# Google Machine Learning and Generative AI for Solutions Architects

Build efficient and scalable AI/ML solutions on Google Cloud

**Kieran Kavanagh**

## Google Machine Learning and Generative AI for Solutions Architects



**Group Product Manager**: Ali Abidi
**Publishing Product Manager**: Sanjana Gupta
**Book Project Manager**: Aparna Nair
**Content Development Editor**: Joseph Sunil
**Technical Editor**: Rahul Limbachiya
**Copy Editor**: Safis Editing
**Proofreader**: Joseph Sunil
**Indexer**: Manju Arasan
**Production Designer**: Prashant Ghare
**DevRel Marketing Coordinator**: Nivedita Singh

First published: June 2024
Production reference: 1120624

Published by Packt Publishing Ltd.
Grosvenor House
11 St Paul's Square
Birmingham
B3 1RB, UK.

ISBN 978-1-80324-527-0
www.packtpub.com

*To my wife, Katelyn, for supporting me through all the hard work, the many late nights, and the busy weekends. I never would have been able to do it without her! To our soon-to-be-born son, whom we can't wait to meet! To my parents, Pat and Michael Kavanagh, and to my siblings, Emmet, Gemma, and Lorraine, who have always supported me through everything in life!*

*– Kieran Kavanagh*

# Foreword

You'll find plenty of books on cloud computing and no shortage of books about artificial intelligence and machine learning (AI/ML), but the industry has long needed a comprehensive book on the vast and awe-inspiring world of AI/ML on Google Cloud, and that book is finally here.

An established leader in AI/ML and generative AI research, Google provides cloud-based tools and models for an enormous variety of business use cases. In this book, you will learn all about these tools, models, and solutions and how to apply them in the real world.

With decades of experience in the technology industry, working on large and complex projects at companies such as Google, Amazon, and AT&T, the author brings a unique perspective and real-life expertise to the pages of this book. With almost 600 such pages, the book covers not only the various tools and offerings available but also the common challenges experienced by most companies when they undertake large-scale, enterprise-grade AI/ML projects, and strategies for addressing those challenges.

The book can be seen as three comprehensive courses all rolled into one:

1. A course on AI/ML and the typical model development life cycle.
2. A course on cloud computing and Google Cloud.
3. A course on building enterprise-grade AI/ML solutions.

While most AI/ML courses are inherently academic in structure, as a solutions architect, the author understands what is required to bring the relevant AI/ML topics out of the realm of academia, including all of the considerations required to build and run business-critical systems for large enterprises. Although the book is intended mainly for people currently working as—or aspiring to be—solutions architects who wish to build their skills in AI/ML and generative AI, this book would also be a valuable tool for current and aspiring data scientists who wish to build workloads on Google Cloud. This is especially true for the practical exercises that accompany each chapter, in which you will build real AI/ML projects on Google Cloud.

Whether you're a seasoned architect or just beginning to embark on your cloud AI journey, this book will guide you on the path from data processing to cutting-edge generative AI models. It unlocks the power of Vertex AI, arming you with the tools to automate, monitor, and continuously improve your AI solutions.

As businesses look to AI to solve increasingly complex problems, the need for solutions architects fluent in these technologies has never been greater. This book is not just a guide; it's an investment in your future and the future of the businesses you will empower.

Let's dive in!

*Priyanka Vergadia*

*Leader, Developer Relations, Google*

# Contributors

## About the author

**Kieran Kavanagh** is a Principal Architect at Google Cloud, working with Google's largest retail customers and driving some of the industry's most challenging digital transformation and generative AI initiatives. Before joining Google, he was a Principal AI/ML Solutions Architect in Strategic Accounts at **Amazon Web Services (AWS)**, building some of the most complex AI/ML systems in the world. He was also a Principal Architect at AT&T, leading their Mobile Internet infrastructure design, and he is a public speaker on the topics of AI/ML, MLOps, and large-scale cloud transformation.

Originally from Cork, Ireland, he now lives in Atlanta, GA, with his wife, Katelyn.

*Thanks to the colleagues who provided early feedback on this book, such as Alex Martin, Brij Dhanda, Elia Secchi, Mikhail Chrestkha, and Vipin Nair, among others.*

## About the reviewer

**Jeremy Wortz** is a Googler and multi-industry data scientist with over twenty years of experience, making the connection between business and tech when it comes to analytics and machine learning. He has spearheaded advanced ML solutions for multiple industries and has assessed over two hundred AI startups. He has also founded his own startup, building one of the first scalable custom AutoML solutions for demand forecasting at the largest retailer in the US.

Jeremy lives in the western suburbs of Chicago with his family and enjoys traveling abroad with them.

Jeremy also is an adjunct professor of Technology Entrepreneurship at Northwestern University, teaching the next generation of students how to form their own tech startups.

# Table of Contents

# 5

## Building Custom ML Models on Google Cloud 135



# 6

## Diving Deeper – Preparing and Processing Data for AI/ML Workloads on Google Cloud 163

# 9

## Neural Networks and Deep Learning 235



# 10

## Deploying, Monitoring, and Scaling in Production 259

# 11

## Machine Learning Engineering and MLOps with Google Cloud 281



# 12

## Bias, Explainability, Fairness, and Lineage 301

# 13

## ML Governance and the Google Cloud Architecture Framework 333



# 14

## Additional AI/ML Tools, Frameworks, and Considerations 369

# Part 3: Generative AI

## 15



## 16

17

## Generative AI on Google Cloud 459



18

## Bringing It All Together: Building ML Solutions with Google Cloud and Vertex AI 487



## Index 501

## Other Books You May Enjoy 526

# Preface

Almost every company nowadays is either using or trying to use AI/ML in some way, especially with the recent revolutions regarding generative AI. While AI/ML research is undoubtedly complex, what is often more complex is actually building and running applications that use AI/ML effectively. This book teaches you how to successfully design and run AI/ML workloads, based on years of experience implementing large-scale and highly complex AI/ML projects at some of the world's leading technology companies.

The early chapters in this book provide an overview of the different categories of artificial intelligence and machine learning (AI/ML), as well as general cloud computing concepts. This is followed by an overview of Google Cloud, including the Google Cloud products related to AI/ML and examples of their intended use cases.

Then, the book progresses through the stages of a typical machine-learning project and model development life-cycle. Each chapter covers an important stage in the life-cycle. You will not only learn those concepts but will put them into action in the practical exercises that accompany each chapter. The process begins with procuring and preparing data and moves on to training ML models. Then, we will deploy the models and get inferences from them. You will also learn about monitoring and updating models after deployment to ensure that they continue to provide the best possible results. Additionally, you will automate all of those steps by building an end-to-end MLOps solution.

The book not only covers all of the steps in the machine-learning model development life-cycle but also covers important topics in implementing and managing machine-learning solutions at enterprise scale. We will dive into considerations such as privacy, compliance, ethics, and many other topics that are necessary to understand for running ML solutions in a real business context.

By the end of this book, you will possess advanced knowledge of cloud computing, Google Cloud, AI/ML, and generative AI. You will have built complex projects, solutions, and models, addressing real-world business use cases, and have learned common challenges that companies often run into when building AI/ML solutions, as well as how to address those challenges, based on many years of experience on some of the industry's largest and most complex AI/ML systems and projects. You will also have learned and implemented important solution architecture considerations such as reliability, scalability, and security, and how they apply to AI/ML use cases.

These are among the most in-demand and high-paying skills in the technology industry and among the most sought-after skills in the world, in general, across all industries. With that in mind, join me on this journey and begin advancing your career today.

## Who this book is for

This book is primarily intended for solutions architects, or aspiring solutions architects, who wish to learn how to design and implement AI/ML solutions on Google Cloud. However, it can also be used by data scientists who wish to learn Google Cloud's AI/ML and generative AI tools and important solution architecture concepts that can help them build robust solutions.

In my career, I've gained a lot of experience in writing technical documents for broad audiences. I've learned that a best practice in that regard is to never assume that the readers of the documents have prior "intrinsic knowledge" of the topic at hand. In this book, you'll find that I start each chapter with the basics that need to be understood before moving on to more complex topics. The benefit of this approach is that if you already have knowledge of the basics, you can skip straight to the more complex topics, whereas if a topic is new to you, you will not be left behind. This means that the book is suitable both for beginners and advanced practitioners because each chapter starts with the basic concepts and progresses towards covering more advanced use cases.

Also, more broadly, the book begins by covering important fundamental AI/ML concepts and then builds in complexity through examples and hands-on activities to eventually dive deep into advanced, cutting-edge AI/ML applications that address real-world use cases in today's market.

## What this book covers

*Chapter 1, AI/ML Concepts, Real-World Applications, and Challenges*, sets the stage for the AI/ML topics that will be used in more depth in the rest of the book.

*Chapter 2, Understanding the ML Model Development Life Cycle*, introduces the common steps that are found in typical, well-structured ML projects.

*Chapter 3, AI/ML Tooling and the Google Cloud AI/ML Landscape*, describes the tools for building AI/ML solutions, specifically on Google Cloud

*Chapter 4, Utilizing Google Cloud's High-Level AI Services*, starts to use the tools to implement real-world AI/ML use cases.

*Chapter 5, Building Custom ML Models on Google Cloud*, gets hands-on with Scikit-learn on Google Cloud.

*Chapter 6, Diving Deeper - Preparing and Processing Data for AI/ML Workloads on Google Cloud*, outlines the procedures to implement a complex data processing workload.

*Chapter 7, Feature Engineering and Dimensionality Reduction*, builds on the data processing themes from the previous chapter, this chapter focuses on one of the most important steps in an ML project: feature engineering.

*Chapter 8, Hyperparameters and Optimization*, outlines the importance of hyperparameter tuning, and how to implement this in Vertex AI.

*Chapter 9, Neural Networks and Deep Learning*, provides an overview of the use of neural networks to solve more complex problems in AI/ML.

*Chapter 10, Deploying, Monitoring, and Scaling in Production*, focuses on how to productionize ML models, the kinds of challenges that are faced at this point in the process, and how to use Vertex AI to address some of those challenges.

*Chapter 11, Machine Learning Engineering and MLOps with Google Cloud,* dives into more detail on deployment concepts and challenges and describes the importance of MLOps in addressing these challenges for large-scale production AI/ML workloads.

*Chapter 12, Bias, Explainability, Fairness, and Lineage*, discusses these concepts in detail, and explains how to effectively incorporate these concepts into the readers' ML workloads.

*Chapter 13, ML Governance and the Google Cloud Architecture Framework*, describes architectural design patterns for AI/ML workloads on Google Cloud.

*Chapter 14, Additional AI/ML Tools, Frameworks, and Considerations*, branches out into additional popular AI/ML frameworks such as PyTorch, Spark ML, and BigQuery ML.

*Chapter 15, Introduction to Generative AI*, outlines the fundamental concepts of Generative AI and focus on its distinctions from "traditional" predictive AI /ML.

*Chapter 16, Advanced Generative AI Concepts and Use Cases*, dives deeper into embeddings, vector databases, and frameworks such as RAG and LangChain.

*Chapter 17, Generative AI on Google Cloud*, explores Google Cloud's Generative AI products and solutions.

*Chapter 18, Bringing It All Together: Building ML Solutions with Google Cloud and Vertex AI,* brings together all of the major elements we've learned throughout the book and helps us in building reference architectures.

## To get the most out of this book

Throughout the book, I provide comprehensive instructions on how to install all required software. I also provide all the code required to build all of the solutions I will outline in this book, as well as detailed descriptions of what the code is doing. Prior knowledge of Python would be an advantage, but it is not a hard prerequisite.

| Software/hardware covered in the book | Operating system requirements |
|---|---|
| Python | Linux |
| Scikit-learn | Linux |
| TensorFlow | Linux |

**If you are using the digital version of this book, we advise you to type the code yourself or access the code from the book's GitHub repository (a link is available in the next section). Doing so will help you avoid any potential errors related to the copying and pasting of code.**

## Download the example code files

You can download the example code files for this book from GitHub at `https://github.com/PacktPublishing/Google-Machine-Learning-for-Solutions-Architects`. If there's an update to the code, it will be updated in the GitHub repository.

We also have other code bundles from our rich catalog of books and videos available at `https://github.com/PacktPublishing/`. Check them out!

## Conventions used

There are a number of text conventions used throughout this book.

`Code in text`: Indicates code words in text, database table names, folder names, filenames, file extensions, pathnames, dummy URLs, user input, and Twitter handles. Here is an example: "Mount the downloaded `WebStorm-10*.dmg` disk image file as another disk in your system."

A block of code is set as follows:

```
import matplotlib.pyplot as plt
import pandas as pd
# Load dataset
data = load_wine()
df = pd.DataFrame(data.data, columns=data.feature_names)
```

When we wish to draw your attention to a particular part of a code block, the relevant lines or items are set in bold:

```
{
  "name": "BUCKET_NAME",
  "location": "BUCKET_LOCATION",
  "storageClass": "STORAGE_CLASS",
  "iamConfiguration": {
    "uniformBucketLevelAccess": {
      "enabled": true
    },
  }
}
```

Any command-line input or output is written as follows:

```
$ cd packt-ml-sa
```

**Bold**: Indicates a new term, an important word, or words that you see onscreen. For instance, words in menus or dialog boxes appear in **bold**. Here is an example: "Select **System info** from the **Administration** panel."

> **Tips or important notes**
> Appear like this.

## Get in touch

Feedback from our readers is always welcome.

**General feedback**: If you have questions about any aspect of this book, email us at `customercare@packtpub.com` and mention the book title in the subject of your message.

**Errata**: Although we have taken every care to ensure the accuracy of our content, mistakes do happen. If you have found a mistake in this book, we would be grateful if you would report this to us. Please visit `www.packtpub.com/support/errata` and fill in the form.

**Piracy**: If you come across any illegal copies of our works in any form on the internet, we would be grateful if you would provide us with the location address or website name. Please contact us at `copyright@packtpub.com` with a link to the material.

**If you are interested in becoming an author**: If there is a topic that you have expertise in and you are interested in either writing or contributing to a book, please visit `authors.packtpub.com`.

## Share Your Thoughts

Once you've read *Google Machine Learning and Generative AI for Solutions Architects*, we'd love to hear your thoughts! Scan the QR code below to go straight to the Amazon review page for this book and share your feedback.

`https://packt.link/r/1-803-24527-1`

Your review is important to us and the tech community and will help us make sure we're delivering excellent quality content.

## Download a free PDF copy of this book

Thanks for purchasing this book!

Do you like to read on the go but are unable to carry your print books everywhere?

Is your eBook purchase not compatible with the device of your choice?

Don't worry, now with every Packt book you get a DRM-free PDF version of that book at no cost.

Read anywhere, any place, on any device. Search, copy, and paste code from your favorite technical books directly into your application.

The perks don't stop there, you can get exclusive access to discounts, newsletters, and great free content in your inbox daily

Follow these simple steps to get the benefits:

1. Scan the QR code or visit the link below

`https://packt.link/free-ebook/978-1-80324-527-0`

2. Submit your proof of purchase
3. That's it! We'll send your free PDF and other benefits to your email directly

# Part 1: The Basics

This part establishes the baseline for the rest of the book. It covers all the basic topics that need to be understood to grasp the more complex concepts (and implement the workloads) that come later in the book. We begin by covering some of the fundamental concepts of AI/ML, and we discuss examples of how AI/ML is used in real-world use cases. Most importantly, we discuss common challenges that companies often run into when implementing AI/ML projects at scale and begin discussing how to address such challenges, forming the basis for deeper discussions throughout this book. Next, we outline the steps in a typical AI/ML project lifecycle, which will be used to form the overall structure of much of this book. We round out this part by introducing Google Cloud and common AI/ML tooling.

This part contains the following chapters:

- *Chapter 1, AI/ML Concepts, Real-World Applications, and Challenges*
- *Chapter 2, Understanding the ML Model Development Life Cycle*
- *Chapter 3, AI/ML Tooling and the Google Cloud AI/ML Landscape*

# 1

# AI/ML Concepts, Real-World Applications, and Challenges

This chapter will introduce basic concepts that will be explored in more detail throughout the rest of the book. We understand that readers of this book may be starting from different stages in their **artificial intelligence/machine learning** (**AI/ML**) journey, whereby some readers may already be advanced practitioners who are familiar with running AI/ML workloads while others may be newer to AI/ML in general. For this reason, we will briefly describe important fundamental concepts as required throughout the book to ensure that all readers have a common baseline upon which to build their understanding of the topics we discuss. Readers who are newer to AI/ML will benefit from learning the important underlying concepts rather than diving straight into the deep end of each topic without a baseline context, and advanced practitioners should find them to be useful knowledge refreshers.

In this chapter, we're going to cover the following main topics:

- Terminology—AI, ML, **deep learning** (**DL**), and **generative AI** (**GenAI**)
- A brief history of AI/ML
- ML approaches and use cases
- A brief discussion of ML basic concepts
- Common challenges in developing ML applications

By the end of this chapter, you will understand the common types of AI/ML approaches and their real-world applications, as well as some historical background on the development of AI/ML concepts. Finally, you will learn about common types of challenges and pitfalls that companies encounter when they begin to implement AI/ML workloads. This is a particularly important part of the book, especially for the solutions architect role, and it provides real-world insights that are not found in academic courses; these insights come from years of experience in the field, working on large-scale AI/ML projects with many different companies.

## Terminology – AI, ML, DL, and GenAI

Here, we describe how the terms *AI* and *ML* relate to each other. It should be noted that these terms are often used interchangeably, as well as the abbreviated term, *AI/ML*, which serves as an umbrella term to encapsulate both AI and ML. We also describe how the terms *DL* and *GenAI* fit in under the umbrella of AI/ML.

We'll begin by briefly including officially-accepted definitions of the terms *AI* and *ML*. We have chosen to include definitions from the *Collins English Dictionary*, in which AI is defined as "*a type of computer technology concerned with making machines work in an intelligent way, similar to the way that the human mind works*" and ML is defined as "*a branch of artificial intelligence in which a computer generates rules underlying or based on raw data that has been fed into it.*" The term *DL* has not yet been officially included as a dictionary term, but the *Collins English Dictionary* lists it as a new word suggestion, with the proposed definition of "*a type of machine learning concerned with artificial neural networks allowing advanced pattern recognition.*" We understand that official dictionary definitions don't always explain the concepts completely, but it's important to include them for reference, and we will cover these concepts in more detail as we progress through the book. All of those terms constitute what we are now beginning to refer to as "Traditional AI", to distinguish it from GenAI, which is a much newer, and quite different concept. There will be an entire section of this book dedicated to GenAI, so this distinction will become clearer.

As a general rule, DL is considered to be a sub-field of ML, and ML is considered to be a sub-field of AI. GenAI can be seen as a sub-field within DL, because it uses Deep Neural Networks and concepts from Natural Language Processing in its application. You will often see them graphically represented in literature as a set of concentric circles, whereby AI is the broadest field, ML is nested as a sub-field within AI, and DL is nested as a sub-field within ML. I'm adding to this conceptual representation by including GenAI within the field of DL, although it is more of an association rather than a strict sub-category:

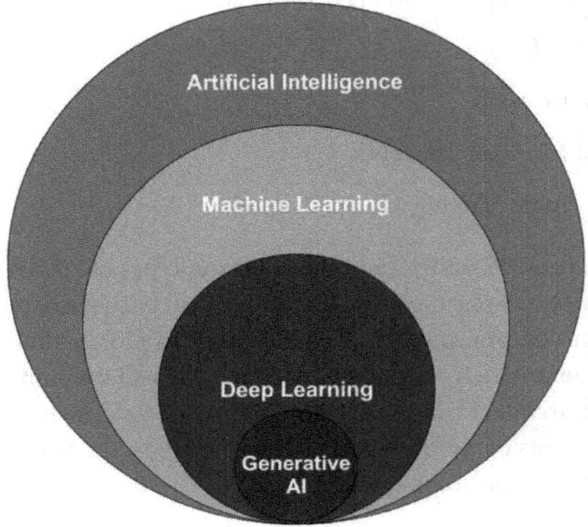


Figure 1.1: Depicting the relationship between the terms AI, ML, DL, and GenAI

Now that we have covered some basic terminology regarding AI/ML, let's briefly discuss its history and understand how the AI/ML industry has developed so far.

## A brief history of AI/ML

If we traveled back in time by only a few years — to the year 2015 — and compared the state of the AI/ML industry to what it is today, we would see that relatively few companies had commercially implemented large-scale AI/ML use cases at that point. Although we would find academic research being performed in this space, we wouldn't regularly hear AI/ML being discussed in mainstream media, and successful commercial or industrial implementations had mainly been achieved only by some of the world's largest, industry-leading technology or niche companies. Jumping forward by just 2 years, we find that by the end of 2017, the tech industry is abuzz with discussions of AI/ML, and it seems to be the main topic—or at least one of the main topics—on everybody's mind.

Based on our time-traveling adventure, one would not be faulted for believing that AI/ML is a brand-new term that suddenly emerged only in the past few years. In reality, however, these concepts have been developing over many decades. As the next step in our time-traveling journey, we travel further back in time to the 1950s. The first use of the term *artificial intelligence* is credited to Professor John McCarthy in 1955 (McCarthy et al. 1955), and a number of other important developments that contributed significantly to this field of science took place during the 1950s, such as Alan Turing's 1950 paper, *Computing Machinery and Intelligence*, in which he posed the question, "*Can machines think?*" (Turing, 1950), and Frank Rosenblatt's work on the "Perceptron" (Rosenblatt, 1957), which we'll look at in more detail later in this book.

As an extension of our time-traveling journey, it should be noted that AI/ML algorithms today use mathematical concepts that were originally discovered and formulated centuries or millennia ago. For example, many of the algorithms we will explore in this book use concepts from linear algebra and calculus, which have been in use for centuries, and when we learn about cost functions, training, and evaluation, we will be using concepts from Euclidean geometry, such as the Pythagorean theorem, which dates back millennia. Interestingly, although Pythagoras is believed to have lived around 2,500 years ago, there is some evidence that concepts from the "Pythagorean theorem" were already understood and used by previous civilizations such as the Babylonians of Mesopotamia more than 1,000 years before his birth (Götze, 1945, 37-38). How fascinating it is that some of our most cutting-edge DL algorithms today use the same mathematical constructs that were used by ancient civilizations in the Bronze Age! In the 1960's and 1970's, the practice of using computers to perform statistical analysis and modeling on data began to grow significantly, and specialized software such as **Statistical Analysis System** (**SAS**) and **Statistical Package for the Social Sciences** (**SPSS**) emerged for these purposes. These tools were generally used for **in-memory** processing, meaning that all of the data used with these tools would be loaded into memory on a single computing machine. In the next section, you will see why it's important to call this out.

Next, we travel forward in time, to the present day, and one of the questions that come to mind is this: If all of this started back in the 1950s, why does it seem that AI and ML are concepts that everybody has suddenly become enthusiastic about only in the past few years? Why did we not see such widespread adoption and success of AI/ML implementations before now? A number of factors have contributed to the gap in time that has been experienced between when these concepts originally began to be researched and when they finally started to gain publicly noticeable traction in the industry during the past few years (see the "AI Winter," for example). As we will see in later chapters of this book, one such factor is that AI/ML use cases generally require large amounts of data and extensive computing resources. This is one of the reasons why—until recently—AI/ML research was usually being performed only by entities that could afford to amass these required resources, such as large technology companies, well-established research institutions, and government bodies.

What has changed in recent years to help AI/ML break out beyond the exclusive realm of large corporations and research institutions? How have smaller companies, and even amateur hobbyists, suddenly obtained access to the resources required to train, host, and evaluate ML models and experiment with new ideas on how to apply AI/ML to an ever-increasing plethora of interesting new use cases? One of the primary contributors to this sudden revolution has been "cloud computing," as well as iterative tooling development for building and running AI/ML workloads, and advances in DL approaches.

## AI/ML and cloud computing

To understand in more detail how cloud computing has helped to suddenly revolutionize AI/ML research and real-world application, let's consider the types of resources that are required to train, host, evaluate, and manage ML models. While we could use a laptop or a home computer to train and evaluate a relatively simple model on a small dataset, as we want to scale out our research and use cases, we would quickly find that the computing resources on our personal computer would not be sufficient to train larger models and the hard drive(s) in our personal computer would not be large enough to store the required datasets. These were also the limitations experienced by the statistical modeling tools such as SAS and SPSS that we mentioned in the previous section, which processed data "in-memory" on a single machine.

To illustrate this concept, we're going to scale up our use cases in a series of steps. If we were to scale up only slightly beyond the resources of the most powerful personal computer on the market, we would need to run our workload on a hardware "server," which contains more powerful computing resources, and we could attach multiple large-capacity hard drives to this server, potentially as a **Redundant Array of Independent Disks** (**RAID**) array (formerly **Redundant Array of Inexpensive Disks**). This is still something that an individual person could perform in their home, but powerful servers—especially the "latest and greatest" servers on the market—can be quite expensive to purchase, and it would require a bit more technical knowledge to set up the server and configure the RAID array. At this point, we are already extending outside the realm of all but the most dedicated amateur hobbyists.

Scaling beyond the resources of the most powerful hardware server on the market would require us to create a cluster of servers. Apart from the additional expense of purchasing multiple servers and their attached disks, this would also require more advanced technical knowledge to build and configure a network that links the servers together appropriately. This would not be an economically viable approach for most hobbyists, but it may still make some sense for a small company.

Next, let's scale all the way up to some of today's most advanced DL use cases, which can take weeks or months to train, on hundreds of high-powered and very expensive servers. If we wanted to run these kinds of workloads completely by ourselves, we would need to build a data center, hire teams of experts to install hundreds of servers, build and configure a complex network to link them all together appropriately, and perform multiple other supporting activities to get our infrastructure set up. Let's imagine for a moment that we want to create a start-up company that will use DL to implement a new breakthrough idea that we've devised. Simply building the data center could take a couple of years and would cost many millions of dollars, before we could ever start experimenting with our idea. This would not be a viable option.

With cloud computing, however, we could simply write a script or click some links and buttons on a cloud computing provider's website, and it would spin up all of the servers we need within minutes. We could then perform our experiments — iteratively train and evaluate our models — and then simply shut down the servers when our work is done. As you can imagine, this is infinitely easier, cheaper, and more achievable than trying to build and manage our own data centers. This now provides small companies and limited-funding researchers or hobbyists with access to compute power and resources that were previously only available to very large organizations.

Note that it was not only the ability to more easily create and access the required hardware infrastructure that helped revolutionize the AI/ML industry. Relevant software tools and frameworks have also been evolving over time, as well as the amount of data that has become available. While the tools such as SAS and SPSS that were developed in the 1960's and 1970's were sufficient for performing statistical modeling on datasets that could fit in memory on a single machine, the rapid growth in popularity of the Internet caused a dramatic increase in the amount of data that companies could gather and produce. In parallel, libraries were developed in languages such as Python, which made activities such as data processing, analysis, and modeling much easier. I can confidently say that it is a lot easier to perform many kinds of modeling use cases with libraries such as scikit-learn, PyTorch, and Keras, than it was with the earlier tools mentioned above. Still, many of today's machine learning algorithms can be seen as an evolutions of traditional statistical modeling techniques, which have been enhanced to handle larger and more complex use cases. In addition to this, tools such as Apache Hadoop and Apache Spark, which I will discuss in more detail later in this book, made it possible to implement use cases that could span beyond the limits of a single machine, and therefore handle much larger datasets.

So far, in our discussion of scaling out our use cases, we have mainly focused on training and evaluating ML models, but those activities are only a subset of what's required to create an ML application that is actually used in the real world. Throughout this book, we will often use the term *in production* to refer to the concept of creating, hosting, and serving AI/ML applications that are used in the real world, outside of a laboratory testing environment.

Even large, well-established organizations with teams of experienced data scientists often find that successfully hosting an ML application in production can be more complex and challenging than the model training and evaluation process. Let's now take a look at how cloud computing can provide additional value for addressing these challenges beyond simply spinning up the required compute and storage resources. In later chapters in this book, we will discuss all of the steps in a typical ML project in great detail, but for now, let's consider at a high level what kinds of resources and infrastructure would be required to host an ML model in production if cloud computing did not exist.

In addition to the activities required for model training and evaluation, such as constructing the data centers, installing the servers, building and configuring a complex network, and maintaining all of the hardware over time, we would need to perform many other activities to actually host and serve an ML model for production usage. For example, it would be necessary to create an interface to expose our model to end users or other systems. The most likely approach would be to use a web-based interface, in which case we would need to build a cluster of web servers and configure and manage those web servers on an ongoing basis. We would need to develop and build an application on those web servers to expose our model to web clients, and then install and configure load balancers and distribute load across our web servers. All of this infrastructure would, of course, need to be secured appropriately. *Figure 1.2* shows an example of the kind of infrastructure you may need to set up; bear in mind that you may need to duplicate that infrastructure for each layer in your solution—for example, you would need to duplicate that infrastructure multiple times for your web server layer, your application server layer, and your model serving layer. The diagram shows two load balancers and routers for redundancy, in case one of those components fails:

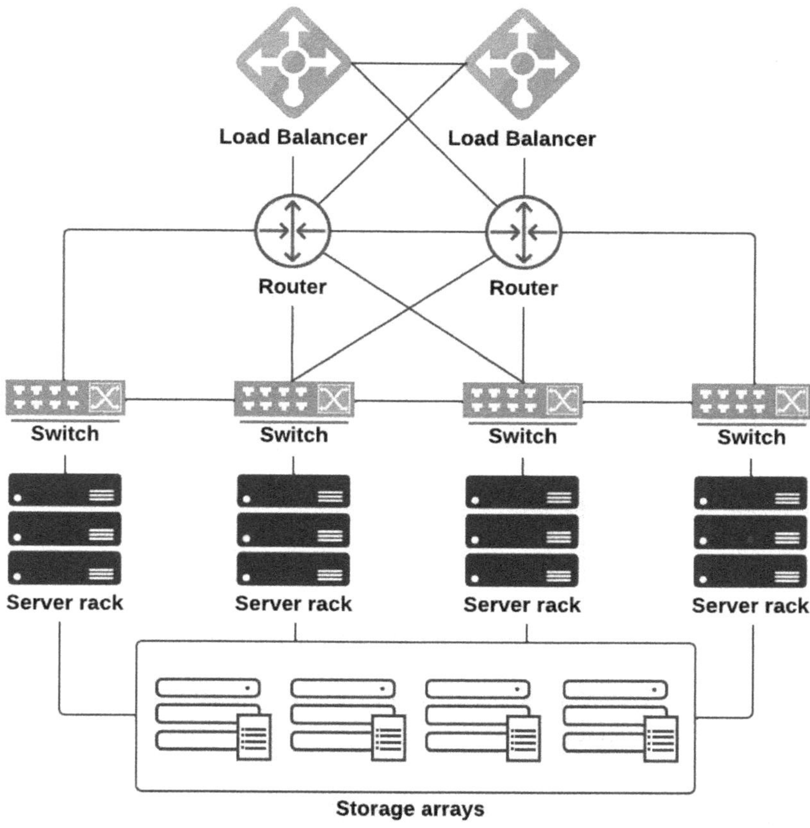


Figure 1.2: Example infrastructure for model hosting

Unfortunately, most data scientists are not also networking experts, web server configuration experts, and security experts, so even if we had a team of the best data scientists who had created a breakthrough model, we would need many other teams of dedicated experts just to build and maintain the infrastructure required to expose that model to clients. On the other hand, if we wanted to use a service such as Vertex AI on Google Cloud, it would automatically build and manage all of this infrastructure for us, and we could go from laboratory testing to production hosting within minutes.

It's important to note that without cloud computing, companies wouldn't just find it less convenient to build their AI/ML workloads, but it would also be prohibitive – that is, large companies wouldn't invest in building the required infrastructure unless they were sure their application was going to be successful in advance, which is a very difficult thing to forecast. Smaller companies wouldn't be able to get started without significant up-front investments in infrastructure expenses, which most would not be able to obtain. As such, AI/ML research, experimentation, and eventual implementation in the real world would be far less prevalent and achievable without cloud computing.

Speaking of implementing AI/ML in the real world, let's take a look at the different kinds of AI/ML approaches and some of their real-world use cases.

## ML approaches and use cases

AI/ML applications are usually intended to make some kind of prediction based on input data, with perhaps the exception of Generative AI, because Generative AI is intended to generate content rather than simply making predictions. In order to make predictions, ML models first need to be **trained**, and how they are trained depends on the approach being used. While ML is a broad concept that encompasses many different fields of research, with endless new use cases being created almost every day, the industry generally groups ML approaches into three high-level categories:

- **Supervised learning** (SL)
- **Unsupervised learning** (UL)
- **Reinforcement learning** (RL)

### SL

SL is the most commonly used type of ML in the industry and perhaps the easiest to describe. The term *supervised* indicates that we are informing the ML model of the correct answers during the training process. For example, let's imagine that we want to train a model to be able to identify photographs of cats. In this case, we would use thousands or millions of photographs in our training set, and we would tell the model which photographs contain cats and which ones do not. We do this via a process called **labeling**, which we will describe in more detail in later chapters. If trained correctly, our model would learn how to distinguish input features that identify a cat in each photograph. If we then presented new photographs that the model had never seen before (that is, that were not included in the training set), our model would be able to identify whether those photographs contained cats. More specifically, for each photograph, our model would be able to predict the probability that it contains a cat, based on observed features in the photograph.

There are two subcategories of SL: classification and regression.

#### Classification

Our cat-identification model described previously is an example of a classification use case in which our model can classify whether our photo contains a cat. Classification is further broken down into binary classification or multi-class classification. Binary classification provides a "yes" or "no" prediction. For example, in this case, we would ask the model, "Does this photograph contain a cat?", and the model would respond with either "Yes" or "No." If we were to train our model to identify many different types of objects, then it would be capable of performing multi-class classification, and we could ask a broader question such as, "What do you see in this photograph?". In this case, the model could respond with multiple different object classifications, including "cat" (if it sees a cat in the photograph), among other objects that it predicts to exist in the photograph (see *Figure 1.3*):

Figure 1.3: Classification of cat and flowers in a photograph

## Real-world applications of classification

Of course, classification can be used for much more important use cases than identifying pictures of cats. An example of an important real-world classification use case is in medical diagnoses, whereby an ML model can predict the presence of a medical condition in a patient based on input data such as physical symptoms or a radiology image.

### Regression

While classification is useful when there are discrete answers to our questions, regression is used when we deal with "continuous variables," whereby the answer to our question could be any value in a continuum, such as 0.1, 2.3, 9894.6, 105, or 0.00000487. To introduce some terminology, the inputs that we provide to our model are referred to as "input variables," and the variables that we wish to predict are referred to as "target variables" or "dependent variables."

> **Note**
>
> When we use the term *regression* here, we are referring to linear regression. This is not to be confused with logistic regression, which is actually a type of classification that we will cover later in this book.

The aim of linear regression is to define a linear function that maps input variables to an output target variable. For example, we may want to predict the grade a student will achieve on an exam based on the number of hours they spent studying, from data we have regarding previous students' grades and how many hours they spent studying. We could graph the data as follows (see *Figure 1.4*), where the stars represent each student in the dataset; that is, they represent each student's grade and the number of hours they spent studying:

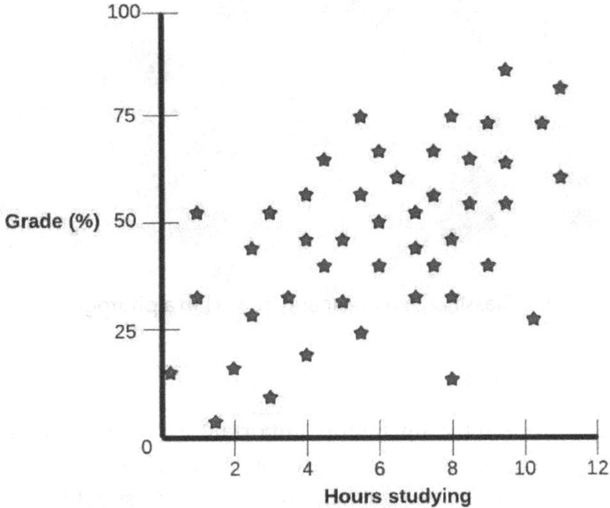


Figure 1.4: Student grades and hours studying

As we can see in the diagram, there appears to be a relationship or correlation between the grades achieved and the number of hours spent studying; that is, students who studied for longer hours generally got higher grades.

A linear regression model trained on this dataset would try to find the function or line that best represents that relationship. You may remember from school that a simple line function is often represented by the formula $y = ax + b$, where $a$ is a multiple of $x$, and $b$ is where the line intercepts the $y$ axis. To find the most accurate function, the linear regression process tries to define a line that minimizes the distance between that line and each of the data points (stars), which may look something like *Figure 1.5*. The distances between the line and some of the data points are shown in red for reference. In later sections, we'll discuss in more detail how those distances are calculated:

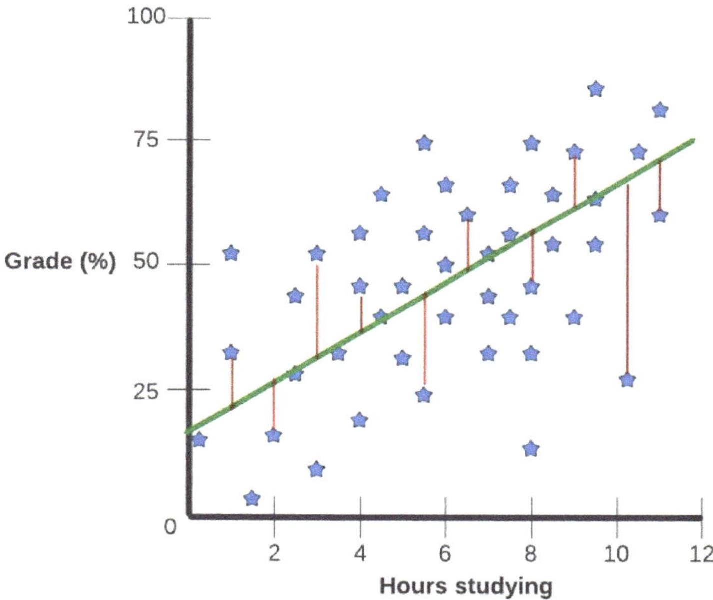


Figure 1.5: Linear regression function

With this function, we could now estimate or predict future student grades based on the number of hours they spend studying. For example, if a student spends 10 hours studying, we would predict that they would achieve a grade of approximately 70% based on what we see in *Figure 1.6*:

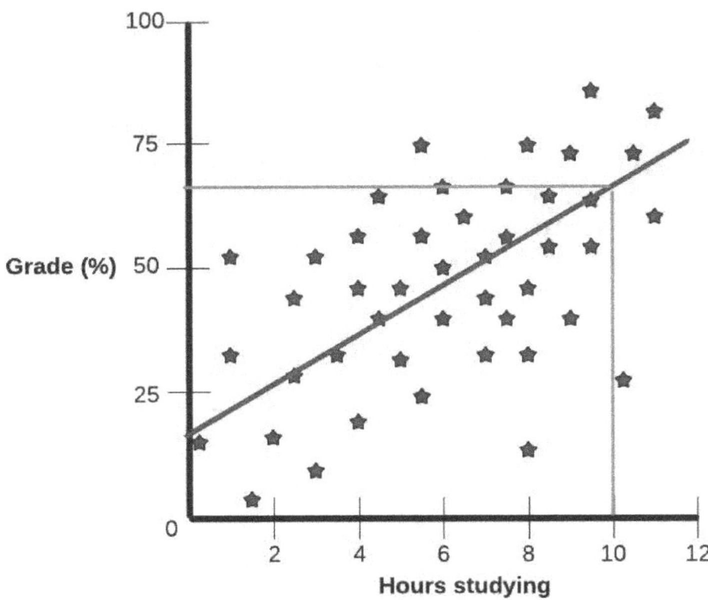


Figure 1.6: Applying linear regression function

### Real-world applications of linear regression

Regression is one of the most widely used types of ML, and it is useful for many different business use cases. It's the quintessential example of "predicting the future" that is often attributed to the power of ML. Business leaders often want to predict numbers that relate to the performance of their business, such as predicting sales for the next quarter, based on historical sales and other market data. Whenever you have numerical metrics that you can track, and sufficient historical features relating to those metrics, you can try to predict or "forecast" the future values of those metrics, from stock market prices and housing prices to blood pressure measurements in various medical scenarios.

## UL

With UL, we do not train the model on a dataset in which the entries are labeled with the correct answers. Instead, we ask the model to find unknown or non-predetermined patterns in the data. An analogy we could use here is that in SL, we are teaching the model about what exists in the data, whereas in UL, the model is teaching us about what exists in the data, such as underlying trends among various data points in our dataset.

The most common type of UL is something known as "clustering," whereby data points are grouped together based on some kinds of similarities that are observed by the model. *Figure 1.7* provides a visual representation of this concept, showing the input data on the left and the resulting data clusters on the right:

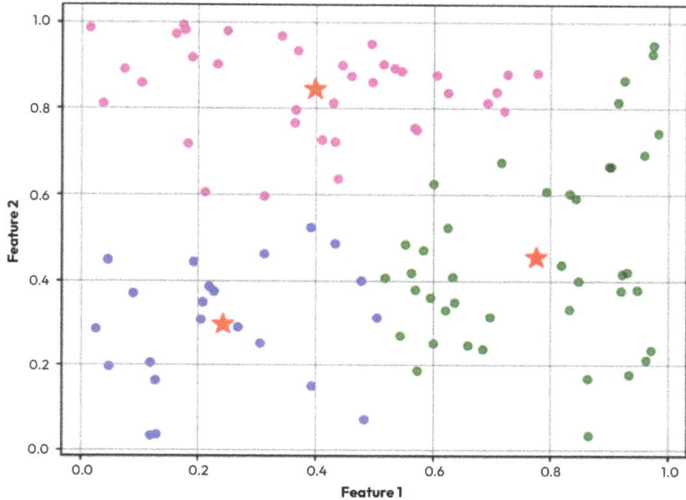


Figure 1.7: Clustering

### *Real-world applications of UL*

An example of a real-world application of clustering would be for categorizing groups of customers with similar purchasing preferences. You may have noticed this being utilized when you are purchasing an item online and you see recommendations for other items that may interest you, accompanied by a message such as "people who purchased this item also purchased these other items."

Another important real-world use case is fraud detection, in which case one of the clusters could represent legitimate transactions and another cluster could represent unusual or potentially fraudulent transactions, or anything that does not match the characteristics of legitimate transactions could be flagged as potentially fraudulent. As new transactions occur, the model could group them accordingly based on their input characteristics and could trigger a warning response if a transaction appears to be fraudulent. Have you ever received notifications or questions from your bank when you tried to use your credit card on the first day of vacation in a new location? That's because the bank's ML models determined that the characteristics of the transaction were abnormal in some way; in this case, it was a transaction from a location that is far from where you usually use your card.

> **Note**
>
> Clustering can actually be considered as a type of unsupervised classification.

## RL

In RL, the mechanism for training a model is quite different from the previous two approaches. To introduce some terminology, we say that the model uses an **agent** that has an overall goal it wants to achieve (the desired model output). The agent interacts with its **environment** by sending **actions** to the environment. The environment evaluates the actions and provides feedback in the form of a **reward** signal, which indicates whether or not the actions contribute to achieving the overall goal, and **observations**, which describe the current state of the environment. See *Figure 1.8* for a visual representation of this process. To make a very broad analogy, this is similar to how we train some animals, such as dogs. For example, if the dog performs a desired action, then the trainer rewards it with a tasty treat. Conversely, if the dog does something undesirable, the trainer may reprimand it in some way. In the case of RL, the model randomly attempts different actions in its environment. If an action or set of actions is deemed to contribute to achieving the overall goal, then the environment provides a positive reward as feedback to the agent, whereas if the action is considered to be detrimental to achieving the overall goal, then the environment provides a negative reward as feedback to the agent. In this case, the reward is usually just a numeric value, such as 0.5 or -0.2, rather than a tasty treat, because unfortunately for ML models, they aren't yet complex enough to enjoy tasty treats:

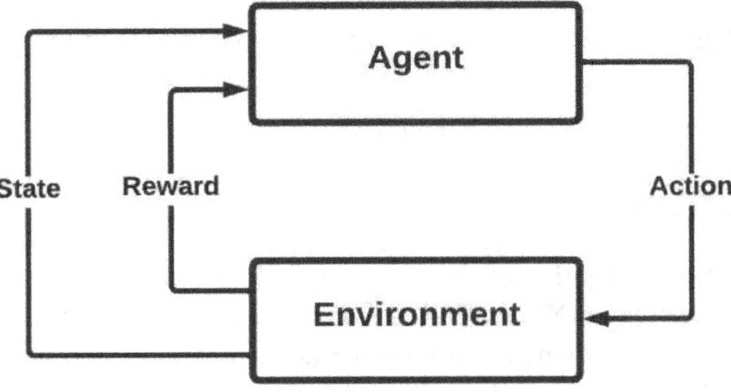


Figure 1.8: RL

The model's environment is the space within which the goal and all possible actions exist, and the observations are features of the environment. This can be a physical environment, such as when a robot is moving around in a physical space, or something abstract, based on the problem that the model is trying to address. For example, you could create a model that becomes an expert in playing a game such as chess or a video game. The model will begin by trying all kinds of random actions, most of which may seem silly or bizarre at first, but based on feedback from the environment, the model's actions will gradually become more relevant and may eventually outperform the actions of human experts in that task.

RL could actually be considered a type of SL because feedback is provided to the model when it makes a prediction, and the model learns and makes improvements based on that feedback. However, it differs from the standard concept of SL, which we described previously, because we are not providing labeled correct answers as part of the training process. Instead, the model is provided with signals that help it understand what kinds of actions it should perform in order to progress toward achieving the required goal.

### *Real-world applications of RL*

RL is not yet as widely adopted in the industry as "traditional" SL and UL, but some interesting applications are emerging. In addition to the gaming use cases mentioned in a previous paragraph, one of the most recognizable applications of RL is in robotic navigation and self-driving cars. In this application, the car could be considered as the model agent, whereby it performs actions such as accelerating, braking, and steering the wheels. Sensors on the car, such as cameras and lidar sensors, provide observations on the state of the environment. If the actions performed by the car help it to achieve a goal such as navigating a course or self-parking without hitting any obstacles, then it would receive positive rewards, whereas if it collided with an obstacle, then it would receive a negative reward. Over time, it could learn to navigate the course or self-park and avoid obstacles.

Another important real-world application of RL is in healthcare, where it has shown promising results for use cases such as medical imaging diagnoses (Zhou et al., 2021, 1-39) and determining what kinds of medical treatments work for individual patients based on their conditions, via mechanisms such as **dynamic treatment regimes** (**DTRs**).

Now that we've discussed the different types of ML approaches and examples of their real-world use cases, let's take a look at some of the underlying concepts that form the basis for how these ML implementations work.

## A brief discussion of ML basic concepts

Mathematics is the hidden magic behind ML, and pretty much all ML algorithms function by using mathematics to find relationships and patterns in data. This book focuses on practical implementations of AI/ML on Google Cloud; it is not a theoretical academic course, so we will not go into a lot of detail on the mathematical equations upon which ML models operate, but we will include mathematical formulae for reference where relevant throughout the book, and here we present some basic concepts that are widely used in AI/ML algorithms. There are plenty of academic materials available for learning each of these concepts in more detail. As an architect, understanding the mathematical concepts could be considered an extracurricular credit rather than a requirement; you usually would not need to dive into the mathematical details of ML algorithms in your day-to-day work, but if you want to have a better understanding of how some of the algorithms work, you can review these concepts in more detail.

## Linear algebra

In ML, we make frequent use of vectors and matrices to store and represent information. To briefly cover some definitions, *Collins Dictionary* defines a vector as "*a variable quantity, such as force, that has size and direction*" and a matrix as "*an arrangement of numbers, symbols, or letters in rows and columns which is used in solving mathematical problems.*" Let's take a look at what this really means. Representing information in matrices is most easily demonstrated if we use tabular data as an example. Consider the information in *Table 1.1*, which represents house sales in King County, Washington (excerpted from a Kaggle dataset: `https://www.kaggle.com/datasets/harlfoxem/housesalesprediction`):

| price | bedrooms | bathrooms | sqft_living | sqft_lot | floors | waterfront | view | condition | grade | yr_built | yr_renovated |
|---|---|---|---|---|---|---|---|---|---|---|---|
| 221900 | 3 | 1 | 1180 | 5650 | 1 | 0 | 0 | 3 | 7 | 1955 | 0 |
| 538000 | 3 | 2.25 | 2570 | 7242 | 2 | 1 | 0 | 3 | 7 | 1951 | 1991 |
| 257500 | 3 | 2.25 | 1715 | 6819 | 2 | 0 | 0 | 3 | 7 | 1995 | 0 |
| 291850 | 3 | 1.5 | 1060 | 9711 | 1 | 0 | 0 | 3 | 7 | 1963 | 0 |
| 229500 | 3 | 1 | 1780 | 7470 | 1 | 0 | 0 | 3 | 7 | 1960 | 0 |
| 323000 | 3 | 2.5 | 1890 | 6560 | 2 | 0 | 0 | 3 | 7 | 2003 | 0 |
| 650000 | 4 | 3 | 2950 | 5000 | 2 | 0 | 3 | 3 | 9 | 1979 | 0 |

Table 1.1: King County house sales

The dataset depicted in *Table 1.1* has 7 rows (not including the title row) and 13 columns, where each row represents a single house sale, which we consider a data point or an observation in our dataset, and each column represents individual features of the data points. We can consider each row and each column to be vectors. Bear in mind that a vector can also be considered as a matrix with one row or one column (that is, a one-dimensional vector). So, for each individual house purchase in our dataset, we have a vector that contains each of the features of that house. Let's imagine that we want to predict the price of houses based on the features (other than the price) of each house. We would want to find the function that best describes the relationship between the price and all of the other features, and linear regression is one way in which we could do this. In this case, we would want to find the set of values by which we should multiply each feature and then add all of the results of those multiplications together in order to correctly estimate the price of each house. This means that each feature would have a corresponding multiplier (or "coefficient"). In order to efficiently compute the multiplications of the features and the coefficients and then add all of the results together, we could represent all of the coefficients also as a vector and calculate the dot product of the feature vector and the coefficient vector. We'll take a minute here to clarify what it means to calculate the dot product. If we have two vectors, $A$ and $B$, where $A = [a\ b\ c]$, and $B = [d\ e\ f]$, the dot product is calculated as follows:

$$a*d + b*e + c*f$$

To illustrate, let's take the first row of *Table 1.1* (without the price) as a feature vector; it would look like this:

```
[3 1 1180 5650 1 0 0 3 7 1955 0]
```

Now, let's create an initial vector of random coefficients (we can just create random coefficients at first and improve our guesses later during the model training process), which needs to have the same number of elements as our preceding feature vector:

```
[1 5 0.3 0.001 2 7 2.5 108.67 14.234 0.103 8]
```

> **Note**
>
> When calculating the dot product, there are rules regarding the shapes of each vector, but we are omitting those details here for simplicity. We will go deeper into those details in later chapters.

The dot product of our feature vector and coefficient vector is shown here:

```
3*1 + 1*5 + 1180*0.3 + 5650*0.001 + 1*2 + 0*7 + 0*2.5 + 3*108.67 +
7*14.234 + 1955*0.103 + 0*8 = 996.663
```

From our first guess at what the coefficients should be, we estimate that the price of the house in *row 1* of *Table 1.1* would be $996.663. However, we can see in *Table 1.1* that the actual price of that house was $221,900. We can now calculate the error resulting from our guess as follows:

```
221900 - 996.663 = 220903
```

We often refer to this as the **loss** or the **cost** of our linear function, and it is similar to what was represented by the red lines in *Figure 1.5* earlier, where this value represents the "distance" from the correct answer; that is, how far away our guess is from the correct answer. This is the first step in the learning process, and in later sections, we will want to find coefficients that minimize this error.

## Calculus

One common use of calculus in ML is in the error minimization process mentioned previously. In later chapters, we will define something called a **loss function** (or a **cost function**), and we will use mechanisms such as "Gradient descent" (described later) to minimize that loss function. In that case, we will use calculus to derive the slope at various points on the curve that represents the loss function, and we will use that information to work toward minimizing the cost function (see *Figure 1.9*):

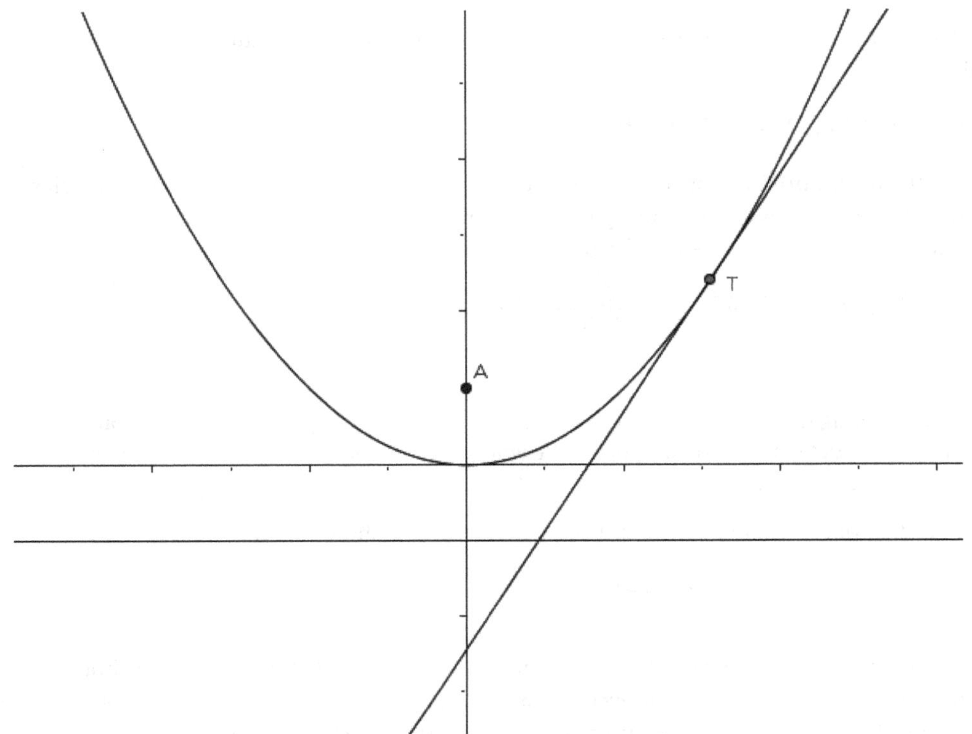


Figure 1.9: The slope at a point on a function curve (source: https://commons.wikimedia.org/wiki/File:Parabola_tangent.png)

## Statistics and probability

ML models don't provide answers as definite facts. Instead, the results from an ML model are often provided as approximations, probabilities, or inferences. We most commonly call the results from an ML model invocation an **inference**. Referencing our cat classification model from earlier in this chapter, a model that we would use to identify cats in a photograph would usually respond to us with the "probability" of a cat existing in the photograph. For example, the model may tell us that it is 97.3% sure that it sees a cat in the photograph. One of the main goals of ML is to ensure that these probabilities are as accurate as possible. A model would not be effective if it says it's 100% sure it sees a cat, but there is actually no cat in the photograph. In the case of binary classification, where the response is either true or false, there would generally be a threshold above which we consider the probability response to be true or below which we would consider it to be false. For example, we could determine that anything above 72.3% probability is deemed to be positive, and anything below that threshold is negative. The threshold value would vary based on the use case and is one of the parameters that need to be determined when building such models.

If we break the process down a bit further, in the case of the cat classification model, it has observed some features in the photograph, and based on previous training in which it has seen those kinds of features (or features similar to those), it estimates the probability of a cat being present in the photograph.

We will also see later in this book that statistical analysis plays an important role in the early stages of an ML project when data scientists are exploring how datasets can be used to address a business problem. In such data exploration activities, data scientists usually analyze the statistical distributions of values for each of the variables or features in the dataset. For example, when exploring a dataset, data scientists often want to see statistical information regarding each of the numeric variables in the data, such as the mean, median, mode, and the minimum and maximum range in the values; see *Figure 1.10*, in which the statistical distributions of some features from our house sales dataset are shown:

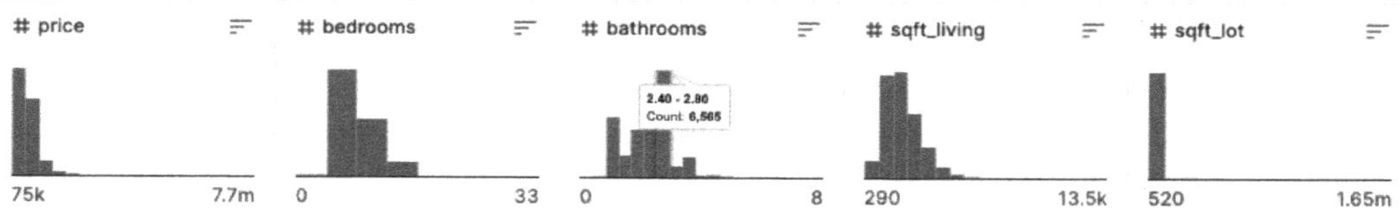


Figure 1.10: Statistical distributions of dataset features

## Metrics

> **Note**
>
> We also introduce the term *data science* here. While data science is a broad scientific field, for the purposes of this book, we use the term *data science* to incorporate all of the steps required to create an ML model, including all data preparation and processing steps.

Data science and ML are fields in which we constantly strive for improvement, whether it's to improve the accuracy of our models, how quickly they train and perform, or how much compute power they use. There's a well-known saying, "*What isn't measured cannot be improved*" (this is actually an approximation of slightly different observations from Peter Drucker and Lord Kelvin), and this saying holds a lot of truth; in order to improve something in a methodical way, you need to be able to measure some attribute of that thing. For this reason, metrics are an essential component of any ML project, and selecting the correct metric to monitor can have a critical impact on the success of an ML implementation.

Apart from operational metrics, such as measuring the latency of responses from your ML models, there are also various metrics for measuring how accurate an ML model's inferences are.

For example, in linear regression, it's common to measure the **Mean Absolute Error** (MAE), **Mean Squared Error** (MSE), or **Root Mean Squared Error** (RMSE), while for classification use-cases, we often use metrics such as Accuracy and Precision. We will explore all of these metrics, and many others, in later chapters.

Having discussed some of the underlying theory and mathematical concepts that are used in ML, let's bring the discussion back to the real world again, and let's take a look at what kinds of challenges exist for companies when they try to implement ML workloads.

# Common challenges in developing ML applications

Companies typically run into common kinds of challenges when they embark on an AI/ML development journey, and it is often a key requirement of an architect's role to understand common challenges in a given problem space. As an architect, if you are not aware of challenges and how to address them, it's unlikely that you will design an appropriate solution. In this section, we introduce the most frequently encountered challenges and pitfalls at a high level, and in later sections of this book, we discuss ways to address or alleviate some of these hurdles of AI/ML development.

## Gathering, processing, and labeling data

Data is the key ingredient in ML because, in general, ML models cannot function without data. There's an often-quoted adage that data scientists spend up to 80% of their time working on finding, cleaning, and processing data before they can begin to make use of it for analytical or data science purposes. This is an important concept to understand; that is, data scientists are not tasked simply with finding relevant data, although that is, in itself, often a difficult task; they also need to convert that data to a state that can be used efficiently by ML algorithms. Not only may data be unusable for many kinds of ML models in its raw format, but a data scientist may also need to combine data from many different sources, each with different formats and different problems that need to be addressed in the raw data before it can be used by an ML model. Also, the available data may not be sufficient to make the kinds of predictions that we'd like to get from an ML model, and data scientists often need to invent ways to generate new data by cleverly using what's available from their existing data sources. We will cover this in more detail when we discuss a practice known as "feature engineering" later in this book.

### *Data quality impact on model performance*

The effectiveness of a data scientist in performing the aforementioned tasks can have drastic impacts on how well or poorly the resulting ML models function because the data that is fed into ML models usually has a direct influence on the model's output accuracy. Bear in mind that for some business applications, a tiny difference in the ML model's accuracy can result in a difference of millions of dollars in revenue for business owners. Another well-known expression that describes this process well is "garbage in, garbage out." The concept is quite simple; if the data you feed into the model doesn't accurately represent what you're trying to predict, the model will not be able to make accurate predictions.

It's not just a model's outputs that are affected by the quality or contents of the data. Large ML models can be expensive and time-consuming to train, and inadequately prepared data can increase the time and expense required to train a model. As an architect or data scientist, these factors play a fundamental role in how we design our workloads because an architect's purpose is not just to design solutions that address technical challenges, but often what will be equally or even more important will be the cost of implementing the solution. If we designed a solution that would be too expensive to implement, then the project may not get approval to proceed, or the company may lose money by implementing the solution.

### Bias and fairness

Another important concept that we will cover in more detail in this book is the concept of bias and fairness. The goal and challenge in this regard is to ensure that the data we use to train and evaluate ML models represents a fair distribution of all relevant classes for our dataset. For example, if our model will make predictions that will affect people's lives, such as approving a loan or a credit card application, we need to ensure that the dataset used to train the model fairly represents all relevant demographic groups and does not become inadvertently biased in relation to any particular demographic groups.

### Data labeling

In addition to the challenges described previously, another substantial challenge exists specifically for **supervised ML** (**SML**) applications. As we discussed earlier in this chapter, SML models learn from labels in the data that provide the "correct" answer for each data entry. See *Figure 1.11* for an example, in which the dataset contains labels describing whether or not students passed their exams, in addition to other details regarding those exams, such as the grade received and the number of hours spent studying. However, usually, these datasets and the related labels need to be generated or created somehow, and considering that some datasets may contain millions of data points, it can be difficult, time-consuming, and error-prone to label all of the data accurately:

| Hours studying | Grade received | Passed exam |
|---|---|---|
| 8 | 76 | Y |
| 10 | 84 | Y |
| 5 | 35 | N |
| 3 | 22 | N |
| 7.5 | 75 | Y |

Figure 1.11: Example of labels (highlighted in green) in a dataset

### *Data governance and regulatory compliance*

It's important to control how data is being stored and processed within your company and who has access to the data. Special care must be taken with regard to sensitive data — for example, data that contains customers' personal details such as their address, date of birth, or credit card number. There are specific regulations that need to be upheld in this regard, such as the **California Consumer Privacy Act** (**CCPA**), the **Children's Online Privacy Protection Act** (**COPPA**), the **General Data Protection Regulation** (**GDPR**), and the **Health Insurance Portability and Accountability Act** (**HIPAA**), which outline detailed rules on how specific types of data must be handled. For companies that operate on an international scale, abiding by all of the varying regulations in different countries can be quite complicated. When data scientists are gathering, storing, exploring, processing, and labeling data, they need to keep these security requirements in mind, and as an AI/ML solutions architect, you will need to ensure that the data storage and processing infrastructure facilitates adhering to these regulations and other important information security practices. We will cover each of Google Cloud's relevant data storage and processing infrastructure options in this book and provide additional guidance on data governance concepts where appropriate.

### *Data and model lineage*

Data science contains the word "science" for a reason. As with most scientific fields, it involves iterative experimentation. When data scientists create new models, they usually go through a complex process in which they need to experiment with different datasets, different transformations on the datasets, different algorithms and parameters, and other supporting activities and resources. A team of data scientists could try hundreds of different combinations of steps before they create the desired model, and each of those steps has inputs and outputs. If a data scientist has a breakthrough discovery and creates a killer new model, and then they leave the company or something happens to them, we would not be able to recreate their work unless they kept detailed notes of all of the steps they took to create that model, including all input and output artifacts that were used and created by each step.

This is also important during the experimentation process, in which data scientists may want to collaborate with other scientists on their team or on other teams. If a data scientist gets some promising results from an experiment, they could share the details with their peers, who could validate the results or build on top of them by combining the outputs of other experiments they had performed. This kind of collaboration is fundamental to many kinds of scientific research and is often required for significant progress to occur.

Data and model lineage refers to this process of tracking all of the steps and their associated inputs and outputs that were required to create a model. It's not only important for collaboration and progress but also for governance purposes and fairness in AI/ML development; it's also important to understand how a model was created and which data artifacts, algorithms, and parameters were used along the way.

As companies begin to perform AI/ML research, they often do not have robust lineage tracking mechanisms in place, and collaboration at scale can be hampered as a result. Even worse, companies sometimes find themselves using models for which nobody has a good understanding of how they work or how they were created. This is not a good position to be in if you want to update those models or they need to be audited for compliance reasons. Later in this book, we'll see how Google Cloud's Vertex AI platform can help to ensure that data and model lineage are being tracked appropriately.

## Organizational challenges

Most large companies have evolved over time and generally consist of multiple organizations that are loosely connected to each other. As large companies begin to experiment with AI/ML, research often takes place organically in each organization without coordination between the different parts of the company. When this happens, knowledge and data are often not shared adequately — or at all — across the company, and this leads to the formation of silos within each organization, which in turn create obstacles that impede the company's overall success regarding AI/ML solution development. As an AI/ML solutions architect, you will need to advise company leadership on how to structure their organizations and their corporate policies to make their AI/ML journey as successful as possible.

Let's imagine that we own a large company, and a data science team in one organization — let's call it "Organization A" — in our company has spent the past year gathering, cleaning, and experimenting with a large dataset, and they finally had some success in training an ML model that is providing promising results. Let's also imagine that — similar to most companies — the organizations that make up our business operate mainly independently of each other, with little communication between them unless it is required as part of regular business operations. Now, another organization in our company, named "Organization B," starts exploring AI/ML, and they have a similar use case to Organization A. Because the organizations operate independently and do not regularly communicate with each other, Organization B will start from scratch and will spend the next year wasting their time doing work that has already been completed elsewhere in the company.

Now, let's imagine that our company consists of 20 large organizations, each with hundreds of product development teams. Consider how much time would be wasted if even just 20% of those product development teams started creating AI/ML workloads without communicating with each other. It may be hard to believe, but this is how most large companies operate when they begin to experiment with AI/ML.

There are mainly four different types of silos that form in the scenarios described previously, and they are related to the following four topics:

- Knowledge
- Data
- AI/ML models
- Tooling and development

### Knowledge silos

This one is pretty straightforward: if organizations are not sharing knowledge effectively with each other, teams all across the company will waste time trying to solve similar problems from scratch again and again.

### Data silos

We've already talked about the importance and difficulty of getting access to data; especially getting access to clean, processed data that is ready to be used for training ML models. In most companies, each organization (and, possibly, each team) will build its own datasets. If a team in Organization B wanted to get access to a dataset that was built by Organization A, they would first need to learn of the existence of that dataset (which requires some knowledge sharing to occur). Then, they would need to request access to the data, which can often go through months of escalations through upper management just to get the required approvals. Next, a multi-month project would need to be carried out in order to actually set up the integration between the Organization A and Organization B systems. In an industry where AI/ML use cases and opportunities are evolving so quickly, these are the kinds of obstacles and processes that kill a company's ability to rapidly innovate in this space. *Figure 1.12* shows an example of data silos in a company. You will soon learn about ways to effectively and securely share datasets between organizations in order to break down data silos:

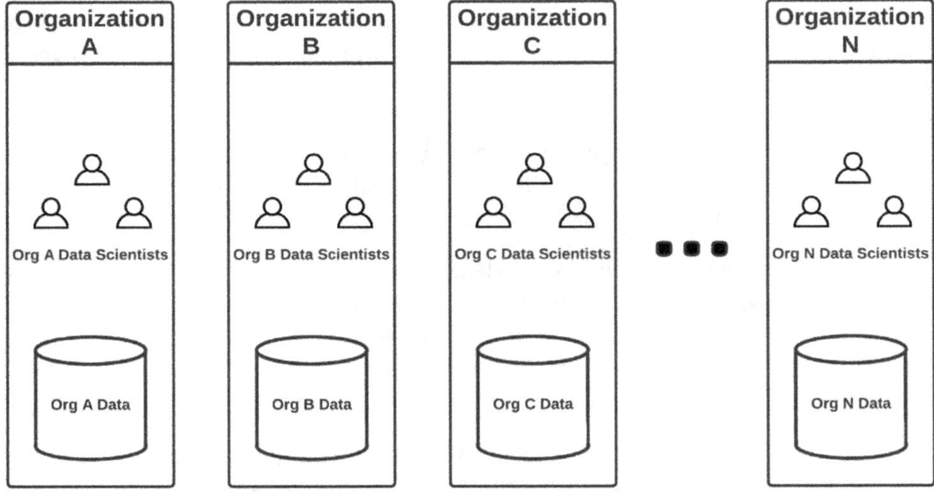


Figure 1.12: An example of data silos

### *Model silos*

This is an extension of the knowledge and data silo concepts. Just as with knowledge and datasets, some kinds of models can be reused once they've been developed. If a team in Organization A has created a useful model, and that model could be reused by other teams in the company, then we should ensure that such sharing is enabled not only by our corporate structure, culture, and policies but also by our AI/ML development infrastructure. To understand this in more detail, you will learn how to share models, what kinds of requirements that entails, and how our AI/ML development tools and infrastructure can help or hinder this process.

### *Tooling and development silos*

In large companies, various development teams may use different tools and methodologies to build their AI/ML workloads. The selection of those tools and methodologies is often based on arbitrary factors such as what kinds of tools the employees used at previous companies. Employees in each organization or team will install their chosen tools on their machines and start developing in an ad hoc manner. For example, Employee A in the Organization B organization will install and use a tool named `scikit-learn` for their development and will use MySQL databases to store their application data, while Employee B in the Organization B organization will install and use PyTorch for their development and will use Oracle databases to store their application data. Next, Employee A may leave the company, and a new employee, Employee C, will be hired, and they will prefer using TensorFlow and some other type of database. This "wild west" approach makes it very difficult for employees and teams to collaborate and share artifacts at scale across the company.

In later chapters, we will go into detail on how to prevent, fix, and design around these pitfalls, but for now, it's critical to highlight the importance of standardization. As companies begin to build their data science strategies, they should standardize as much as possible. Standardize the toolsets that will be used for AI/ML development and the types of data systems and formats that will be used. Establish company practices that encourage knowledge sharing and simplify data and model sharing across teams and organizations in a secure manner. Without these strategies, it will be difficult to collaborate and innovate rapidly at scale. One caveat is that you will need to find the balance between standardization and flexibility. Lack of standardization leads to the problems mentioned previously, but if your standardization strategy is too rigid, it could hinder your developers' productivity. For example, it would be too rigid to force all of your developers to use only one type of database and only one specific programming language or framework. Different tools are best suited to different use cases, and your company should provide guidelines to your employees on what tools are recommended for which use cases.

## Operationalization and ongoing management of AI/ML models

By now, it's hopefully pretty clear that AI/ML model development can be complicated and challenging. However, even when you've successfully created a model that makes useful predictions, your work is still not done. Companies often find it difficult to bring a model out into the real world even though it has been working well in the lab. We already covered some of the infrastructural and logistical activities that need to be performed in order to host a model, but what introduces additional complexity is that most models need to evolve over time, because the environment in which they operate will almost inevitably evolve and change over time. This is similar to regular software development, in which we need to update our applications in order to provide new functionality or to react to changes in how our customers are using our products.

Another important factor is knowing when we may need to update our models. When our models are running in the real world, we need to monitor them on an ongoing basis in order to determine whether they continue to adequately meet the business needs they were created to address.

In this book, you'll learn about the unique requirements for monitoring and updating AI/ML models and how traditional software DevOps mechanisms are not sufficient by themselves for these purposes, but how we can build upon those mechanisms to suit the needs of AI/ML workloads.

## Edge cases

The term *edge cases* is being used as a pun with a double meaning here. In traditional software development, edge cases are abnormal or extreme use cases that can cause anomalous behavior. However, in this case, we also refer to the concept of edge computing, which is a sub-field of cloud computing that focuses on providing compute resources as close as possible to customers with low-latency requirements (see *Figure 1.13*). We refer to the locations of these compute resources as "edge locations" because they exist outside the core cloud computing infrastructure locations, and they generally have limited resources in comparison to core cloud computing infrastructure locations.

ML models often require powerful compute resources in order to function, and this can present challenges for edge computing use cases due to limited resources at edge locations.

However, some ML models need to operate at or close to the "edge." For example, consider a self-driving car, which needs to perform actions to navigate in its environment. Before and after each action, it needs to consult an ML model to determine the best next actions to perform. In this case, it cannot use a model that is hosted in a far-away data center because it cannot wait for an API request to travel over the internet to a server in the cloud, and then wait for the server to provide a response before it decides what to do next. Instead, it needs to make decisions and react to its environment within milliseconds. This is a clear use case for edge computing.

In later chapters, we explore some of the requirements and solutions for these scenarios and how to address them for AI/ML workloads:

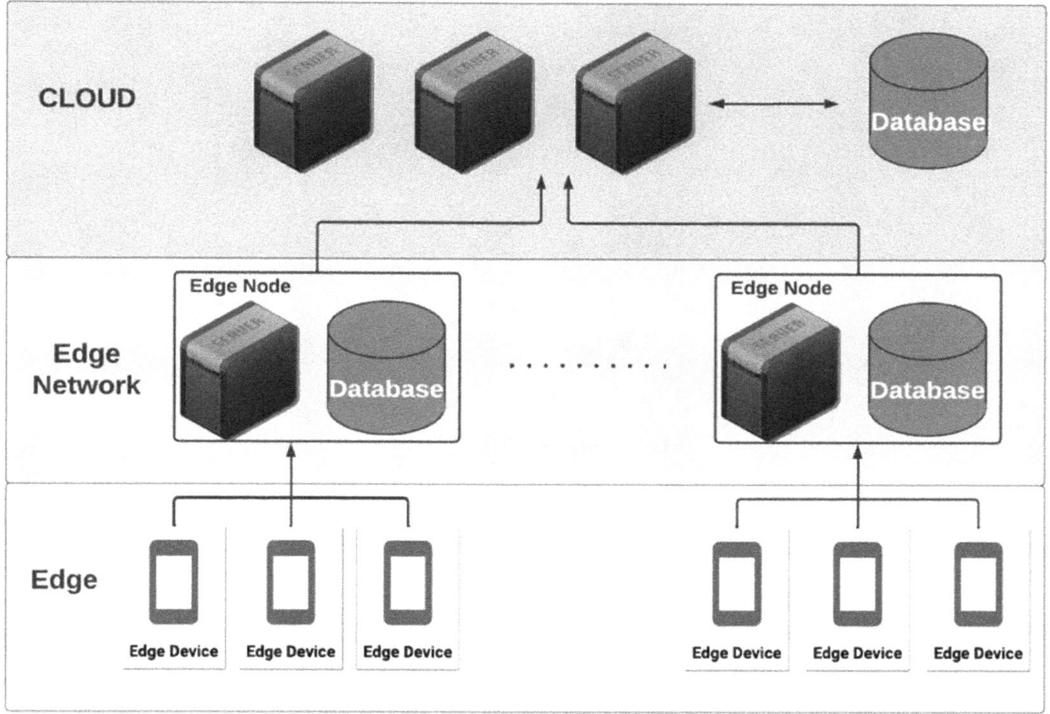


Figure 1.13: Edge computing

# Summary

In this chapter, we introduced basic terminology related to AI/ML and some background information on how AI/ML has developed over time. We also explored different AI/ML approaches that exist today and some of their applications in the real world. Finally, and perhaps most importantly, we summarized common challenges and pitfalls that companies typically run into when they begin to implement AI/ML workloads.

In the coming chapters, we will dive deeper into the model development process.

# 2

# Understanding the ML Model Development Life Cycle

In this chapter, we will explore the different steps that exist in a typical AI/ML project. This information is an important foundation for an AI/ML solutions architect role because you will need to advise companies on how to implement these steps efficiently. It is also a foundation for the rest of the contents of this book, as in later chapters, you will create your own machine learning projects, and it's important that you understand the steps in the process. We will also explore the concept of MLOps in this book and how the ML model development life cycle serves as the basis for the MLOps paradigm.

This chapter covers the following topics:

- An overview of the ML model development life cycle
- Common challenges encountered in the ML model development life cycle
- Best practices for overcoming common challenges

## An overview of the ML model development life cycle

You may be familiar with the concept of the **software development life cycle** (**SDLC**), which is taught in computer science classes in schools all over the world. The SDLC concept began to be formulated in the 1960s and early 1970s, and by now it is a well-established and well-understood process that is used in various formats by pretty much every company that develops software. Without formalized processes for people to follow when developing software, it would be difficult for companies to efficiently produce high-quality software and the software development industry would be quite chaotic. In fact, that's how the software development industry was in its early years, and that's how the machine learning industry currently is for most companies. Only in the past couple of years has the industry started to establish some structure around how companies should develop ML models and their related applications.

In this section, we provide a high-level overview of the ML model development life cycle, outlining each of the steps that you will encounter in most machine learning projects. Let's begin with a quick recap of the SDLC. We will make references to this more well-established set of processes where relevant.

> **Note**
>
> The term **MDLC**, representing **model development life cycle**, was originally coined by a friend and colleague of mine named Fei Yuan. He and I worked on building an MLOps process at Amazon before the term "MLOps" began to be used in the industry. That's not to say that we were the only people trying to automate the steps in data science projects. For example, the technical reviewer for this book shared with me that he and some colleagues had implemented MLOps-type workloads in the early 2000's with SAS and a process model called CRISP-DM, which stands for CRoss Industry Standard Process for Data Mining. You can learn more about CRISP-DM at the following URL: `https://www.datascience-pm.com/crisp-dm-2/`. Fortunately, in recent years, MLOps has become an important and popular concept in ML model development, and now there are lots of MLOps tools available, which we will cover later in this book.

## SDLC – a quick recap

One of the first and simplest versions of the SDLC is known as the Waterfall model because the flow of activities in the process is sequential, where the deliverables from each activity serve as dependencies for the next activity in the flow. *Figure 2.1* shows the original Waterfall diagram from a paper titled *Managing the development of large software systems* by Winston Royce in 1970.

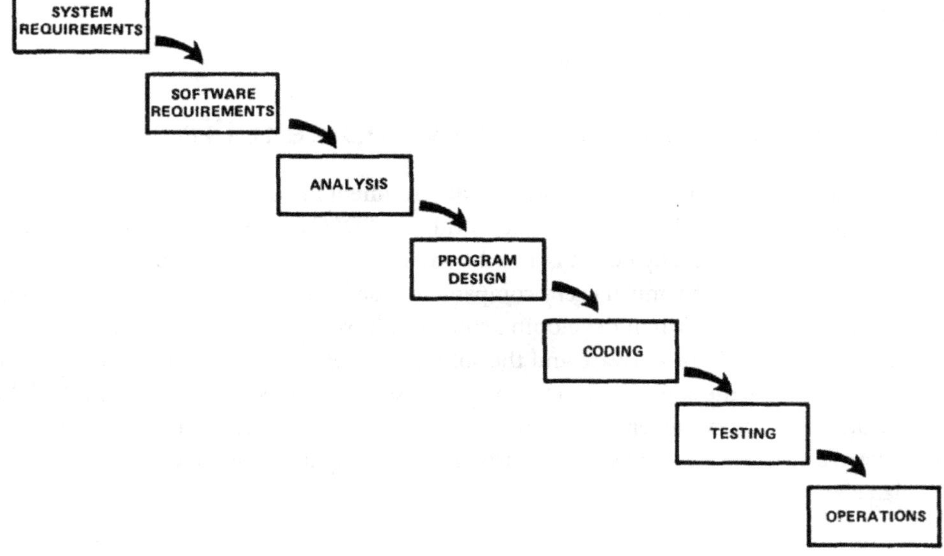


Figure 2.1: SDLC Waterfall model

As we can see, the process starts by gathering and analyzing the requirements that the system needs to satisfy and then designing, coding, and testing the software before deploying it for use. After the resulting software has been deployed, you then need to manage it via ongoing operations activities. This model was later updated to include feedback loops between the various stages. For example, the feedback from testing could result in an updated coding step, which in turn could result in an updated program design step, and so on.

A well-known limitation of the Waterfall model is that the process doesn't facilitate the rapid innovation or flexibility that is required in today's fast-paced software development industry, whereby new requirements often come to light during the development, testing, and deployment phases. The software design needs to be updated frequently, and the various stages in the process are more cyclical in nature to allow updates to occur more flexibly (see *Figure 2.2*). As a result, newer development methodologies such as Agile have emerged. Nevertheless, the methodical sequence of events from gathering requirements through to designing, coding, testing, deploying, and monitoring software still exists in various forms in system design projects, and this extends to ML systems and projects.

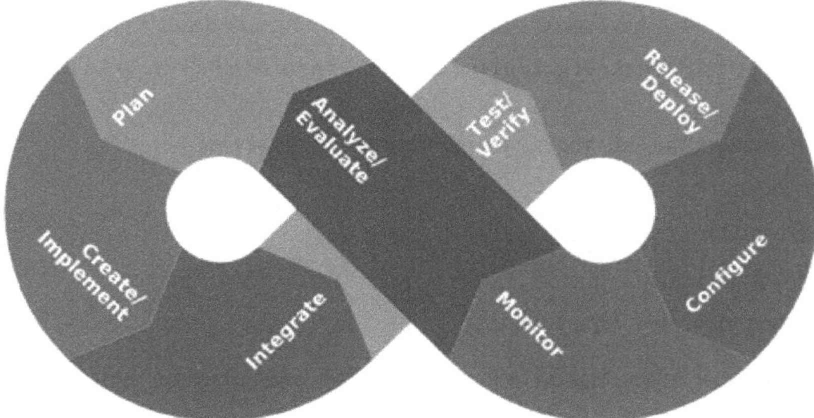


Figure 2.2: A cyclical approach to software development, often referred to as DevOps (source: https://openclipart.org/download/313185/1546764098.svg)

## Typical ML project stages

Interestingly, it took a while for the lessons from the traditional software development industry to be applied to ML model development. When ML development suddenly experienced a huge uptick in popularity during the past few years, many companies dived into the race without formalized processes in place, and as a result, companies ran into their own random unexpected issues without much ability to standardize across the industry. Fortunately, lessons have been learned from the early pioneers in this process, and standardized project activities have emerged. The following are the kinds of steps that you can expect to take in most ML development projects:

1. Gather, analyze, and understand the business requirements for which the model will be developed.
2. Find and gather relevant data.
3. Explore and understand the contents of the data.
4. Transform or manipulate the data for ML model training, which may include feature engineering and storing features for use in later steps. This step is often also closely linked to *step 6* because the selected algorithm may have specific requirements regarding how the data needs to be presented.
5. For supervised learning models, label the data if the required labels are not already present in the dataset.
6. Pick an algorithm that suits the requirements of the business case.
7. Train a model.
8. Configure and tune hyperparameters.
9. Deploy the model.
10. Monitor the model after deployment.

*Figure 2.3* shows a visual representation of these steps, and we will dive into these steps in detail in the coming sections:

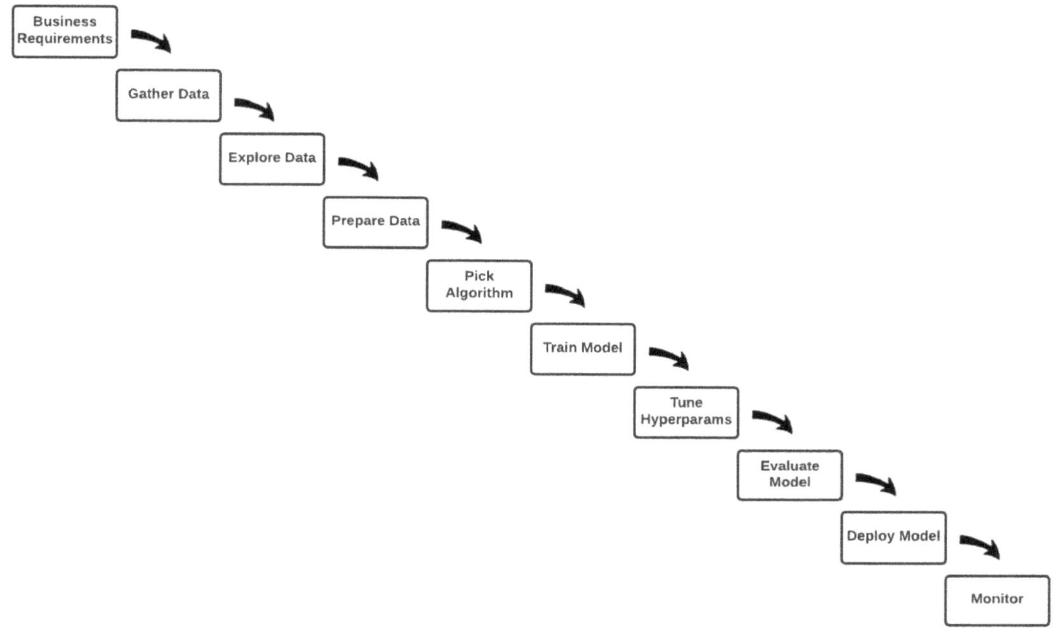


Figure 2.3: Typical ML project stages

As you can see, there are some similarities between the ML model development process and the traditional SDLC, but there are also some differences that are unique to ML model development. Most notable is the inclusion of data in the process. The fact that we now need to include the manipulation of data in our overall process adds a lot of complexity, as you will see when we go through each of the process steps in more detail. It should be noted that each step in the overall life cycle is often cyclical in nature, whereby the data science team may need to perform each task—or combinations of tasks—multiple times, with different inputs and outputs, using a trial-and-error methodology until they find the optimal approach to use for each step.

### Gathering, analyzing, and understanding the business requirements

This step is often omitted from ML life cycle diagrams because such diagrams usually focus on the technical steps, which follow later in our project. This can be considered as step zero because this generally needs to happen before any of the technical steps in our project begin. Just like in traditional software development, the overall process must begin with gathering and understanding the business requirements that the model will be built to address. For example, will the models produced by our project be used to forecast sales revenue for the next year or are we setting out to build an application that will monitor people's health data and provide them with health-related recommendations based on that data? The business requirements influence the decisions we make in the later steps of our project, such as what kinds of data we need to gather, what ML algorithms we will use to train our models, and what kinds of metrics we will measure in relation to our models' performance.

In this part of an AI/ML project, the solutions architect will work with business leaders to understand what they want to achieve from a business perspective and will then work with technical personnel to translate the business requirements into technical requirements. Defining the technical requirements is one of the first steps in defining the overall strategy to meet the business objectives outlined by the business leadership. This includes identifying any constraints that may exist, such as working with data scientists to determine what kinds of data would be required to address the business goals and whether that data can be gathered, generated, or procured from somewhere.

### *Finding and gathering relevant data*

We briefly touched on this topic in *Chapter 1*. Data is what ML models learn from, so without data, there is no machine learning. If the project team—including the data scientists and data engineers (we will explain these roles in more detail later)—cannot figure out how to get the data that would be required to meet the business objective, then the project could be a non-starter right from the beginning, so this is a critical step in the process. Sources of data vary based on the type of project, but the following are some examples of data that could be used for various AI/ML use cases:

- Historical data that contains the details of customer credit card transactions and/or banking transactions
- Data relating to what customers are purchasing online
- Housing sales data in a particular region
- Log entry data that contains details of a technical system's operational events
- Health data that is tracked by wearable devices such as watches or fitness trackers
- Data gathered from people filling in forms or surveys
- Data streamed from **Internet of Things** (**IoT**) devices such as factory conveyor belts or a fleet of construction vehicles

As you can see, there are different types of data that can be used for many different purposes. The data science team's first major task will be to define and locate the data that needs to be used for the project. This is not an atomic activity in that it does not need to happen all at once at the beginning of a project. Often, the science team begins with some idea of the data they need, and then they may refine the data requirements based on testing and feedback in later steps of the project.

### *Exploring and understanding the data*

When the data science team has gathered data that they believe could be used to address the business requirements, they usually don't just dive into training ML models on that data. Instead, they usually need to inspect the data to assess whether it really can be used adequately for the purposes of the project. Raw data is often not in an optimal state to be used by some ML algorithms. Let's take a couple of examples from our list of potential data sources. If we're using data gathered from people filling in forms or surveys, people may input the details incorrectly. They may leave some fields blank or misspell some of the details during input. As another example, if we're using data from sensors such as wearable health trackers or other IoT devices such as mechanical machinery sensors, those sensors may malfunction and record corrupted data. As such, data scientists often need to inspect the data and look for errors, anomalies, or potentially corrupted data. In *Chapter 1*, we also mentioned that data scientists may want to get statistical details regarding the data, such as the range of values that are generally seen for particular variables in the data or other statistical distribution details. In the hands-on activities later in this book, we will be using data visualization tools and other data inspection tools to explore and understand the contents of our datasets.

### *Transforming or manipulating the data for ML model training*

Missing or corrupted data can cause problems when training ML models. Some algorithms that need to operate on numeric data will produce errors if they encounter non-numeric values, including null and garbled/corrupted characters. Even for algorithms that can handle such values gracefully, the values may skew the learning process in unexpected ways, thus affecting the performance of the resulting models.

When data scientists find that a dataset isn't perfectly ready for use in training ML models, they don't usually just give up but rather try to make changes to the data to bring it closer to the desired state. We call this process **feature engineering**.

> **Note**
>
> Some literature publications use the term "feature engineering" only to refer to the process of creating new features from existing features (such as our price per square foot example), while other literature uses the same term to describe all activities related to manipulating features in our dataset, including replacing missing values.

This could include data cleaning (or cleansing) techniques, such as replacing missing data with something more meaningful. As an example, let's imagine that some medical condition is more likely to occur as a person gets older and we want to build a model that predicts the likelihood of this condition occurring. In this case, a person's age would be an important input feature in our dataset. During our data exploration activities, if we discover that some of the records in our dataset are missing the age value, we could compute the average age of all of the people in our dataset and replace each missing age value with the average age value. Alternatively, we could replace each value with the modal (i.e., the most frequently occurring) age value. Either of these cases would at least be better than having missing or corrupted values in our dataset during the training process.

Also, the optimal variables and values for addressing specific business requirements may not be readily available in whatever raw data we can access. Instead, data scientists often need to combine data from different sources and come up with clever ways to derive new data from the available data. A very simple example would be if we specifically need a person's age as an input variable but the dataset only contains their date of birth. In that case, the data scientist could add another feature, age, to the dataset, and subtract the date of birth from the current date in order to calculate the person's age. A slightly more complex example would be if we wanted to predict housing prices and we determined that price per square foot would be an important input feature for our model but our dataset only contains the total price for each house and the total area of each house in square feet. In that case, to create our price per square foot input feature for each house, we could divide the total cost of each house by the total area of that house in square feet and then add that as a feature in our dataset.

It's important to understand that when a data scientist has created the features that are important for training their model, they will often want to store those features somewhere for later use, rather than needing to re-create them again and again. Later in this book, we will explore tools that have been developed for this purpose.

### Data labeling

As we discussed in *Chapter 1*, supervised learning algorithms rely on labels in the dataset during the training process, which tell the models what the correct answers are for the types of data relationships that the model is trying to learn. *Figure 2.4* shows our example of a labeled dataset.

| Hours studying | Grade received | Passed exam |
|---|---|---|
| 8 | 76 | Y |
| 10 | 84 | Y |
| 5 | 35 | N |
| 3 | 22 | N |
| 7.5 | 75 | Y |

Figure 2.4: An example of labels (highlighted in green) in a dataset

If you're lucky, you will find a dataset that can be used to address your business requirements and already contains the requisite labels or "correct answers" for the variables that you want to predict. If not, you will need to add the labels to the dataset. Again, considering that your dataset could contain millions of data points, this could be a significantly complex and time-consuming task to take on. And, just like any of the other features in your dataset, the quality of your labels directly impacts how reliable your model's predictions will be. With this in mind, you will need access to a labor force that can accurately label your dataset and other tools that facilitate the labeling.

Another data preparation step that is used for supervised learning algorithms is to split the dataset into three different subsets, which are used for the training, validation, and testing of the model, respectively. We will describe the use of these subsets in the model training section later in this chapter.

An important thing to call out at this point is the concept of **data leakage**, which refers to a scenario in which information from outside the training dataset is used to create the model. This can cause the model to perform well on the training data (because it has information it wouldn't have in a real-world scenario) but perform poorly in production due to these unintentional hints.

There are various causes that can lead to data leakage, such as how and when we split our datasets during our data science project, or how we label our data. For example, during data preparation activities such as labeling or feature engineering, we could accidentally include knowledge that would not be available to the model in a real-world application when the model needs to make predictions. Consider a scenario in which we are using historical data to train our model. We may accidentally include information that only became available after the events that are represented in the dataset actually occurred. While this data may be relevant and may help influence the outcome, it will harm our model performance if that information would not be available to our model in a real world scenario at the time when our model needs to make a prediction.

#### *Picking an algorithm and model architecture*

There are lots of different types of ML algorithms that can be used for various purposes, with new algorithms and model architecture patterns emerging regularly. In some cases, choosing your approach is an easy decision because there are some algorithms and model architectures that are particularly suited to specific use cases. For example, if you want to implement a computer vision use case, then something such as a convolutional neural network architecture would be a good starting point. On the other hand, choosing what kind of ML algorithm and implementation to use for a specific problem can be a difficult task that often depends on the experience of the data science team. For example, an experienced team of data scientists will have worked on many different projects and developed a working understanding of what kinds of algorithms work best for various circumstances, whereas a less experienced data science team may need to perform a lot more experimentation with various algorithms and model architectures.

In addition to direct business requirements, such as "we need a computer vision model to identify manufacturing defects", the chosen algorithm can also depend on less-tangible business requirements, such as "the model needs to run on limited computing resources" or "the explainability of the model is extremely important in this use case." Each of the aforementioned requirements puts different constraints on the types of algorithms the data science team could select for the given use case.

As with most of the steps in the overall AI/ML project life cycle, selecting the best algorithm and model architecture to use can require the data science team to implement a cyclical trial-and-error approach, whereby they may experiment with different algorithms, architectures, and inputs/outputs until they find the optimal implementation. We will be exploring various algorithms and their unique characteristics in the hands-on activities later in this book, but overall, it is best to start with a simple baseline model, so that we have a starting point to compare metrics and understand the base dataset. Then, we can test out more complex models and assess whether they perform better.

### Training a model

This is probably the most well-known activity in the AI/ML project life cycle. It's where the model actually learns from the data. For unsupervised algorithms, this is where they may form the clusters that we talked about in *Chapter 1*, for example. For supervised algorithms, this is where our training, validation, and testing datasets come into the picture. In *Chapter 1*, we briefly talked about how linear algebra and calculus can be used in machine learning. If we take our linear regression example, this is exactly where those concepts would come into play. Our model would first try to find the relationships between the features and the labeled target outputs. That is, it would try to find the coefficients for each of our features, which, when used in combination (e.g., by adding them all together), would produce the labeled target output. It tries to calculate the coefficients that would work for all data points in the dataset, and to do this, it needs to scan through all of the items in the training dataset.

The model usually starts this process with a random guess, so it inevitably is incorrect on the first try. However, it then calculates the errors and makes adjustments to minimize those errors in future iterations through the dataset. There are a number of different methods and algorithms that can be used to minimize errors, but a very popular method is called gradient descent. We briefly mentioned gradient descent in *Chapter 1*, but we'll talk about it in more detail here. In gradient descent, the algorithm works on finding the minimum point of what we call the loss function, which is a representation of the errors that are produced when our model tries to guess the coefficients of the features that produce the labeled outputs for each data point in our dataset. *Equation 2.1* shows the equation for calculating the **mean squared error** (**MSE**) as an example for a loss function for linear regression:

$$MSE = \frac{1}{n}\sum_{i=1}^{n}(y_i - \hat{y}_i)^2$$

Equation 2.1: Mean squared error formula

In *Equation 2.1*, *n* represents the number of data points in our dataset.

To understand what this formula is saying, let's start with the section in parenthesis: $(y_i - \hat{y}_i)$

$y_i$ represents our model's predicted target variable for each datapoint and $\hat{y}_i$ represents the true target variable's value for each datapoint.

Within the parentheses in *Equation 2.1*, we subtract the true value from our model's predicted value in order to calculate the error of our model's prediction, similar to what we described in *Chapter 1*, and we then square the result. In this case, we are calculating what's referred to as the **Euclidean distance**, which is the distance between the predicted value and the true value in two-dimensional space. Squaring the result also has the effect of removing negative values from the results of our subtractions.

The summation symbol, Σ (sigma), in the equation represents adding up all of the calculated errors for all data points in our training dataset. Then, we divide the final result—i.e., the total error for all predictions—by the number of data points in our dataset in order to calculate the average (or mean) error for all of our predictions.

Remember that we want to minimize the error in each training iteration, by finding the minimum point of this loss function (also referred to as the **objective function**). To get an understanding of what it means to find the minimum point in a loss function, it helps if we can graph the function. *Figure 2.5* shows an example of a two-dimensional loss function graph for the MSE:

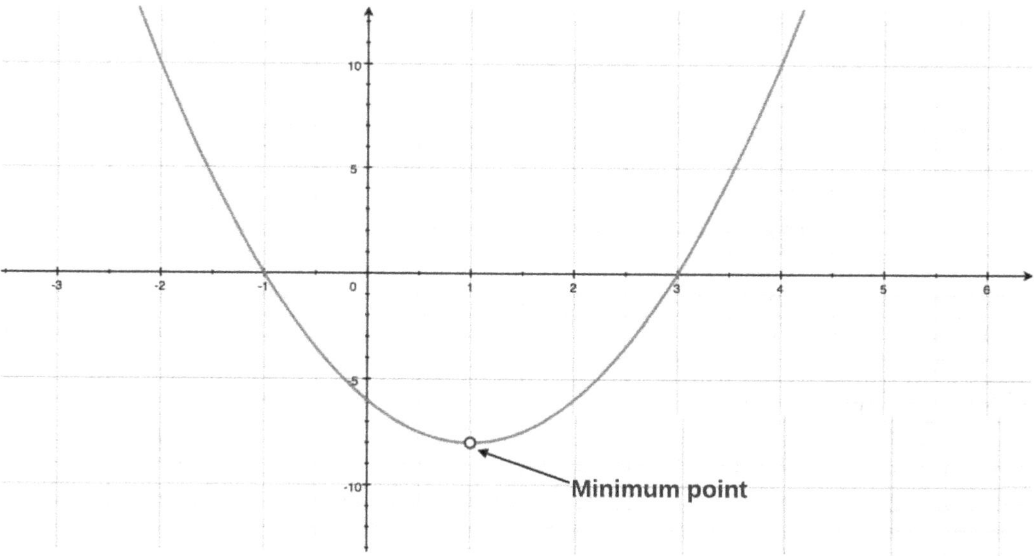


Figure 2.5: MSE loss function graph showing minimum point

Each time the algorithm calculates the loss from each training iteration, that loss value can be represented as a point on the graph. Considering that we want to move toward the minimum point in order to minimize the loss, we want to take a step downward on the graph. Whenever we want to move from one point to another (even when moving our bodies around in real life), there are two aspects of the movement that we need to determine: direction and magnitude, i.e., in which direction do we want to move and how far? This is where gradient descent comes into the picture. It helps us to determine the direction in which we should move in order to progress towards the minimum point. Let's take a look at how it works in more detail.

Imagine the graph in *Figure 2.5* is a valley between two mountains and we are standing at a point that represents the loss calculated in our most recent training iteration. That point is somewhere on the side of the valley, such as the location shown in *Figure 2.6*.

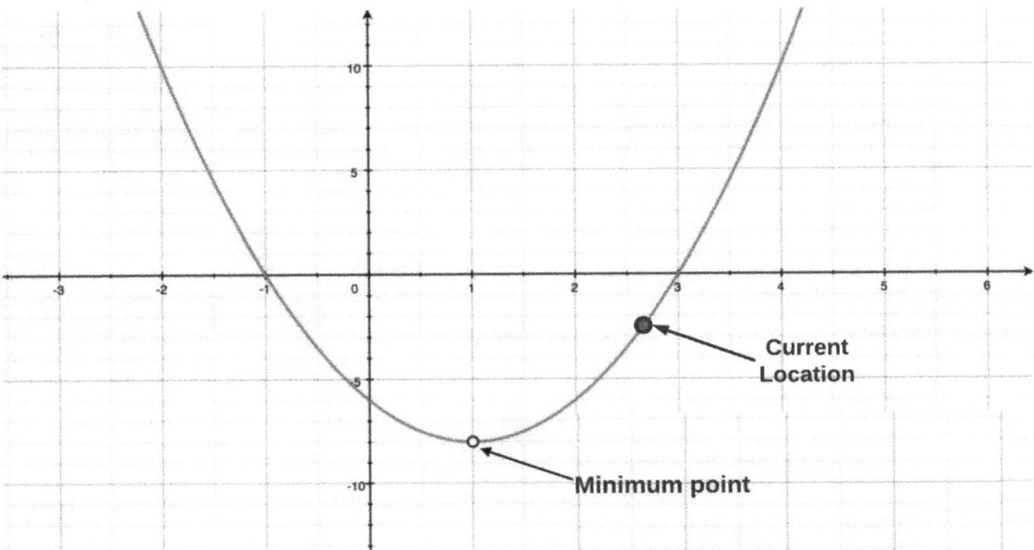


Figure 2.6: MSE loss function graph showing current location

For a human, walking downhill is somewhat instinctive because we have sensory input that tells us which way is downhill. For example, our feet can feel the slope of the hill at our current location, we can feel the pull of gravity downwards, and we may also be able to see our surroundings and therefore see which way is downhill. However, our gradient descent algorithm does not have these sensory inputs and it can only use mathematics to find out which way is downhill. Then, it needs to use programmatic methods to define what it means to take a step in that direction. Our algorithm knows the current location on the graph and it can calculate the derivative of the function (a concept from differential calculus) to determine the slope of the graph at the current location. *Figure 2.7* shows an example of the slope of a line at a point on the graph, which represents the derivative of the function at that point.

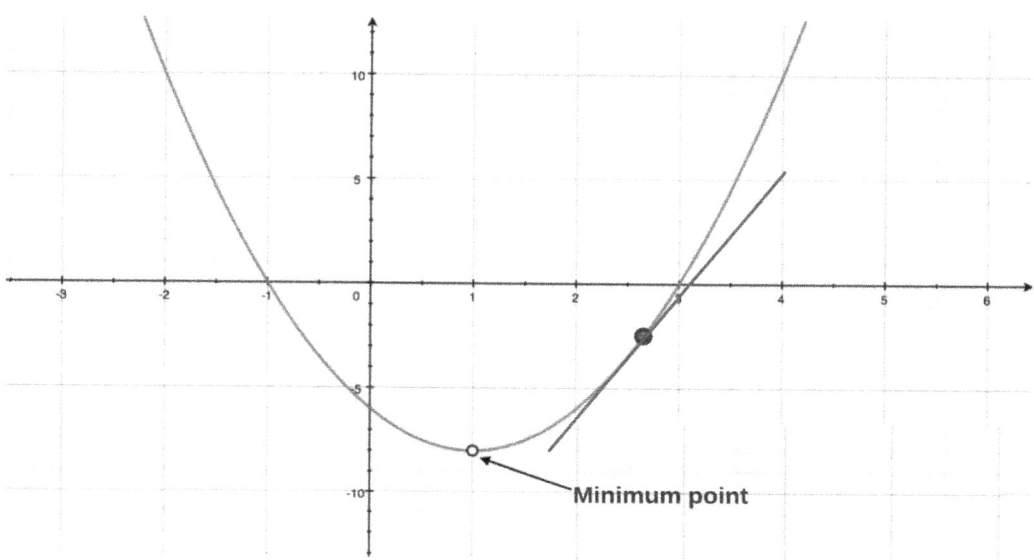


Figure 2.7: Derivative of the loss function at a particular point

When the derivative has been calculated, this information can be used by our algorithm to take a step toward the minimum point. *Equation 2.2* shows how each next step is calculated for gradient descent in the context of linear regression:

$$\theta_j = \theta_j - \alpha \frac{1}{m} \sum_{i=1}^{m} \left( h_\theta(x^{(i)}) - y^{(i)} \right) x_j^{(i)}$$

Equation 2.2: Gradient descent for linear regression

In *Equation 2.2*, $\theta_j$ represents the location on the graph, and Θ represents the vector of coefficients for the features of each data point in our dataset. Remember that we are trying to find the set of coefficients for our features that results in the least error between our model's predictions and the true values of our data points' target variables:

- $h_\theta(x^{(i)})$ then represents the predicted target variable for each data point
- $y^{(i)}$ represents the true target value for each data point

As we can see, in the broader set of parentheses, we are again subtracting the true target value from the predicted target variable value for each data point. This is because *Equation 2.2* is derived from *Equation 2.1* (the mathematical proof of this derivation is omitted here for simplicity).

$m$ represents the number of features we have for each data point in our dataset.

$\alpha$ is what we refer to as the **learning rate**. It's one of the hyperparameters of our algorithm and it determines the size of the step we should take; i.e., the magnitude by which we should move in the selected direction.

Altogether, $\frac{1}{m}\sum_{i=1}^{m}\left(h_\theta(x^{(i)}) - y^{(i)}\right)x_j^{(i)}$ represents the derivative of the loss function at our current location on the graph. Therefore, *Equation 2.2* says that our next location will be equal to our current location minus the derivative of the loss function at our current location on the graph multiplied by the learning rate. This then takes a step toward the minimum point, whereby the derivative of the loss function determines the direction and the combination with the learning rate defines the magnitude.

It's very important to note that *Equation 2.2* represents just one step in the gradient descent process. We iterate over this process many times, in each iteration scanning through the dataset and inputting our estimated coefficients, calculating the error/loss, and then reducing the loss by using gradient descent to step toward the loss function's minimum point. We may never exactly reach the minimum point but even if we can get quite close, our model's estimates and predictions could be acceptably accurate.

We should note that there are different configurations for gradient descent. In batch gradient descent, we would go through the entire training set in each iteration. Alternatively, we could implement mini-batch gradient descent, in which case each iteration would process subsets of the training dataset. This approach is less thorough but can be more efficient. A popular implementation is known as stochastic gradient descent (the word "stochastic" means "random"). In stochastic gradient descent, we take a subset of random samples from our dataset in each iteration, which could be as little as one data point in each sample. The key here is that because we take a random subset in each iteration, we start from a different point in our feature space each time. At first, this seems somewhat erratic, as we jump around to different points in our feature space. However, this approach has been shown to be quite effective in minimizing the overall loss function. It can also help to avoid something referred to as **local minima**, which refers to the fact that some loss functions are not as simple as the one we showed in *Figure 2.5*. They may have multiple peaks and valleys, in which case the bottom of any of the valleys could be considered a type of minimum point, but a local minimum may not be the overall minimum of the function, which is referred to as the **global minimum**. *Figure 2.8* shows an example of multiple minima and maxima:

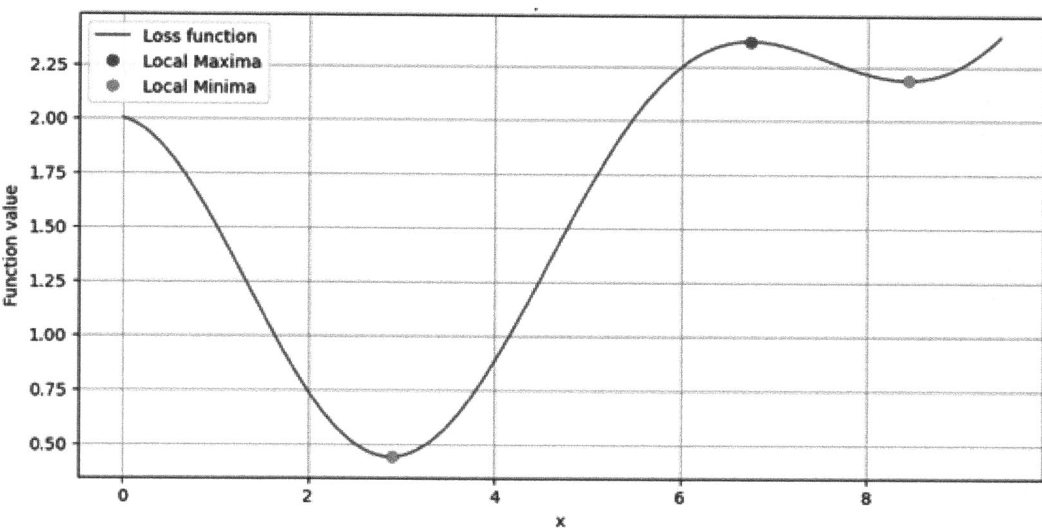


Figure 2.8: Local and global minima and maxima

Although we focused on two-dimensional loss functions in this section, loss functions can have more than two dimensions, and the same concepts also apply in higher-dimensional space.

> **Note**
>
> In this section, we chose a specific algorithm (linear regression) and type of dataset (tabular) to illustrate an example of the model training process. There are, of course, other algorithms and types of data for different use cases, and the training processes for those use cases would have their own unique implementations. However, the overall model training process generally involves processing the input data to try to find some kind of useful pattern or relationship and then incrementally honing the accuracy of that pattern or relationship in a repetitive way until some determined threshold of accuracy is met or until the training is deemed to be ineffective (if the model is failing to effectively learn any useful patterns or relationships) and therefore is terminated.

### *Configuring and tuning hyperparameters*

Hyperparameters are parameters that define aspects of how your model training jobs run. They are not the parameters in your dataset from which your models learn but rather external configuration options related to how the model training process is executed. To start with a simple example: we've discussed that training jobs often need to cycle through the training dataset multiple times in order to learn patterns in the data. One of the hyperparameters you may configure for a model training job would be to specify how many times it should cycle over the dataset. This is often referred to as the number of **epochs**, where an epoch represents one iteration through the training dataset.

Choosing the optimal values for your hyperparameters is another activity that usually requires a lot of trial-and-error attempts, in which you may need to experiment with different combinations of hyperparameter values in order to find the optimal settings to maximize your model training performance. Let's continue our simple example of configuring the number of times the training job should process through the training dataset. Considering that the model could potentially learn more information each time it runs through the dataset, we may at first think that deciding on the number of epochs would be a simple choice, i.e., that we should just set this value to be very high so that the model learns more from the data. However, in reality, this is often not the best option because it's not always true that a model learns more useful information every time it processes through the dataset. There are concepts in machine learning called **underfitting** and **overfitting**, which we will explore in the *challenges* section of this chapter. They relate to problems in which continued training on the existing dataset will fail to yield desired results.

Even in cases when the model does learn useful information each time it processes the dataset, it often reaches a point where the rate at which it is learning new information will taper off or reach a plateau. When this happens, it would be inefficient to continue running through the dataset again and again. Bear in mind that it can be very expensive to train a model on a large dataset, so you do not want to keep processing the data when the model is reaching a plateau in its learning. We can measure the rate of learning by generating a **learning curve** graph, which graphs the training errors during the training process. *Figure 2.9* shows an example of a learning curve:

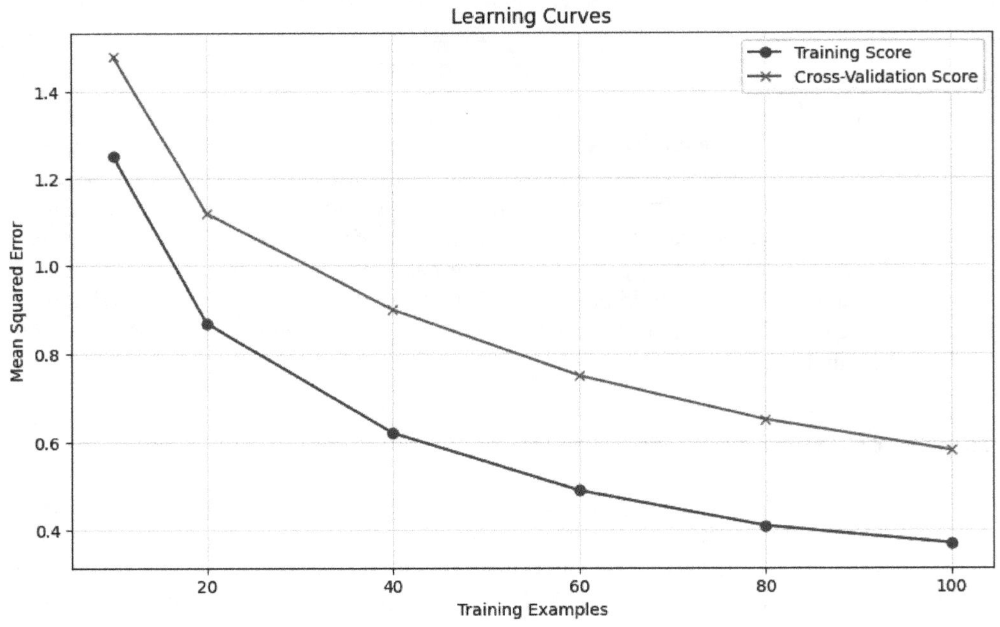


Figure 2.9: Learning curve graph

In *Figure 2.9*, the gap between the blue and red lines represents the error between our model's predictions on the training dataset and the testing dataset. When the graph shows that the gap between the blue and red lines is not significantly decreasing, then we know that continued training will not make our model significantly more accurate, and it may be a good point at which to stop the training process.

An analogy would be if we had a student who had read a book so many times that they memorized every word and thoroughly understood every concept in the book. At that point, it would be unnecessary to continue instructing the student to read the book again and again because they would no longer learn anything new from that book. Now let's also imagine that we needed to pay the student every time they read the book, which is analogous to paying for the computing resources required to train the model on the dataset.

Configuring the number of epochs is just one example of trying to find the best configuration for a particular hyperparameter. Different types of algorithms have different types of hyperparameters that pertain to them, and trying to test all of the various combinations of hyperparameter values can be a painstaking and time-consuming task for data scientists. Fortunately, Google Cloud has a tool that performs this task for us in an automated fashion, which we will explore later in this book.

### Deploying the model

Finally! We have found the right combinations of data, algorithms, and hyperparameter values and we've gotten to a point where our model is ready to be used in the real world, which we refer to as **hosting** or **serving** our model. Getting to this point has required a lot of work. We may have trained hundreds of different versions of our model to get it to a point at which it's ready to be deployed in production. Our data science team may have had to experiment with lots of different dataset versions and transformations and lots of different algorithms and hyperparameter values in order to finally get some meaningful results or insights. Up until only around five years ago, performing all of those steps and tracking their results would have been a very slow, manual, and painstaking process. It still can be a lot of work, but at least now there are tools that automate a lot of these steps and they can be completed much more quickly, perhaps taking weeks instead of months. Later in this book, we'll talk about something called AutoML, which can reduce this entire process down to a few minutes or hours in just a few short commands!

Deploying our model can be as simple as packaging it into a Docker container and deploying the container on a server, although we will usually want to create some kind of web-based API to facilitate access to the model by applications. We will do this in a hands-on activity later. Also, when we cover MLOps, we'll see how it may make sense for us to create pipelines to automate the deployment of our model, in much the same way as we use CI/CD pipelines to build and deploy regular software applications.

### *Monitoring the model after deployment*

You might think that once you've successfully tested and deployed the model to production then your job is done. However, the fun doesn't stop there! Just like regular software, you need to monitor the performance of your model on an ongoing basis. This includes traditional monitoring, such as keeping track of how many requests your model is serving in a given timeframe (per second, for example), how long it takes for your model to respond to a request (latency), and whether these metrics are changing over time. However, ML models also have additional requirements that need to be monitored, such as ML-specific metrics such as those mentioned in *Chapter 1* (MAE, MSE, accuracy, precision, etc.). These metrics help us to understand how our models are performing from an inference perspective, so we need to monitor them and ensure they continue to meet our business requirements.

> **Note**
>
> The stages in the ML model development life cycle that you've learned about in this section form the basis for understanding MLOps and AutoML. We dedicate an entire chapter of this book to ML engineering and MLOps on Google Cloud, but for now, at a high level, you can consider the goals of MLOps and AutoML to be the automation of all of the steps in the ML model development life cycle. You'll see in the chapter on MLOps that there are tools we can use to create pipelines that automate all of the outlined steps. We can have complex combinations of pipelines within pipelines, which would automate everything from preparing and transforming the input data to training and deploying the models, monitoring the models in production, and automatically kicking off the entire process all over again if we detect that our model has stopped providing desirable results and we want to retrain the models on updated data. This would provide a self-healing model ecosystem, which helps to keep our models up to date on an ongoing basis.

## Roles and personas in AI/ML projects

Throughout this book, we mention various roles such as data scientist, data engineer, and ML engineer. We also mention more traditional roles such as software engineer, project manager, stakeholder, and business leader. The traditional roles have been defined in the industry for decades now, so we will not define them here, but there is often confusion about the newer roles that are specific to AI/ML projects, so we'll take some time to briefly describe them here. In small teams, it should be noted that a single person may perform all of these roles:

- **Data engineer**: Data engineers are usually involved in the earlier stages of the data science project—specifically the data gathering, exploration, and transformation stages. A data engineer will often be tasked with finding relevant data and cleaning it up to be used in later stages of the project.

- **Data scientist**: A data scientist is usually the role we associate with actually training ML models. They will usually also perform data gathering, exploration, and transformation activities as they iterate through various model training experiments. In some cases, data scientists will be the more senior members working on the project and they may provide direction to the data engineers and ML engineers. They will often be responsible for the resulting ML models that are created, although those models are created and deployed with help from the data engineers and ML engineers.
- **ML engineer**: The ML engineer role usually refers to a software engineer who has ML or data science expertise. They understand ML concepts and they are experts in the stages of the model development life cycle. They are usually the bridge that brings DevOps expertise to an ML project in order to create an MLOps workload. When a data scientist creates a model that they believe is ready to be used in production, they may work with an ML engineer to put all of the mechanisms in place to actually deploy the model into production via an MLOps pipeline.

Now that we've covered the major steps and concepts found in a typical AI/ML project, let's take a look at the kinds of pitfalls that companies often run into when trying to implement such projects.

# Common challenges encountered in the ML model development life cycle

For some of the stages in the ML model development life cycle, we've already discussed various challenges that you are likely to encounter in those stages. However, in this section, we specifically call out major challenges that you need to be aware of as an AI/ML solutions architect interacting with companies who are implementing AI/ML workloads. In the *Best practices for overcoming common challenges* section later in this chapter, we'll look at ways to overcome many of these challenges.

## Finding and gathering relevant data

One of our first major challenges is finding relevant data for the business problem our models are being built to address. We presented some examples of potential data sources in the previous section, and in some cases, the data you need may be readily available to you, but finding the relevant data is not always straightforward for data scientists and data engineers. The following are some common challenges with regard to finding and accessing the right data:

- If you work in a large company, the data may exist somewhere in your company owned by another team or organization, but you may not know about it or you may not know how to find it.
- The data might need to be created from a combination of different data sources that are dispersed throughout your company, all owned by disjointed organizations.
- The data could contain sensitive information, and therefore be subject to regulations and restrictions regarding how it needs to be stored and accessed.

- You may need to consult with experts in order to find, validate, and understand the data, e.g., financial data, medical data, atmospheric data, or other data related to fields of specific expertise.
- The data may be stored in databases that are restricted for use only in production transaction operations.
- You may not know whether you can trust the contents of the data. For example, is it accurate?
- The data could contain inherent biases or other unknown challenges.

## Picking an algorithm and model architecture

When it comes to picking an algorithm or model architecture to use, one of the biggest challenges initially can be just figuring out where to start. You don't want to spend months just experimenting with different options before finding a useful implementation or never finding a useful implementation.

## Data labeling

Data labeling can be a very manual task, requiring humans to go through enormous amounts of data and add the labels for each data point. Tools have been built in recent years that automate some labeling tasks or make the tasks easier for humans to perform, but there is still a need to have humans involved in labeling datasets. So, a major challenge that can exist for companies is finding and hiring a strong data-labeling workforce. Bear in mind that some data labeling tasks may require specific expertise. For example, consider a dataset consisting of medical images. A specific characteristic in the image could indicate the presence of a particular medical condition. It often requires special training to be able to read the medical images and identify the specific characteristics in question, so this is not a task for which you could hire any random person. As we've discussed previously, if your labels are not accurate, then your models will not be accurate, and in the example of medical imaging described here, this could have critical implications for a medical facility that is tasked with diagnosing life-threatening medical conditions.

## Training models

Two classic challenges with regard to training models are the problems of **underfitting** and **overfitting**. These challenges relate to how well your model learns a relationship or pattern in your data.

In the case of supervised learning, we generally split our dataset into the three subsets mentioned earlier: training, validation, and testing. The validation set is usually used in hyperparameter tuning, the training dataset is what the model is trained on, and the testing dataset is how we evaluate the trained model. We evaluate the model based on the metrics that we've defined for that model, such as accuracy, precision, or MSE. In this context, the test dataset is new data that the model did not see during the model training process and we want to determine if the model's predictions are accurate when it sees this new data—i.e., we want to determine how well the model **generalizes** to new data.

If a model provides very accurate predictions on the training dataset but inaccurate or less accurate predictions on the testing dataset then we say that the model is overfitting. This means it fits too closely to the training data, and it cannot perform well when it is exposed to new data.

On the other hand, if we find that the model is not performing well on either dataset (training or testing), then we say it is underfitting.

*Figure 2.10* shows an example of fitting, overfitting, and underfitting for a classification model that is trying to determine the difference between the blue and red data points. The black boundary line represents overfitting because it fits much too precisely to the dataset, and the purple line represents underfitting because it does not do a good job of capturing the differences between the blue and red dots. The green line does a pretty good job of separating the blue and red dots; it's not completely perfect, but it could be an acceptable model that generalizes well to the characteristics of the data points.

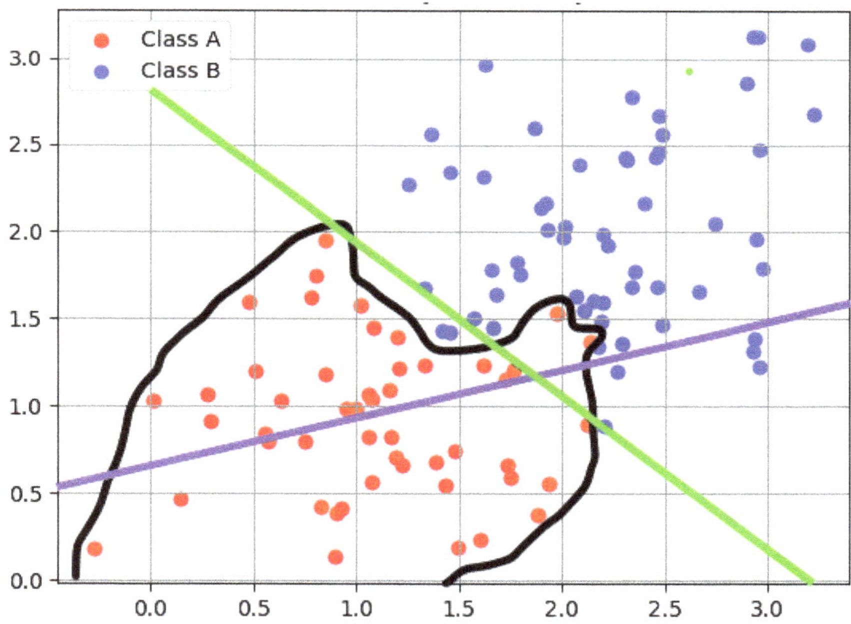


Figure 2.10: Example of fitting, overfitting, and underfitting

In addition to classic training challenges such as those mentioned, there are other challenges that relate to the overall process of how training is performed as part of a broader AI/ML project, such as lineage tracking, as we mentioned in *Chapter 1*. In large AI/ML projects, there may be multiple teams of data scientists performing experiments and training hundreds of models. Keeping track of their results and sharing them with each other can be very challenging in large projects.

## Configuring and tuning hyperparameters

Finding the optimal set of hyperparameter values can be almost as challenging as curating a usable dataset or choosing the correct algorithm to use. Considering that hyperparameters affect how our model training jobs operate, each job can take a long time to run, and there can be thousands of combinations of hyperparameter values to explore, doing this manually can be very challenging and time-consuming.

## Evaluating models

Although we generally perform some validation and testing during the training and hyperparameter tuning processes, we should thoroughly evaluate our model before deploying it in production. In the section titled "Common challenges in developing machine learning applications" in Chapter 1, we discussed the challenge of delivering business value from data science projects. We called out the need for data scientists to work with business stakeholders to thoroughly understand the business requirements that the target AI/ML system is intended to address. The evaluation step in our data science project is where we check to ensure that the models and solutions we've created adequately address those business requirements, based on the metrics we defined for measuring success. In addition to the data science team evaluating the models, it may also be relevant at this point to review the results with the business stakeholders to ensure they align with expectations. If we find that the results are not satisfactory, we usually need to retry the process from earlier in the data science lifecycle, perhaps with new or different data, different algorithms, and/or different hyperparameter values. We may need to repeat these steps in an interactive process until we appropriately satisfy the business requirements.

## Deploying models

When deploying our model, we'll need to select the kinds of computing resources that are required to serve our model adequately. Depending on the model architecture, we may need to include GPUs/TPUs, usually also in combination with CPUs, and, of course, RAM. A really important activity at this point in the project is to "right-size" these components. To do this, we'll need to estimate how much of each type of component will be required to serve our model as accurately as possible given the traffic we expect to receive in terms of requests per second to our model. Why is this so important? Well, model hosting is often reported by companies to be by far their number one cost when it comes to their AI/ML. It can account for up to 90% of a company's AI/ML costs. So, if we configure our servers with more resources than we need, it will increase those costs. On the other hand, if we don't configure enough resources, we will not be able to handle the number of requests coming from our clients, resulting in disruption of service from our models. Fortunately, cloud AI/ML services such as Vertex will auto-scale our model-hosting infrastructure to meet increased demand, but we still need to get the sizing as accurate as possible for each server in order to control our costs.

Something we also need to keep in mind when deploying our models is how quickly our applications need to get responses for the inference requests. We will need to keep this in mind as we architect our model hosting infrastructure.

## Monitoring models after deployment

In addition to monitoring the various metrics associated with our models, an important concept to call out at this point is something referred to as **drift**, which can be represented in various formats, such as model drift, data drift, or concept drift. To explain the concept of drift, we will first dive a bit deeper into the relationship between the model training process and how models operate in production.

Note that during model training, the model was exposed to data in a specific format and with specific constraints, and that's how it learned. Let's refer to the state of the input data as its **shape**, which refers to the format and constraints of that data. When we deploy our model to production and we expose it to new data in order to get predictions from the model, we want to ensure that the shape of the new data matches the shape of the training data as much as possible. We are not referring to the contents of the data but rather how the data is represented to the model.

During our training process, we may have performed transformations on raw data to make it more suitable for training the model. If so, we need to perform the same kinds of transformations on any new data that we send to the model to make predictions. In the real world, however, data can change over time, and therefore the shape of the data can change over time. We refer to this as drift, which is when the raw data out in the real world has fundamentally changed in some way since we trained our model. Let's look at a couple of examples of drift in order to clarify the point:

**Example one**:

We trained our model on data that was gathered from customers filling in forms online. Our model tries to predict whether those customers would likely respond well to a specific marketing campaign in which we would send them targeted emails with discounts on shoes. Recently, an administrator decided that they wanted to capture some additional information from customers and that some of the previously captured data is no longer relevant, so they added some fields to the form and removed some other fields. Now, when new customers fill in the form and that data is sent to our model, there will be extra fields in the input that the model has never seen before and the other fields that it expects to see will no longer be there. This could affect our model's ability to interpret and use the input data effectively, causing errors or incorrect predictions.

**Example two**:

We have built a model that estimates how quickly we can deliver products to customers. The model uses inputs from a number of different sources. One of the sources is a dataset that contains historical data on how long it took to deliver products to customers in past deliveries. We receive an update of that dataset every day, which contains the details of all of the orders that were delivered the previous day. A critical feature in that dataset is the delivery time, which is measured in days. This dataset is created by a system that is owned by a different organization in our company, so we do not have control over that dataset. Our delivery process has become a lot more efficient recently, and some products are being delivered the same day, so the delivery time has now been updated to be measured in hours instead of days. However, nobody informed our data science team about this change. Now, our model looks at the delivery time feature, and because the unit of measurement has changed, its predictions for new delivery times are incorrect.

Another interesting example is something with which we all are somewhat familiar. Many large retailers use AI/ML models to forecast what to stock in their inventory based on data related to what customers are purchasing and they try to look for emerging trends to identify changes in consumer behavior. In the first few weeks of the COVID-19 pandemic, there was an enormous and sudden shift in what everybody wanted to buy. The models were probably surprised to find that everybody was suddenly very interested in one thing in particular, which up until then, was generally sold at a very predictable rate. What was this thing that the models predicted everybody was suddenly interested in? Toilet paper!

Figure 2.11: Empty toilet paper shelves in a store (source: https://commons.wikimedia.org/wiki/File:COVID-19_-_Toilet_Paper_Shortage_%2849740588227%29.jpg)

There are also much more subtle changes that can happen to our data over time. We've already discussed how data scientists often want to inspect data before training a model, and one of the aspects they often want to inspect is the statistical distribution of the values of each of the features (mean, mode, maximum, minimum, etc.). These are important details because they give us an understanding of what kinds of values our features are generally expected to contain. During inspection, this can help us to identify outliers that could indicate erroneous data or that may inform us of some other characteristic of the data that we were previously unaware of. We can also apply this knowledge to make predictions during production. We can analyze the data that is sent to our models for inference purposes, and if we see that the statistical distributions have changed in a consistent way, then it could alert us to potential data corruption or to the fact that the data has genuinely changed, which could indicate a need to update our models by training them on new data that matches the updated shape that we're observing in the real world.

As we can see, drift can lead to our models becoming less accurate or providing erroneous results, and we therefore need to monitor specifically for this by inspecting the data that we're observing in production, as well as monitoring the expected metrics for our models. If we see that our model's metrics are declining over time—or suddenly—this could be an indication of drift between the kind of data the model was trained on and the kind of data it observes in production.

# Best practices for overcoming common challenges

This section contains pointers and best practices that companies have developed over time to address many of the challenges discussed in the previous section.

## Finding and gathering relevant data

We discussed data silos being a common challenge in large companies, as well as restrictions with regard to how data is stored and accessed, especially sensitive data that could be subject to various regulations and compliance requirements. A key to overcoming these challenges is to break down the silos by creating centralized data lakes and data discovery mechanisms, such as a searchable data catalog that contains metadata describing the various datasets in our data lake. To ensure that our data is stored and accessed securely, we need to implement robust encryption and permission-based access mechanisms. We will explore these topics in more detail in later chapters, and we will also perform some hands-on activities regarding detecting and addressing bias and other problems in our datasets.

In cases where our data exists in databases that are restricted specifically for transactional business operations, we could implement a **change data capture** (**CDC**) solution that replicates our data to a data lake that can then be used for data analytics and AI/ML workloads.

Considering that the data-gathering process happens at the very beginning of our AI/ML workload, it's critical that we implement data quality checks at this point in order to prevent issues later in our workload. For example, we know that training our models on corrupt data will result in errors or less accurate model outputs. Bear in mind that in most ML use cases, we are periodically training our models on updated data that's coming from some source, perhaps from a system that is owned and operated by another team or organization. Therefore, if we create data quality checks as this data is coming into our workload and we detect data quality issues, we should implement mechanisms that prevent further steps in our process from proceeding. Otherwise, performing the downstream steps in our process, such as data transformations and model training, would be a waste of time and money and could lead to worse consequences in production, such as malfunctioning models.

## Data labeling

If your company is having difficulties finding a workforce to perform your labeling tasks, Google Cloud's data labeling service can help you to get your data labeled appropriately.

## Picking an algorithm and model architecture

Where should we start when picking an algorithm? This is a common challenge, and as a result, data scientists have been constructing solutions to try to make this easier. In this section, we'll describe a tiered framework for approaching a new AI/ML project in this context:

- **Tier 1**: You can see if a packaged solution already exists for your business problem. For example, Google Cloud has created packaged solutions for lots of different types of use cases such as computer vision, NLP, and forecasting. We'll be covering these solutions in more detail in the coming chapters.
- **Tier 2**: If you want to create and deploy your own model without doing any of the work required to do so, check out Google Cloud's AutoML functionality to see if it meets your needs. We'll also explore this in a hands-on activity in a later chapter in this book.
- **Tier 3**: If you want to get started with a model that has been trained by somebody else, there are numerous hubs and "model zoos" that exist for data scientists to share models they've created. Analogous to software libraries in traditional software development, these are assets that have been created by other data scientists for specific purposes, which you can reuse rather than starting from scratch to implement the same functionality. For example, you can find pre-trained models for various use cases in Google Cloud's AI Hub (`https://cloud.google.com/ai-hub/docs/introduction`).
- **Tier 4**: If the previous options do not meet your specific needs, you can create your own custom models. In this case, Google Cloud Vertex AI provides built-in algorithms for common use cases such as linear regression, image classification, object detection, and many more, or you can install your own custom model to run on Vertex AI. Vertex AI provides many tools for each step in the AI/ML project life cycle, and we will explore most of them in this book.

## Training models

There are some established methods to address overfitting and underfitting. One cause for overfitting can be that the model did not get access to a sufficient amount of different data points from which to learn an appropriate pattern. Looking at a very extreme case of this, let's imagine that we have just one data point in our dataset and our model processes this data point over and over again until it finds a set of coefficients that it can use to relate the input features accurately to the target output for that data point. Now, whenever it sees that same data point, it can easily and accurately predict the target variable. However, if we show it a new datapoint with a similar structure—i.e., the same number and types of features—but different values for those features, it's unlikely that our model will accurately predict the output for the new datapoint because it has only learned the specific characteristics of a single data point during training. This is a case of overfitting, whereby the model works really well on the specific data point on which it was trained but does not make accurate predictions for other data points.

One method that can help to address overfitting in this regard is to provide more data points when training our model. If our algorithm has seen thousands or millions of data points during training, it's much more likely to have built a more generalized model that has a broader understanding of the feature space and the relationship between the features and the target variables. Therefore, when it sees a new data point, it may be able to make more accurate predictions for the target variable of that new data point.

There is a trade-off that we need to keep in mind in this context: although our training process is likely to build a more generalized model as we provide more and more data points, we need to keep in mind that training models on large datasets can be expensive. We may find that after the model has seen millions of data points, each new training iteration is only increasing the model's generalization metrics by very small amounts. For example, if our model currently has a 99.67% accuracy rate and each training iteration increases its accuracy by 0.0001% but it costs thousands of dollars to do so, it may not make sense from a financial perspective to keep training the model on more and more data points, especially if our business considers 99.5% to be accurate enough to meet the business needs. This last point is important—the trade-off in training costs versus increases in accuracy is dependent on the business requirements. If we're building a model for a medical diagnosis use case or a model whose forecasting accuracy can cost our company millions of dollars if it is incorrect by 0.001%, then it may be worth it to keep training the model on more data points. In any case, what you will generally need to do is define a threshold at which the business considers the model's metrics to be sufficient and measure the increase in that metric as the model is trained on more data points. If you see that the metric begins to plateau after a certain amount of data points, it may be time to stop the training process.

It should be noted that adding more data points is not always an option because you may only have a limited dataset to begin with and it may be difficult to gather more real-world data for your specific use case. In these scenarios, you may be able to generate synthetic data with similar characteristics as your real-world data or use mechanisms to optimize the use of your existing dataset during the training process, such as maximizing the training dataset by using cross-validation, which we'll explore in a hands-on activity later in this book.

Another potential cause of overfitting is if the model is too "complex", by which we mean that too many features may be used for each data point in the training dataset. Again, taking an extreme example, if each data point has thousands of features, the model will learn very specific relationships between the features and the training dataset, which may not generalize well to other data points. In this case, a solution could be to remove features that are not deemed to be critical to figuring out the optimal relationship between the features and the target variables. Selecting the relevant features can be a challenge in itself, and we will explore mechanisms such as **principal component analysis** (**PCA**) to help select the most relevant features.

The opposite of overfitting, then, is underfitting, and one potential cause of underfitting is if the model is too simple, in which case there may not be a sufficient number of features for each data point in the dataset for the model to determine a meaningful relationship between the features and the target variables. Of course, in this case, we would want to find or generate additional features than can help our model to learn more meaningful relationships between those features and the target variables.

To address the challenge of keeping track of experiments and their results in large-scale ML projects, we will use Vertex ML Metadata, which tracks all of our experiments and their inputs and outputs for us (i.e., the lineage of our data and ML model artifacts).

## Configuring and tuning hyperparameters

There are some methodical ways in which to explore what we call the "hyperparameter space", which means all of the different possible values for our hyperparameters. The following are some popular methods:

- **Random search**: The random search approach uses a subsampling technique in which hyperparameter values are selected at random for each training job experiment. This will not result in all possible values of every hyperparameter being tested, but it can often be quite an efficient method for finding an effective set of hyperparameter values.
- **Grid search**: The grid search approach to hyperparameter tuning is the most exhaustive because it will try out every possible combination of values of each hyperparameter. This means that it will generally take a lot more time than a random search approach. Also, bear in mind that each training job costs money, so if you have a large hyperparameter space, this can be very expensive or even infeasible.
- **Bayesian optimization**: Earlier in this chapter, we talked about using gradient descent to optimize a function by finding its minimum point. Bayesian optimization is another type of optimization technique. It's quite a complex process, and it is usually more efficient than the other approaches mentioned previously. Fortunately, Google Cloud's Vertex AI Vizier service will perform Bayesian optimization for you, so if you use that tool, you won't need to implement it yourself.

  If you are interested in diving into the inner workings of Bayesian optimization, I recommend referring to the following paper: `https://arxiv.org/abs/1807.02811`

Google Cloud's Vertex AI Vizier service will run lots of training job experiments for you, trying out many different combinations of hyperparameter values for each experiment, and will find the optimal hyperparameter values to run your ML training jobs efficiently.

> **Key tip**
>
> To save yourself a lot of painstaking work and time when doing hyperparameter optimization, use a cloud service that has been built for that purpose, such as Google Cloud's Vertex AI Vizier service.

## Deploying models

Latency is often a key factor in the deployment of our models, in that we need to ensure our model hosting infrastructure meets the latency requirements expected by our client applications.

One decision point that comes into view in this context is whether we have a batch or online use case. In an online use case, a client sends a piece of input data to our model and it waits to receive an inference response. This usually happens when our client application needs an answer quickly, such as when a customer is performing a transaction on our website and we want to see if the transaction seems fraudulent. This is a real-time use case, and therefore the latency generally needs to be very low; perhaps a few milliseconds. You will usually need to work with business leaders to define the acceptable latency.

In a batch use case, our model can process large amounts of data at a time. For example, as input to our model at inference time, we may provide a large file containing thousands or millions of data points for which we want our model to make predictions, and our model could work for hours on processing those inputs and save all of the inference results as outputs in another file, which we could reference later.

> **Tip**
>
> Batch use cases are usually associated with scenarios in which we do not require low latency. However, ironically, there is also a scenario in which batch use cases can actually help provide lower latency at inference time. Consider a scenario in which we're running a retail website and we want to get insights from our users' purchasing histories in order to recommend products that customers may be interested in buying when they visit our website. Depending on the amount of historical data we have, it could take a long time to process that data. Therefore, we don't want to do this in real time when a customer visits our website. Instead, we could periodically run a batch inference job every night and store our results in a file or a key–value database. Then, when customers visit our site, we can fetch the pre-computed inferences from our file or database. In this scenario, fetching a value from a file or database will usually be much quicker than performing an online inference in real time. Note that this would only be suitable for certain use cases. For example, it would not work for a transactional fraud evaluation use case because we would need real-time characteristics from the ongoing transaction for that scenario. Consequently, as a data scientist or AI/ML solutions architect, you will need to determine what kind of inferencing approach works best for each of your use cases.

## Monitoring models after deployment

If you detect that drift is occurring in production, this is a sign that you may need to update your models. We recommend putting mechanisms in place to automate the retraining of your models with updated data if you detect drift, especially if you're managing large numbers of models. I've worked with organizations that are running hundreds of models simultaneously in production, and it's not feasible to manually monitor, curate, and retrain all of those models on an ongoing basis. In that scenario, we've implemented MLOps frameworks that would retrain the models on updated data whenever a model's metrics consistently dropped below a preconfigured threshold that we considered to be acceptable. The MLOps framework would then test the new model, and if the new model's metrics outperformed the current model's metrics in production, it would replace the production model with the new model.

When it comes to defining and monitoring the model's operational metrics in production, you can use Google Cloud Monitoring for that purpose.

# Summary

In this chapter, we explored a quick recap of the traditional SDLC and introduced the concept of the ML model development life cycle. We discussed each of the steps that we usually encounter in most AI/ML projects, and then we dived into specific challenges that generally exist in each step. Finally, we covered approaches and best practices that companies have learned over time to help them address some of those common challenges.

In the next chapter, we'll begin to explore the various different services in Google Cloud that can be used to implement AI/ML workloads.

# 3

# AI/ML Tooling and the Google Cloud AI/ML Landscape

In this chapter, we take a look at the various tools in Google Cloud that can be used to implement AI/ML workloads. We start off with a quick overview of some of the fundamental Google Cloud services that function as the building blocks for almost all workloads on Google Cloud. We then progress toward more advanced services that are used specifically for data science and AI/ML workloads. This is the final chapter in the *Basics* part of this book, and like the previous two chapters, it provides foundational information that we build upon throughout the book. If you already have knowledge of Google Cloud's services, this chapter may serve as a refresher for that knowledge. If you are new to Google Cloud, this chapter is an essential part of your learning process, because it introduces concepts that are assumed to be known in the rest of the book.

To describe how each tool is used in the context of data science projects, we will refer to the steps of the ML model lifecycle, as laid out in *Chapter 2*. *Figure 3.1* shows a simplified diagram of the ML model lifecycle. In reality, combinations of these steps could be repeated in cycles throughout the model lifecycle, but we will omit those details for simplicity at this point. Our simplified workflow example assumes that the outputs from each step are satisfactory, and we can move on to the next step in the process. It also includes hyperparameter optimization in the **Train Model** step.

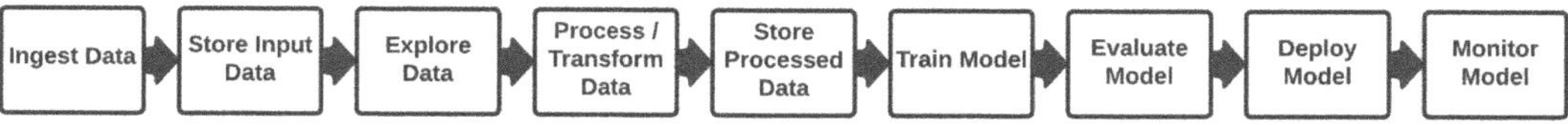


Figure 3.1: Simplified ML model lifecycle

This chapter will cover the following topics:

- Why Google Cloud?
- Prerequisites for using Google Cloud tools
- Google Cloud services overview

- Google Cloud tools for data processing
- Google Cloud VertexAI
- Standard industry tools on Google Cloud
- Choosing the right tool for the job

Let's begin with a discussion of why we would want to use Google Cloud for data science and AI/ML use cases in the first place.

# Why Google Cloud?

Google has been a well-known leader in the AI/ML space for a very long time. They have contributed a lot to the AI/ML industry, through countless research papers, publications, and donations of AI/ML libraries to the open source community, such as TensorFlow, one of the most widely used ML libraries of all time. Their search and advertising algorithms have been leading their respective industries for years, and their peer organizations, such as DeepMind, dedicate their entire existence to pure AI/ML research.

Google has also been spearheading initiatives such as **Ethical AI**, championing the concepts of fairness and explainability to ensure that AI is held accountable and used only for purposes that are beneficial to humans. AI/ML is not something that Google is trying to use, but rather it is a core tenet of Google's business.

A significant testament to Google Cloud's leadership in this space was when Gartner officially recognized them as a leader in the 2022 Gartner® Magic Quadrant™ for Cloud AI Developer Services. Google Cloud provides a wide array of services and tools for implementing AI/ML use cases and embraces open source and third-party solutions in order to provide the broadest selection possible to their customers. By using Google Cloud for AI/ML workloads, you can benefit from the decades of AI/ML research performed, and expertise gained, by Google in this space.

# Prerequisites for using Google Cloud tools and services

This section is going to be pretty simple because Google Cloud makes it very easy to get started in trying out its services. If you have a Gmail account, then you pretty much already have everything you need to get started on Google Cloud. As a generous bonus, Google Cloud gives USD $300 in credits to new customers, plus additional free credits to new customers who verify their business email addresses. You can use those free credits to explore and evaluate Google Cloud's various services. Also, many of Google Cloud's services provide a **Free Tier**, which allows you to use those services free of charge up to their specified free usage limit, the details of which you can find in the Google Cloud documentation (`https://cloud.google.com/free/docs/free-cloud-features`).

If you need to create a new Google Cloud account, you can sign up at `https://console.cloud.google.com/freetrial`, and when you need to go beyond free usage, you can upgrade to a paid Cloud Billing account.

After you've created and logged into your account, you can start using the Google Cloud services that we'll be using in this book, and many more. When you first try to use a Google Cloud service, you may need to enable the API for that service. This is a simple, one-click action that you only need to perform once in each Google Cloud project. *Figure 3.2* shows an example of the page that's displayed when you first try to use the Google Filestore service (we will describe Filestore in more detail later in this chapter). You can simply click the **Enable** button to enable the API.

> **Definition**
>
> A Google Cloud project organizes all your Google Cloud resources. It consists of a set of users; a set of APIs; and billing, authentication, and monitoring settings for those APIs. All Google Cloud resources, along with user permissions for accessing them, reside in a project.

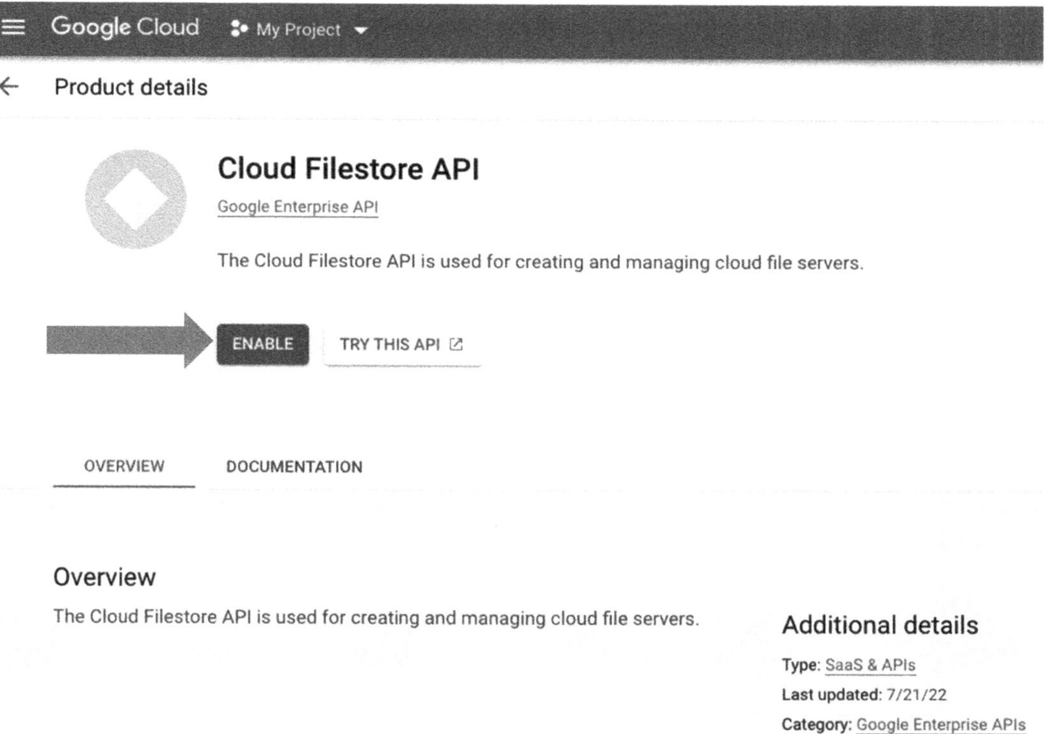


Figure 3.2: Enabling a Google Cloud API on first use

## Security, privacy, and compliance

The first thing we need to think about when we've created our Google Cloud account is security. This also extends to privacy and compliance, and these are hot topics in the data analytics and AI/ML industries today because your customers want to know that their data is being handled securely. Fortunately, these topics are major priorities for Google Cloud, and as a result, Google Cloud provides a plethora of default controls and dedicated services to facilitate and uphold these priorities. We will introduce some of the important concepts and related services briefly here, and we will dive deeper into these topics in later chapters of this book.

### *Who has access to what?*

The first topic to discuss in the context of security, privacy, and compliance is identity and access management; that is, identifying and controlling who has access to which resources in your Google Cloud environments. Google Cloud provides the **Identity and Access Management** (**IAM**) service for this purpose. This service enables you to define identities such as users and groups of users, and permissions with regard to accessing Google Cloud resources. For every attempted action on Google Cloud, whether it's to read an object from storage or to run a piece of code as a cloud function, the permissions associated with that action, the resource upon which the action would act, and the invoking identity would be evaluated by Google Cloud IAM, and the action would only be permitted if the correct combination of permissions has been applied to all relevant identities and resources. For additional convenience, you can integrate Google Cloud IAM with external **Identity Providers** (**IdPs**) and directories, such as Active Directory.

### *Data security*

Google Cloud encrypts all data at rest by default. You can control the keys that are used to encrypt your data by using **Customer-Managed Encryption Keys** (**CMEKs**), or you can let Google Cloud manage all that functionality for you. Regarding data in transit, Google has built global networks with stringent security controls, and TLS encryption is used to protect data that is being transported throughout these global networks. Google Cloud also enables you to encrypt your data even when the data is actively in use, through Confidential Computing, which uses a hardware-based **Trusted Execution Environment** (TEE). TEEs are secure and isolated environments that prevent unauthorized access or modification of applications and data while they are in use.

### *Infrastructure security*

In addition to Google Cloud's state-of-the-art infrastructure security controls, Google Cloud provides tools to help prevent and detect potential security threats and vulnerabilities. For example, you can use Cloud Firewall and Cloud Armor to prevent **Distributed Denial-of-Service** (**DDoS**) and common OWASP threats. You can use Chronicle, Security Command Center, and Mandiant, for **Security Incident and Event Monitoring** (SIEM), **Security Orchestration Automation and Response** (SOAR), intrusion detection, and threat intelligence. In addition to all of these Google Cloud services, you can use third-party observability and reporting services such as Splunk on Google Cloud.

## Compliance

Google Cloud provides audit data that tracks the actions being performed by identities on resources in your environments, which is important for compliance reasons. Google Cloud participates in formal compliance programs such as FedRamp, SOC2, and SOC3, and supports compliance standards such as **Payment Card Industry Data Security Standard** (**PCI DSS**), and multiple ISO/IEC international standards. You can see additional details regarding Google Cloud's compliance program participation at `https://cloud.google.com/security/compliance`.

## Interacting with Google Cloud services

There are numerous ways in which you can interact with Google Cloud services. At a high level, you can either use the **Graphical User Interface** (**GUI**), the **Command-Line Interface** (**CLI**), or the API. We explore each of these options in more detail in this section.

## Console

One of the most straightforward ways to interact with Google Cloud services is via the Google Cloud console, which provides a GUI. You can access the console at `https://console.cloud.google.com/`.

The console enables you to perform actions in Google Cloud by clicking around in a web-based interface in your browser. For example, you create a **virtual machine** (**VM**) by clicking **Compute Engine** in the products menu, and then going to the VM instances page and clicking **Create an instance** to specify the desired properties of your VM, as shown in *Figure 3.3*.

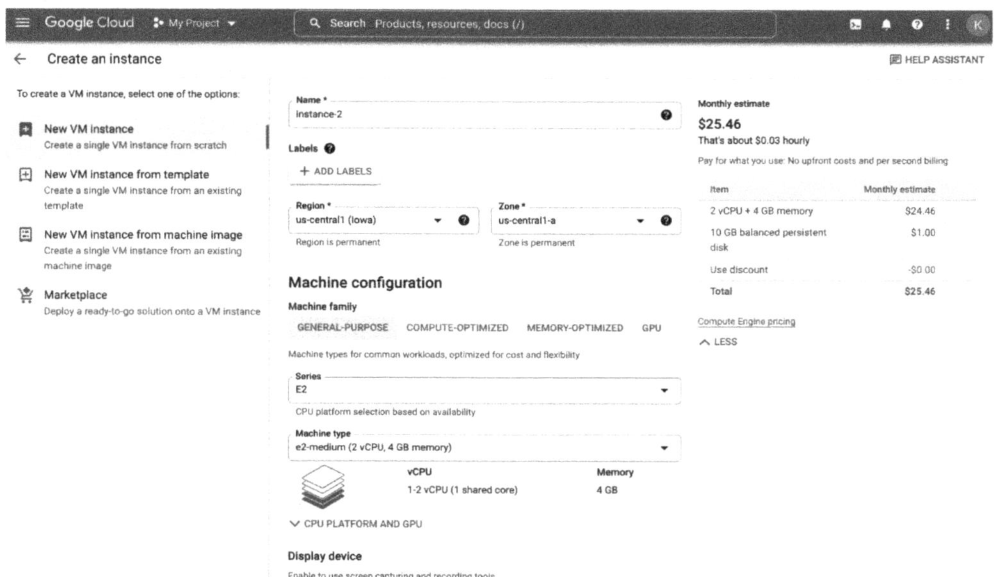


Figure 3.3: Creating a VM in the Google Cloud console

#### *The gcloud CLI*

If you prefer to use a CLI, Google Cloud has built a tool named **gcloud**, which enables you to interact with Google Cloud services by executing text-based commands. This is particularly useful if you wish to automate sequences of Google Cloud service API actions by composing scripts that run multiple commands in order. For example, you could create a Bash script that contains multiple commands, and you could execute that script either manually or on a periodic schedule if it contains actions that frequently need to be repeated. This approach would be suitable for ad hoc automation that can be implemented with little effort. There are other ways to automate more complex sequences of actions on Google Cloud, which we will explore in later chapters.

The following is an example of a `gcloud` command. This command will enable the API for `SERVICE_NAME`, which is a placeholder for the name of the Google Cloud service with which we want to interact:

```
gcloud services enable SERVICE_NAME
```

For example, to enable the Filestore API, rather than clicking the **Enable** button in the Google Cloud console, as we described earlier in this chapter, we could instead use the following `gcloud` CLI command:

```
gcloud services enable file.googleapis.com
```

In this case, `file` is the command-line name for the Google Cloud Filestore service (the full service name is `file.googleapis.com`).

To use the gcloud CLI, you can install it on any machine on which you wish to run it, as it supports many different operating systems, such as Linux, macOS, and Windows, or you can use Google Cloud Shell, described next.

### Google Cloud Shell

Google Cloud Shell is a very convenient way to use the gcloud CLI and interact with Google Cloud APIs. It's a tool that provides a Linux-based environment in which you can issue commands to Google Cloud service APIs.

You can open the Cloud Shell by clicking on the >_ symbol in the top-right corner of the Google Cloud console screen, as shown in *Figure 3.4*:

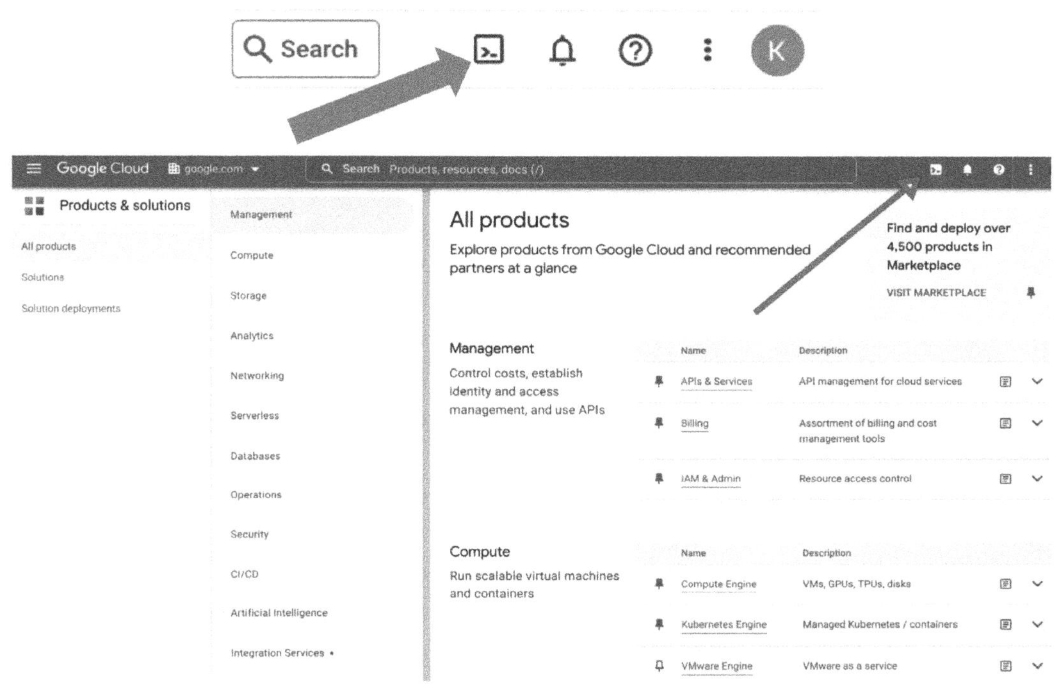


Figure 3.4: Activating Google Cloud Shell

The terminal will then appear at the bottom of the screen, as shown in *Figure 3.5*:

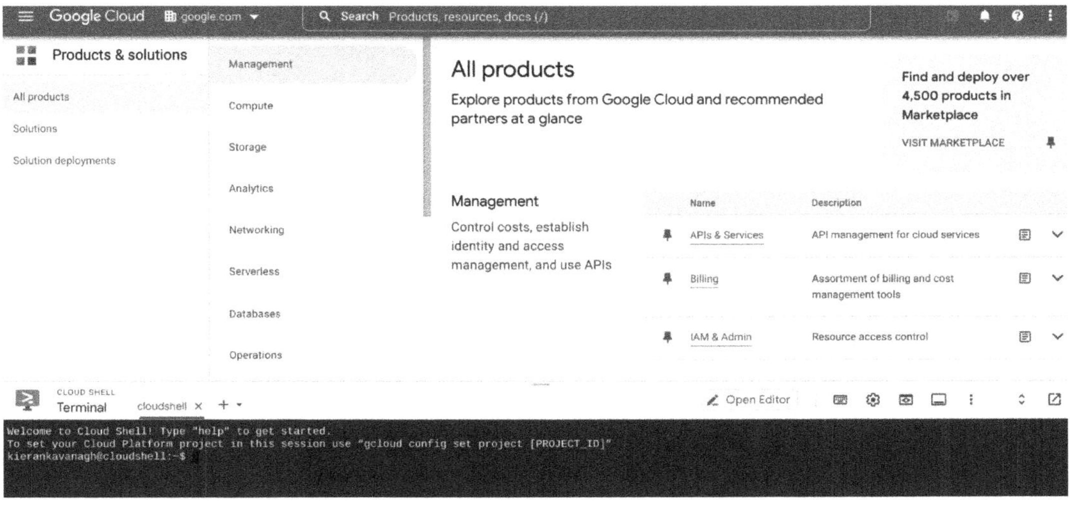


Figure 3.5: Google Cloud Shell

When you first try to use Cloud Shell, you need to authorize it to interact with Google Cloud service APIs, as depicted in *Figure 3.6*.

Authorize Cloud Shell

Cloud Shell needs permission to use your credentials for the gcloud command.

Click Authorize to grant permission to this and future calls.

Figure 3.6: Authorizing Google Cloud Shell

### *API access*

The most low-level method for interacting with Google Cloud services is by programmatically invoking their APIs directly. This method differs from the GUI and CLI access because it is not intended for direct human interaction, but rather it is suitable for more advanced use cases, such as interacting with Google Cloud services via your application software. As an example, let's consider an application that saves users' photos in the cloud. When new users sign up, we may wish to create a new Google Cloud Storage bucket to store their photos, among other signup-related activities (we will describe the Google Cloud Storage service later in this chapter). We could create the following REST API request for that purpose:

```
curl -X POST --data-binary @JSON_FILE_NAME \
     -H "Authorization: Bearer OAUTH2_TOKEN" \
     -H «Content-Type: application/json» \
     https://storage.googleapis.com/storage/v1/b?project=PROJECT_
IDENTIFIER
```

Let's break this down:

- `JSON_FILE_NAME` is the name of the required JSON file that specifies the bucket details
- `OAUTH2_TOKEN` is an access token that is required to invoke the API
- `PROJECT_IDENTIFIER` is the ID or number of the project with which our bucket will be associated, for example, `my-project`

The required JSON file is structured as follows:

```
{
  "name": "BUCKET_NAME",
  "location": "BUCKET_LOCATION",
  "storageClass": "STORAGE_CLASS",
  "iamConfiguration": {
    "uniformBucketLevelAccess": {
      "enabled": true
    },
  }
}
```

Here's the breakdown of this:

- `BUCKET_NAME` is the name we want to give our bucket.
- `BUCKET_LOCATION` is the location where you want to store your bucket object data. For more information regarding Google Cloud locations, refer to the Google Cloud documentation here: `https://cloud.google.com/compute/docs/regions-zones`.
- `STORAGE_CLASS` is the default storage class of your bucket. For more information regarding Google Cloud Storage classes, refer to the Google Cloud documentation here: `https://cloud.google.com/storage/docs/storage-classes`.

In practice, it's most common to use Google Cloud's client **Software Development Kits** (**SDKs**) to create such API calls programmatically. For example, in the following Python code, we import the Google Cloud Storage client library, and we then define a function to create a new bucket, specifying the bucket name, the location, and the storage class:

```
# import the GCS client library
from google.cloud import storage


def create_bucket_class_location(bucket_name):
    """
    Create a new bucket in the US region with the coldline storage
    class
    """
    # bucket_name = "your-new-bucket-name"

    storage_client = storage.Client()

    bucket = storage_client.bucket(bucket_name)
    bucket.storage_class = "COLDLINE"
```

```
    new_bucket = storage_client.create_bucket(bucket, location="us")

    return new_bucket
```

Now that we've covered some of the basics of how to interact with Google Cloud services, let's discuss the types of Google Cloud services we will use in this book.

# Google Cloud services overview

Having covered the basics of how to set up a Google Cloud account and how to enable and interact with the various services, we will now introduce the services that we are going to use in this book to create AI/ML workloads. We will first cover the fundamental cloud services upon which almost all workloads are built, and then we will cover the more advanced services related to data science and AI/ML.

## Google Cloud computing services

Considering that the word *computing* is included directly in the term **cloud computing**, and the fact that computing services form the basis of all other cloud services, we will start this section with a brief overview of Google Cloud's computing services.

### *Google Compute Engine (GCE)*

A few years ago, the term cloud computing was pretty much synonymous with the term **virtualization**. Traditionally, companies had physical servers on their own premises, and this was then contrasted with creating virtual servers in the cloud, either public or private. Hence, perhaps the easiest concept to understand in cloud computing is virtualization, where we simply create a virtual server instead of a physical server by introducing an abstraction layer called a hypervisor between the hardware and our server's operating system, as depicted in *Figure 3.7*. For most companies, if they're already running physical servers, then their first foray into the world of the cloud is usually implemented by using VMs because this is the simplest step to transition from the physical paradigm to the cloud paradigm. **Google Compute Engine** (**GCE**) is Google Cloud's service for running VMs in the cloud. It provides some useful features, such as auto-scaling based on demand, which is one of the well-established benefits of cloud computing.

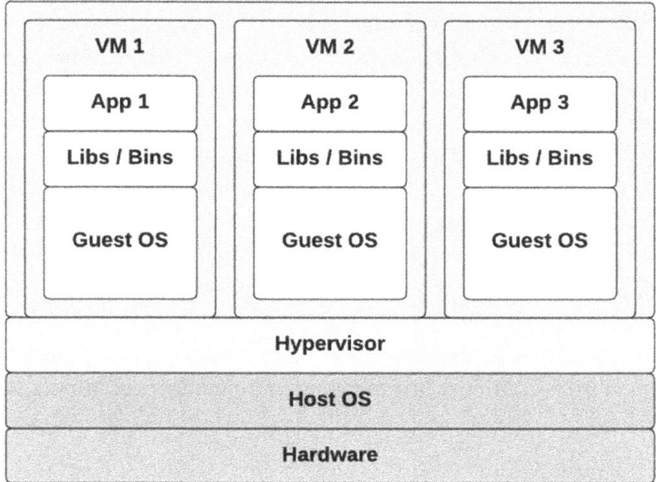


Figure 3.7: Example VM implementation

### Google Kubernetes Engine (GKE)

A newer type of virtualization was created in the 2000s by using Linux **cgroups** and **Namespaces** to isolate computing resources such as CPUs, RAM, and storage resources, for specific processes within a running operating system. With containerization, the abstraction layer moves higher in the stack, whereby it exists between the operating system and our applications, as depicted in *Figure 3.8*.

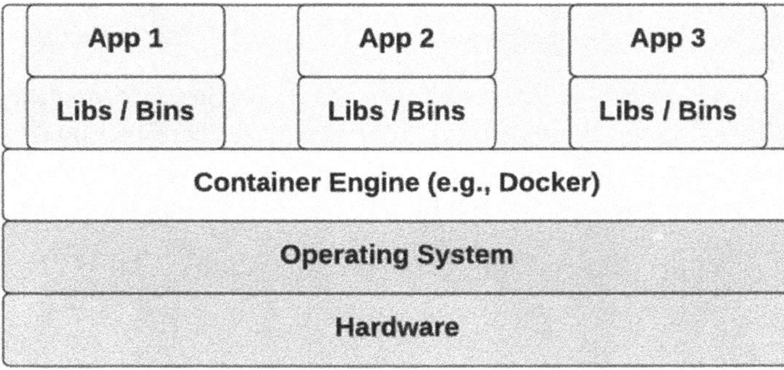


Figure 3.8: Example container implementation

This gives us some interesting benefits beyond those afforded by hypervisor-based virtualization. For example, containers are generally much smaller and more *lightweight* than VMs, meaning they contain much fewer software components. While a VM has to boot up an entire operating system and lots of software applications before it becomes usable, which can take a few minutes, a container usually only contains your application code and any required dependencies and can therefore be loaded in seconds. This makes a big difference when it comes to auto-scaling and auto-healing cloud-based software workloads. Starting up new VMs in relation to a sudden increase in traffic may not happen quickly enough and you may lose some requests while the VMs boot up and load your application. The same applies to restarting a VM due to some kind of problem. In both of those cases, a container would usually start much more quickly. Containing fewer components also means that containers can be deployed much more quickly, and this makes them a perfect environment for microservices with DevOps CI/CD pipelines. There are also many other benefits of containers, such as portability and easy manageability.

However, one of the challenges introduced by containerization also stems from their lighter footprint. Because they are generally smaller than VMs, it's common to have more of them in a single application deployment. Managing lots of tiny containers can be challenging, especially in terms of application lifecycle management and orchestration; that is, determining how and where to run your workloads, and assigning adequate computing resources to them. This is where Kubernetes comes into the picture. The following are the official definitions for Kubernetes, in general, and **Google Kubernetes Engine** (**GKE**), in particular.

Kubernetes, also known as **K8s**, is an open source system for automating the deployment, scaling, and management of containerized applications. It groups containers that make up an application into logical units for easy management and discovery. GKE provides a managed environment for deploying, managing, and scaling your containerized applications using Google infrastructure.

*Figure 3.9* shows an example of how Kubernetes organizes and orchestrates applications. It deploys your applications as Pods, which are groups of containers with similar functionality, and it deploys agents on your hardware servers or the host operating systems that keep track of resource utilization and communicate that information back to the Kubernetes master, which uses that information to manage Pod deployments.

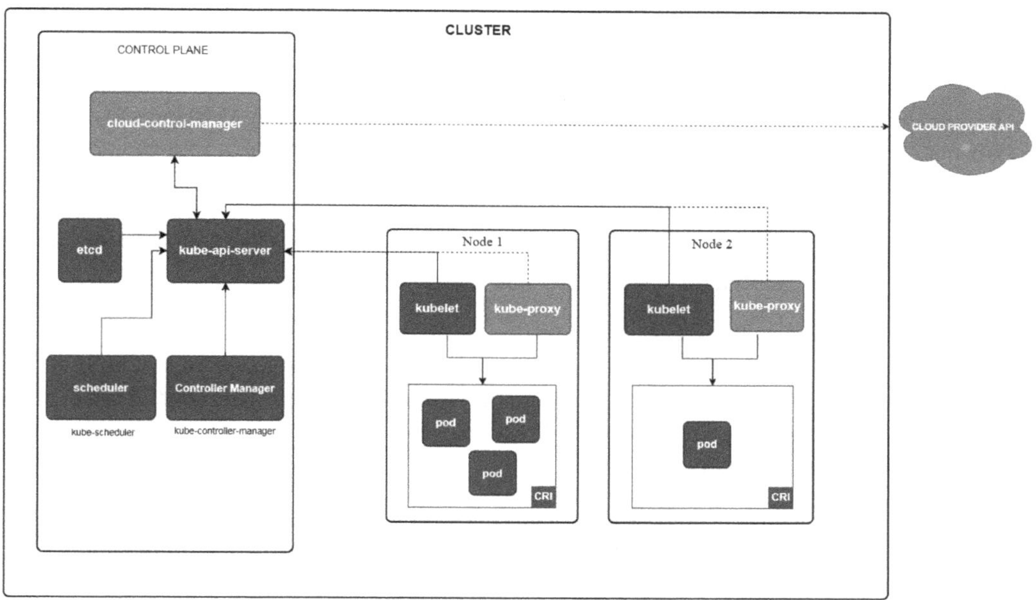


Figure 3.9: Example GKE implementation
(source: https://kubernetes.io/docs/concepts/architecture/)

### *Google Cloud serverless computing*

The term **serverless** in the context of cloud computing refers to the concept of running your code on a cloud provider's infrastructure without needing to manage any of the servers that will be used to run your code. In reality, there are still servers that are being used, behind the scenes, but the cloud provider creates and manages them on your behalf so that you don't need to perform those actions. Google Cloud has two primary services that relate to serverless computing, which are named **Cloud Functions**, and **Cloud Run**. Another Google Cloud service, named **App Engine**, is also often bundled under the serverless umbrella, and we will describe that service, and how it differs from Cloud Functions and Cloud Run, later in this section.

> Note
>
> Many other Google Cloud services also run in a serverless fashion, whereby the actions they perform on your behalf run on servers that are managed in the background, without any need for you to manage those servers. However, Cloud Functions and Cloud Run are the two Google Cloud services that relate to *serverless computing*, which specifically refers to running your code without the need to explicitly manage servers.

### Cloud Functions

With Cloud Functions, you simply write small pieces of code — for example, a single function — and Google Cloud will run that code for you in response to events that you specify as triggers to run that code. You don't need to manage any containers, servers, or infrastructure on which your code executes, and there are many types of triggers that you can configure. For example, whenever a file is uploaded to your Google Cloud Storage bucket, that could trigger your piece of code to execute. Your code could then process that file in some way, feed it into another Google Cloud service to be processed, or simply send a notification to inform somebody that the file has been uploaded.

This concept is referred to as **Functions as a Service** (**FaaS**) because it is usually used to execute a single function for each event trigger. This approach is suitable for when you want to simply write and run small code snippets that respond to events that occur in your environment. You can also use Cloud Functions to connect with other Google Cloud or third-party cloud services to streamline challenging orchestration problems.

In addition to sparing you the trouble of managing servers, another advantage of using Cloud Functions is that you don't have to pay for servers when no events are happening in your environment.

### Cloud Run

Cloud Run is a different type of serverless computing service that is more suitable for long-running application processes. While cloud functions are intended to run small pieces of code in response to specific events that occur, Cloud Run can run more complex applications. This also means that it provides more flexibility and control with regard to how your code executes. For example, it runs your code in containers, and you have more control over what executes in those containers. If your application requires custom software package dependencies, for example, you can provision those dependencies to be available in your containers.

Cloud Run abstracts away all infrastructure management by automatically scaling up and down from zero almost instantaneously, depending on traffic, and it only charges you for the exact resources you use.

## App Engine

While App Engine can also be considered as a serverless service, because it manages the underlying infrastructure for you, its use cases differ from those of Cloud Functions and Cloud Run. App Engine comes in two levels of service, referred to as **Standard** and **Flexible**. In the standard environment, your application runs on a lightweight server inside a sandbox. This sandbox restricts what your application can do. For example, the sandbox only allows your app to use a limited set of software binary libraries, and your app cannot write to a permanent disk. The standard environment also limits the CPU and memory options available to your application. Because of these restrictions, most App Engine standard applications tend to be stateless web applications that respond to HTTP requests quickly. In contrast, the flexible environment runs your application in Docker containers on Google Compute Engine VMs, which have fewer restrictions.

Also, note that the standard environment can scale from zero instances up to thousands very quickly, but the flexible environment must have at least one instance running and can take longer to scale up in response to sudden traffic increases.

App Engine is generally suited to large web applications. Its flexible environment can be more customizable than Cloud Run. However, if you want to deploy a long-running web application without managing the underlying infrastructure, I recommend first evaluating whether Cloud Run could meet your application's needs and comparing the costs of running your app on Cloud Run versus App Engine.

### *Google Cloud Batch*

Some workflows are intended to run for a long time without the need for human interaction. Examples of such workloads include media transcoding, computational fluid dynamics, Monte Carlo simulations, genomics processing, and drug discovery, among others. These kinds of workloads usually require large amounts of computing power and can be optimized by running tasks in parallel. Creating and running these jobs by yourself can incur overheads such as managing servers, queueing mechanisms, parallelization, and failure logic. Fortunately, the Google Cloud Batch service has been built to manage all these kinds of activities for you. As a fully managed job scheduler, it automatically scales the infrastructure required to run your batch jobs up or down and handles parallelization and retry logic that you can configure in case any errors occur during execution.

Now that we've covered the primary computing services on Google Cloud, let's review some of the services you can use to integrate between different Google Cloud services.

## Google Cloud integration services

In addition to Google Cloud infrastructure services such as compute and storage, we often need to implement integrations between the services in order to create complex workloads. Google Cloud has created tools specifically for this purpose, and we will briefly discuss some of the relevant tools in this section.

### *Pub/Sub*

Google Cloud Pub/Sub is a messaging service that can be used to pass data between components of your system architecture, whether those components are other Google Cloud services, third-party services, or components you've built yourself. It's an extremely versatile service that can be used for a wide array of system integration use cases, such as decoupling microservices, or streaming data into a data lake.

Pub/Sub relates to the system architecture concept of publishing and subscribing, whereby one system can publish a message or a piece of data to a shared space, or **Topic**, and other systems can then receive that piece of data by subscribing to that topic. The messages can be delivered via either a **Push** or **Pull** mechanism. In the case of a push approach, the Pub/Sub service initiates the communication with the subscriber systems and sends the message to those systems. In the case of a pull model, the subscriber systems initiate the communication to the Pub/Sub service and then request or pull the information from the Pub/Sub service.

Pub/Sub also caters to nuanced messaging needs such as publishing messages in order (if required) and retrying failed message transmissions. Google Cloud also provides an offering called Pub/Sub Lite, which is a lower-cost option with fewer features than the regular Pub/Sub product.

### Google Cloud Tasks

Similar to Pub/Sub, Google Cloud Tasks is a service that can be used to implement message passing and asynchronous system integration. With Pub/Sub, the publishers and subscribers are completely decoupled, and they have no control over each other's implementations. On the other hand, with Cloud Tasks, the publisher (or **task producer**) fully controls the overall execution of the workload. It can be specifically used for cases where a task producer needs to control the execution timing of a specific webhook or remote procedure call. Cloud Tasks is included in this section for completeness because it's an alternative to Pub/Sub for some use cases, but we will not use Cloud Tasks in this book.

### Eventarc

Eventarc is a Google Cloud service that enables you to build *event-driven* workloads. This is a common pattern for companies that want their workloads to execute in response to events that happen in their environment. We touched on this topic briefly when we introduced Cloud Functions. Cloud Functions can be triggered directly by certain event sources, but Eventarc provides much more flexibility and control for implementing complex event-driven architectures, in conjunction with Cloud Functions and other Google Cloud services, as well as some third-party applications.

Eventarc uses Pub/Sub to route messages from **Event Providers** to **Event Destinations**. As the names suggest, event providers send events to Eventarc, and Eventarc sends events to event destinations. It provides a way to standardize your event processing architectures, rather than building random, ad hoc event-driven implementations between your various system components.

### Workflows

While Eventarc provides a mechanism for standardizing your event-driven workloads, Google Cloud Workflows, as the name suggests, is a service that has been specifically built to orchestrate complex workflows, in which the coordination of activities among various systems needs to be implemented in a specific order. With this in mind, Google Cloud Workflows and Eventarc make a great pair when used together to implement complex, event-driven workloads. Workflows can either be triggered by events, or you can create batch workflows that can be triggered in different ways.

Workflows can orchestrate activities between various microservices and custom or third-party APIs. The Workflows service maintains the state of each step in your workload during execution, meaning that it tracks the inputs and outputs of each step, and it knows which steps have already been completed, which steps are currently executing, and which steps remain to be invoked in the workflow. It allows you to visualize all of your workflow's steps and their dependencies, and if any step in the process fails, you can use the Workflows service to figure out which one and determine what to do next. *Figure 3.10* shows an example of a workflow process for an online retail system, in which a customer purchases an item. Various Google Cloud computing products are used to run each of the software services in the workflow, and the coordination of each of the steps in the process is managed by the Workflows service. While this is an example of a simple order processing workflow, note that most large retail companies work with supply chains consisting of extremely complex webs of interconnected systems and partners.

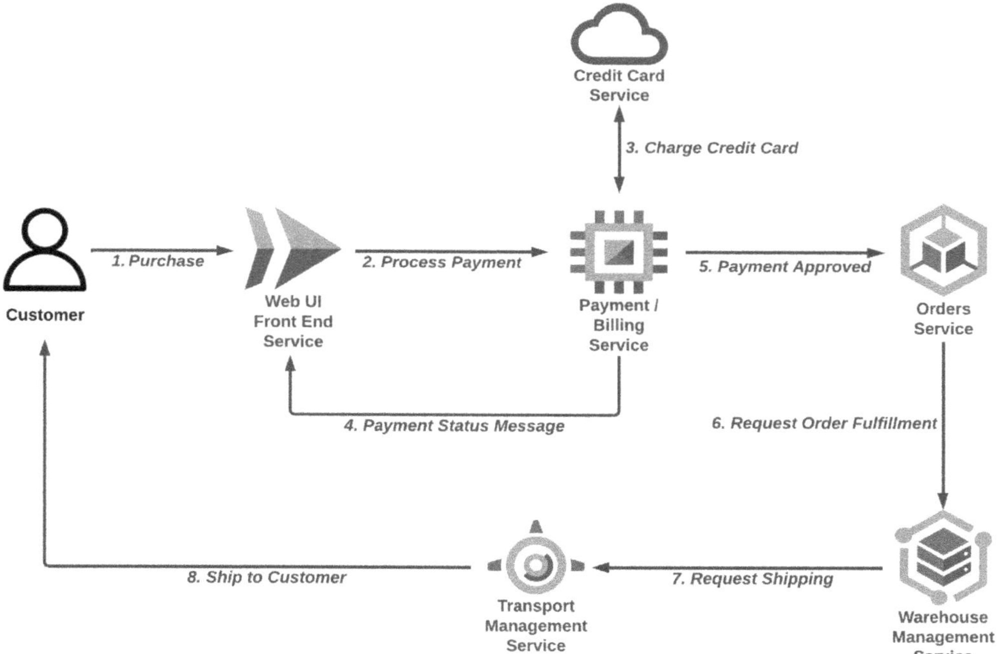


Figure 3.10: Example workflow for online retail system

The Workflows service is best suited to orchestrating activities between services. If you want to implement an orchestration workflow for data engineering, then Google Cloud Composer may be more suitable. We discuss Google Cloud Composer later in this chapter.

### Scheduler

Google Cloud Scheduler is a relatively simple but very useful service that can be used to execute workloads according to a schedule. For example, if you want a process to run at the same time every day, every hour, or every month, you could use Cloud Scheduler to define and kick off those executions. Any of you who are familiar with Unix-based operating systems may see a similarity with the cron service.

Google Cloud Scheduler can be used in conjunction with many of the integration services we've described in this section. For example, you could schedule a message to be sent to a Pub/Sub topic every 15 minutes, and that could then be sent to Eventarc and used to invoke a cloud function.

## Networking and connectivity

Very few workloads exist without the need to set up some kind of network connectivity. For example, even if you have only a single server, you generally need to connect to it in some way in order to perform any actions on it. As you scale beyond a single server, those servers usually need to communicate with each other. In this section, we discuss the fundamental networking and connectivity concepts upon which we will build our workloads in later chapters of this book.

### Virtual Private Cloud (VPC)

The first concept we introduce in this section is the **Virtual Private Cloud** (**VPC**) concept. A VPC is a virtual network that can span all Google Cloud regions. The reason it is called a virtual private cloud is that it defines the boundaries of your networking infrastructure, and therefore where you run your workloads within Google Cloud. You can, however, peer or share connectivity with other VPCs in order to communicate across VPC boundaries.

### Hybrid networking

If you're working for a company that has its own on-premises servers and networks, and you want to connect them to the cloud, this is referred to as **Hybrid Connectivity**. It's a common need for many companies, and therefore, Google Cloud has created specific solutions to facilitate this kind of connectivity, which consist of the following offerings.

#### Dedicated Interconnect

Dedicated Interconnect provides direct physical connections between your on-premises network and Google's network. It offers a guaranteed uptime of 99.99% and can connect either one or two links that each can support up to 100 **gigabits per second** (**Gbps**) in bandwidth. It requires hardware connectivity to be set up at specific Dedicated Interconnect locations, and therefore it can require non-trivial effort to set it up. This option is for companies who need high-bandwidth networking between their premises and Google Cloud for long-lived connectivity.

### Partner Interconnect

If you don't have your own infrastructure built at one of the Dedicated Interconnect locations, there are Google Cloud partners that offer connectivity with up to 99.99% availability through Partner Interconnect. This option also requires some effort in working with the partners to set it up, but it doesn't require the same amount of investment as Dedicated Interconnect. A trade-off is that the partners generally share the connections among many customers, so the bandwidth is less than that of Dedicated Interconnect.

### Private Google Access (PGA) for on-premises

This is a basic connectivity option that provides direct access to Google services such as Cloud Storage and BigQuery from your on-premises locations.

### Virtual Private Network (VPN)

Perhaps the easiest way to connect your on-premises resources to your Google Cloud VPC is via a **Virtual Private Network** (**VPN**), which uses **IP security** (**IPsec**) mechanisms to offer a low-cost option that delivers 1.5 – 3.0 Gbps of throughput over an encrypted public internet connection. Unlike the Interconnect services mentioned previously, this option does not require any special, hardware-related network connectivity in any specific location.

Now that we've covered the fundamental Google Cloud services that underpin the workloads that we'll build in this book, it's time to dive into the services that we will directly use to create our data processing workloads to prepare for our AI/ML use cases.

## Google Cloud tools for data storage and processing

Since gathering data is the first major step in an AI/ML project (after establishing the business objectives of the project), we begin our exploration of Google Cloud's AI/ML-related services by first reviewing the tools for storing and processing data. *Figure 3.2* shows the steps in the life cycle that relate to ingesting, storing, and processing data. It should be noted that the **Train Model**, **Evaluate Model**, and **Monitor Model** steps would also usually create outputs that need to be stored somewhere.

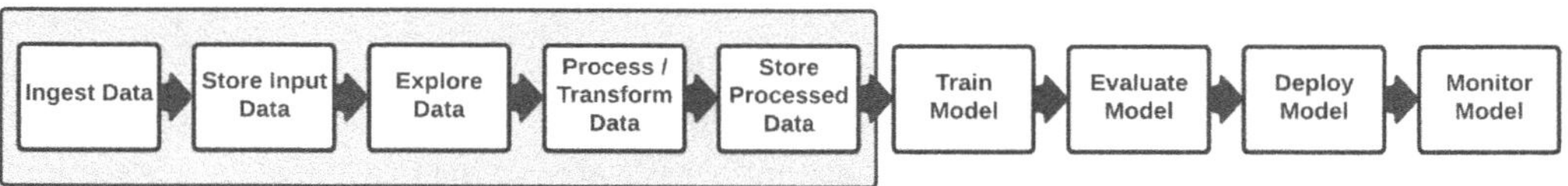


Figure 3.11: Ingesting, storing, exploring, and processing data

As can be seen in *Figure 3.11*, and as we've discussed previously, working with data is a very prominent part of any AI/ML project.

## Data ingestion

Before we can do anything with data in Google Cloud, we need to get access to the data, and we often want to ingest that data into some kind of storage service on Google Cloud. In this section, we'll discuss some of the tools that exist on Google Cloud for ingesting data, and in the next section, we will cover the Google Cloud storage systems into which the ingestion services ingest data. We will not focus on Google Cloud's database services in this chapter, nor the related **Database Migration Service** (**DMS**), because for machine learning purposes, we would usually extract data from databases and place it into one of the storage systems described in this section. An exception to this may be Google Cloud Bigtable, but we will discuss that service separately in a later chapter.

### *gsutil*

Perhaps the simplest way to transfer data to **Google Cloud Storage** (**GCS**), or between GCS buckets, is via the gsutil command-line tool, which can be used to transfer up to 1 TB of data with a simple command.

### *The Data transfer service*

If you want to transfer more than 1 TB of data, you can use the data transfer service, which transfers data quickly and securely from on-premises systems or other public cloud providers. For large data migration projects, in which you may wish to run multiple data transfer jobs, it lets you centralize your job management to monitor the status of each job. You can transfer petabytes of data consisting of billions of files, at up to tens of Gbps of bandwidth, and the data transfer service will optimize your network bandwidth to accelerate transfers. You can ingest your data into GCS and then have other Google Cloud services access it from there.

### *The BigQuery Data Transfer Service*

The BigQuery Data Transfer Service automates data movement specifically into BigQuery, on a scheduled, managed basis. You can access the BigQuery Data Transfer Service using the Google Cloud console, the **bq** command-line tool, or the BigQuery Data Transfer Service API. It supports lots of data sources, such as GCS, Google Ads, YouTube, Amazon S3, Amazon Redshift, Teradata, and many more.

## Data storage

There are many different ways to store data in Google Cloud, and the types of tools and services you select for data storage will depend on your use case and what you're trying to achieve. In this section, we'll take a look at the different products and services provided by Google Cloud in the data storage space, and the kinds of workloads to which they best relate.

### Concepts – data warehouses, data lakes, and lake houses

Before diving into each of the major Google Cloud data storage services, it's important to discuss the concepts of data warehouses, data lakes, and lake houses, which are all terms that have become quite popular in the industry in recent years.

A **data warehouse** usually contains structured data in a format that is optimized for analytics purposes, such as columnar data formats like **Parquet** or **Optimized Row Columnar** (**ORC**). This is because data analytics queries often operate on database columns rather than rows. For example, we might run a query to find out the average age of customers who buy our products, and this query would therefore focus on the *age* column in our customer database table. Columnar data formats store all of the elements of each column near each other on the physical storage disks so queries that operate on database columns run more efficiently.

A **data lake**, as the name suggests, serves as a reservoir in which you can store huge amounts of data in a variety of formats, both structured and unstructured. Because there are no specific requirements regarding query optimization, data lakes can usually store much more data than data warehouses. Data lakes are a key ingredient in breaking down the problematic data silos that we described in *Chapter 2*, and they can serve as the foundation of your data management strategy.

The term, **data lake house**, refers to more recently emerging patterns, in which companies utilize a combination of data warehouses and data lakes in order to get the best of both worlds and support a broader set of use cases, such as real-time analytics, batch data processing, machine learning, and visualization, all from the same source.

### Google Cloud Storage (GCS)

In addition to Google Compute Engine, **Google Cloud Storage** (**GCS**) is one of the most fundamental services in Google Cloud. It supports what's referred to as **object** storage, and it is perhaps the most versatile of all the storage services in Google Cloud, because you can store pretty much any type of data in GCS, and it can be directly accessed from most Google Cloud services that process data. It is especially suitable for large amounts of data, and it can be used as the basis for building an enterprise data lake.

GCS provides different storage classes to optimize your usage based on cost and frequency of access. For objects that you don't access frequently, you can put them in a lower-cost storage class, and you can even configure GCS to automatically move your objects between storage classes based on criteria such as the age of each object. For more information on each of the different storage classes, and which one works best for different use cases, reference the table at `https://cloud.google.com/storage/docs/storage-classes`.

### Filestore

The Google Cloud Filestore service is a high-performance, fully managed file storage service, used for workloads in which a structured file system is needed. This is the concept of **Network Attached Storage**, in which your VMs and containers can *mount* a shared filesystem, and can access and operate on the files in the shared directory structure. It uses the **Network File System version 3** (**NFSv3**) protocol and supports any NFSv3-compatible clients.

Filestore is available in three different formats:

- Filestore Basic, which is best for file sharing, software development, and web hosting
- Filestore Enterprise, which is best for critical applications such as SAP workloads
- Filestore High Scale, which is best for high-performance computing, including genome sequencing, financial services trading analysis, and other high-performance workloads

Access to the shared file system depends on the permissions you've configured, and the networking connectivity that you have set up by using the products that we discussed in the *Networking and connectivity* section of this chapter.

### Persistent Disk

So far, we've covered object storage and file storage. Another type of storage is referred to as block storage. This type of storage may be most familiar because it's the traditional type of storage used by disks that are directly attached to computers; that is, **Direct Attached Storage** (**DAS**). For example, the disk drive in your laptop uses this type of storage. In companies' own on-premises data centers, many servers may be connected to shared block storage devices referred to as a **Storage Area Network** (**SAN**), using the types of shared RAID array configurations that we briefly discussed in *Chapter 1*. In either case, these block storage devices appear to our server operating systems as if they are directly attached disks, and they are used as such by the applications running on our servers or in our containers.

SANs can require a lot of effort to set up and maintain, but when using Google Cloud Persistent Disk, you can simply define what kind of disk storage you want, and the required capacity, and all of the underlying infrastructure is managed for you by Google.

At a high level, Persistent Disk provides two different storage types, which are **Hard Disk Drives** (**HDDs**) and **Solid State Drives** (**SSDs**). HDDs offer low-cost storage when bulk throughput is of primary importance. SSDs offer high performance and speed for both random-access workloads and bulk throughput. Both types can be sized up to 64 TB.

### BigQuery

Google Cloud BigQuery is a serverless data warehouse, meaning that you can use it without needing to configure or manage any servers. As a data warehouse, it straddles both storage and processing. It can store your data in a format that's optimized for data analytics workloads, and it provides tools that allow you to run SQL queries on that data. You can also use it to run queries on other storage systems such as GCS, Cloud SQL, Cloud Spanner, Cloud Bigtable, and even storage systems on AWS or Azure. Additionally, it provides built-in machine learning that enables you to get ML inferences from your data via SQL queries, without needing to use other services. On the other hand, if you explicitly want to use other services such as Vertex AI, which we will describe later in this chapter, it integrates easily with many other Google Cloud services.

It supports geospatial analysis, so you can augment your analytics workflows with location data, and it supports real-time analytics on streaming data when you integrate streaming solutions such as Dataflow with BigQuery BI Engine, which is an in-memory analysis service that provides a sub-second query response time. BI Engine also natively integrates with Looker Studio and works with many business intelligence tools. BigQuery is an extremely popular service on Google Cloud, and you will learn how to use many of its features in this book.

After you've ingested and stored data in Google Cloud, you will often want to organize and manage it so that it can easily be discovered and utilized effectively. In the next section, we will discuss Google Cloud's data management tools.

## Data management

The services we describe in this section enable you to organize your data and make it easier to manage how your data can be discovered and accessed by users in your organization, thus breaking down or preventing data silos. These tools act as a supporting layer between the data storage services we discussed in the previous section and the data processing services we will discuss in subsequent sections in this chapter.

### BigLake

BigLake is a storage engine that unifies data warehouses and data lakes by enabling BigQuery and open source frameworks such as Spark to access data with fine-grained access control. For example, you could store data in GCS and make it available as a BigLake table, and then you could access that data from BigQuery or Spark. Fine-grained access control means that you can control access to the data at the table, row, and column level. As an example, you could ensure that your data scientists can see all columns except the credit card information column, or you could ensure that the sales department for a particular geographical location can only see the rows that pertain to that location, and cannot see data in any rows that relate to other geographical locations.

BigLake allows you to perform analytics on distributed data regardless of where and how it's stored, using your preferred analytics tools – open source or cloud native – over a single copy of the data. This is important because it means you don't need to move the data around between your data lakes and data warehouses, which has traditionally been laborious and expensive to do. BigLake also supports open source engines such as Apache Spark, Presto, and Trino, and open formats such as Parquet, Avro, ORC, CSV, and JSON, serving multiple compute engines through Apache Arrow. You can centrally manage data security policies in one place and have them consistently enforced across multiple query engines, and across multiple clouds when using BigQuery Omni. It can also integrate with Google Cloud Dataplex, which we will describe next, to enhance this functionality and provide unified data governance and management at scale.

### Dataplex

Google refers to Dataplex as an "*intelligent data fabric that enables organizations to centrally discover, manage, monitor, and govern their data across data lakes, data warehouses, and data marts, with consistent controls*." This relates to the concept of breaking down data silos. In *Chapter 2*, we talked about data silos being a common challenge that companies run into when they wish to perform data science tasks and the complexities of managing who can access the data securely when you have many datasets owned by various organizations throughout your company. Dataplex helps to overcome these challenges by enabling data discovery and providing a single pane of glass for data management across data silos, and centralized security and governance. This means that you can define security and governance policies in Dataplex, and have them applied to data that is stored and accessed by other systems, in a consistent manner. It integrates with other Google Cloud data management services such as BigQuery, Cloud Storage, and Vertex AI.

With Dataplex, the idea is to create a *data mesh*, in which there are logical connections between your various data stores and data processing systems, rather than disjointed data silos. It also uses Google's AI/ML capabilities to provide additional features such as automated data life cycle management, data quality enforcement, and lineage tracking (you may remember that, in *Chapter 2*, we also talked about lineage tracking being a difficult, common challenge that companies face when implementing data science workloads).

Google Cloud originally had a standalone service named Data Catalog, which could be used to store metadata about your various datasets, and therefore provide discoverability by allowing you to view and search through the metadata to understand what datasets are available. This service is now provided within Dataplex, and it can even automate data discovery, classification, and metadata enrichment. It can then logically organize data that exists across multiple storage services into business-specific domains using the concepts of Dataplex lakes and data zones.

Dataplex also provides some data processing functionality via its *serverless data exploration workbench*, which provides one-click access to Spark SQL scripts and Jupyter notebooks, allowing you to interactively query your datasets. The workbench also allows teams to publish, share, and search for datasets, therefore enabling discoverability and collaboration across teams.

Speaking of data processing, our next section will cover some of the primary tools and services available for processing data on Google Cloud.

## Data processing

When you have used some of the services we discussed in the previous sections to store, organize, and manage data, you may then wish to process that data in some way. Fortunately, there are a number of tools and services in Google Cloud that can be used for this purpose, and we will explore them here.

### *Dataproc*

Dataproc is a fully managed and highly scalable service for running Apache Hadoop, Apache Spark, Apache Flink, Presto, and 30+ open source tools and frameworks. It is therefore very popular for data processing workloads on Google Cloud; especially when open source tools are preferred. You can either manage the servers that process your data yourself, or Dataproc also provides a serverless option, in which Google will manage all of the servers for you. People sometimes want to manage the servers themselves if they want to use customized configurations or customized tools. Dataproc also integrates with other Google tools such as BigQuery, and Vertex AI to cater to flexible data management needs and data science projects, and you can enforce fine-grained row and column-level access controls with Dataproc, BigLake, and Dataplex. You can also manage and enforce user authorization and authentication using existing Kerberos and Apache Ranger policies, and it provides the built-in Dataproc Metastore, which eliminates the need to run your own Hive Metastore or catalog service.

Managing your own on-premises Hadoop or Spark clusters can require a lot of work. One of the great things about Dataproc is that you can easily spin up clusters on demand in order to run a data processing workload, and then automatically shut them down when you're not using them, and clusters can automatically scale up and down to meet your needs, which helps to save costs. You can also use Google Compute Engine Spot instances to further save costs for workloads that can tolerate being interrupted. You can run your workloads in VMs or containers, and it also supports GPUs if you need to use those in your data processing workloads.

### *Dataprep*

Dataprep by Trifacta is a tool for visually exploring, cleaning, and preparing structured and unstructured data for analysis, reporting, and machine learning. It's serverless, so there is no infrastructure to deploy or manage. It's a very useful tool for the data exploration phase of a data science project, enabling you to explore and understand data with visual data distributions, and automatically detecting schemas, data types, possible joins, and anomalies such as missing values, outliers, and duplicates. You can then define a sequence of transformations to clean up and prepare your data for training ML models. You can do all of this visually, without needing to write any code, and it even suggests what kinds of transformations you may wish to implement, such as aggregation, pivot, unpivot, joins, union, extraction, calculation, comparison, condition, merge, regular expressions, and more. You can also apply data quality rules to ensure that your data meets your quality requirements, and it allows teams to collaborate on datasets by sharing or copying them as needed.

### Dataflow

In order to discuss Google Cloud Dataflow, we will first take a minute to introduce Apache Beam. In previous sections and chapters of this book, we've referred to *batch* data processing, in which large amounts of data are processed by long-running jobs, and *streaming* data processing, in which small pieces of data are processed very quickly, usually in real time or near real time. There are generally slightly different tools for each of those processing types. For example, you might use Hadoop for batch processing and you might use Apache Flink for stream processing. Apache Beam provides a unified model that can be used for both batch and streaming workloads. In the words of the Apache Beam project management committee, this allows you to "*write once, run anywhere.*" This can be very useful as it enables your data engineers to simplify how they code their data processing workloads by using this unified model instead of using completely different tools and code for their batch and streaming use cases.

Google Cloud Dataflow is a fully managed, serverless service for executing Apache Beam pipelines. Once you have defined your data processing steps as an Apache Beam pipeline, you can then use your preferred data processing engines, such as Spark, Flink, or Google Cloud Dataflow, to execute the pipeline steps. As a fully managed service, Dataflow can automatically provision and scale the resources required to run your data processing steps, and it integrates with other tools and services such as BigQuery, enabling you to use SQL to access your data, and Vertex AI notebooks for ML model training use cases.

### Looker

Looker is Google Cloud's business intelligence platform with embedded analytics. One of its main advantages is LookML, which is a powerful SQL-based modeling language. You can use LookML to centrally define and manage business rules and definitions as a version-controlled data model, and LookML can then create efficient SQL queries on your behalf. As a business intelligence tool, Looker then provides a user interface in which you can visualize your data in graphs, charts, and dashboards.

It comes in a few different service tiers, providing different levels of business intelligence functionality. Google originally had a business intelligence tool named Data Studio, and Looker was created by a separate company, but Google acquired that company and has integrated Looker into its Cloud Service portfolio. They also integrated their original Data Studio product into Looker, creating Looker Studio, and added an enterprise version of that tool, named Looker Studio Pro, which provides additional functionality as well as customer support.

### Data Fusion

In order to discuss Data Fusion, let's first talk about the concepts of **ETL** and **ELT**, which stand for **Extract, Transform, Load**, and **Extract, Load, Transform**, respectively. The ETL concept has been around since the 1970s, and it's a common pattern that's used in data science and data engineering when you need to perform transformations on your data. The pattern used in this case is to extract your data from its source storage location, transform it in some way, and then load it into your desired destination storage location. Examples of transformations could be to change the format of the data from CSV to JSON, or to remove all rows that contain missing information. A complex data engineering project may require the creation of ETL pipelines that define multiple transformation steps. ELT, on the other hand, has gained popularity in recent years, especially in relation to cloud-based data processing workloads. The idea with ELT is that you can extract your data from source locations and load it into a data lake or data warehouse, and then perform different kinds of transformations based on your project needs. This is also often referred to as *data integration*. Using this approach can enable analysts to use SQL to get insights and value from the data without requiring data engineers to create complex ETL pipelines.

With this in mind, Data Fusion is Google Cloud's serverless product that allows you to run your ETL/ELT workloads without having to manage any servers or infrastructure. It provides a visual interface that enables you to define your data transformation steps without having to write code, which makes it easy for less technical analysts to process the data, and it even tracks that ever-important data lineage that we discussed in *Chapter 2*, as your data progresses through each transformation step.

Data Fusion integrates with other Google Cloud services such as BigQuery, Dataflow, Dataproc, Datastore, Cloud Storage, and Pub/Sub, making it easy to perform data processing workflows across those services.

### Google Cloud Composer

Composer is a Google Cloud orchestration service built on the open source Apache Airflow project, and it is particularly useful for data integration or data processing workloads. Like Data Fusion, it integrates with other Google Cloud services such as BigQuery, Dataflow, Dataproc, Datastore, Cloud Storage, and Pub/Sub, making it easy to orchestrate data processing workflows across those services.

It is highly scalable, and it can also be used to implement workloads across multiple cloud providers and on-premises locations. It orchestrates workflows by using **Directed Acyclic Graphs** (**DAGs**), which represent the tasks your workflow needs to execute, as well as all of the relationships and dependencies between them.

Similar to Google Cloud Workflows, Composer monitors the execution of the tasks in your workload and tracks whether each step is completed correctly or whether any problems have occurred. *Figure 3.12* shows an example of a workflow being orchestrated by Google Cloud Composer, in which data regarding customer orders is ingested periodically from Cloud Bigtable into Cloud Dataproc and enriched with data regarding broader retail trends from third-party providers, which is provided via Google Cloud Storage. The outputs are then stored in BigQuery for analysis. This data could then be used to build business intelligence dashboards in Looker or to train machine learning models in Vertex AI, for example.

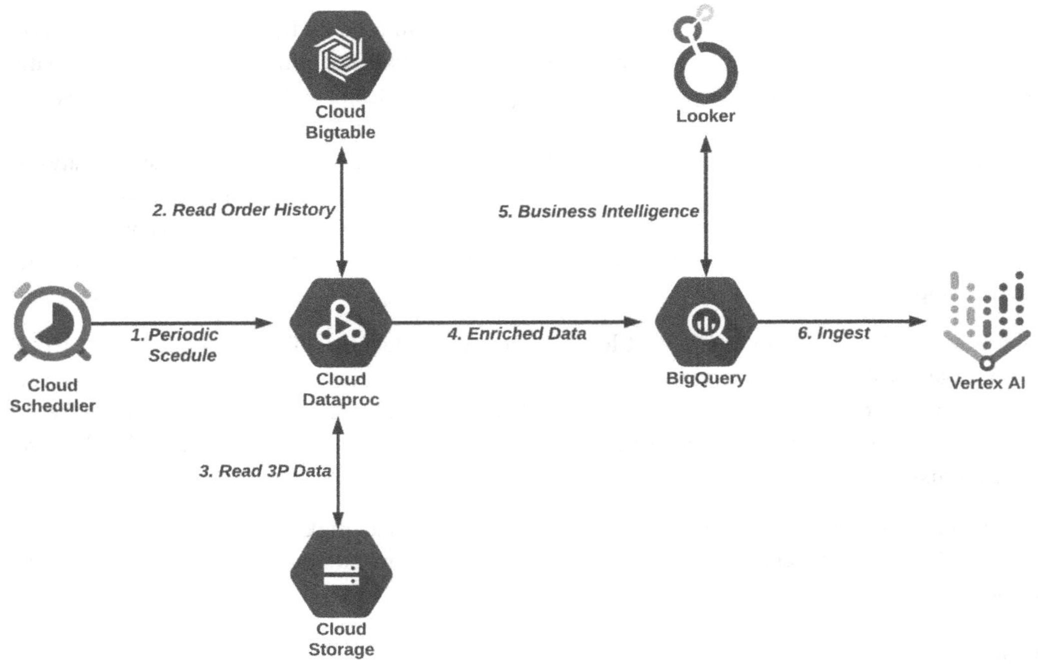


Figure 3.12: Data processing workflow

As depicted in *Figure 3.12*, when we've processed and stored our data, we may want to use it to train a machine learning model. Before we get into training our own models, let's explore some of the AI/ML capabilities that we can use on Google Cloud that provide models trained by Google.

# Google Cloud AI tools and AutoML

In this section, we will cover Google Cloud's AI tools that can be used to implement AI use cases without the need to understand the underlying machine learning concepts. With these services, you can simply send a request to an API, and get a response from ML models that are created and maintained by Google. There's no need to manually preprocess data, train or hyper-tune machine learning models, or manage any infrastructure.

We will group these services into the following categories: **Natural Language Processing** (**NLP**), **Computer Vision**, and **Discovery**.

> **Note**
>
> At the time of writing this, in January 2023, Google has also announced a preview of a service named **Timeseries Insights API**, which can be used to perform analysis and gather insights in real time from your time series datasets, for use cases such as time series forecasting, or detecting anomalies and trends in your data while they are happening. This is an interesting new service because running forecasting and anomaly detection workloads over billions of time series data points is computationally intensive, and most existing systems implement these workloads as batch jobs, which limits the type of analysis you can perform online, such as deciding whether to alert based on a sudden increase or decrease in data values.

## NLP

Natural language processing, in the context of AI, refers to the use of computers to understand and process natural human language. It can be further broken down into **Natural Language Understanding** (**NLU**) and **Natural Language Generation** (**NLG**). NLU is concerned with understanding the content and meaning of words and sentences, as they are understood by humans. NLG takes this one step further and attempts to create or generate words and sentences in a way that can be understood by humans. In this section, we'll discuss some of Google Cloud's NLP-related services.

## The Natural Language API

You can use the Google Cloud Natural Language API to understand the contents of textual inputs. This can be used for purposes such as sentiment analysis, entity analysis, content classification, and syntax analysis. With sentiment analysis, it can tell you what kinds of emotions are suggested by the content. For example, you could feed all of your product reviews into this API and get an understanding of whether people are responding positively to your product, or whether they may be frustrated by something about the product. Entity analysis can identify what kinds of content exist in the text, such as people's names, place names, locations, addresses, and phone numbers. The content classification feature is useful if you have large amounts of textual data and you want to organize and categorize it based on the contents.

### *Text-to-Speech*

The name of this service is somewhat self-explanatory. The service will take a textual input that you provide, and it will convert it to an audible spoken output. This can be very useful for accessibility use cases, whereby if somebody is visually impaired and cannot read text content, it could be automatically spoken to them. It provides the option to use many different voices to personalize the user experience, and you can even create custom voices using your own recordings. As of January 2023, it supported more than 40 languages and variants. It also supports **Speech Synthesis Markup Language** (**SSML**), which enables you to have more control over how words and phrases are pronounced.

### Speech-to-Text

This service does pretty much the opposite of the previous service. In this case, you can provide audio inputs, and the service will transcribe any spoken language into a textual output. This is useful for dictation purposes, accessibility use cases such as closed captioning, and other use cases such as quality control. For example, you could provide recordings of your customer service calls and it would convert them to text. Then you could feed that text into the Natural Language API to understand whether your customers are frustrated or happy with the service they are receiving. As of January 2023, the service supported an impressive 125 languages and variants.

### Translation AI

Another service whose name is self-explanatory, this service can be used to translate from one language to another. It can help you to internationalize your products, and engage your customers with localization of content. It can detect more than 100 languages, and you can customize the translations with industry- or domain-specific terms. It provides **Translation Hub**, which allows you to manage translation workloads at scale, as well as Media Translation API, which can deliver real-time audio translation directly to your applications.

### Contact Center AI (CCAI)

**Contact Center AI** (**CCAI**) provides human-like AI-powered contact center experiences. It consists of a number of different components, such as Dialogflow, which can be used to create chatbots that can have intelligent, human-like conversations with customers.

Have you ever been on hold with a company's customer service line for an hour before anybody helps you with your concerns? This, of course, happens when customer service centers are overloaded with calls. Using chatbots can offload a significant amount of cases with simple questions, freeing up humans to focus on more complex customer interactions.

When a human does need to get involved, CCAI has another feature, named Agent Assist, that provides support to human agents while they handle customer interactions. It can recommend ready-to-send responses to customers, provide answers to customer questions from a centralized knowledge base, and transcribe calls in real time.

CCAI also includes CCAI Insights, which uses NLP to identify customer sentiment and reasons for calls, which helps contact center managers learn about customer interactions to improve call outcomes.

There's also the option to use CCAI Platform, which provides a **Contact Center as a Service** (**CCaaS**) solution.

### *Document AI*

Document AI goes beyond understanding the content of textual inputs, and also incorporates structure. It provides pre-trained models for data extraction, or you can use Document AI Workbench to create custom models, and you can use Document AI Warehouse to search and store documents. For example, if you collect information from people via forms, Document AI could be used to extract the data from those forms and store it in a database, or send it to another data processing system to process the data in some way or feed it to another ML model to perform some kind of other task. It could also be used to categorize and organize documents based on their contents. Some companies process millions of forms and contracts per year, and before these kinds of AI systems existed, all of those documents had to be processed by humans, which led to extremely laborious and error-prone work. With Document AI, you can automate that work, and you can also enrich the data using Google **Enterprise Knowledge Graph** (**EKG**), or you can enhance the functionality of Document AI with human inputs by using its **Human-in-the-Loop** (**HITL**) AI functionality. With HITL AI, experts can verify the outputs from Document AI, and provide corrections if needed. You can either use your own workforce of experts for this purpose or, if you don't have such experts in your employment, you can use Google's HITL workforce.

Document AI's **Optical Character Recognition** (**OCR**) functionality flows over into the computer vision realm, which we will discuss next.

## Computer vision

You can use Google Cloud Vision AI to create your own computer vision applications or get insights from images and videos with pre-trained APIs, AutoML, or custom models. It enables you to spin up video and image analytics applications in minutes, for use cases such as detecting objects, reading handwriting, or creating image metadata. It consists of three main components: Vertex AI Vision, Vision API, and custom ML models. Vertex AI Vision includes **Streams** to ingest real-time video data, **Applications** to let you create an application by combining various components, and **Vision Warehouse** to store model output and streaming data. The Vision API offers pre-trained ML models that you can access through REST and RPC APIs, allowing you to assign labels to images and classify them into millions of predefined categories. If you need to develop more specialized models, you can use AutoML or build your own custom models.

Vision AI can be used to implement interesting use cases such as image search, in which you could use Vision API and AutoML Vision to make images searchable based on topics and scenes detected in the images, or product search, in which you could enable customers to find products of interest within images and visually search product catalogs using the Vision API.

Also in the computer vision space is Google Cloud Video AI, which can analyze video content for use cases such as content discovery. Video AI can recognize over 20,000 objects, places, and actions in video, it can extract metadata at the video, shot, or frame level, and you can even create your own custom entity labels with AutoML Video Intelligence. It consists of two main components: the Video Intelligence API, which provides pre-trained ML models, and Vertex AI for AutoML video, which provides a graphical interface to train your own custom models to classify and track objects in videos, without the need for ML experience. Vertex AI for AutoML video can be used for projects requiring custom labels that aren't covered by the pre-trained Video Intelligence API.

### Discovery AI

Google Cloud Discovery AI includes services such as search and recommendation engines. These kinds of features are essential for today's online businesses, in which it's important to ensure that your customers find what they want on your website as quickly and easily as possible. If this doesn't happen, they'll go to your competitor's website. Imagine if your company could integrate Google-quality search into its website. This includes functionality such as image-based product searches, which we discussed in the previous section, making it easier for customers to search for products with an image, by using object recognition to provide real-time results of similar items from your product catalog.

In addition to your customers finding products by directly searching for them, companies see a significant amount of business being driven by recommendation engines, which can make personalized recommendations for products based on customers' previous purchasing behavior. I've experienced this many times myself as a consumer, whereby I'm purchasing something on a website, I see a recommendation for something else that I may be interested in purchasing, and I think "*Actually, yes I do also want one of those*," and I go ahead and add it to my cart before checking out. These kinds of personalized experiences help to maintain customer loyalty, which is also extremely important in today's online business world.

Bringing these two concepts together – that is, search and personalization – makes a lot of sense, because not only will the search results intelligently match what the customer searched for, but the ranking of the results can be catered to what the specific customer is more likely to find relevant to their preferences.

A third component of Discovery AI is Browse AI, which extends the search functionality to work on category pages in addition to text queries. Without this, retailers would mostly sort products on their category and navigation pages by historic bestsellers. This ordering doesn't adapt well to new product additions, changes in product availability, and sales. With Browse AI, retailers can sort products on these pages in an order that is personalized to each user, and based on predicted, rather than historic, bestsellers. This ordering can rapidly adapt to new products, product stockouts, and price changes, without needing to wait for backward-looking bestseller lists to catch up.

Now that we've taken a look at the high-level AI services on Google Cloud, which you can use to get inferences from ML models that are trained and maintained by Google by simply calling an API, let's discuss how you can start to train your own models on Google Cloud, starting with AutoML.

# AutoML

Some of the services we discussed in the previous sections use completely pre-trained models, and others allow you to bring your own data to either train or *up-train* a model based on your data. Pre-trained models are trained on datasets provided by Google or other sources, and the term, up-train, refers to augmenting a pre-trained model with additional data. If you want to create more customized use cases than those supported by the high-level API services, you may want to train your own models. Vertex AI, which we will describe later in this chapter, provides a plethora of tools for implementing every step in the model development process. However, before we get to the level of customizing every step in the process, one way in which you can easily start getting inferences from ML models that are trained on your data in Google Cloud is to use AutoML, which enables developers with limited ML expertise to train models specific to their business needs in as little as a few minutes or hours. The actual amount of time depends on the algorithms being used, how much data is used for training, and some other factors, but in any case, AutoML saves you a lot of work and time, considering that data scientists can spend weeks on these tasks when not using AutoML. Looking at our ML model development lifecycle diagram in *Figure 3.13*, AutoML automatically performs all of the steps in the process for us, as indicated by everything encapsulated within the blue box.

Figure 3.13: ML model lifecycle managed by AutoML

How does AutoML work? If we think back to all of the steps in a typical data science project that we discussed in *Chapter 2*, you may remember that each step – especially in the early stages of a project – required a lot of trial and error. For example, you might try a number of different data transformation techniques, and then try out a few different algorithms during training, as well as different combinations of hyperparameter values for those algorithms. Each of those steps could take many days or weeks of work before you find a good candidate model (i.e., a model that we believe may satisfy our business objectives and metrics). AutoML automatically runs many trial jobs very quickly – generally much more quickly than a human could achieve – and evaluates the outcomes against desired metric thresholds. Jobs whose outcomes do not meet the desired criteria are not selected as candidates, and AutoML continues to try other options until suitable candidates are found, or some other threshold is met, such as all options being exhausted.

How can you start using AutoML? Some of the AI services we discussed in the previous sections in this chapter already use AutoML in order to train the models that you access via those APIs. Examples include AutoML image, in which you can get insights from object detection and image classification, AutoML Video for streaming video analysis, AutoML Text to understand the structure, meaning, and sentiment of text, AutoML Translation to translate between languages, and AutoML Forecasting to provide forecasts based on time series data.

Another AutoML use case that we haven't explored yet is AutoML for tabular data (structured data that's stored in tables). This is a very common format for storing business data, as it provides a way to organize information that is easy for humans to read and understand. AutoML Tabular supports multiple ML use cases with tabular data, such as binary classification, multi-class classification, regression, and forecasting.

We can use AutoML to automate a lot of the trial and error steps and develop candidate models, and then we can customize further from that point if we wish. For example, we could use Vertex AI Tabular Workflows, which creates a *glassbox*-managed AutoML pipeline that lets us see and interpret each step in the model building and deployment process. We can then tweak any steps in the process as we see fit, and automate any updates via MLOps pipelines. We will be performing exactly these kinds of activities in later chapters of this book.

Next, we're going to dive deeper into ML model customization, and we will explore Vertex AI in more detail, as it can be used to customize every step in a data science project.

# Google Cloud Vertex AI

This is where we start getting to expert-level AI/ML. Think back on all of the data science concepts we've covered so far in this book; all of the different types of ML approaches, algorithms, use cases, and tasks. Vertex AI is where you can accomplish all of them, and many more that we will yet explore in this book. You can use Vertex AI as your central command center for performing everything AI/ML-related. Our ML model lifecycle diagram in *Figure 3.14* illustrates this point graphically. Traditionally, we would expect an AI/ML platform to mainly take care of training, evaluating, deploying, and monitoring models. This is what's represented by the blue box on the right-hand side of *Figure 3.14*, and all of these activities are of course supported by Vertex AI. However, with additional features such as notebooks and MLOps pipelines, Vertex goes beyond just those traditional ML activities, to also enable us to perform all of the tasks in our lifecycle, including data exploration and processing, as represented by the light blue dashed box on the left-hand side of the following diagram:

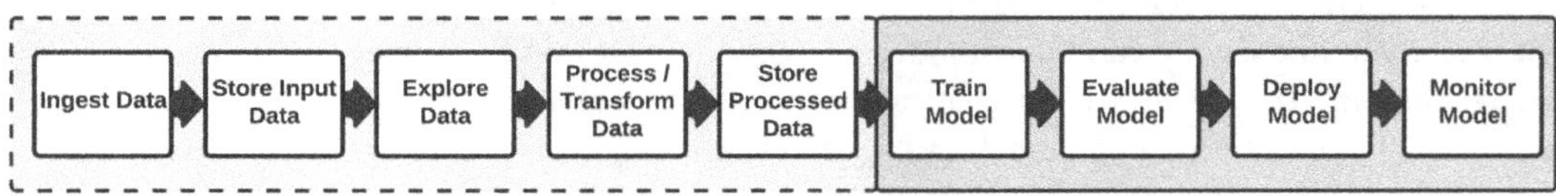


Figure 3.14: ML model lifecycle with Vertex AI

Let's take a look at Vertex AI's features in a bit more detail. Starting with the basics, we can use Vertex AI's Deep Learning VM Images to instantiate a VM image containing the most popular AI frameworks on a Compute Engine instance, or we can use Vertex AI Deep Learning Containers to quickly build and deploy models in a portable and consistent containerized environment.

The VMs and container images provide building blocks on which we can develop customized ML workloads, but we can also use Vertex AI in other ways to perform and manage all of the steps in our model development lifecycle. We can use Jupyter notebooks to explore data and experiment with each of the steps in the process, while Vertex AI Data Labeling enables us to get highly accurate labels from human labelers for creating better supervised ML models, and Vertex AI Feature Store provides a fully managed feature repository for serving, sharing, and reusing ML features. Vertex AI Training provides pre-built algorithms and allows users to bring their custom code to train models in a fully managed training service for greater flexibility and customization, or for users running training on-premises or in another cloud environment. Vertex AI Vizier automates all of our hyperparameter tuning jobs for us, finding an optimal set of hyperparameter values for us, and saving us from having to do a lot of painstaking work manually! Optimized hyperparameter values lead to more accurate and more efficient models.

When we need to deploy our models, we can use Vertex AI Predictions, which will host our models on infrastructure that's managed by Google, which auto-scales to meet our models' traffic needs. It can host our models for batch or online use cases, and it offers a unified framework to deploy custom models built on any framework, such as TensorFlow, PyTorch, scikit-learn, or XGB, as well as BigQuery ML and AutoML models, and on a broad range of machine types and GPUs. After deployment, we can use Vertex AI Model Monitoring to provide automated alerts for data drift, concept drift, or other model performance incidents that may require supervision. We can then automate the entire process as MLOps pipelines by using Vertex Pipelines, which allows us to trigger retraining of our models when needed, and to manage updates to our models in a version-controlled way. We can also use Vertex AI TensorBoard to visualize ML experiment outcomes and compare models and model versions against each other to easily identify the best-performing models. This visualization and tracking tool for ML experimentation includes model graphs that display images, text, and audio data.

What's really great is that we can perform and manage all of these activities from Vertex AI Workbench, which is a Jupyter-based fully managed, scalable, enterprise-ready compute infrastructure with security controls and user management capabilities. All the while, the lineage details of our models as they progress through every step in the process can be tracked by Vertex Experiments and the Vertex ML Metadata service, which provides artifact, lineage, and execution tracking for ML workflows, with an easy-to-use Python SDK.

Vertex AI provides even more functionality beyond all of the features we mentioned previously, such as Vertex AI Matching Engine, which is a massively scalable, low latency, and cost-efficient vector similarity matching service. Vertex AI Neural Architecture Search enables us to build new model architectures targeting application-specific needs and optimize our existing model architectures for latency, memory, and power, in an automated service, and we can use Vertex Explainable AI to understand and build trust in our model predictions with actionable explanations integrated into Vertex AI Prediction, AutoML Tables, and Vertex AI Workbench. Explainable AI provides detailed model evaluation metrics and feature attributions, indicating how important each input feature is to our predictions.

## Standard industry tools on Google Cloud

In addition to Google Cloud's own data science tools that we've been describing so far in this chapter, you can also use other data science tools such as open source frameworks or other popular industry solutions. There are lots of great libraries out there that make it easy to perform the various tasks in the model development life cycle. When it comes to data exploration and processing, for example, the beloved pandas library is a staple of any ML and data analysis course. You can use it for handling missing data, slicing, subsetting, reshaping, merging, and joining datasets. Matplotlib is right up there with pandas for data exploration as it allows you to visualize your data via customizable and interactive plots and charts that can be exported into various file formats. NumPy allows you to easily manipulate and play around with the kinds of $n$-dimensional arrays and vectors we find in so many ML implementations. Learning NumPy also sets you up to start using frameworks such as scikit-Learn, TensorFlow, and PyTorch.

Speaking of scikit-learn, it is as much of a staple in any machine learning course as pandas. If you take an ML course, you will almost certainly use scikit-learn at some point in your learning process, and for good reason; it's a framework that's easy to use and understand and contains lots of built-in algorithms and datasets that you can use to implement ML workloads. And, it's more than just an easy framework for learning purposes; many companies also use scikit-learn for their production ML workloads.

While we're still on the topic of general-purpose ML frameworks, the next framework we'll discuss is the wildly popular TensorFlow, which was originally created by Google before they open sourced it, and is therefore very well supported on Google Cloud. TensorFlow can be used for everything from processing data to NLP and computer vision. You can use it to train, deploy, and serve models, and with **TensorFlow Extended** (**TFX**), you can implement end-to-end MLOps pipelines to automate all of those steps. We'll certainly be exploring TensorFlow in more detail in this book, as well as Keras, which is an API that provides access to TensorFlow via a Python interface that's popular for its ease of use, and advertises itself as "*an API designed for human beings, not machines.*"

We'll round out our discussion of general-purpose ML frameworks with PyTorch, which was originally developed by Facebook (Meta AI) and is now open sourced under the Linux Foundation. PyTorch has been rapidly gaining popularity in recent years, especially among Python developers, and it has become a very widely used framework in addition to TensorFlow. In this book, we're not going to get into the argument of which framework is better. There are staunch supporters on each side of this debate, and if you Google "TensorFlow versus PyTorch," you'll find no shortage of websites and forums highlighting how one is better than the other for particular types of use cases. We will also be using PyTorch in later chapters of this book.

Switching gears from general-purpose ML frameworks to more specialized frameworks, you might want to use something such as OpenCV for computer vision workloads, or SpaCy for NLP. OpenCV has a broad selection of algorithms for applying ML and deep learning to images and video content, to perform many of the tasks we discussed earlier in this chapter, such as object recognition, and tracking objects and actions through video frames. SpaCy has lots of pre-trained word embeddings and pipelines supporting multiple languages, and it supports custom models written in TensorFlow and PyTorch and lots of Python packages for NLP.

The good news is that all of the tools and frameworks we've discussed in this section can easily be used on Google Cloud. In addition to open source tools, there are many popular third-party data science solutions that can be used on Google Cloud, providing flexibility for people to use whatever tools they prefer for the objectives they want to achieve. We've already talked about using Spark on Google Cloud through services such as Dataproc and Vertex AI, and you can use third-party Spark offerings such as Databricks via the Google Cloud Marketplace. The marketplace allows you to find thousands of solutions that run on Google Cloud, including deep learning solutions from companies such as Hugging Face.

With all of these tools and all of the services provided by Google Cloud, you might wonder how to choose the right tool for your data science workloads. Let's take a look at that discussion in more detail in the next section.

# Choosing the right tool for the job

Your choice of data processing tools will depend heavily on what kind of data processing tasks you need to accomplish. If you have a bunch of raw data that you need to transform in bulk, as in an ETL/ELT task, Data Fusion would be a good place to start your assessment, whereas if you want to perform relatively simple transformations using SQL syntax, then start with BigQuery, and if you want to visualize and transform data via an easy-to-use GUI, then go with Dataprep. If you prefer to stick to using open source tools, then you might want to use something like pandas or Spark. We discussed pandas being a good starting point for people who are beginning to learn about data exploration and preprocessing, and how it's also more than an educational tool. pandas is really great for initial data exploration and data processing at a moderate scale. However, for large-scale data processing projects, Spark's highly parallelized functionality will be a lot more efficient, and if you want to use it without managing the infrastructure yourself, then you can run it on Dataproc. Dataflow is the recommended choice for streaming data, and it has the additional benefit of also working well for batch data processing with its unified programming model.

When it comes to choosing a tool for AI/ML workloads, the decision may be a bit more straightforward. If you or your company has a pre-existing affinity for a specific third party such as Hugging Face, then you may be guided in that direction. The general best practice is that you should start with the highest level of abstraction available to you because then you won't have to spend a lot of time and effort building and maintaining something that is already available as a service for you to use. If that does not meet your needs for any reason, such as specific business requirements that require some level of customization, then move to a solution that provides more control over how the workload is implemented. For example, if you're building an application that needs to include some kind of NLP functionality, start by evaluating the Google Cloud NLP API. If, for any reason, you cannot achieve your desired objective with that solution, move on to evaluating the use of AutoML to automate training a model on your specific data. If your objective is still not met by the models created during that process, then it's time to increase your level of customization, and potentially use Vertex AI to build a completely customized model.

Another very important factor in your decision process is your budget. Services that do more work for you – such as the higher-level AI services – may cost more than the lower-level services, but it's very important to factor in how much you are paying your employees to manage infrastructure and perform the tasks that could be performed on your behalf by the higher-level services.

## Summary

In this chapter, we covered all of the foundational and primary services that you can use for implementing AI/ML workloads on Google Cloud. We started with the basic services such as Google Compute Engine, and Google Cloud networking services, upon which all other services are built. We then explored how you could use those services to set up connectivity with systems outside of Google Cloud. Next, we discussed the services that you can use to import or transfer data to Google Cloud and the various storage systems that you can use for storing that data in the cloud. Having covered the primary storage systems, we moved on to discuss Google Cloud's data management and data processing services. The final step in our journey was to understand all of the different AI/ML services that exist in Google Cloud.

At this point, you have already learned a lot, and with this knowledge, you could now have an intelligent discussion about AI/ML and Google Cloud. This is a significant achievement because people spend a very long time trying to learn about AI/ML, and a very long time trying to learn about Google Cloud. Right now, you know more than a lot of people out there, and you should feel proud and give yourself a pat on the back for coming this far.

This concludes the *Basics* part of our book, and all of the knowledge you have gained in the previous chapters will form the basis of what you will learn in the rest of this book. In the next chapter and beyond, we will start performing hands-on activities, diving deeper into the services and concepts we have covered thus far, and actually start to build data science workloads on Google Cloud!

# Part 2: Diving in and building AI/ML solutions

At this point, we have equipped ourselves with the prerequisite knowledge to start building AI/ML solutions on Google Cloud. Therefore, this part focuses on actually building those solutions. We begin this part by exploring Google Cloud's high-level AI services. These are the services you can use without requiring any expertise in AI/ML. We then move on to building custom ML models on Google Cloud. After building your first custom model, the remaining chapters in this part focus on diving deep into each step in the typical AI/ML project, such as data preparation, feature engineering, hyperparameter optimization, and deploying, monitoring, and scaling in production. We then dive into the concept of MLOps to automate all of the steps in the machine learning model development lifecycle. Next, we go beyond the technology realm and incorporate broader business and societal concepts that apply to AI/ML models, such as bias, explainability, and fairness. We then broaden the discussion further by discussing ML system governance, and how to use the Google Cloud Architecture Framework to build robust, enterprise-grade AI/ML systems and solutions. Finally, we wrap up this section by exploring additional AI/ML use cases, frameworks, and technologies.

This part contains the following chapters:

- *Chapter 10, Deploying, Monitoring, and Scaling in Production*
- *Chapter 11, Machine Learning Engineering and MLOps with Google CLoud*
- *Chapter 12, Bias, Explainability, Fairness, and Lineage*
- *Chapter 13, ML Governance and the Google Cloud Architecture Framework*
- *Chapter 14, Additional AI/ML Tools, Frameworks, and Considerations*

# 4
# Utilizing Google Cloud's High-Level AI Services

Now that you have been equipped with an arsenal of information regarding AI/ML and Google Cloud, you are ready to start diving in and implementing AI/ML workloads in the cloud – that's exactly what we will do in this chapter! You're going to start by using Google Cloud's high-level AI/ML APIs, such as the Natural Language API and the Vision API, which enable you to implement AI/ML functionality using models that are trained and maintained by Google. Then, you'll use Vertex AI AutoML to train your own ML model, all without the need for any AI/ML expertise.

This chapter will cover the following topics:

- Using Document AI to extract information from documents
- Using the Google Cloud Natural Language API to to get sentiment analysis insights from textual inputs
- Using Vertex AI AutoML

## Prerequisites for this chapter

First, we need to perform some setup steps to lay the foundations for the activities we will be performing in this chapter.

## Cloning this book's GitHub repository to your local machine

Throughout the rest of this book, we will perform a lot of hands-on activities that require the use of resources such as code, data, and other files. Many of these resources are stored in the GitHub repository associated with this book, so the easiest way to access them would be to clone the repository to your local machine (that is, your laptop/PC). The exact way to do this will differ based on the operating system you use, but the process is generally to open a command terminal on your system, navigate to a directory of your choice, and run the following commands:

```
git init
git clone https://github.com/PacktPublishing/Google-Machine-Learning-for-Solutions-Architects
```

## The Google Cloud console

In the Prerequisites for using *Google Cloud tools and services* section of *Chapter 3*, we discussed that you would need to create a Google Cloud account to interact with most Google Cloud services. We will use the Google Cloud console to perform many of the activities in this chapter. Once you have created a Google Cloud account, you can log into the console by navigating to `https://console.cloud.google.com`.

Throughout this book, you will need to navigate to different services within the Google Cloud console. Whenever you need to do this, you can click on the symbol with three horizontal lines next to the Google Cloud logo in the top-left corner of the screen to view the menu of Google Cloud services or products (in this context, we'll use the terms "services" and "products" interchangeably throughout this book). See *Figure 4.1* for reference:

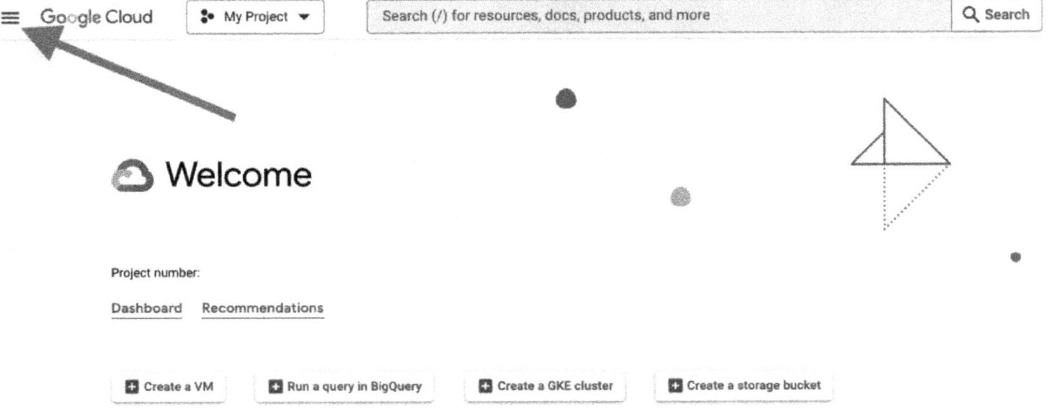


Figure 4.1: Accessing the Google Cloud services menu

You can then scroll through the list of services and select the relevant one. Some of the menu items have sub-menus nested under them, as shown in *Figure 4.2*:

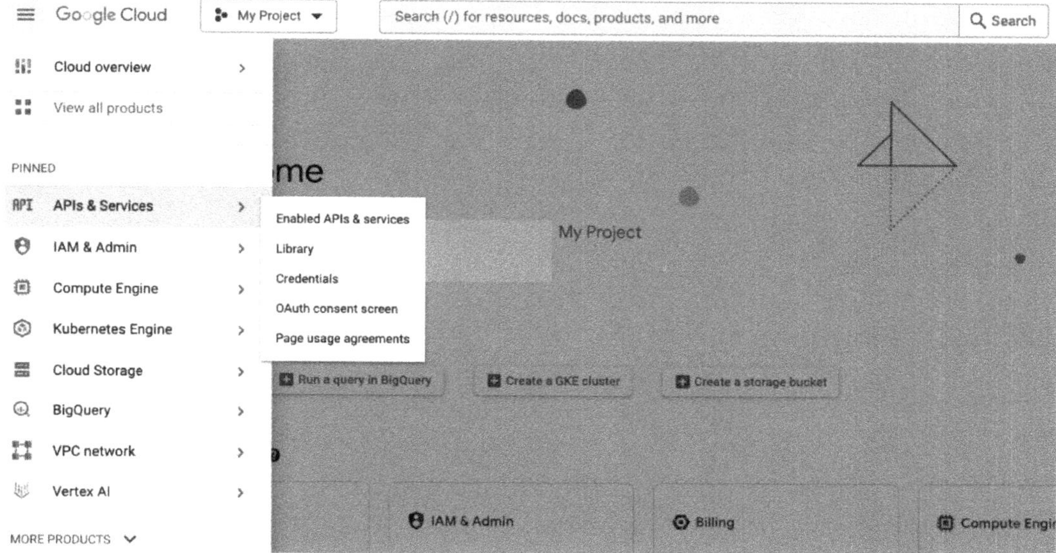


Figure 4.2: Google Cloud services sub-menus

Throughout this book, we will sometimes use shorthand notation to represent the navigation path through the services menus and sub-menus. For example, to navigate to the **Credentials** sub-menu item, we will represent this as the **Google Cloud services** menu → **APIs & Services** → **Credentials**.

Now that we've looked at how to access the Google Cloud console, let's ensure that we have a Google Cloud project to use for the activities in this book.

## Google Cloud project

A Google Cloud project is an environment that contains all of your resources in Google Cloud. Resources include things such as virtual machines, network configurations, databases, storage buckets, identity and access controls, and pretty much everything that you create on Google Cloud.

By default, when you create your Google Cloud account and log into the console for the first time, Google Cloud automatically creates your first project for you, with the name **My First Project**. If, for any reason, this has not happened, you will need to create one by following these steps:

1. In the Google Cloud console, navigate to the **Google Cloud services** menu → **IAM & Admin** → **Create a Project**.

2. Type a name for your project in the **Project Name** text box.

   You could also edit the **Project ID** value if you'd like, but it's more common to leave this at the generated value unless you have a specific naming convention that you'd like to implement. Note that this cannot be changed after the project is created.

3. In the **Location** field, click **Browse** to display potential locations for your project. Then, click **Select**.
4. Click **Create**. The console will navigate to the **Dashboard** page and your project will be created within a few minutes.

Now that we have our Google Cloud project in place, let's ensure that our billing details are set up for using Google Cloud.

## Google Cloud Billing

A Google Cloud Billing account contains all of the details that enable Google Cloud to bill you for using their services.

If you don't already have a Google Cloud Billing account, then you will need to create one by following these steps:

1. In the Google Cloud console, navigate to the **Google Cloud services** menu → **Billing**.
2. On the page that appears, select **Manage Billing Accounts**, then **Add billing account**.
3. Click **Create account**.
4. Enter a name for the Cloud Billing account.
5. Depending on your configuration, you will also need to select one of the following:
   - **Organization**: If you see an **Organization** dropdown, then you must select an organization before you can continue
   - **Country**: If you are prompted to select a **country**, select the country that corresponds with your billing mailing address
6. Click **Continue**.
7. You will then need to fill in all of your billing details.
8. When you have entered all of your details, click **Submit and enable billing**. To learn more about Google Cloud Billing accounts and concepts, visit `https://cloud.google.com/billing/docs/how-to/create-billing-account`.

Now that our Google Cloud project and Billing account have been set up, we're ready to start using Google Cloud products in our project.

## Google Cloud Shell

We will use Google Cloud Shell to perform some of the activities in this chapter.

Open Cloud Shell, as described in the *Interacting with Google Cloud services* section of *Chapter 3*. As a reminder, you can access it by clicking the **Cloud Shell** symbol in the top-right corner of the screen, as shown in *Figure 4.3*:

Figure 4.3: Cloud Shell symbol

Now that we've opened Cloud Shell, we'll perform some basic setup steps to prepare our Cloud Shell environment.

## Authentication

When a Google Cloud service is invoked in some way, it usually wants to know who or what (identity) is invoking it. This is for many reasons, such as billing and checking whether the invoker is allowed to perform the action that is being requested. This process of identification is called "authentication." There are numerous different ways to authenticate with a Google Cloud API. For example, when you log into the Google Cloud console, you are authenticating with the console. Then, as you navigate and perform actions in the console, your authentication details are used to control what kinds of actions you are allowed to perform. One of the simplest authentication mechanisms is referred to as **API keys**. We'll explore this next.

### *Google Cloud API keys*

API keys are a very basic type of authentication supported by some Google Cloud APIs. An API key does not provide any identification or authorization information to the called API; they are only generally used for billing purposes (by linking to a Google Cloud project) or to track usage against Google Cloud quotas.

In this chapter, we will use them to access the Google Cloud Natural Language API.

### Creating an API key

Although we will use Cloud Shell to perform many of the steps in this chapter, Google Cloud currently only supports creating API keys in the Google Cloud console. To create our API key, perform the following steps:

1. In the Google Cloud console, navigate to the **Google Cloud services** menu → **APIs & Services** → **Credentials**.
2. Select **Create credentials**, then **API key**.
3. Copy the generated API key and click **Close**.
4. You will need to paste this API key in a later step. If you need to view or copy the key again, you can select **Show key** in the **API Keys** section of the console (that is, the section you are currently in).
5. With that, we have created an API key that we can use in later activities in this chapter.

## Enabling the relevant Google Cloud APIs

As we discussed in *Chapter 3*, before you can use a Google Cloud service API, you need to enable it. Since we are already logged into Cloud Shell, we can easily do that directly from here by using the `gcloud` command. We will enable the Google Cloud Natural Language API, the Vision API, the Document AI API, and the Google Cloud Storage API as we will be using all of them in this chapter. To do this, run the following commands in Cloud Shell:

```
gcloud services enable language.googleapis.com
gcloud services enable vision.googleapis.com
gcloud services enable documentai.googleapis.com
gcloud services enable storage.googleapis.com
```

The responses should be similar to the following:

```
Operation "operations/..." finished successfully.
```

With the required APIs now enabled, we just have a few more steps to complete before we can start using them.

## Storing authentication credentials in environment variables

We will reference our authentication credentials – that is, the API key that we created earlier – in many commands in this chapter. To make it easier to reference the credentials, we will store them in Linux environment variables in Cloud Shell.

To store the API key, run the following command, but replace `<YOUR_API_KEY>` with the API key you created and copied earlier:

```
export API_KEY=<YOUR_API_KEY>
```

We're almost finished with the environment setup steps now. Next, we will clone our GitHub repository, after which we'll be ready to start using Google Cloud's high-level AI services.

### Creating a directory and cloning our GitHub repository

You already cloned our repository to your local machine. In this section, you will also clone it to your Cloud Shell environment. To do this, perform the following steps in Cloud Shell:

1. Create a directory:

   ```
   mkdir packt-ml-sa
   ```

2. Change location to that directory:

   ```
   cd packt-ml-sa
   ```

3. Initiate the directory as a local `git` repository:

   ```
   git init
   ```

4. Clone the `git` repository that contains the code we will use in the Document AI section of this chapter:

   ```
   git clone https://github.com/PacktPublishing/Google-Machine-
   Learning-for-Solutions-Architects
   ```

Now that we have all of the required code copied to our Cloud Shell environment, it's time to start executing it!

## Detecting text in images with the Cloud Vision API

We're going to begin with a very simple example to show you just how easy it is to start using AI/ML services on Google Cloud. In just a few short steps, you will be able to extract text and associated metadata from an image and save that text in a file. This process is called **optical character recognition** (**OCR**).

To begin this process, download the following image to your computer (the image can also be found in the `Chapter04/images` folder in the git repository you created above): `https://github.com/PacktPublishing/Google-Machine-Learning-for-Solutions-Architects/blob/main/Chapter-04/images/poem.png`.

The image is shown in *Figure 4.4*:

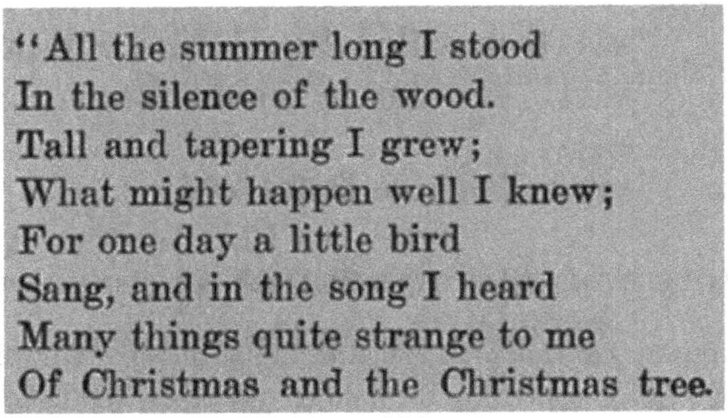

"All the summer long I stood
In the silence of the wood.
Tall and tapering I grew;
What might happen well I knew;
For one day a little bird
Sang, and in the song I heard
Many things quite strange to me
Of Christmas and the Christmas tree.

Figure 4.4: Sign to be used for OCR

Continue with the remaining steps to upload this image to your Cloud Shell environment:

1. Click the symbol with the three dots at the top-right corner of the Cloud Shell area (at the bottom of your Google Cloud console screen), then select **Upload**. See *Figure 4.5* for reference:

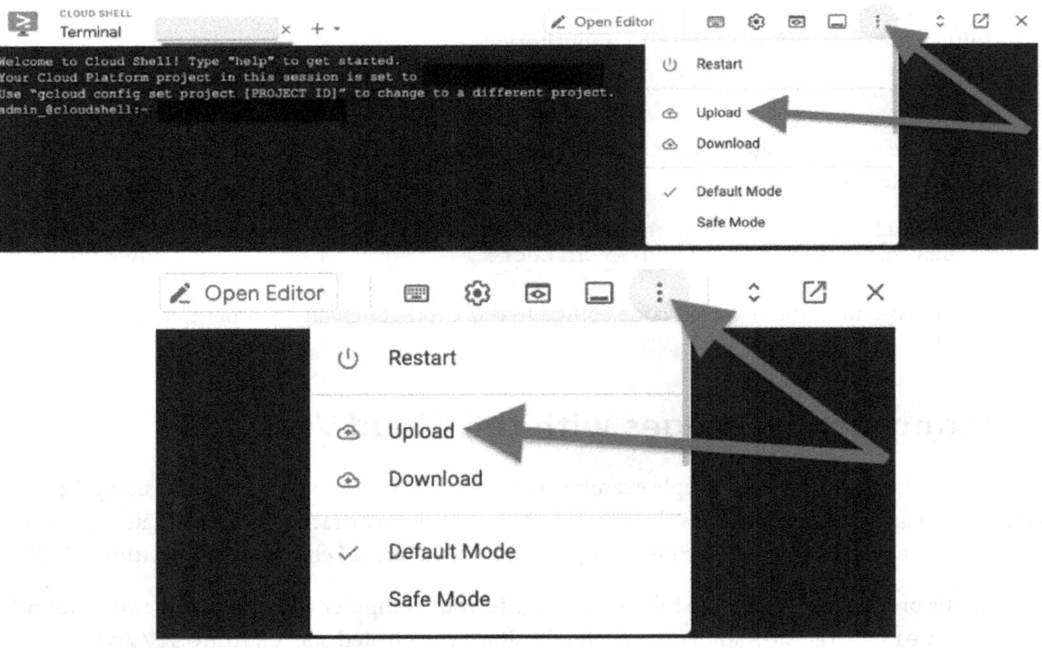


Figure 4.5: Cloud Shell upload

2. You will be presented with a file upload dialogue box (see *Figure 4.6* for reference). Select **Choose files** and browse for the file that you previously downloaded:

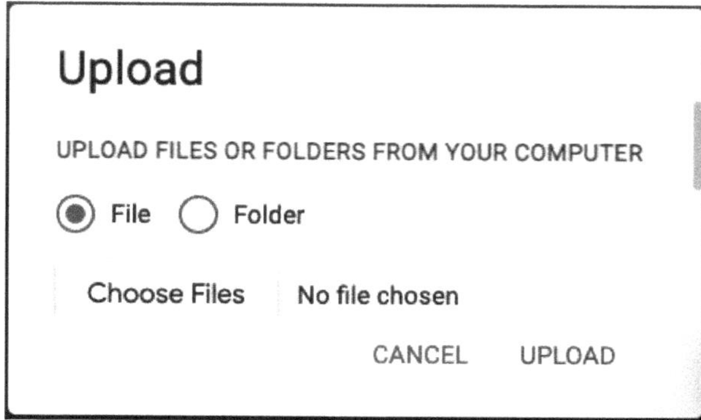


Figure 4.6: Cloud Shell upload prompt

3. Now, invoke the Cloud Vision API to extract the text and save it in a file with the following command (in this case, the filename of the image we're using is `poem.png`). The results are stored in a file named `vision-ocr.json`:

```
gcloud ml vision detect-text poem.png> vision-ocr.json
```

4. Inspect the contents of the file:

```
cat vision-ocr.json
```

> **Note**
> The file is not conducive to being printed in a book, so we will not display it here.

As you can see, each word that is extracted is stored with some associated metadata, such as the *X* and *Y* coordinates of the corners of the bounding boxes that contain that word in the image. This metadata helps us understand where each word appeared in the image, which can be useful information for developers.

5. If you just want to see the detected text, without the other metadata, you can use the following command:

```
cat vision-ocr.json | grep description
```

You will see the entire body of detected text in the first description, and then individual words in the subsequent descriptions. The output will look similar to the following (truncated here for brevity):

```
"description": "\"All the summer long I stood\nIn the silence
of the wood.\nTall and tapering I grew;\nWhat might happen well
I knew;\nFor one day a little bird.\nSang, and in the song I
heard\nMany things quite strange to me",
          "description": "\""
          "description": "All"
          "description": "the"
          "description": "summer"
          "description": "long"
          "description": "I"
          "description": "stood"
```

Congratulations – you have just implemented your first AI/ML workload on Google Cloud! It really is that simple! Next, we'll start looking at some slightly more advanced use cases.

## Using Document AI to extract information from documents

As you may recall, we discussed Document AI at length in *Chapter 3*. We talked about how it goes beyond understanding the content of textual inputs and also incorporates structure. In this section, we're going to take a look at how Document AI does this, and we'll see it in action for some real-world use cases. We'll start by covering some important Document AI concepts.

### Document AI concepts

Document AI uses the concept of "processors" to process documents. There are three main types of processors that you can use:

- General processors
- Specialized processors
- Custom processors

General processors and specialized processors use models that are pre-trained by Google, so you can use them without needing to train your own models, and without any AI/ML expertise. However, to provide additional flexibility, Google Cloud gives you the ability to further train specialized processors with your data to improve their accuracy for use cases that are specific to your business. This process is referred to as "uptraining."

General processors include a processor for performing OCR on images of text documents and forms. While we performed OCR using the Cloud Vision API in the previous section, Document AI provides additional functionality, such as identifying structured information within the input documents. This includes understanding key-value pairs in input forms, which can be very useful for automated data entry use cases.

**Note**

This is an important consideration in the Solution Architect role; there are often multiple different services that could be used to achieve similar business goals, and the Solution Architect's job is to create or select the most suitable solution based on specific business requirements. In the case of selecting between the Cloud Vision API and Document AI for OCR use cases, bear in mind that the Cloud Vision API generally requires less initial effort to use, but it also provides fewer features than Document AI. Therefore, if you have a very simple use case that can be satisfied with the Cloud Vision API, then you can select that option, whereas more complex use cases would steer you toward using Document AI.

Specialized processors provide models for processing specific types of documents, such as US Federal Government forms, identification documents, and invoices. The specialized processors are currently categorized into four different types:

- Procurement
- Identity
- Lending
- Contract

At the time of writing (March 2023), Google Cloud recently announced the general availability of new functionality in Document AI, which is Document AI Workbench. Document AI Workbench enables you to create completely new, custom types of processors, beyond what's provided by Google Cloud, for your specific use cases.

Let's take a look at how some of the Document AI processors work.

## Performing OCR with Document AI

We're going to use the simplest use case, OCR, for demonstration purposes, which will also allow us to directly contrast this functionality against how we performed OCR with the Cloud Vision API.

### *Creating a Document AI OCR processor*

Before we can process any documents, we need to create a processor. We're going to perform this action in the Google Cloud console:

1. In the Google Cloud console, navigate to the **Google Cloud services** menu → **Document AI** → **Processor Gallery**.
2. Under **Document OCR**, click **CREATE PROCESSOR**. See *Figure 4.7* for reference:

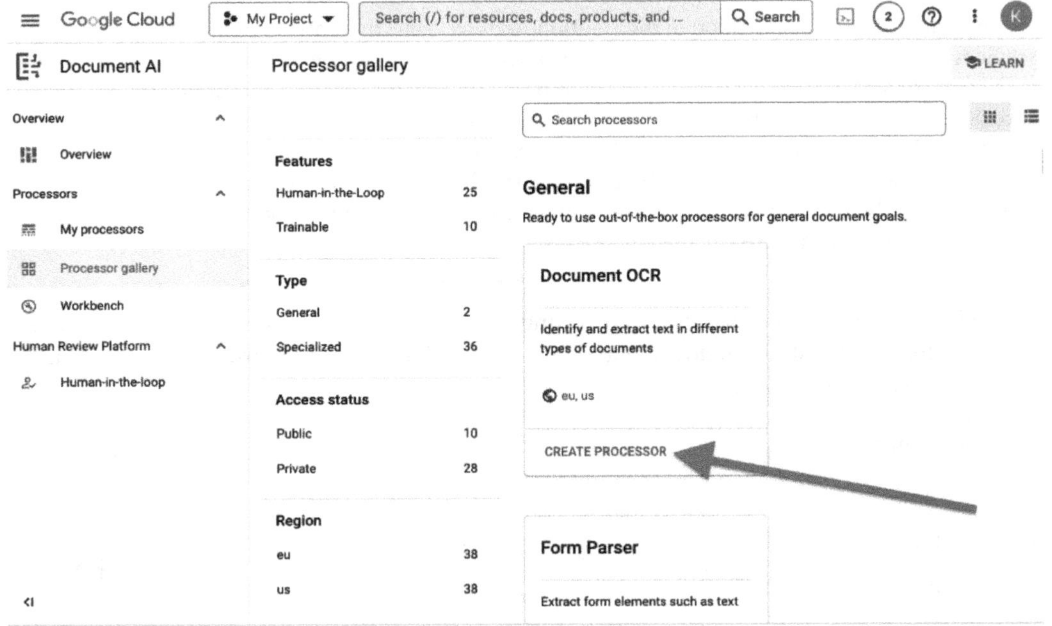


Figure 4.7: CREATE PROCESSOR

3. Input **OCR-Processor** as the name, select the region nearest to you, and click **Create**.

**A note on regions and zones**

For this book, whenever you need to select a region, we recommend selecting the same region every time since we will build resources throughout this book that may need to be used in combination with other resources created in different chapters of this book. Generally, when you need to use some cloud resources with other cloud resources, it's often much easier if those resources are in the same region. Sometimes, it can be difficult or impossible for resources in one region to reference resources in a different region without building a customized solution to do so.

In some activities in this book, you will also have the option of selecting a specific zone within a region. This decision is usually less important for the activities in this book because accessing resources across zones within a region is generally quite easy. In a production environment, you may want to keep resources in the same zone if you have specific business requirements, such as clusters of compute instances that need to work very closely together.

If you don't have such specific requirements, then the best practice is to distribute your resources across multiple zones to improve workload resilience and availability. You can learn more about Google Cloud regions and zones in the documentation at the following URL: `https://cloud.google.com/compute/docs/regions-zones`.

4. When your processor gets created, you should automatically be redirected to the **PROCESSOR DETAILS** page. If not, you can view the details by selecting **My processors** in the menu on the left-hand side of the screen and then selecting your newly created processor. The processor details are shown in *Figure 4.8*:

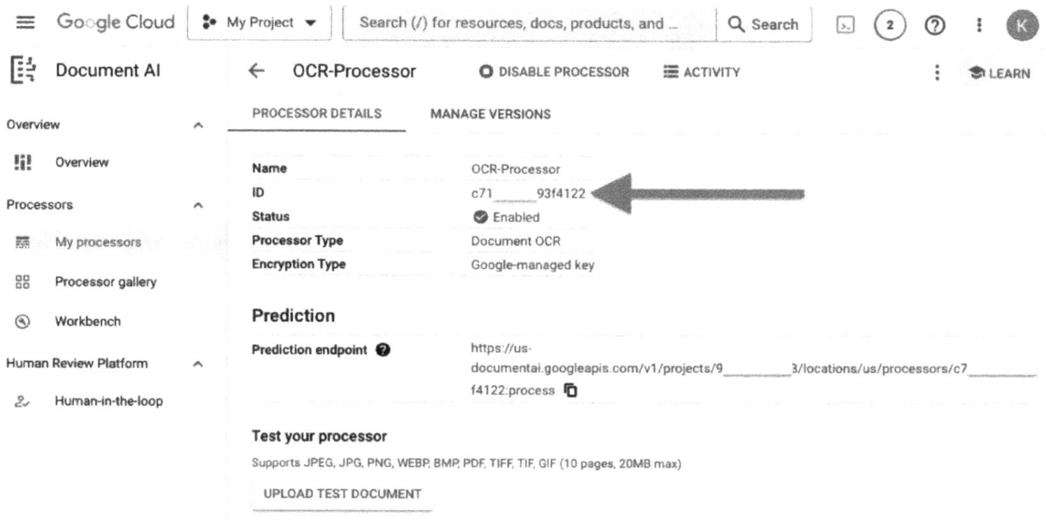


Figure 4.8: OCR processor details

Take note of your processor ID (highlighted by the red arrow in *Figure 4.8*) as you will need to reference it in the code we will use in the next section.

### Invoking the OCR processor

Now that we've created our processor, we can start using it. In this section, we will run a simple piece of Python code that uses the Document AI Python client library to process a file and extract the necessary information. We're going to use the same file that we used with the Cloud Vision API earlier in this chapter.

Perform the following steps in Cloud Shell:

1. Install the latest version of the Document AI client packages:

```
python -m venv env
source env/bin/activate
pip install --upgrade google-cloud-documentai
```

2. Take note of your project ID as you will need to reference it in the Python code that will be used to interact with Document AI. You can use the following command to view it:

```
echo $DEVSHELL_PROJECT_ID
```

3. Ensure that you are in your home directory:

```
cd ~
```

4. Copy the `docai-ocr.py` file to your home directory from the `git` directory we created earlier. We're going to edit this file outside of the branch because we won't need to merge the updates back to the main branch:

```
cp ~/packt-ml-sa/Google-Machine-Learning-for-Solutions-
Architects/Chapter-04/docai-ocr.py ~
```

In this Python file, we'll begin by importing the required libraries. Then, we'll define some variables (we will need to replace the values of those variables with the values of the processor we created in the previous section).

Then, we'll define a function that creates a Document AI client, reads the input file, creates a request object based on the file's contents, and sends this request to be processed by the Document AI service.

Finally, the function will print the resulting document text.

To perform these actions, execute the following steps:

1. Use `nano` to open the file for editing:

```
nano docai-ocr.py
```

2. Replace the variable definitions with the values you saved after creating the processor in the Google Cloud console:

    i. Replace the `project_id` value with your project ID, which you viewed in the `$DEVSHELL_PROJECT_ID` environment variable previously.

    ii. Specify 'us' or 'eu' as the location, depending on which region you used.

    iii. Replace the `processor_id` value with your processor ID, which you can view on the **PROCESSOR DETAILS** page in the console after creating your processor.

    iv. Replace the `file_path` value with the path to the file we want to process. The Cloud Shell upload feature that we used generally stores files in your Cloud Shell home directory, which is usually `/home/admin_`.

    v. In this case, the `mime_type` value is `'image/png'`. You can view the list of supported MIME types at `https://cloud.google.com/document-ai/docs/file-types`.

3. When you have completed all of the edits, press *Ctrl* + *X* to exit `nano`, then *Y* to save the file, and then *Enter* to confirm this.

4. Now, we're ready to execute OCR! The following command will run our Python function and save the results in a file named `docai-ocr.txt`:

```
python docai-ocr.py > docai-ocr.txt
```

5. Inspect the contents of the file:

```
cat docai-ocr.txt
```

The response should look like this:

```
All the summer long I stood
In the silence of the wood.
Tall and tapering I grew;
What might happen well I knew;
For one day a little bird.
Sang, and in the song I heard
Many things quite strange to me
```

Note how the output is formatted differently than the results we received when we used the Cloud Vision API. This further demonstrates how we can use different services to achieve similar results. However, there are subtle differences between the services, and we need to pick the best option based on business requirements. The Cloud Vision API required much less initial effort to use, but bear in mind that Document AI has much more powerful features, such as automated document processing, and customized models for specific use cases.

Let's take a look at some of Document AI's additional features.

#### *Document AI's response*

In our OCR example, we focused on the `document.txt` object in the response. However, the full response that is returned contains a lot more information that we can use in various ways, such as the number of pages, paragraphs, and lines in the document, and many other types of metadata. When using form parsers or specialized processors, it can even highlight structured data types such as tables and key-value pairs.

It should also be noted that in the preceding example, we performed an online inference request, in which we got a response in real time for a single document. Document AI also allows us to perform batch inference requests if we need to process large numbers of documents at a time.

#### *Human in the loop (HITL) AI*

In the preceding use case, the model was able to identify all the words in the image. However, in reality, the sources of our images may not always be clearly legible. For example, we may need to read information from pictures of road signs that could be worn in places. This is something to bear in mind when thinking about how accurate you need the results to be. Document AI provides a **HITL**) feature to enable you to improve the accuracy of the results by having a human review and make updates where necessary.

## Using the Google Cloud Natural Language API to get sentiment analysis insights from textual inputs

**Natural language processing** (**NLP**) and **natural language understanding** (**NLU**) are becoming ever more prominent in our daily lives, and researchers continue to find interesting new use cases almost every day. In this chapter, we'll explore the simplest way to get powerful NLP/NLU functionality on Google Cloud by using the Google Cloud Natural Language API. In later chapters, we will build and use much more complex language use cases.

### Sentiment analysis with the Natural Language API

Sentiment analysis allows us to get insights regarding the primary emotional tone of a piece of text. This is important for many business use cases, especially when it comes to connecting with, and understanding, your customers. For example, if you want to understand how customers feel about a new product you've released, you can analyze various customer interaction channels, such as reviews, social media reactions, and customer service center logs, to find out how people are reacting to the new product. This enables you to answer questions such as the following:

- Are they happy with it?
- Is there a pattern of complaints emerging about a specific feature?

Doing this kind of analysis manually would not be possible if you have thousands of reviews, social media reactions, and service center logs to process.

Fortunately, you can perform sentiment analysis on a piece of text with a simple API call to the Google Cloud Natural Language API.

To do so, perform the following steps in Cloud Shell:

1. Create a JSON file that will contain the piece of text that we want to analyze. Just like when we discussed Document AI, we can use the `nano` command to do this, which will create and open the file for editing:

   ```
   nano request.json
   ```

   Paste the following text into the `request.json` file:

   ```
   {
     "document":{
       "type":"PLAIN_TEXT",
       "content":"This is the best soap I've ever used! It smells
   great, my skin feels amazing after using it, and my partner
   loves it too!"
     },
     "encodingType":"UTF8"
   }
   ```

2. Press *Ctrl* + *X* to exit `nano`.
3. Press *Y* to save the file.
4. Press *Enter* to confirm this.
5. Now, we can send the request to the Natural Language API's `analyzeSentiment` endpoint. To do that, we will use the following `curl` command. Note that we are using the `API_KEY` environment variable that we created earlier in this chapter. This will be used to authenticate our request:

   ```
   curl -X POST -s -H "Content-Type: application/json" --data-
   binary @request.json "https://language.googleapis.com/v1/
   documents:analyzeSentiment?key=${API_KEY}"
   ```

   Your response should look like this:

   ```
   {"documentSentiment": {
    { "magnitude": 1.9, "score": 0.9
     }, "language": "en",
     "sentences": [
       {"text": {"content": "This is the best soap I've ever
   used!", "beginOffset": 0},
         "sentiment": {"magnitude": 0.9,"score": 0.9 }},
   ```

```
    {"text": {"content": "It smells great, my skin feels amazing
after using it, and my partner loves it too!", "beginOffset": 38
},
      "sentiment": {"magnitude": 0.9,"score": 0.9}
    }]}
```

The first thing we can see in the output is the overall document score, which is the score for the entire body of text. After that, we can see the scores for the individual sentences that were detected. The score of the sentiment ranges between -1.0 (negative) and 1.0 (positive), and the magnitude indicates the overall strength of emotion (both positive and negative) within the given text. You can learn more about the response fields and scores at `https://cloud.google.com/natural-language/docs/basics#sentiment_analysis_response_fields`.

In this case, both of the sentences constitute a positive review, so they both have high scores, and the overall document score is therefore also high.

> **Note**
>
> If your scores are slightly different than those in the outputs, that's because the models that serve these requests are constantly being updated with new data.

Retry all of the previous steps in this section, but add the following two sentences at the end of the review: "The only bad thing is that it's just too expensive, and that really sucks. Very annoying!"

The overall review will then look as follows:

*"This is the best soap I've ever used! It smells great, my skin feels amazing after using it, and my partner loves it too! The only bad thing is that it's just too expensive, and that really sucks. Very annoying!"*

When you submit the updated request, the response output will look as follows:

```
{ "documentSentiment": { "magnitude": 3.4,"score": 0.1
  }, "language": "en",
  "sentences": [
    {"text": {"content": "This is the best soap I've ever
used!","beginOffset": 0},
      "sentiment": {"magnitude": 0.9,"score": 0.9}},
    {"text": {"content": "It smells great, my skin feels amazing after
using it, and my pertner loves it too!",
        "beginOffset": 38},
      "sentiment": {"magnitude": 0.9,"score": 0.9}},
    {"text": {"content": "The only bad thing is that it's just too
expensive, and that really sucks.",
        "beginOffset": 122},
      "sentiment": {"magnitude": 0.5,"score": -0.5}},
    {"text": {"content": "Very annoying!",
```

```
        "beginOffset": 197},
      "sentiment": {"magnitude": 0.8,"score": -0.8}
    }]}
```


In the output, notice that the negative sentences at the end have much lower scores, so this brings down the overall score for the entire review. This makes sense because the overall sentiment of the review is frustration, even though it starts with a positive sentiment.

The Natural Language API also provides other types of functionality, such as the following:

- **Entity analysis**: This involves identifying what kinds of entities (for example, people, places, and so on) are present in a piece of text. The supported entities are listed at `https://cloud.google.com/natural-language/docs/reference/rest/v1/Entity#type`.
- **Entity sentiment analysis**: A combination of entity analysis and sentiment analysis.
- **Syntactic analysis**: This involves inspecting the linguistic structure of a given piece of text.
- **Content classification**: This involves categorizing the content of a document or piece of text.

To explore the Natural Language API in more detail, let's take a look at one of its other features content classification.

## Classifying content with the Natural Language API

Let's imagine that we want to build a search engine that can be used to search across large amounts of documents and content objects. One of the first things we will need to do is classify our documents and content objects into categories. Doing this manually on millions of objects would be impossible, or at least extremely laborious and error-prone. This is where the content classification feature of the Natural Language API can be useful.

To demonstrate, we're going to use the text that was produced by our Document AI OCR processor in the previous section of this chapter. This will also demonstrate an important concept, which is that you can combine multiple AI services to create more complex use cases to meet your business needs. In this case, not only can we classify regular text inputs, but we can detect text in images and then categorize the contents of that text.

Perform the following steps in Cloud Shell:

1. Create a JSON file that we will use in our request to the API. This will contain the text from our OCR processor:

   ```
   nano classify-request.json
   ```

2. Paste the following text into the `classify-request.json` file:

```
{
  "document":{
    "type":"PLAIN_TEXT",
    "content":""All the summer long I stood
In the silence of the wood.
Tall and tapering I grew;
What might happen well I knew;
For one day a little bird.
Sang, and in the song I heard
Many things quite strange to me "
  },
  "classificationModelOptions":{
    "v2Model":{
      "contentCategoriesVersion":"V2"
    }
  }
}
```

3. Press *Ctrl* + *X* to exit nano.
4. Press *Y* to save the file.
5. Press *Enter* to confirm this.
6. Now, we can send the request to the Natural Language API's `classifyText` endpoint. To do that, we will use the following `curl` command. Note that we are using the `API_KEY` environment variable that we created earlier in this chapter. This will be used to authenticate our request:

```
curl "https://language.googleapis.com/v1/
documents:classifyText?key=${API_KEY}"   -s -X POST -H "Content-
Type: application/json" --data-binary @classify-request.json
```

Your response should look like this:

```
{
  "categories": [
    {
      "name": "/Books & Literature/Poetry",
      "confidence": 0.45689428
    },
    {
      "name": "/Arts & Entertainment/Music & Audio/Music
Reference",
      "confidence": 0.22331826
    },
```

```
        {
          "name": "/Arts & Entertainment/Music & Audio/Rock Music",
          "confidence": 0.109513044
        }
      ]
    }
```

That's more like it! Note that the response contains entries for multiple potential categories, each with different levels of confidence.. This demonstrates two important realities for the Solution Architect role:

- While we want to automate everything as much as possible via AI/ML, it's often necessary to have mechanisms for humans to review the model outputs and make corrections where needed.
- AI/ML workloads often consist of several steps in which data is passed from one step to the next, and it's important to implement data quality checks at each stage in the process.

> **Note**
>
> When using HITL reviews, we wouldn't need to have a human review every data point (that would be impractical, and would negate the benefits of AI/ML), but we should have mechanisms that define what kinds of values we expect to see in a given use case, if possible, or flag our outputs for human review if our model's confidence levels are below specified thresholds for some data points.

Now that we've explored how to use pre-trained models provided by Google, we're going to look at the next level of complexity when it comes to implementing AI/ML workloads on Google Cloud, which is to train our own models in a managed way, using AutoML.

## Using Vertex AI AutoML

As we discussed in *Chapter 3*, we can use AutoML to automate all of the steps in the model training and evaluation process. In this section, we're going to build an AutoML Model with Vertex AI.

> **Note**
>
> Refer to the Vertex AutoML pricing at the following link before following the instructions in this section: `https://cloud.google.com/vertex-ai/pricing`.

## Use case – forecasting

The use case we will focus on for this workload is forecasting. After all, forecasting is one of the most fundamental business processes that almost all businesses have to perform in some form or other. For example, whether you own a global online retail company or just a single store in a small town, you will need to estimate how much of each product you should purchase each month, or perhaps each day, based on the expected customer demand for those items. And, forecasting is not just for physical goods. For example, if you owned a consulting or services company, you would need to estimate how many people you would need to hire to cater to the expected needs of your customers in the coming months. Being able to predict the future is a pretty important superpower, and in our case, we're going to skip right the to "get-rich-quick" use case of forecasting stock market performance.

Are you ready to train your first ML model? Let's dive in!

### *Preparation – creating a BigQuery dataset for our prediction outputs*

BigQuery is a useful tool for viewing and performing analytical queries on our prediction outputs. For this reason, we will create a BigQuery dataset to store the outputs from our AutoML tests. To create a dataset in BigQuery, perform the following steps:

1. In the Google Cloud console, navigate to the **Google Cloud services** menu → **BigQuery**.
2. In the top-left corner of the screen, you will see your project's name. Click on the three vertical dots to the right of your project's name (see *Figure 4.9* for reference):

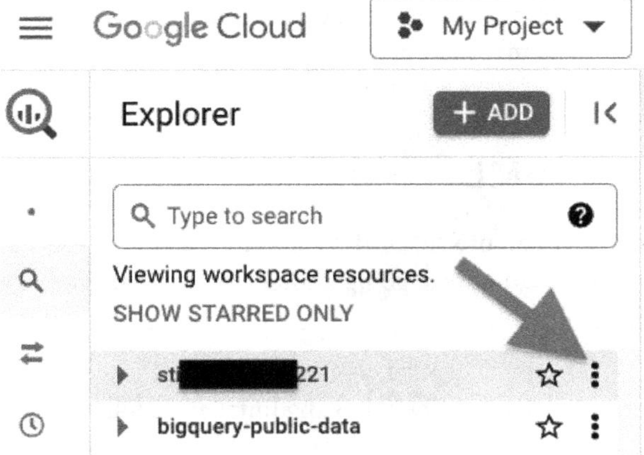


Figure 4.9: BigQuery project menu

3. In the menu that's displayed, select **Create dataset**.
4. Give your dataset a name, such as `forecast_test_dataset` (see *Figure 4.10*).

5. Select your preferred region, then select **Create dataset**:

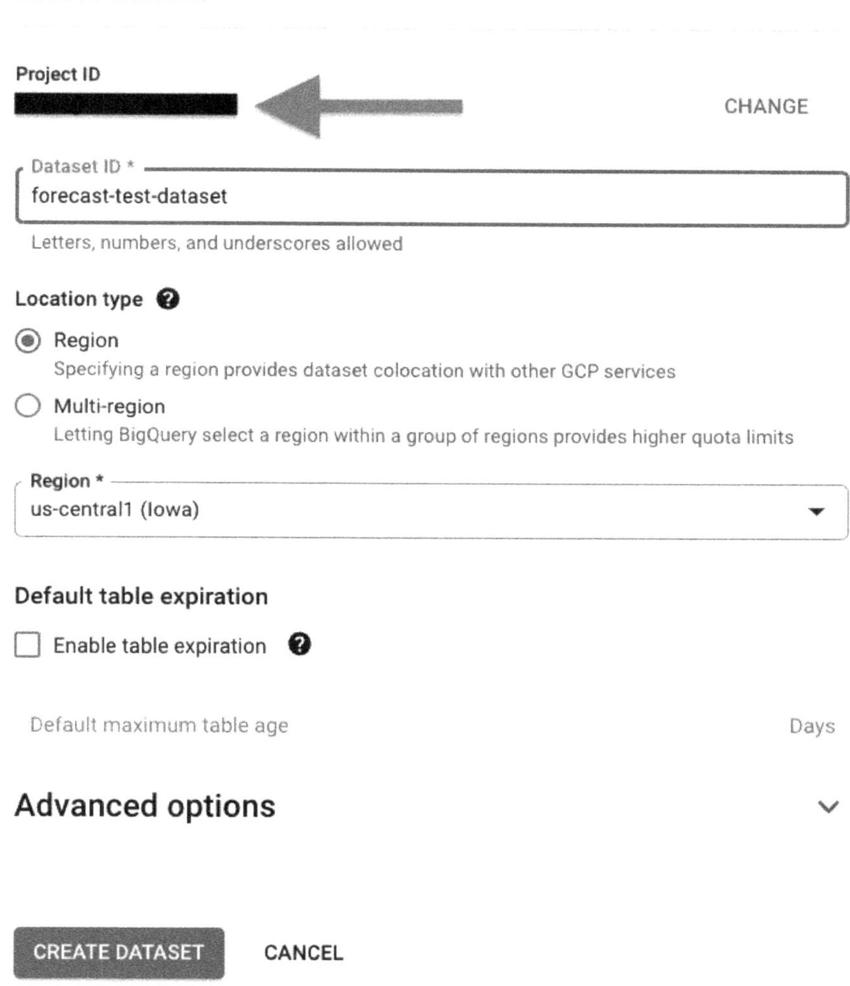


Figure 4.10: Creating a BigQuery dataset

6. That's it – we don't need to do anything else in BigQuery for now.

### Creating the AutoML workload

Now, it's time to define the AutoML job that's going to automate all the steps in our data science projects. As a reminder, the steps in the model life cycle are shown in *Figure 4.11*:

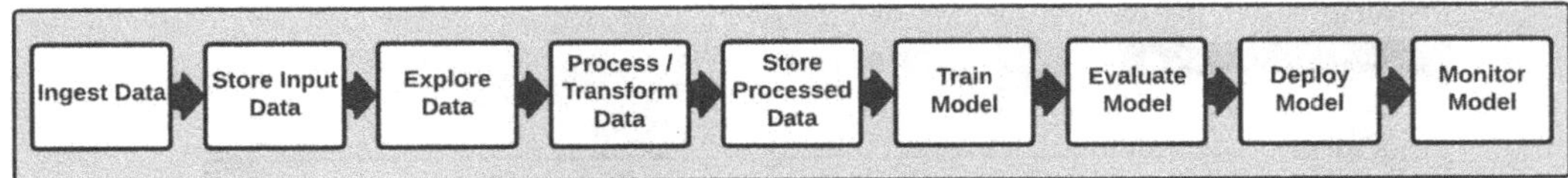


Figure 4.11: ML model life cycle managed by AutoML

As we can see, the first step in the process is to ingest some data. We need to create a dataset for this purpose. We're going to use a public dataset from Kaggle, which is a useful resource for practicing ML concepts. Forecasting use cases generally require **time series** data, which is a sequence of data that is chronologically ordered according to time intervals. For this reason, our dataset will need to include a timestamp field. We will dive into how model training and deployment work later in this book, but for now, we'll focus on how easily Vertex AI AutoML enables us to train a forecasting model, without the need for much or any AI/ML expertise. Note that Vertex AI AutoML can also be used for classification and regression use cases.

In our example, we will use a subset of the *DJIA 30 Stock Time Series* dataset, which can be found at `https://www.kaggle.com/datasets/szrlee/stock-time-series-20050101-to-20171231`.

For reference, the data contains the following fields:

- `Date`: In yy-mm-dd format
- `Open`: The price of the stock at market open (this is NYSE data, so it's all in USD)
- `High`: The highest price reached in the day
- `Low Close`: The lowest price reached in the day
- `Volume`: The number of shares traded
- `Name`: The stock's ticker name

In the clone of our GitHub repository that you created on your local machine earlier in this chapter, you will find a modified version of this file in a directory named `data`, which exists within the directory named `Chapter04`. The name of the modified file is `automl-forecast-train-1002.csv`.

Therefore, you should find the file at the following path on your local machine (the slashes will be reversed if you're using Microsoft Windows): `[Location in which you cloned our GitHub repository]/Chapter-04/data/automl-forecast-train-1002.csv`.

To create our dataset in Google Cloud, perform the following steps:

1. In the Google Cloud console, navigate to the **Google Cloud services** menu → **Vertex AI** → **Datasets**.
2. Select **Create dataset**. A dataset name will be generated automatically, but you can change it to something easier to remember if you wish. In our example, we've called it `my-forecasting-dataset`.
3. You will also be asked to select a data type and objective. Select **Tabular**, and then select **Forecasting**. Your selections should look like what's shown in *Figure 4.12*.
4. Select your preferred region, then select **Create** (note that this must be the same region in which you created the BigQuery dataset in the previous section):

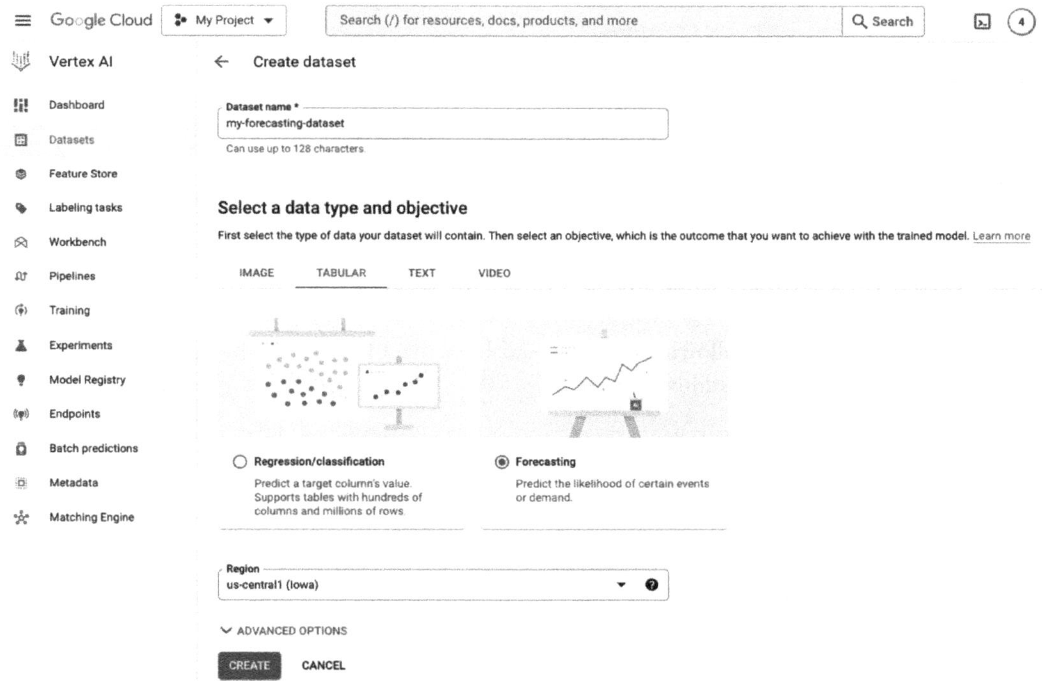


Figure 4.12: Create dataset

5. Next, we need to add data to the dataset. Select **Upload CSV files from your computer**, then click **Select files**. See *Figure 4.13* for reference:

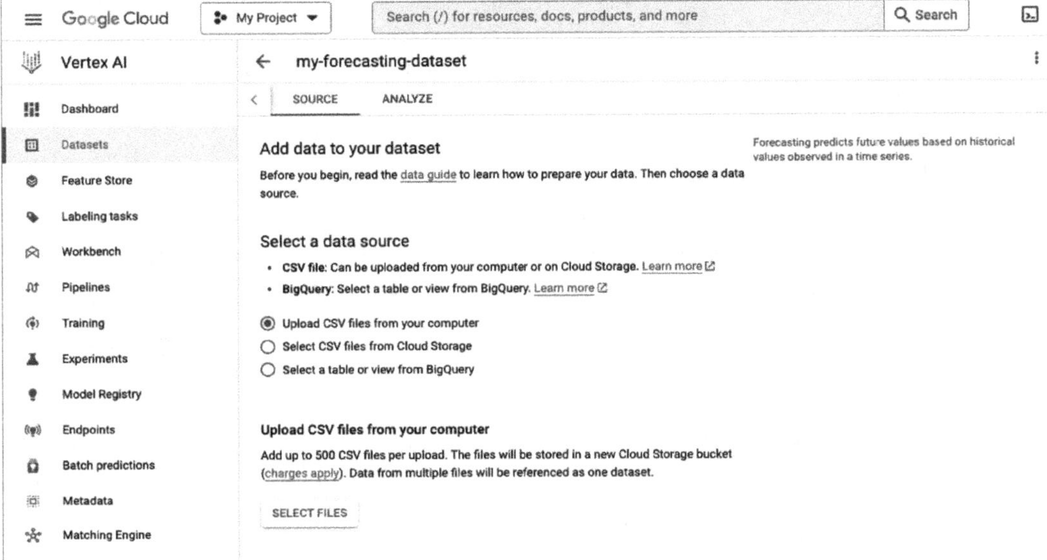


Figure 4.13: Add data to your dataset

6. Select the file from the following path on your local machine (the slashes will be reversed if you're using Microsoft Windows):

```
[Location in which you cloned our GitHub repository]/Chapter-04/
data/automl-forecast-train-1002.csv
```

7. In the **Cloud Storage path** input field, select **Browse**.
8. We're going to create a new Cloud Storage bucket for this use case. To do this, click on the symbol in the top-right corner that looks like a bucket with a plus sign ( ).
9. Give the bucket a unique name and select **Create** (you can leave all of the bucket configuration options at their default values).
10. If you are prompted to prevent public access, select **Confirm**. This prevents any items in your bucket from being made public, which is a best practice unless you specifically want to create publicly accessible content. In this example, we do not need to create publicly accessible content.

11. When your bucket has been created, click **Select** at the bottom of the screen. With that, your bucket is now the storage location for our training data. The resulting screen should look similar to what's shown in *Figure 4.14*:

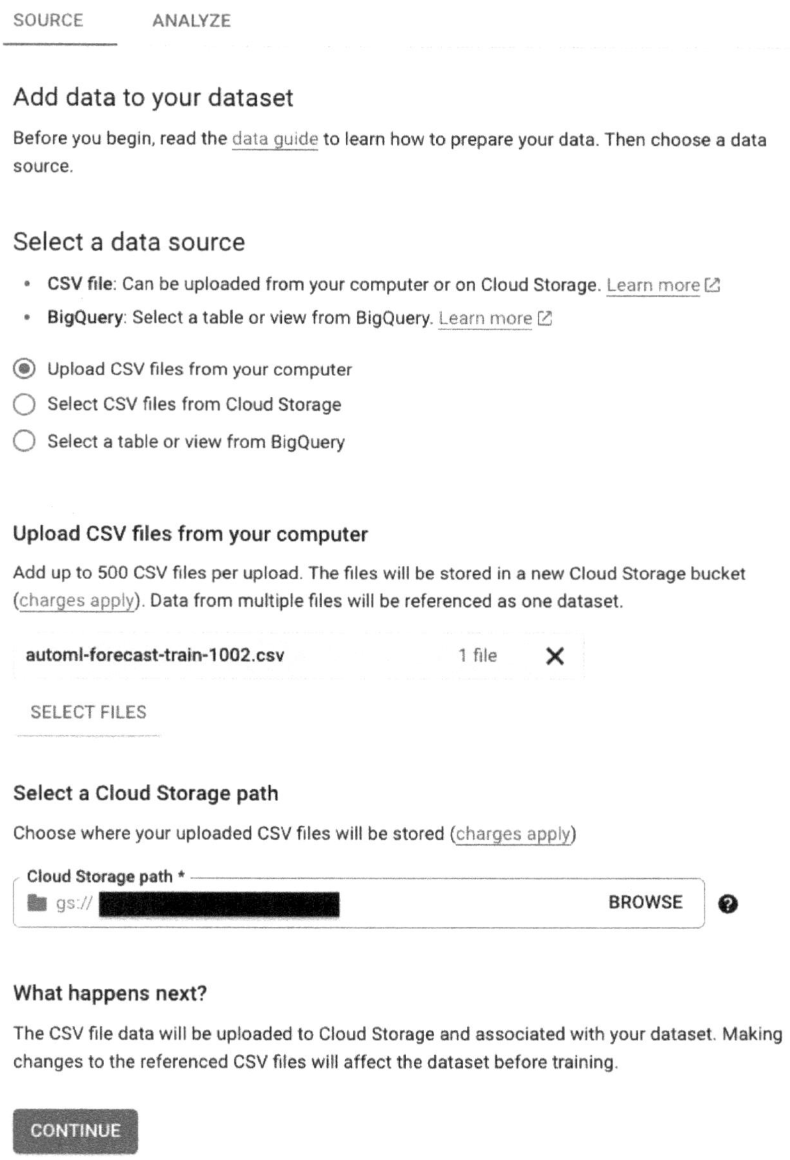

SOURCE ANALYZE

Add data to your dataset

Before you begin, read the data guide to learn how to prepare your data. Then choose a data source.

Select a data source

- **CSV file**: Can be uploaded from your computer or on Cloud Storage. Learn more
- **BigQuery**: Select a table or view from BigQuery. Learn more

Upload CSV files from your computer

Select CSV files from Cloud Storage

Select a table or view from BigQuery

**Upload CSV files from your computer**

Add up to 500 CSV files per upload. The files will be stored in a new Cloud Storage bucket (charges apply). Data from multiple files will be referenced as one dataset.

**automl-forecast-train-1002.csv** 1 file

SELECT FILES

**Select a Cloud Storage path**

Choose where your uploaded CSV files will be stored (charges apply)

**Cloud Storage path ***

gs://

BROWSE

**What happens next?**

The CSV file data will be uploaded to Cloud Storage and associated with your dataset. Making changes to the referenced CSV files will affect the dataset before training.

CONTINUE

Figure 4.14: Add data to your dataset

12. Select **Continue**.
13. When the file has been uploaded, you should see a message saying that the upload has been completed. Now, we're ready to start training our model!
14. In the details screen for your dataset, select the button that says **Train new model**.
15. On the next screen that appears, select **AutoML (Default)**.
16. Next, enter the details for the AutoML workload, as follows (see *Figure 4.15* for the final configuration details):

    I. The dataset name is automatically selected as the name for the workload. You can leave it at the default value, or change the name as you wish.

    II. We're going to try to predict the volume of the company's stocks that will be sold on each date, so select **Volume** as the target column.

    III. The **Name** column is the series identifier column.

    IV. The **Date** column contains our timestamps, so we will select this as the timestamp column.

    V. Our **Date** column contains a timestamp for each day, so the data granularity is **Daily**.

    VI. We can leave holiday regions blank for this use case.

    VII. We're going to set **Forecast horizon** and **Context window** to **10**. You can play around with these values if you'd like to see how they affect the resulting outputs.

    VIII. Select **Export test dataset to BigQuery**.

    IX. In the **BigQuery path** input field, input the details of the BigQuery dataset we created earlier in this section. The format will be `[project_name].[dataset_name].[table_name]`, where we have the following:

    i. `project_name` is the name of your Google Cloud project (hint: remember seeing this in the BigQuery console when you were creating the BigQuery dataset).

    ii. `dataset_name` is the name of the BigQuery dataset you created; for example, **forecast_test_dataset**.

    iii. `table_name` is the name of the table that will be created to store the test outputs. You can type any name here, and that name will be assigned to the table that will be created. *It's important to note that this table will be created by our AutoML job, so it must NOT have been created manually in BigQuery before this point. Only the dataset must have been created; not the table within that dataset.*

    iv. When you have entered the dataset details, click **Select**. The dataset path will appear in the **BigQuery path** input field:

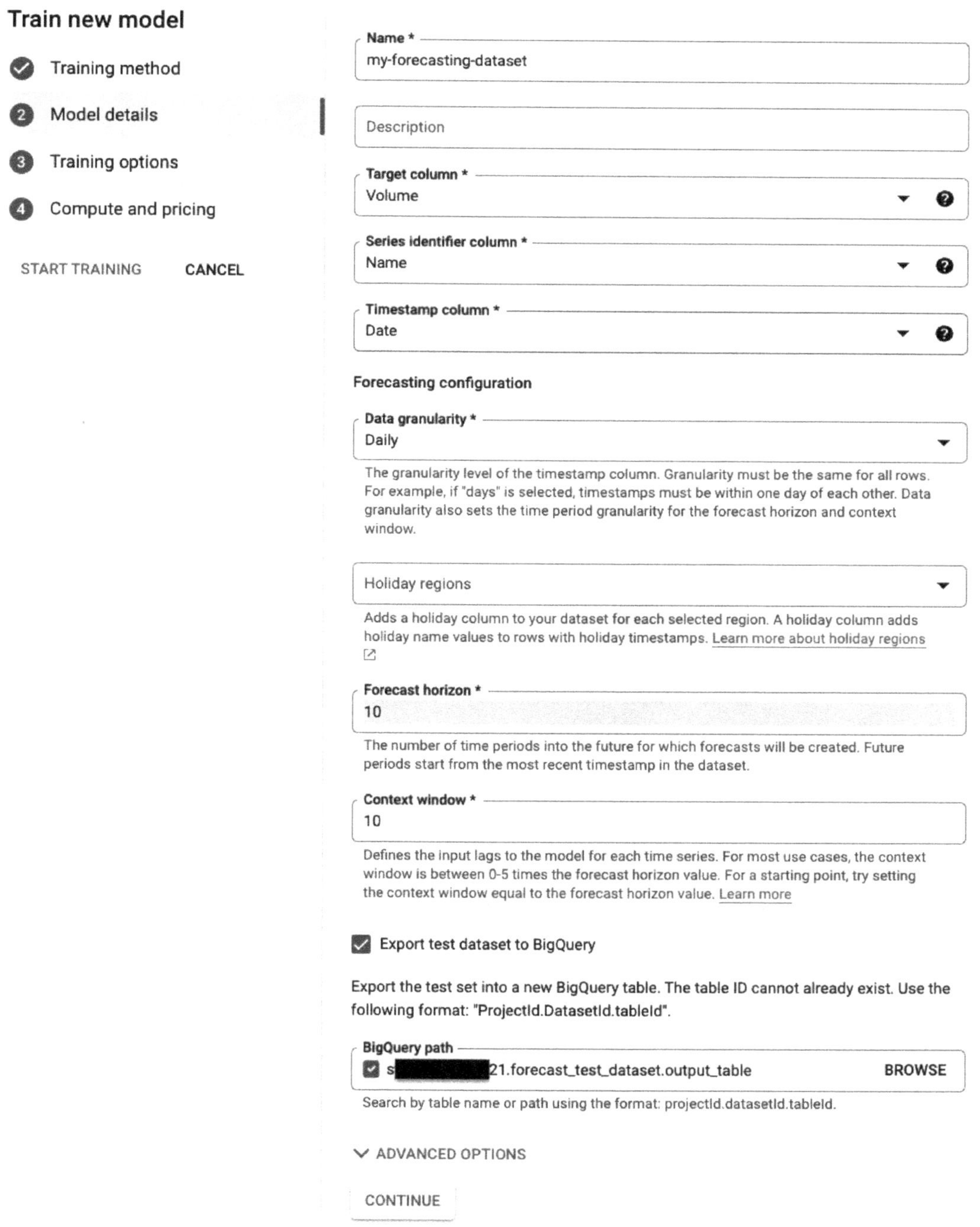

Train new model

Training method

2 Model details

3 Training options

4 Compute and pricing

START TRAINING CANCEL

Name *
my-forecasting-dataset

Description

Target column *
Volume

Series identifier column *
Name

Timestamp column *
Date

Forecasting configuration

Data granularity *
Daily

The granularity level of the timestamp column. Granularity must be the same for all rows. For example, if "days" is selected, timestamps must be within one day of each other. Data granularity also sets the time period granularity for the forecast horizon and context window.

Holiday regions

Adds a holiday column to your dataset for each selected region. A holiday column adds holiday name values to rows with holiday timestamps. Learn more about holiday regions

Forecast horizon *
10

The number of time periods into the future for which forecasts will be created. Future periods start from the most recent timestamp in the dataset.

Context window *
10

Defines the input lags to the model for each time series. For most use cases, the context window is between 0-5 times the forecast horizon value. For a starting point, try setting the context window equal to the forecast horizon value. Learn more

Export test dataset to BigQuery

Export the test set into a new BigQuery table. The table ID cannot already exist. Use the following format: "ProjectId.DatasetId.tableId".

BigQuery path
s█████21.forecast_test_dataset.output_table BROWSE

Search by table name or path using the format: projectId.datasetId.tableId.

ADVANCED OPTIONS

CONTINUE

Figure 4.15: Model details

17. Select **Continue**.
18. On the **Training options** screen that appears, select the checkboxes next to the **Close**, **High**, **Low**, and **Open** column names.
19. When you do that, four menus will appear in blue text near the top of the screen (see *Figure 4.16* for reference). Click on the **Feature type** menu and select **Covariate**. This is a required step, and it indicates that these columns in the training dataset contain values that change over time:

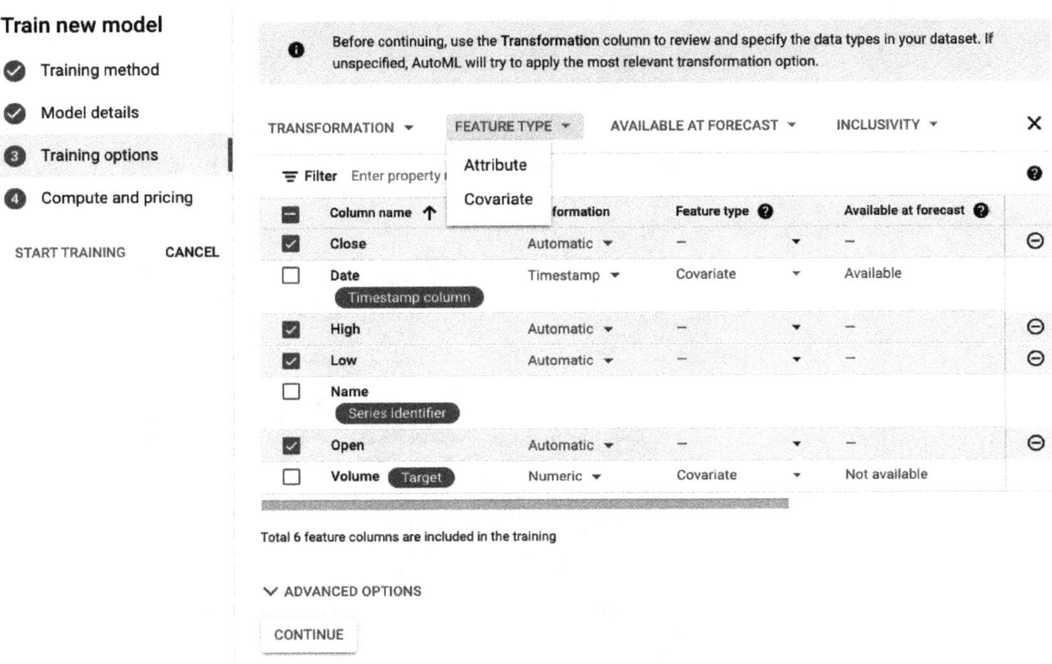


Figure 4.16: Training options

20. Select **Continue**.
21. On the **Compute and pricing** screen that appears, we can specify the budget in terms of the maximum number of node hours for which we want to allow our AutoML job to run. A link is provided on that screen for further details on how this pricing model works. For testing purposes, we will select the minimum possible number of node hours, which is **1**. For this reason, enter `1` in the input field (see *Figure 4.17* for reference):

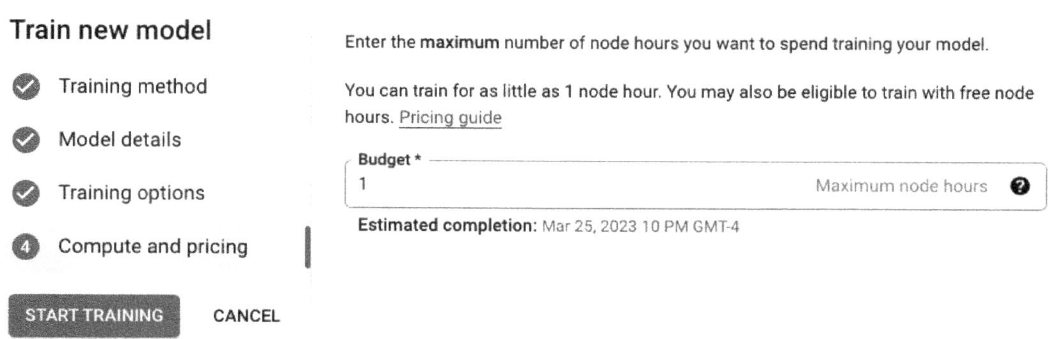


Figure 4.17: Compute and pricing budget

22. Select **Start training**.
23. To monitor our training job, we can select **Training** from the menu on the left-hand side of the screen.
24. When the status of our job changes to **Finished**, our model has been trained, and we can begin using it to get predictions.

> **Note**
>
> Due to the various steps in the AutoML process, and the concept of a "node hour," the job may run for more than 1 hour.

Congratulations! You have trained your first ML model on Google Cloud. Let's take a look at some details regarding our model.

### *Viewing model details*

When the status of our model training job changes to **Finished**, we can click on the name of the training job and see a lot of useful details regarding our model (see *Figure 4.18* for reference). Along the top of the screen, we can see various performance metrics, such as **Mean Absolute Error** (**MAE**), **Mean Absolute Percentage Error** (**MAPE**), and others. You can click the question mark symbol next to each one to learn more about it. We will also discuss these metrics in more detail in later chapters in this book.

Another useful piece of information is **Feature importance**, which shows us how much each input feature appears to influence the model outputs. This is very important for understanding how our model works.

Can you think of how each of these features may influence the predicted sales volume of stocks? For example, when a stock price is low, do people buy more of that stock? If this is true, does it make sense that the **Low** feature – which represents the stock's lowest price point each day – would be important in predicting how much of that stock would be sold on a given day?

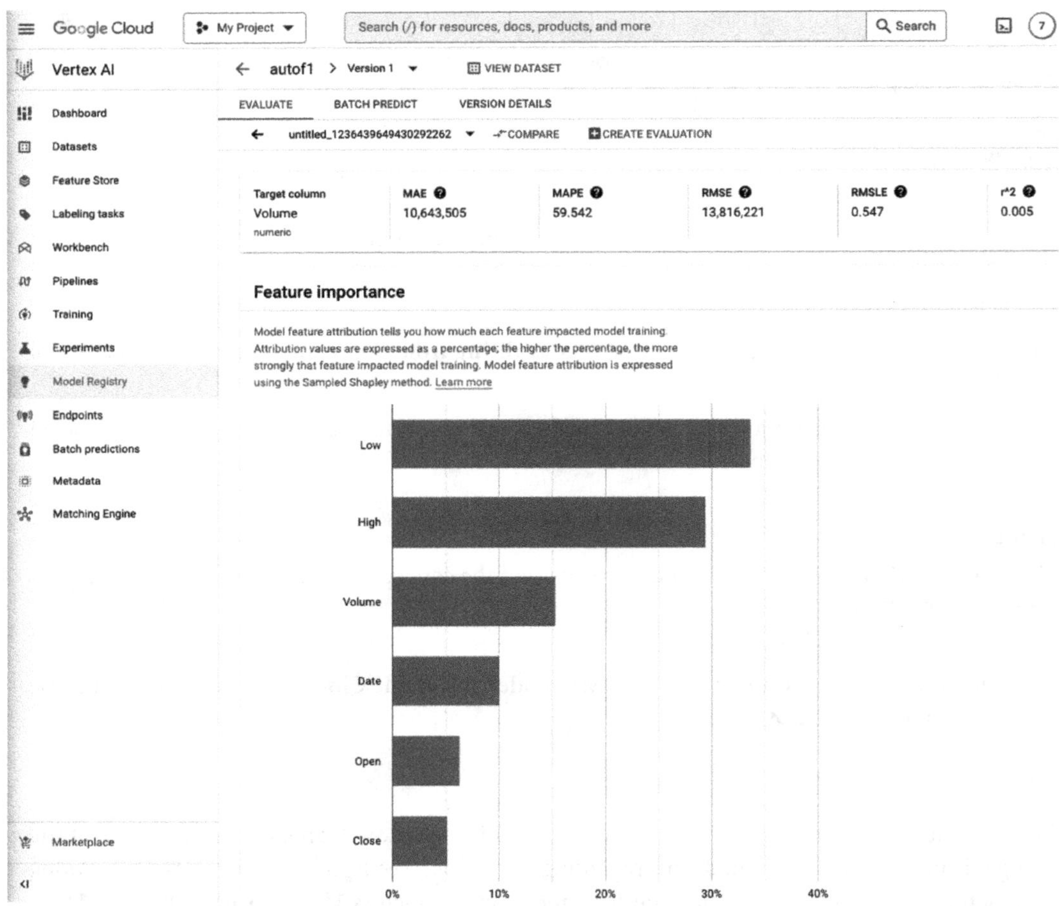


Figure 4.18: Model metrics

In addition to our model's performance metrics, let's take a look at some of the predictions our model made as part of the testing process during our AutoML workload execution.

### *Viewing model predictions*

To take a look at some of the model outputs that were generated by our AutoML job, we need to view the BigQuery table that we created to store those outputs. To do this, perform the following steps:

1. In the Google Cloud console, navigate to the **Google Cloud services** menu → **BigQuery**.
2. In the top-left corner of the screen, click on your project's name, then click on your dataset's name, and then click on the table's name.
3. You will see the schema of the table.
4. Click on the **Preview** tab; a preview of the output data will be displayed.
5. Scroll to the right; you will see columns named **predicted_Volume.value** and **predicted_on_Date**. These show us the prediction outputs from our model.

It's important to note that we only allowed our job to run for 1 node hour, so our prediction values may not be accurate at this point. An AutoML job would usually need to run for a longer time to find the best model. This is something to bear in mind from a cost perspective. If your budget allows, try running the AutoML job for longer periods, and see how it affects the model's performance.

## Summary

In this chapter, you learned how to use Google Cloud's high-level AI/ML APIs to implement AI/ML functionality by using models that are trained and maintained by Google. Then, you moved on to train your own ML model using Vertex AI AutoML. All of this was performed with the help of fully managed services on Google Cloud. In the next chapter, and beyond, we're going to dive in deeper, and you will build your own models from scratch so that you get to see how each of the steps in the model development life cycle works in much more detail.

# 5

# Building Custom ML Models on Google Cloud

In the last chapter, we implemented AI/ML workloads by letting Google do all of the work for us. Now is the point at which we're going to elevate our knowledge and skills to an expert level by building our own models from scratch on Google Cloud.

We will use popular software libraries that are commonly used in data science projects, and we will start implementing some of the concepts we discussed in previous chapters, such as **unsupervised ML (UML)** and **supervised ML (SML)**, including clustering, regression, and classification.

This chapter covers the following topics:

- Background information – libraries
- UML with scikit-learn on Vertex AI
- Implementing a regression model with scikit-learn on Vertex AI
- Implementing a classification model with XGBoost on Vertex AI

# Background information – libraries

Before we dive in and start building our own models, let's take a moment to discuss some of the software libraries we will use in this chapter.

**Definition**

A software library is a collection of code and data that includes useful functions and tools for specific types of programming tasks. When common types of programming tasks are identified in a given industry, such as data manipulation or implementing complex mathematical equations, then usually somebody will eventually create a library that contains the code and other resources required to perform those tasks. The library can then easily be used by others to achieve those same tasks and potentially extended to add more functionality over time, rather than everybody needing to write the code to perform those common tasks over and over again. Without libraries, programmers would have to build everything from scratch all of the time and would waste a lot of time on rudimentary programming tasks. In this chapter, we will use libraries such as scikit-learn, Matplotlib, and XGBoost. Later in this book, we will use other libraries such as TensorFlow and PyTorch, and we will describe those libraries in more detail in their respective chapters.

## scikit-learn

scikit-learn, also referred to as `sklearn`, is an open source Python library that has lots of useful data science tools and practice datasets built in. It's like a "Swiss Army knife" for data science, and it's a popular starting point for budding data scientists to begin working on ML projects because it's relatively easy to start using, as you will see in this chapter.

## Matplotlib

In *Chapter 2*, we discussed typical stages that exist in almost all data science projects, and how one of those stages focuses on data exploration. We also discussed that "visualization" is typically an important part of the data exploration phase, in which data scientists and engineers use tools to create visual representations of the characteristics of their datasets. These visual representations, such as graphs, can help data scientists and engineers build a better understanding of the datasets, as well as how those datasets are affected by experiments and transformations that are performed during data science projects. Data scientists and engineers also often want to build visualizations that represent other aspects of their data science project activities, such as increases or decreases in metrics that help determine how a model is performing. In the words of Matplotlib's developers, "*Matplotlib is a comprehensive library for creating static, animated, and interactive visualizations in Python.*" As such, Matplotlib is a commonly used library for creating visual representations of data.

## pandas

pandas is a Python library for data manipulation and analysis. It is used for both data exploration and for performing transformations on data. For example, with pandas, we could read data in from a file or other data source, preview a subset of the data to understand what it contains, and perform statistical analysis on the dataset. Then, we could also use pandas to make changes to the data to prepare it for training an ML model.

## XGBoost

In *Chapter 1*, we discussed how the concept of Gradient Descent is commonly used in the ML model training process. XGBoost is a popular ML library that is based on the concept of **Gradient Boosting**, which uses an **ensemble** approach that combines many small models (often referred to as **weak learners**) to create a better overall prediction model. In the case of Gradient Boosting, each iteration in the training process trains a small model. And, as is the case in almost every model training process, the resulting model will make some incorrect predictions. The next iteration in the Gradient Boosting process then trains a model on those **residual errors** made by the previous model. This helps to "boost" the training process in each subsequent iteration, with the intention of creating a stronger prediction model overall. XGBoost, which stands for Extreme Gradient Boosting, overcomes the sequential limitations of previous Gradient Boosting algorithms, and it can train thousands of weak learners in parallel. We will describe how XGBoost works in more detail later in this chapter.

> **Note**
>
> In general, the concept of "ensemble" model training refers to the combination of many weak models to create a better or "stronger" overall prediction mechanism, often referred to as a "meta-model." Boosting is just one example of an ensemble approach. Other examples include "Bagging" (or "Bootstrap Aggregation"), which trains many weak learners on subsets of the data, and "Stacking," which can be used to combine models trained with completely different algorithms. The intent, in each case, is to build a more useful prediction mechanism than could be achieved by any of the models individually.
>
> Depending on the ensemble implementation, the predictions of each model in the ensemble could be combined in different ways, such as being summed or averaged, or in classification use cases, a voting mechanism may be implemented in order to determine the resulting best prediction.

## Prerequisites for this chapter

Just like in the previous chapter, we will perform some initial activities that are required before we can start to perform the primary activities in this chapter.

Google Cloud makes it even easier to get started with the data science libraries we described in the previous sections because we can use Vertex AI Workbench notebooks that already have these libraries installed. If you wanted to use these libraries outside of Vertex AI, you would need to install them yourself.

Considering that we're going to use Vertex AI notebooks, and we will not need to explicitly install the libraries, we will not include details on how to do that in this book. If you would like to install the libraries in another environment, including **Google Compute Engine** (GCE) or **Google Kubernetes Engine** (GKE), you can find installation instructions on the respective websites for each of those libraries, such as:

- `https://scikit-learn.org/stable/install.html`
- `https://matplotlib.org/stable/users/installing/index.html`
- `https://pandas.pydata.org/docs/getting_started/install.html`
- `https://xgboost.readthedocs.io/en/stable/install.html`

Next, we will discuss Vertex AI Workbench notebooks in a bit more detail, and how to create a notebook.

> **Note**
>
> Later in this chapter, you will see pieces of code that import these libraries. This is not the same as installing. The libraries are already installed in our Vertex AI Workbench notebooks, but we just need to import them into our notebook context in order to use them.

### Vertex AI Workbench

Vertex AI Workbench is a development environment in Google Cloud that enables you to manage all of your AI/ML development needs within Vertex AI. It is based on Jupyter notebooks, which provide an interactive interface for writing and running code. This makes notebooks extremely versatile because you can codify interactions not only with the broader Vertex AI ecosystem but also with other Google Cloud service APIs.

### *Vertex AI Workbench notebooks*

There are three types of notebooks you can use within Vertex AI Workbench:

- Managed notebooks, which, as the name suggests, are managed for you by Google. These are a good default option for many use cases because they have a lot of useful tools built in and are ready to go on Google Cloud. They run in a JupyterLab environment, which is a web-based interactive development environment for notebooks.
- User-managed notebooks, which, as the name suggests, are managed by you. These are suitable if we require significant customization of our environment. They are customizable instances of Google Cloud's Deep Learning VMs, which are **virtual machine** (**VM**) images that include tools for implementing **deep learning** (**DL**) workloads. However, these notebooks can be used for more than just DL use cases.
- Vertex AI Workbench **instances**, which are the newest option released by Google Cloud in late 2023. These can be seen as a hybrid of the previous two options, and will likely become the primary option over time. At the time of writing this in September 2023, this option is only available in preview mode, so we will stick with *options 1* and *2* for now.

Google also provides an offering called Colab, which is a very popular service that allows you to run free notebooks over the public internet. In late 2023, Google Cloud also released an option named Colab Enterprise, which enables customers to use Colab within their own Google Cloud environment. At the time of writing this, Colab Enterprise is also only available in preview mode.

Managed notebooks and user-managed notebooks are being deprecated, so in this book we will primarily use the newest option: Vertex AI Workbench instances. Let's go ahead and create one now.

## Creating a Vertex AI Workbench instance

To create a Vertex AI Workbench instance, perform the following steps:

> **Note**
>
> For configuration parameters that are not specifically called out in these instructions, leave those parameters at their default values

1. In the Google Cloud console, navigate to the Google Cloud services menu → **Vertex AI** → **Workbench**.
2. Select the **Instances** tab and click **Create New**.
3. In the side panel that appears (see *Figure 5.1* for reference), enter a name for your notebook. Alternatively, you can use the default name that's automatically generated in that input field:

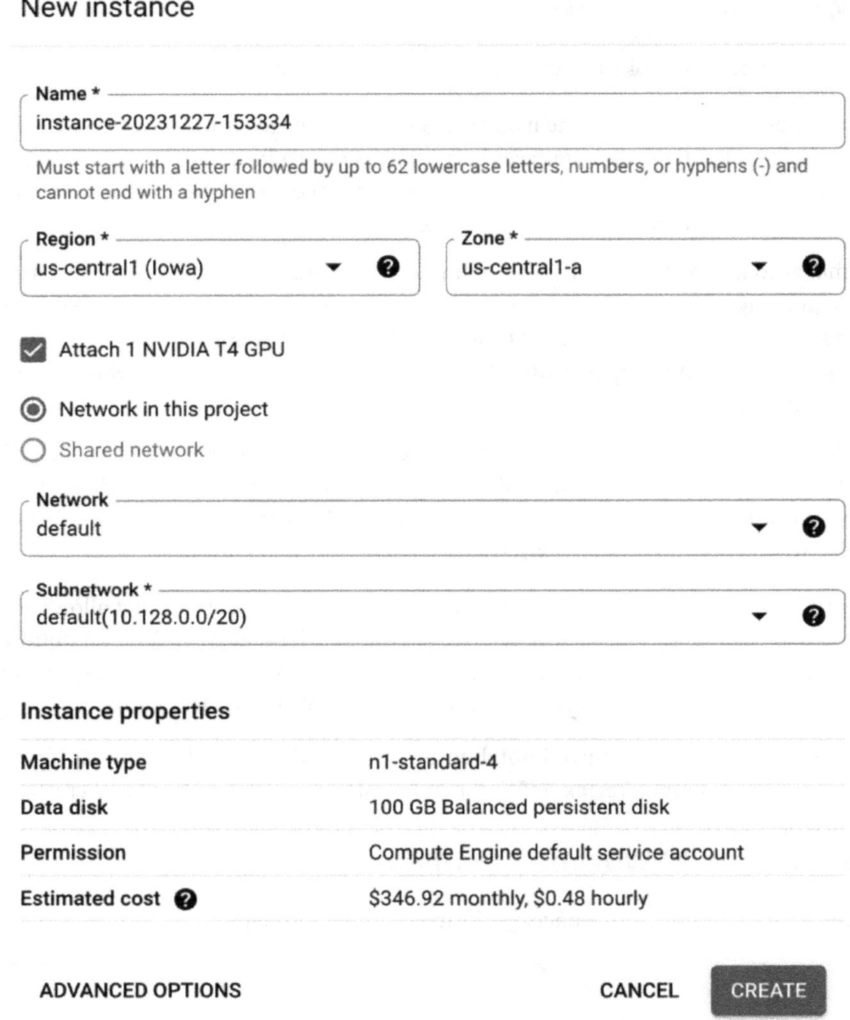


Figure 5.1: The New instance dialogue

4. Select your preferred region.
5. Select the option to attach a GPU (at the time of writing, you should select the box that says **Attach 1 NVIDIA T4 GPU**). See *Figure 5.1* for reference.
6. Leave the networking configuration details at their default values for now.
7. At the bottom of the side panel, click **Advanced Options**.
8. In the next screen that appears (see *Figure 5.2* for reference), ensure that the option to **Enable Dataproc** is selected and click **Continue** (you may need to scroll down to see the **Continue** button):

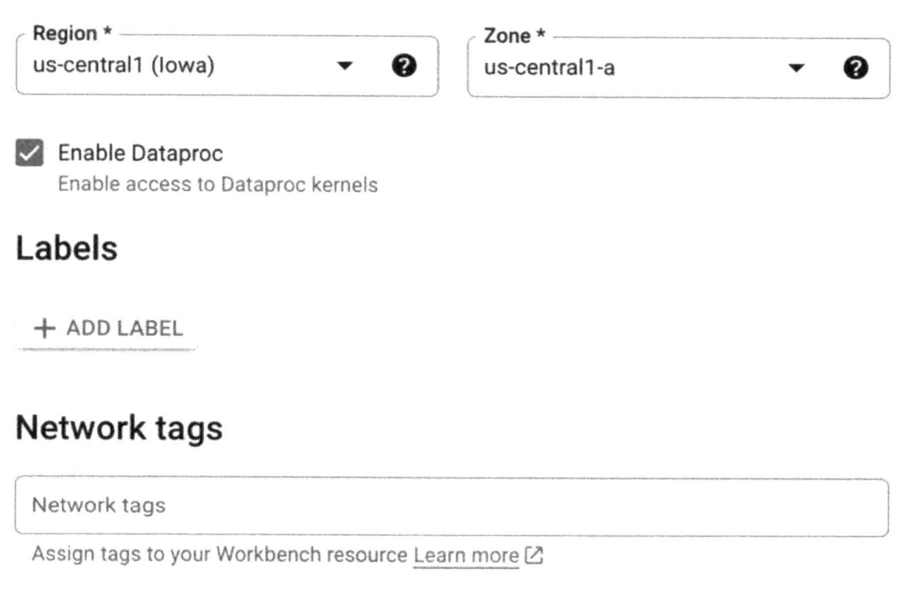

Figure 5.2: The New instance dialogue (continued)

9. On the next screen that appears (that is, the environment configuration screen), select **Use the latest version** and click **Continue**.
10. On the next screen that appears, you can use all of the default values (unless you have preferences to change anything) and click **Continue**.
11. You can click the **Continue** button two more times (that is, to accept the default values on the next two screens, unless you have preferences to change anything) until you reach the **Networking** configuration screen.
12. On the **Networking configuration** screen, ensure that the **Assign external IP address** option is selected.

13. Unless you have any specific networking configuration needs, you can simply use the default network in your project.
14. Click **Continue**.
15. On the next screen that appears (that is, the **IAM & Security** screen), ensure that all the options in the **Security options** section are selected.
16. Select **CREATE** (at the bottom of the screen) and wait a few minutes for the notebook to be created.

    When the notebook is up and running, a green checkmark will appear to the left of the notebook's name.
17. Finally, click **Open JupyterLab**.

At this point, I'd like to highlight an important set of integrations and features that Google Cloud has added to the JupyterLab interface in Vertex AI Notebooks.

## Vertex AI Notebook JupyterLab integrations

In the JupyterLab interface in your Vertex AI Notebook instance, you will notice a set of icons on the far left-hand side of the screen, as shown in *Figure 5.3*:

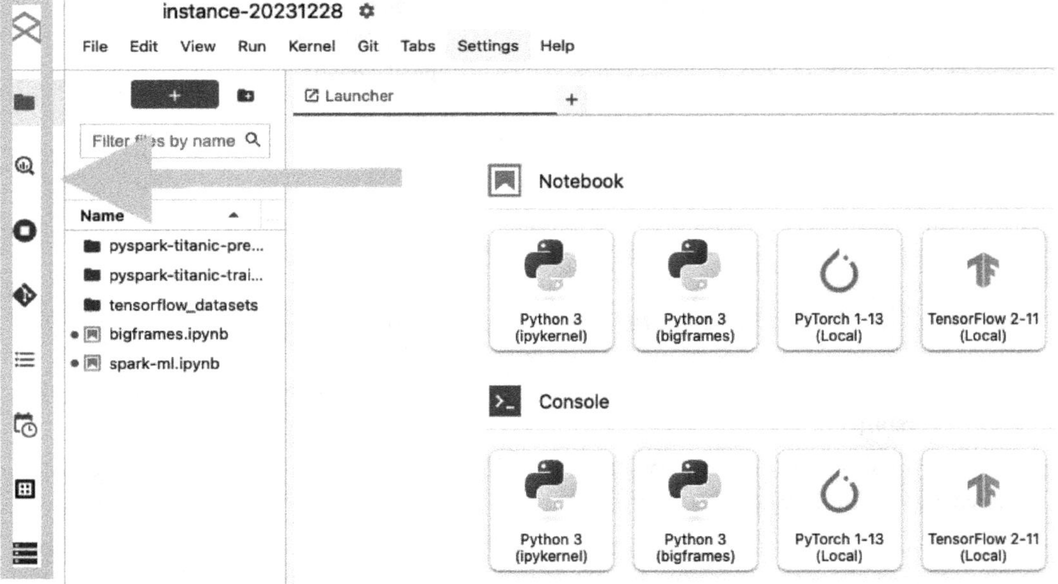


Figure 5.3: Google Cloud JupyterLab integrations

These icons represent integrations that we can use directly within our JupyterLab Notebook instances in Vertex AI. For example, if we click on the BigQuery icon, we can see our BigQuery datasets, and we can even use the integrated SQL Editor to run SQL queries on our BigQuery datasets directly from the JupyterLab Notebook interface.

I recommend clicking on the various icons to learn more about what you can do with them. Other useful features include the ability to integrate with Google Cloud Storage and an option that enables us to schedule the execution of our notebooks. The latter is a very useful feature for workloads that need to be automatically repeated periodically, which is a common need in data science (for example, re-training, evaluating, and deploying a model on new data every day).

Now that the notebook instance has been created, we will clone our GitHub repository so that we can access our notebook code and follow along with the activities in this chapter.

## Cloning the GitHub repository

Cloning our GitHub repository is the easiest way to quickly import all of the resources for the hands-on activities in this chapter into your notebook instance in Google Cloud Vertex AI. To clone our repository into your notebook, perform the following steps:

1. Click on the **Git** symbol in the menu on the left of the screen. The symbol will look like the one shown in *Figure 5.4*:

Figure 5.4: Git symbol

2. Select **Clone Repository**.
3. Enter our repository URL: `https://github.com/PacktPublishing/Google-Machine-Learning-for-Solutions-Architects`.
4. If any options are displayed, leave them at their default values.
5. Select **Clone**.
6. You should see a new folder appear in your notebook, named `Google-Machine-Learning-for-Solutions-Architects`.

Now, we're ready to start using our notebook instance! Let's move on to the next section, in which we will perform some unsupervised training.

## UML with scikit-learn on Vertex AI

In this section, we will start using our Vertex AI Workbench notebook to train models. We will begin with a relatively simple use case in which we will create an unsupervised model to discover clustering patterns in our data. Before we dive into the code, we will first take a minute to discuss the clustering algorithm we will use in this section, which is called K-means.

### K-means

You may remember that we discussed **unsupervised learning** (UL) mechanisms such as clustering in *Chapter 1*. Remember that in clustering, data points are grouped together based on similarities between features or characteristics that are observed by the model. *Figure 5.5* provides a visual representation of this concept, showing the input data on the left and the resulting data clusters on the right:

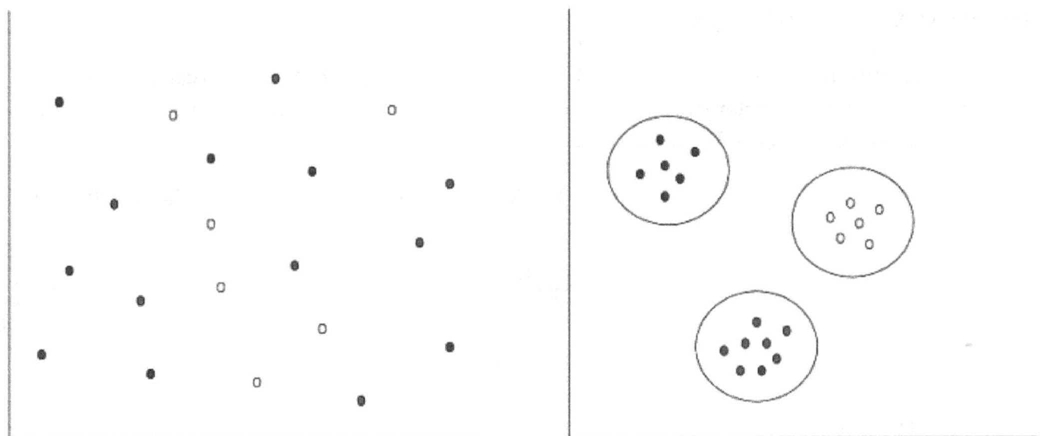

Figure 5.5: Clustering

K-means is an example of a clustering algorithm, and it is categorized as a centroid-based clustering algorithm. What this means is that it chooses a **centroid**, which is a point that represents the center of each of our clusters. The members of each cluster are data points from our dataset, and their membership in each cluster will depend on a mathematical evaluation of how close or far they are from each centroid. This proximity or distance from each centroid is generally calculated in terms of the Euclidean distance, represented by *Equation 5.1*:

$$d\left(x,y\right) = \sqrt{\sum_{i=1}^{n}(x_i - y_i)^2}$$

Equation 5.1: Euclidean distance

Don't worry—we will discuss this in more detail. The graphs shown in *Figure 5.5* represent what's called the feature space, and they show where each of our data points exists in the feature space. In the case of the aforementioned graphs, they represent a two-dimensional feature space, and each data point has an $x$ and $y$ coordinate that represents where the data point exists in the feature space. That is to say, the $x$ and $y$ coordinates are the features of each data point. If we take any two points on the graph, the Euclidean distance is simply the distance in a straight line between those two points, which is calculated by putting the $x$ and $y$ coordinates (that is, the features of each data point) into the equation represented in *Equation 5.1*.

Data points can have more than $x$ and $y$ coordinates as their features, and the concept applies in higher-dimensional feature spaces also.

Let's take a more concrete example than just $x$ and $y$ coordinates. Our dataset could consist of information regarding a retail company's customers, and the features of each customer could include their age, the city they live in, and what they purchased in their most recent visit to the company's store. Clustering algorithms such as K-means may then group those customers by finding similarities between their features and the centroids in each group.

One of the first questions that often comes up when discussing K-means is: How are centroids chosen? For example, how does the algorithm know how many centroids to use, and where to place them in the feature space? Let's address the number of centroids first. This is specified as a "hyperparameter" to the algorithm. This means that you can tell the algorithm how many centroids to use (and therefore how many clusters to form). In fact, this is reflected in the name, "K-means," where K represents the number of centroids or clusters. You would generally try different numbers of centroids until you find a value that maximizes the amount of information gained and minimizes the amount of variance, in each cluster. A mechanism that is often used to find the optimal value for K is referred to as "the elbow method." Using the elbow method, we run K-means clustering on the dataset for a range of different values of K (such as 1-10). Then, for each value of K, the average score is computed for all clusters. The default computed score is called the **distortion**. The distortion represents the sum of the squared distances from each point to the centroid in their assigned cluster, which again relates back to *Equation 5.1*. If you graph the value of K against the distortion score for each run, you will usually notice the distortion score going down each time you increase the value of K. The distortion score will usually go down sharply in the beginning and eventually start to go down in smaller increments. When adding new clusters (that is, increasing the value of K) stops significantly reducing the distortion score, then you can usually consider the optimal value of K to have been found.

Now, let's discuss how the algorithm knows where to place the centroids in the feature space. It begins by placing them at random locations in the feature space and randomly assigning data points to each centroid. It then repeats the following steps until no new data points are added to each cluster, at which point the optimal cluster positioning is believed to have been calculated:

1. Calculate which data points are closest to the centroid and assign them to that centroid.
2. Calculate the average (mean) between those points.
3. Move the centroid.

Now that we've covered the theory of how K-means clustering works, let's move on to the fun part, which is to actually implement the algorithm.

## Implementing a UML workload in Vertex AI

We're going to use the K-means algorithm within scikit-learn to start implementing our UML workload in Vertex AI. One of the quintessential clustering examples using K-means in scikit-learn is to use what's referred to as the iris dataset, to find cluster patterns in the data. The iris dataset, as the name implies, contains data on various iris flowers.

**Business case**

Our company, *BigFlowers.com*, recently acquired a smaller flower company, and in doing so, we acquired all of their digital assets, including the datasets that contain their flower inventory. Unfortunately, they did not do a good job of documenting their datasets, so we don't have a good idea of what's in their inventory.

We've been tasked with finding out as much as we can about the contents of the flower inventory, so we're going to use some data analytics tools and ML models to learn more about the dataset. One of the first things we're going to do is try to find any patterns such as logical groupings, which could help us to understand if there are distinct categories of objects in the dataset and additional information such as how many distinct categories exist.

*Note*: At the time of writing this, the company *BigFlowers.com* does not exist, and our exercises here refer to a fictitious company for example purposes only.

To get started with this task, navigate into the `Google-Machine-Learning-for-Solutions-Architects` folder in the Vertex AI notebook you created in the previous section. Then, navigate into the `Chapter-05` folder, double click on the `Chapter-5.ipynb` file (if prompted, select the `Python` kernel), and perform the following steps.

1. The first thing we need to do in our notebook is to import the resources that we need, such as the scikit-learn K-means class, and the pandas and Matplotlib libraries. We will also import a function to load the iris dataset from scikit-learn. To perform these actions, enter the following code into the interactive prompt in our notebook (or use the notebook you cloned from GitHub) The following code in the notebook performs those steps:

```
from sklearn.cluster import KMeans
from sklearn.datasets import load_iris
import pandas as pd # for exploring our data
import matplotlib.pyplot as plt # for plotting our clusters
from mpl_toolkits.mplot3d import Axes3D # Specifically for
creating a 3-D graph
```

2. To execute the code, hold the *Shift* key on your keyboard, and press the *Enter* key. Considering that the code is simply importing libraries, you will not see any output displayed back to the screen unless an error occurs.

> **Note**
>
> From here onward, when we are performing activities in Vertex AI Workbench notebooks, we will just provide the code samples for each step, and it will imply that you need to enter that code into the next available empty cell in the notebook (unless you are using the cloned notebook that already contains the code) and execute the cell, just as you did with the previous piece of code.

> **Note**
>
> On the left side of each cell in a Jupyter notebook, there is a square bracket symbol that looks like this: `[ ]`
>
> This can be used to understand the status of the cell, and the indicators are as follows:
>
> `[ ]` (Empty): The cell has not yet been executed.
>
> `[*]` (Asterisk): The cell is currently executing.
>
> `[1]` (Any number): the cell has completed executing.

3. Next, we will read the iris dataset in:

```
# Load the iris dataset:
iris = load_iris()
# Assign the data to a variable so we can start to use it:
iris_data = iris.data
```

4. Let's use pandas to get some information about our dataset:

```
# Convert the dataset to a pandas data frame for analysis:
iris_df = pd.DataFrame(iris_data)
# Use the info() function to get some information about the
dataset
iris_df.info()
```

The output should look like what is shown in *Figure 5.6*:

```
<class 'pandas.core.frame.DataFrame'>
RangeIndex: 150 entries, 0 to 149
Data columns (total 4 columns):
 #   Column  Non-Null Count  Dtype
---  ------  --------------  -----
 0   0       150 non-null    float64
 1   1       150 non-null    float64
 2   2       150 non-null    float64
 3   3       150 non-null    float64
dtypes: float64(4)
memory usage: 4.8 KB
```

Figure 5.6: pandas.DataFrame.info() output

5. The `info()` function output shows us what kinds of data our dataset contains. In this case, we can see that it contains 150 rows (indexed from 0 to 149) and 4 columns (indexed from 0 to 3), and the data values in each cell are floating-point numbers. Each row in our dataset is a data point, which represents a particular iris flower, and each column is a feature in our dataset. If we look at the description for the iris dataset in the scikit-learn documentation, which can be found at `https://scikit-learn.org/stable/datasets/toy_dataset.html#iris-dataset`, we can see that the features are:

   A. Sepal length in cm

   B. Sepal width in cm

   C. Petal length in cm

   D. Petal width in cm

   From this, we can understand that the floating-point numbers in each cell in our dataset are measurements of those four aspects of each flower.

6. We can also preview a subset of the data to see the actual values in each cell, like so:

```
iris_df.head()
```

   The output should look like what is shown in *Figure 5.7*:

| | 0 | 1 | 2 | 3 |
|---|---|---|---|---|
| **0** | 5.1 | 3.5 | 1.4 | 0.2 |
| **1** | 4.9 | 3.0 | 1.4 | 0.2 |
| **2** | 4.7 | 3.2 | 1.3 | 0.2 |
| **3** | 4.6 | 3.1 | 1.5 | 0.2 |
| **4** | 5.0 | 3.6 | 1.4 | 0.2 |

Figure 5.7: pandas.DataFrame.head() output

7. Now, we're going to use K-means to group similar data points in the dataset together, based on their features. First, we're creating an instance of a K-means model, and we're specifying that it will have three clusters because in this case, the dataset documentation told us that there are three different categories of iris in our dataset. If we didn't already know how many clusters we needed to use, then we could try different numbers of clusters and use the elbow method to find the best number:

```
kmeans_model = KMeans(n_clusters=3)
```

8. At this point, we have defined our model, but it has not yet learned anything from our data. Next, we instruct the model to "fit" to our dataset. The term *fit* is used by many algorithms to refer to the training process because, during training, this is exactly what the algorithm is trying to do; it is trying to create a model that fits as accurately as possible (without overfitting) to the given dataset:

```
kmeans_model.fit(iris_data)
```

9. When our model has completed the training process, we can now get it to cluster our data. We will feed our original dataset into the model, and according to the patterns it learned during training, it will place the data points into each of the clusters it defined:

```
kmeans_model.predict(iris_data)
```

10. Finally, we store the model's labels in a variable so that we can visualize the clusters in the next cell:

```
labels = kmeans_model.labels_
```

11. Next, we will visualize the clusters that K-means has created:

```
# Create a figure object:
fig = plt.figure()
# Define the axes (note: the auto_add_to_figure option will
default to False from mpl3.5 onwards):
axes = Axes3D(fig, auto_add_to_figure=False)
# Add the axes to the figure:
fig.add_axes(axes)
# Create the scatter plot to graph the outputs from our K-means
model:
axes.scatter(iris_data[:, 2], iris_data[:, 3], iris_data[:, 1],
    c=labels.astype(float))
# Set the labels for the X, Y, and Z axes:
axes.set_xlabel("Petal length")
axes.set_ylabel("Petal width")
axes.set_zlabel("Sepal width")
```

12. The resulting graph should look similar to the graph depicted in *Figure 5.8*. Notice how we can see three distinct clusters of data points in the graph, where each distinct cluster is color-coded as either purple, green, or yellow:

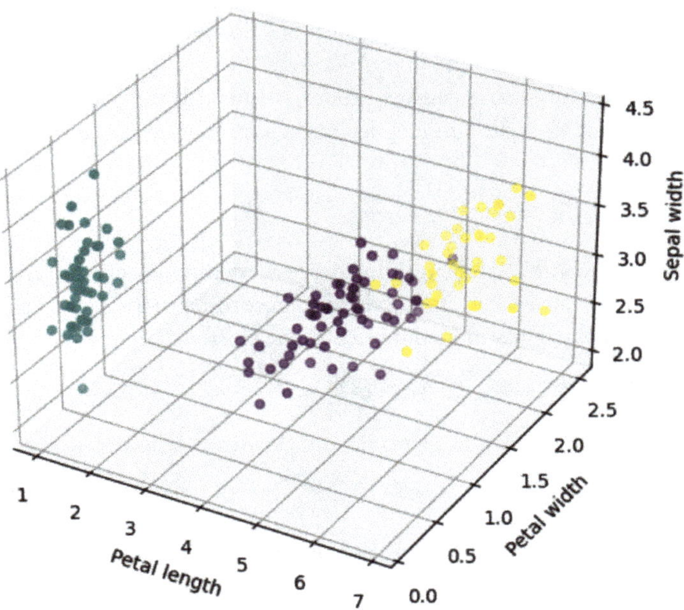


Figure 5.8: K-means cluster graph

This gives us some useful information. We can see the three distinct categories or clusters of data points in our dataset, in which the data points share similar characteristics with other points in their same cluster but differ from the data points in other clusters. We get these insights without ever needing to label our dataset, which is quite useful, considering that labeling is a time-consuming and error-prone task.

> **Note**
>
> Although our dataset contains four features (also referred to as "dimensions"), humans can only see/visualize things in three dimensions. Therefore, we only used three of the four features to create our graph. The following line is where we defined the features to use in our graph, in which the numbers represent the features to be used:
>
> ```
> axes.scatter(iris_data[:, 2], iris_data[:, 3], iris_data[:, 1],
> c=labels.astype(float))
> ```
>
> You can try playing around with the graph by changing each of those numbers to anything between 0 and 3, as long as each entry is a unique value (that is, don't repeat any of the numbers more than once in the code).
>
> Each time you make a change, you can execute the cell again to update the graph.

Now that we've seen what it takes to implement a UML workload in Vertex AI, let's move on and learn how to implement an SML workload in the next section.

# Implementing a regression model with scikit-learn on Vertex AI

The first SML model we're going to build in our Vertex AI Workbench notebook is a linear regression model. You may remember that we described linear regression in *Chapter 1*.

> **Business case**
>
> Our boss at *BigFlowers.com* is running a fun competition at work. Employees are asked to predict the length of an iris flower's petal when given some other measurements related to that flower, such as the sepal length, sepal width, and petal width. The person with the most accurate estimation will get a big prize, and employees are allowed to use technology to help them in their estimations, so we're going to build an ML model to help us make these predictions.

In the previous section, we used the K-means algorithm within scikit-learn. In this section, we will use the linear regression algorithm within scikit-learn. As such, we will need to import the `LinearRegression` class into our notebook context. We will also import a function that will calculate the **Mean Squared Error** (**MSE**) metric, which we described in *Chapter 2* and which is commonly used to evaluate linear regression models.

We can use the same iris dataset for our linear regression model, so we won't need to repeat any of the previous dataset importing steps because it is already loaded into our notebook context. This is an important concept to note; we can use different types of models on the same data, depending on the business use case and desired results. However, as we discussed in *Chapter 2*, SML model training introduces some additional requirements regarding how the dataset is used, which we describe next.

> **Remember**
>
> In **supervised learning** (**SL**) use cases, we need to define which column in the dataset is designated as the "target" feature. This is the feature that we're trying to predict based on the other features in the dataset. During training, some of the elements in this column are used as labels that represent known, correct answers from which the model learns.
>
> The values of the other features in the dataset are used as inputs. During the prediction process, the model uses the relationships it has learned between all of the input features and the target feature to predict the value of the target feature based on the values of the input features.
>
> Also, remember that in SL use cases, we generally split our dataset into subsets such as training, validation, and testing. The training dataset, as the name implies, is what the model is trained on. The testing dataset is how we evaluate the trained model (based on the metrics that we've defined for that model), and the validation set is usually used in hyperparameter tuning.

> *Note*:
>
> We will explore hyperparameter tuning in a later chapter. In our current chapter, we will train a single regression model and we will then test it directly, so we will not need to split out a validation subset of our data. Therefore, we will just split the dataset into subsets for training and testing.

To start building our linear regression model, perform the following steps:

1. Import the `LinearRegression` class from scikit-learn and import a function that will make it easy for us to split our dataset into train and test datasets, as well as a function that will calculate the MSE metric that we will use later to evaluate our model's performance:

```
from sklearn.linear_model import LinearRegression
from sklearn.metrics import mean_squared_error
from sklearn.model_selection import train_test_split
```

2. Define which column in our dataset we want to use as the target feature to predict:

```
target = iris_df[[2]]
```

> **Remember:**
>
> In the previous section of this chapter, we consulted the documentation for the iris dataset, and we saw that the columns in the dataset are as follows: sepal length in cm, sepal width in cm, petal length in cm, petal width in cm. These columns are indexed from 0 to 3. Considering that we want to predict the petal length, we have selected column 2 of our `iris_df` dataframe as our target column.

3. Next, define our input features. In this case, we use all feature columns except column 2 (because we've defined that column 2 is our target):

```
input_feats = iris_df[[0, 1, 3]]
```

4. Split our dataset into separate subsets for training and testing:

```
input_train, input_test, target_train, target_test = \
    train_test_split(input_feats,target,test_size=0.2)
```

> **Note**
>
> Specifying a value of 0.2 for the `test_size` variable means that 20% of the original dataset will be separated to create the test dataset. The remaining 80% then forms the training dataset. The `input_train` and `input_test` datasets contain the input features used during training and the input features that will be used to test the trained model, respectively. The `target_train` and `target_test` datasets contain the respective labels (that is, the "correct answers") for the training and test datasets.

5. Now, let's use the training dataset to train our linear regression model, and then use the test dataset to generate predictions from the model:

```
# Create an instance of a LinearRegression model
lreg_model = LinearRegression()

# Train the model by fitting it to the training data
lreg_model.fit(input_train,target_train)

# Use the test set to generate predictions
target_predictions = lreg_model.predict(input_test)
```

**Remember**

To test the trained model, we send the `input_test` features to it, and we ask it to predict what the corresponding target feature values should be, based on the values of the `input_test` features. We can then compare the model's predictions to the `target_test` values (which are the known, correct answers) in order to see how close its predictions were to the correct, expected values.

6. Finally, it's time to see what kinds of predictions our model has made based on the `input_test` data!

```
pred_df = pd.DataFrame(target_predictions[0:5])
pred_df.head()
```

The output should look similar to what is shown in *Figure 5.9*, although the numbers may differ (the predictions are in the right-hand column, disregarding the number 0 at the top, which is the column header):

| | 0 |
|---|---|
| 0 | 1.726248 |
| 1 | 4.420581 |
| 2 | 1.161504 |
| 3 | 4.531149 |
| 4 | 1.584222 |

Figure 5.9: Linear regression model predictions

7. Now, let's take a look at the corresponding known, correct values from the `target_test` dataset:

```
target_test.head()
```

The output should look similar to what is shown in *Figure 5.10*, although the numbers may differ (the values are in the right-hand column, disregarding the number 2 at the top, which is the column header):

| | 2 |
|---|---|
| **14** | 1.2 |
| **78** | 4.5 |
| **37** | 1.4 |
| **51** | 4.5 |
| **40** | 1.3 |

Figure 5.10: The known, correct values from the target_test dataset

8. As we can see, the numbers are pretty close in some cases. However, let's use metrics to evaluate the model's overall performance. To do this, we use the `mean_squared_error()` function to compare the predictions against the correct values from the `target_test` dataset and generate the MSE metric value:

```
mean_squared_error(target_test,target_predictions)
```

9. The value should be something like 0.08871798773326277.

This is a pretty good value because the error is quite low, which means that our model is doing a good job of predicting the target values. At this point, I think we might win that big prize in the competition to predict the length of the iris petals!

**Diving deeper into the various datasets mentioned in the previous section**

The `train_test_split` function in our hands-on activity created the following subsets from our source dataset:

`input_train`: The input features used during training.

`input_test`: The input features that will be used to test the trained model.

`target_train`: The target labels used during training. During the training process, the model uses these values as the known, correct answers that it is trying to predict. These are key to the training process because, during training, the model tries to learn relationships between the input features that will help it predict these answers as accurately as possible.

`target_test`: The target labels used during testing. These are known, correct answers from the original dataset that were separated out from the training set, so they were not included in the training process. Therefore, the model has never seen these values during training. We then use these values to test the performance of the trained model.

Now that we've seen linear regression in action, let's move on to our first SL classification task.

# Implementing a classification model with XGBoost on Vertex AI

By now, you've started to become familiar with many of the popular libraries that are commonly used in data science projects. In this section, we will start using another very popular library, XGBoost, which can be used for either classification or regression use cases.

While we briefly introduced XGBoost at the beginning of this chapter, we will dive further into how it works here, starting with the concept of decision trees.

## Decision trees

When we discussed the topic of Gradient Boosting earlier in this chapter, we mentioned that one of the components of Gradient Boosting is the concept of weak learners. Decision trees are one example of what could be used as a weak learner. Let's start with a simple example of what a decision tree is. Refer to *Figure 5.11*, which shows a decision tree that is used for estimating whether a bank customer is likely to purchase a house, based on their age group and income:

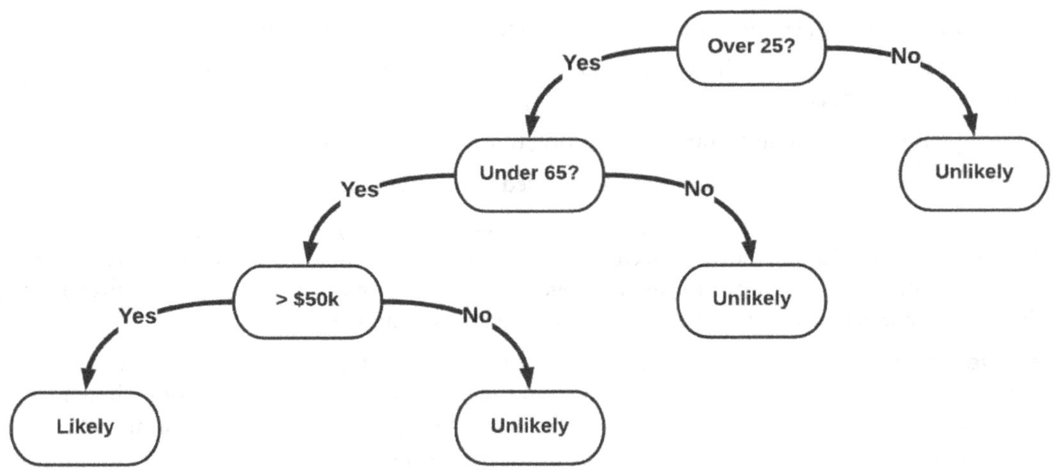


Figure 5.11: A decision tree

A decision tree consists of a sequence of decisions. The process starts at what's referred to as the **root node**, and at each point in the tree, a decision is made, which then guides us on the next step to take. The sequence of decisions therefore results in a path through the tree, until we reach a final point that contains no further decisions. Such a point is referred to as a **leaf node**. Following the steps in the tree in *Figure 5.11*, it determines that a bank customer is likely to purchase a house if they are between 25 years old and 65 years old and earn more than $50,000 per year. All of the other factors in the decision tree indicate that they are otherwise unlikely to purchase a house.

While the example in *Figure 5.11* is quite simple, this kind of process can be used algorithmically in ML applications, whereby the decision at each point is based on some kind of threshold. For example, if the value is less than or greater than a certain threshold, then move on to evaluate feature *A*; otherwise, evaluate feature *B* or stop evaluating features. When using decision trees, one of the goals is to find which features in the dataset can help make the best decisions and how many decisions need to be made, which is referred to as the **tree depth**, in order to get to the best eventual prediction.

Two key concepts in the decision tree process are **entropy** and **information gain**. Entropy, in this context, is the measure of impurity in a given grouping of entities. For example, a grouping that perfectly captures identical entities in the same group would have zero entropy, whereas a grouping that has lots of variability between the entities would have high entropy. Information gain, in this context, can be seen as how much the entropy is decreased by each decision in our decision tree. In our bank-customer example in *Figure 5.11*, the initial set of customers (before any decisions are made) consists of all of our customers, and this set would have a lot of entropy because it would include people of all ages and all kinds of economic situations and other characteristics.

Now, as we try to figure out which customer features could help our algorithm decide whether they would be likely to purchase a house, we find a pattern that suggests that people within a certain age group (let's say, between 25 and 65) and who earn over a certain amount of money per year (let's say, more than $50,000) are more likely to purchase a house than other customers. Therefore, in this case, a customer's age and income are examples of features that help to maximize the information gain and reduce the entropy and would therefore be good decision point features in our decision tree. On the other hand, features such as their name or music preferences are unlikely to have any correlation with their probability of purchasing a house, so those features would not result in significant information gain in the decision points in our decision tree, and the decision tree model would likely learn to ignore those features during training.

Decision tree algorithms can be quite effective for some use cases, although they come with some limitations such as being prone to overfitting. This is where the concept of ensembles comes into play because combining many trees together can make a more powerful prediction model—and one that is also a lot less likely to overfit—than any individual tree by itself. This could be done by using a Bagging approach, such as the Random Forest algorithm (see *Figure 5.12* for reference), which trains each tree in the ensemble on a random sub-sample (with replacement) from the training feature space, or by using a Boosting approach, as we described in the case of Gradient Boosting:

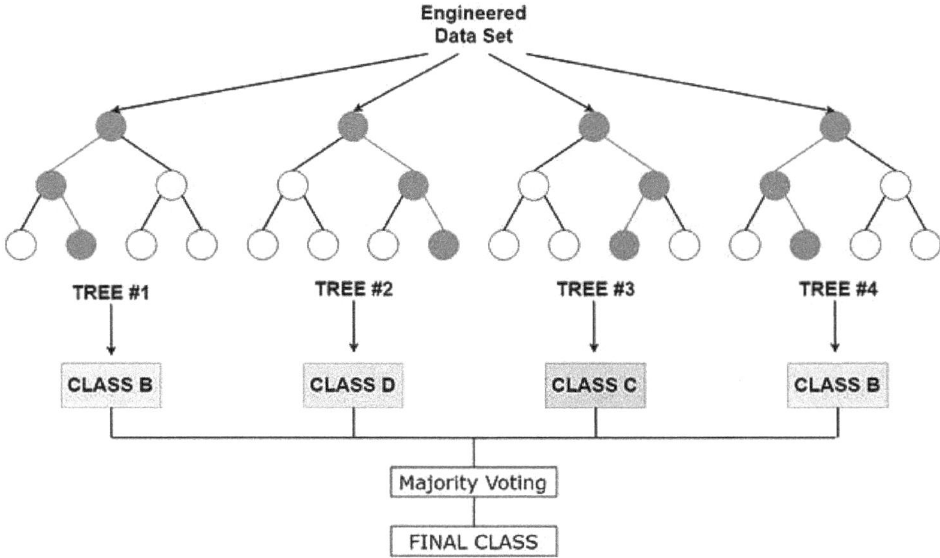


Figure 5.12: Random Forest

When we use Gradient Boosting with decision trees, we refer to this as **gradient-boosted trees**. One of the inherent challenges with a simple gradient-boosted tree implementation is the sequential nature of the algorithm, whereby the errors of a tree in one training iteration need to be used in the next iteration to train the next tree in the ensemble. However, as we mentioned earlier in this chapter, XGBoost overcomes this limitation, and it can train thousands of trees in parallel, which vastly speeds up the overall training time.

With this background information in mind, let's see how we can use XGBoost to build a classification model in Vertex AI. The business case here does not require its own dedicated callout section because it's pretty straightforward: our model will need to predict the class for each iris, based on its petal and sepal measurements.

> **Note**
>
> The iris dataset contains details about irises that fit into one of three classes:
>
> - `Iris-Setosa`
> - `Iris-Versicolour`
> - `Iris-Virginica`
>
> In our linear regression example in the previous section of this chapter, we decided to use the `petal length` feature as the target that we were trying to predict. However, in this section, we will try to predict the class (or category) of each iris flower based on all of the other features in the dataset. Because there are more than two classes, this will be a multi-class classification task.
>
> The class is a new feature of the iris dataset that we will introduce in this section, and we have not interacted with this feature yet. Next, we will talk about how to access this feature.
>
> The iris dataset in scikit-learn contains multiple objects. One of those objects is the `data` object, and that's what we've been using so far in this chapter.
>
> Another object in the dataset is the `target` object, which contains the `class` feature column. The `class` feature column represents the classes as follows:
>
> `0 = Iris-Setosa`
>
> `1 = Iris-Versicolour`
>
> `2 = Iris-Virginica`
>
> Remember the following two lines of code from earlier in this chapter:
>
> ```
> iris = load_iris()
> iris_data = iris.data
> ```
>
> Those lines of code loaded the iris dataset in and specifically assigned the `data` object from that dataset to the `iris_data` variable. We did not reference the `target` object at that time because we did not need to do so.
>
> We will start using the `target` object in this section, and in order to do so, we will assign it to a variable named `iris_classes`.

To train the classification model, perform the following steps in your Vertex AI Workbench notebook:

1. Just as with the libraries we used in the previous section, we will need to import the XGBoost library before we can start using it. More specifically, we will import the `XGBClassifier` class from the `XGBoost` library:

   ```
   from xgboost import XGBClassifier
   ```

2. Considering that we will use it for a classification use case, we will also import a function that will calculate a metric that can be used to evaluate a classification model, which is called `accuracy`:

   ```
   from sklearn.metrics import accuracy_score
   ```

3. Assign the `target` object from the iris dataset to the `iris_classes` variable so that we can begin to reference it:

   ```
   iris_classes = iris.target
   ```

4. Create our dataset splits for training and testing, just like we did in our regression example in the previous section:

   ```
   xgb_input_train,xgb_input_test,xgb_target_train,xgb_target_test=
       train_test_split(iris_data, iris_classes, test_size=.2)
   ```

5. Create a model instance and specify the hyperparameters:

   ```
   xgbc = XGBClassifier(n_estimators=2, max_depth=2,
       learning_rate=1, objective='multi:softmax')
   ```

> **Diving deeper**
>
> The hyperparameters we're specifying in the previous piece of code represent the following:
>
> `n_estimators`: The number of decision trees to use in the ensemble.
>
> `max_depth`: The maximum number of decision points (or maximum depth) for each tree.
>
> `learning_rate`: The learning rate to use during the optimization of the cost function.
>
> `objective`: The type of prediction we want our model to perform. In this case, we want it to use the `softmax` function to perform multi-class classification.

6. Let's train our model:

   ```
   xgbc.fit(xgb_input_train, xgb_target_train)
   ```

   You will see an output similar to what is shown in *Figure 5.13*, which summarizes the values (including default values) of the model's parameters:

```
XGBClassifier(base_score=0.5, booster='gbtree', colsample_bylevel=1,
              colsample_bynode=1, colsample_bytree=1, gamma=0, gpu_id=-1,
              importance_type='gain', interaction_constraints='',
              learning_rate=1, max_delta_step=0, max_depth=2,
              min_child_weight=1, missing=nan, monotone_constraints='()',
              n_estimators=2, n_jobs=0, num_parallel_tree=1,
              objective='multi:softprob', random_state=0, reg_alpha=0,
              reg_lambda=1, scale_pos_weight=None, subsample=1,
              tree_method='exact', validate_parameters=1, verbosity=None)
```

Figure 5.13: XGBoost parameters

7. Now, it's time to get predictions from our model by using the `xgb_input_test` dataset:

   ```
   xgb_predictions = xgbc.predict(xgb_input_test)
   ```

8. Let's see what it predicted for each item in the `xgb_input_test` dataset:

   ```
   xgb_predictions
   ```

   The output should be an array of predicted classes, which are represented by either 0, 1, or 2, similar to the output shown in *Figure 5.14*, although the values you get may differ:

```
array([1, 2, 1, 0, 1, 2, 1, 0, 1, 2, 2, 0, 1, 0, 0, 2, 1, 1, 1, 0, 0, 1,
       2, 0, 2, 0, 1, 0, 0, 2])
```

Figure 5.14: XGBoost iris classification predictions

9. Did our model get it right? In order to find out, we can check what the known correct answers are:

   ```
   xgb_target_test
   ```

   The output of this piece of code should also be an array of classes, which are represented by either 0, 1, or 2, similar to the output shown in *Figure 5.15*, although the values you get may differ:

```
array([1, 2, 1, 0, 1, 2, 1, 0, 1, 2, 2, 0, 1, 0, 0, 2, 1, 1, 1, 0, 0, 1,
       2, 0, 2, 0, 1, 0, 0, 2])
```

Figure 5.15: The known, correct answers

10. They look pretty similar to me, but to be sure, let's test the accuracy by comparing our predictions to the known, correct answers:

    ```
    accuracy_score(xgb_target_test,xgb_predictions)
    ```

    The score will be presented as a floating-point number representing the accuracy percentage. If the result is 1.0, it means that our predictions were 100% accurate!

**Diving deeper**

The accuracy metric quantifies the ratio of correct predictions from all of the predictions performed by our model, as represented by the following formula:

*Accuracy = Correct Predictions / Total Number of Predictions*

There are many other metrics for assessing a model's prediction performance, and scikit-learn has built-in functions to calculate many of those metrics. As additional learning, we recommend you consult the scikit-learn metric documentation at the following link: `https://scikit-learn.org/stable/modules/classes.html#module-sklearn.metrics`.

Great work! You have officially trained multiple of your own models on Google Cloud Vertex AI! Let's recap on what we've learned in this chapter.

## Summary

In this chapter, we took many of the ML concepts from *Chapters 1* and *2* and put them into practice. We used clustering to find patterns in our data in an unsupervised manner, and you specifically learned a lot more about how K-means is used for clustering and how it works.

We then dived into SL, and you explored the linear regression class within scikit-learn and learned how to use metrics to measure the performance of a regression model.

Next, you learned how to use XGBoost to build a classification model and classify items in the iris dataset based on their features.

Not only did you put all of those important concepts into practice, but you also learned how to create and use Vertex AI Workbench-managed notebooks.

Additionally, you learned other important concepts in the ML industry, such as how decision trees work, how Gradient Boosting works, and how XGBoost enhances that functionality to implement one of the most effective ML algorithms in the industry.

That was a lot of stuff to learn in one chapter, and you should be proud that you are significantly elevating your skills and knowledge in areas that are in high demand in today's business world.

In the next chapter, you will learn another set of extremely important skills as we will focus on data analysis and data transformation for ML workloads.

# 6

# Diving Deeper – Preparing and Processing Data for AI/ML Workloads on Google Cloud

In the previous chapter, we did some very rudimentary data exploration by looking at a few details relating to our dataset, using functions such as `pandas.DataFrame.info()` and `pandas.DataFrame.head()`. In this chapter, we will dive deeper into the realm of data exploration and preparation for data science workloads, as represented by the section highlighted in blue in the data science life-cycle diagram shown in *Figure 6.1*:

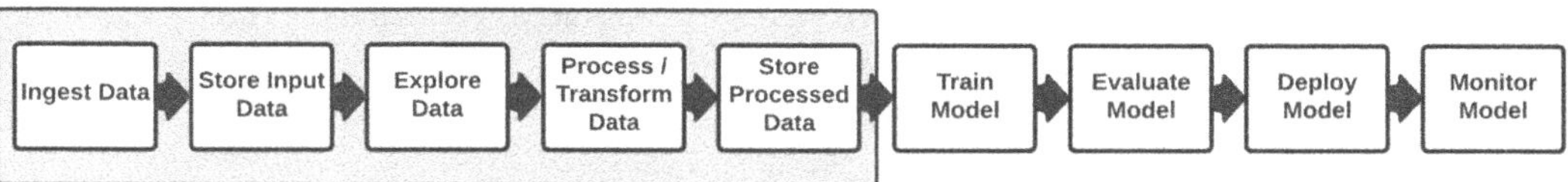


Figure 6.1: Data exploration and processing

In the early stages of a typical data science project, you would likely perform many of the data exploration and preparation steps in Jupyter notebooks, which, as we have seen, are useful for experimenting with small datasets. When you bring your workload into production, however, you are likely to use much larger datasets, in which case you would usually need to use different tools for processing your data. Google is seen as an industry leader when it comes to large-scale data processing, analytics, and AI/ML. For example, Google Cloud BigQuery is well established as one of the most popular data warehouse services in the industry, and Google Cloud has many additional industry-leading services for data processing and analytics workloads.

In this chapter, we will learn how to explore, visualize, and prepare data for ML use cases with tools such as Vertex AI, BigQuery, Dataproc, and Cloud Composer. Furthermore, we will dive into processing streaming data on the fly with Dataflow, and introduce the fundamentals of data pipelines. By the end of this chapter, you will be able to create, build, and run data pipelines on Google Cloud, equipping you with the necessary skills to take on complex data processing tasks in today's fast-paced, data-driven world.

This chapter covers the following topics:

- Prerequisites and basic concepts
- Ingesting data into Google Cloud
- Exploring and visualizing data
- Cleaning and preparing the data for ML workloads
- An introduction to data pipelines
- Processing batch and streaming data
- Building and running data pipelines on Google Cloud

Let's begin by discussing the prerequisites for this chapter.

# Prerequisites for this chapter

The activities in this section need to be completed before we can start to perform the primary activities in this chapter.

## Enabling APIs

In addition to the methods we discussed in earlier chapters to enable Google Cloud APIs, such as the Google Cloud Shell or being prompted in the Google Cloud console, you can also proactively search for an API in order to enable it in the console. To do that, you would perform the following steps, in which *[service/API name]* is the name of the Service/API you wish to enable:

1. In the Google Cloud console, navigate to the **Google Cloud services** menu → **APIs & Services** → **Library**.
2. Search for *[service name]* in the search box.
3. Select the API from the list of results.
4. On the page that displays information about the API, click **Enable**.

Perform the preceding steps for each of the following service/API names:

- Compute Engine API
- Cloud Scheduler API
- Dataflow API
- Data Pipelines API
- Cloud Dataproc API
- Cloud Composer API
- Cloud Pub/Sub API

After you have enabled the required APIs, we're ready to move on to the next section.

## IAM permissions

In this section, we will set up the **identity and access management** (**IAM**) permissions required to enable the activities that come later in this chapter.

### *Service accounts*

In previous chapters, I mentioned that there are multiple ways to authenticate with Google Cloud services and that the API keys you used in *Chapter 4* were the simplest authentication methods. Now that we're advancing to more complex use cases, we will begin to use a more advanced form of authentication, referred to as service accounts.

Google Cloud service accounts are special accounts used by Google Cloud services, applications, and **virtual machines** (**VMs**) to interact with and authenticate to other Google Cloud resources. They are an interesting concept because, in addition to being resources, they are also considered to be identities or principals, just like people, and just like people, they have their own email addresses and permissions associated with them. However, these email addresses and permissions apply to the machines and applications that use them, rather than to people. Service accounts provide a way for machines and applications to have an identity when those systems want to perform an activity that requires authentication with Google Cloud APIs and resources.

We're going to create a service account with the required permissions for the activities we will perform in this chapter.

#### Data processing service account

Considering that we're going to use multiple Google Cloud services to process data in various ways in this chapter, we will create a service account with permissions mainly related to Google Cloud data processing services, such as BigQuery, Cloud Composer, Dataflow, Dataproc, **Google Cloud Storage** (**GCS**), and Pub/Sub.

Perform the following steps to create the required service account:

1. In the Google Cloud console, navigate to the **Google Cloud services** menu → **IAM & Admin** → **Service accounts**.
2. Select **Create service account.**
3. For the service account name, enter `data-processing-sa`.
4. In the section titled **Grant this service account access to project**, add the roles shown in *Figure 6.2*:

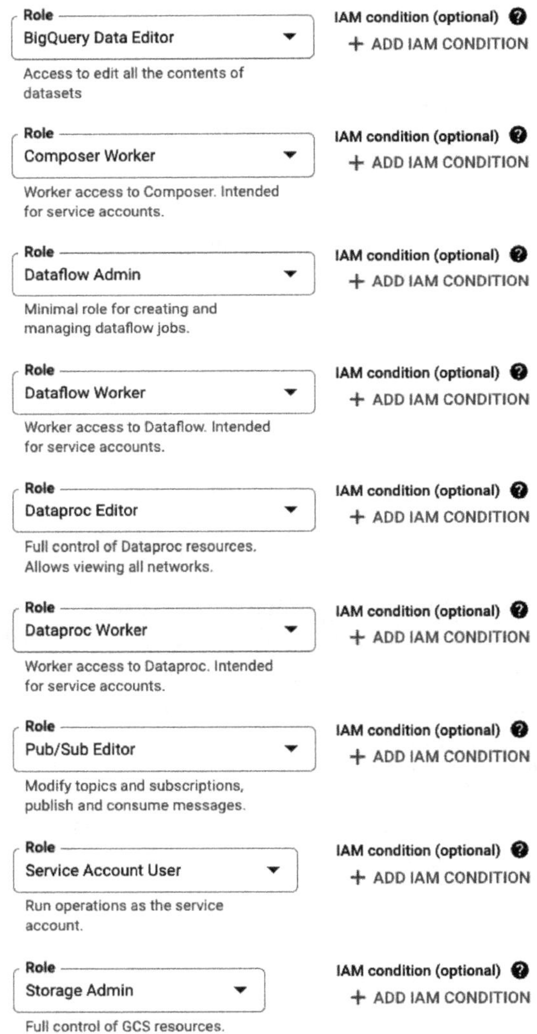


Figure 6.2: Dataflow worker service account permissions

5. Select **Done**.

Our service account is now ready to be used later in this chapter.

> **Note**
>
> In the service account we just created, we also added the **Service Account User** permission. This is because, in some of the use cases we will implement, our service account will also need to temporarily impersonate or "act as" other service accounts or identities. For more information on the concept of impersonating service accounts, see the following documentation: `https://cloud.google.com/iam/docs/service-account-permissions#directly-impersonate`.

## Cloud Storage bucket folders

We will use a Cloud Storage bucket to store data for the activities later in this chapter. We already created a bucket in *Chapter 4*, so we can simply add some folders to the bucket for storing our data. Perform the following steps to create the folders:

1. In the Google Cloud console, navigate to the **Google Cloud services** menu → **Cloud Storage** → **Buckets**.
2. Click on the name of the bucket you created in *Chapter 4*.
3. Select **Create folder**.
4. Name it `data`.
5. Select **Create**.

Repeat the preceding steps with the following additional folder names:

- `code`
- `dataflow`
- `pyspark-airbnb`

Our folders are now ready to store our data.

## Uploading data

We will use a file named `AB_NYC_2019.csv` as a dataset for some of the activities in this chapter. In the clone of our GitHub repository that you created on your local machine in *Chapter 4*, you will find that file in a directory named `data`, which exists within the directory named `Chapter-06`.

Therefore, you should find the file at the following path on your local machine (the slashes will be reversed if you're using Microsoft Windows):

```
[Location in which you cloned our GitHub repository]/Chapter-06/
data/AB_NYC_2019.csv
```

In order to upload this file, perform the following steps:

1. Navigate into the `data` folder you created in GCS in the previous section.
2. Click **Upload files**.
3. Now, select the `AB_NYC_2019.csv` file by navigating to the `Chapter-06/data` directory in the clone of our GitHub repository that you created on your local machine.

Now that we've completed the prerequisites, let's discuss some important industry concepts that we will dive into in this chapter.

# Fundamental concepts in this chapter

In this book, we aim to provide you with knowledge not only regarding how to use the relevant Google Cloud services for various types of workloads but also regarding important industry concepts that relate to each of the relevant technologies. In this section, we briefly cover concepts that provide additional context for this chapter's learning activities.

## Ingesting data into Google Cloud

In previous chapters, you already performed steps to upload data to GCS and Google Cloud BigQuery. In addition to performing bulk uploads to GCS and BigQuery, it's also possible to stream data into those services. You will see this in action in this chapter.

This section provides a more holistic overview of Google Cloud's data ingestion options. We will only cover Google Cloud's services here, but there are also countless options of third-party database and data management services that you can run on Google Cloud, such as MongoDB, Cassandra, Neo4j, and many others that are available through the Google Cloud Marketplace.

For streaming data use cases, which we will describe in more detail later in this chapter, you can use Cloud Pub/Sub to ingest data from sources such as IoT devices or website clickstream feeds. Dataflow can also be used to ingest data from various sources, transform it, and write it to other Google Cloud services such as BigQuery, Bigtable, or Cloud Storage. As we will discuss later in this chapter, Dataflow also supports batch data processing use cases.

Google Cloud Dataproc can be used to ingest data from **Hadoop Distributed File System** (**HDFS**)-compatible sources or other distributed storage systems, and Google Cloud Data Fusion can also be used to ingest data from various sources, transform it, and write it to most Google Cloud storage and database services.

For relational database data, you can use the Google Cloud **Database Migration Service** (**DMS**) to ingest data into Google Cloud SQL, which is a fully managed relational database service for MySQL, PostgreSQL, and SQL Server. You can also use standard SQL clients, import/export tools, and third-party database migration tools to ingest data into Cloud SQL instances.

For non-relational database data, there are multiple options, including the following:

- Google Cloud Bigtable, which is a fully managed, scalable NoSQL database for large-scale, low-latency workloads. You can use the Bigtable HBase API or the Cloud Bigtable client libraries to ingest data into Bigtable.
- Google Cloud Firestore, which is a fully managed, serverless NoSQL document database for web and mobile applications. Firestore provides client libraries, REST APIs, or gRPC APIs to ingest data.
- Some of you may also be familiar with Google Cloud Datastore, which was a separate Google Cloud non-relational database offering but has been somewhat merged with Google Cloud Firestore. To see more details regarding how these database options relate to each other, please see the Google Cloud documentation at the following link: `https://cloud.google.com/datastore/docs/firestore-or-datastore`.

Now that we've covered many of the options for ingesting data into Google Cloud, let's discuss what kinds of things we may want to do with that data after ingesting it, starting with data transformation approaches, such as **Extract, Transform, Load** (**ETL**) and **Extract, Load, Transform** (**ELT**).

## ETL and ELT

ETL and ELT are two approaches for data integration, processing, and storage. In ETL, data is first extracted from source systems and then transformed into another format. Such transformations could include cleansing, enrichment, aggregation, or deduplication. These transformations usually take place within an intermediary processing engine or staging area. Once the data is transformed, it is loaded into the target system, which is typically a data warehouse or a data lake. ETL is a traditional approach that is well suited for environments where data consistency and quality are critical. However, it can be time-consuming and resource-intensive due to the processing steps happening before the data is loaded.

On the other hand, ELT reverses the order of the last two steps. Data is first extracted from source systems and then directly loaded into the target system, such as a modern data warehouse or data lake. The transformation step is performed within the target system itself, using the processing capabilities of the destination system. ELT has gained popularity in recent years due to the increased scalability and processing power of modern cloud-based data warehouses, as it enables faster loading and allows users to perform complex transformations on demand.

Choosing between ETL and ELT depends on the specific requirements of the data processing environment, data quality needs, and the target system's capabilities.

# Batch and streaming data processing

At a high level, there are two main ways in which we can process data: batch and streaming. In this section, we explain each of these approaches, which will equip us with the knowledge we need for the activities that follow in this chapter.

## Batch data processing

Batch data processing usually refers to performing a series of transformations on large datasets in a bulk manner. These kinds of pipelines are suitable for use cases in which you need to process large amounts of data in parallel over a period of hours or even days. As an example, imagine that your company runs an online retail business that has separate deployments in different geographical regions, such as a *www.example.com* website in North America, a *www.example.co.uk* website in the UK, and a *www.example.cn* website in China. Every night, you may want to run a workload that takes all of the items purchased by all customers in each region that day, merges that data across regions, and then feeds that data into an ML model. This would be an example of a batch workload. Other examples of batch data processing use cases include the following:

- Processing daily sales transactions to generate reports for business decision-making
- Analyzing log files from a web server to understand user behavior over a specific time window (for example, a day or week)
- Running large-scale data transformation jobs, such as converting raw data into a structured format for downstream data systems

In addition to data processing tasks that only need to be performed periodically, companies may also need to process data in real time or near real time, which brings us to the topic of streaming data processing.

## Streaming data processing

Streaming data processing, as the name implies, involves working on a stream of data that continuously flows into a system, usually on an ongoing basis. Rather than processing an enormous dataset as a single job, the data in a streaming processing use case usually consists of small pieces of data that are processed in flight. Examples of streaming data processing include the following:

- Analyzing social media feeds to identify trends or detect events (for example, **sentiment analysis** (**SA**), hashtag tracking)
- Monitoring and analyzing IoT sensor data in real time to detect anomalies, trigger alerts, or optimize processes
- Processing financial transactions in real time for fraud detection and prevention

Now that we've discussed the two main high-level categories of data processing use cases, let's start diving into how we actually implement these use cases.

## Data pipelines

Data pipelines are how we automate the concepts we just described, such as large-scale ETL/ELT or streaming data transformations. Data pipelines are essential for organizations that handle large volumes of data or require complex data processing workloads, as they help to streamline data management at scale. Without such automation pipelines, employees would spend a lot of time performing mundane and repetitive but complex and error-prone data processing activities. There are multiple Google Cloud services that can be used for batch and streaming data pipelines, which you will learn to use in this chapter.

Now that we've covered some of the fundamental concepts, it's time to start diving into some practical data processing activities.

Before we start to build automated data processing pipelines, however, we will first need to explore our data so that we can understand what kinds of transformations we would want to implement in our pipelines.

> **Note**
>
> While, in previous chapters, we included the code directly in the pages of this book, we are now advancing to more complex use cases that require a lot of code that would not be suitable to include directly in the pages of this book. Please review the code artifacts in the GitHub repository related to this chapter to understand the code we are using to implement these steps: `https://github.com/PacktPublishing/Google-Machine-Learning-for-Solutions-Architects`.
>
> Throughout the rest of the book, I will continue to include code directly where it makes sense.

# Exploring, visualizing, and preparing data

*Use case*: We're planning a trip to New York City, and we want to get an idea of what the best accommodation options would be. Rather than perusing through and evaluating lots of individual Airbnb postings, we're going to download lots of the reviews and do some bulk data analysis and data processing to get some insights.

We can use the Vertex AI Workbench notebook that we created in *Chapter 5* for this purpose. Please open JupyterLab on that notebook instance. In the directory explorer on the left side of the screen, navigate to the `Chapter-6` directory and open the `Chapter-6-Airbnb.ipynb` notebook. You can choose **Python (Local)** as the kernel. As you did in *Chapter 5*, run each cell in the notebook by selecting the cell and pressing *Shift* + *Enter* on your keyboard.

In the notebook, we use markdown cells to describe each step in detail so that you can understand each step in the process. We use libraries such as `pandas`, `matplotlib`, and `seaborn` to summarize and visualize the contents of our dataset, and then we perform data cleaning and preparation activities such as filling in missing values, removing outliers, and removing features that are not likely to be useful for training a regression model to predict accommodation prices. *Figure 6.3* shows an example of one of our data visualization graphs, in which we view the range and distribution of prices for the listings in our dataset:

```
plt.figure(figsize=(10, 5))
sns.histplot(data['price'], bins=100)
plt.title('Price Distribution')
plt.xlabel('Price')
plt.ylabel('Count')
plt.xlim(1,1000)
plt.show()
```

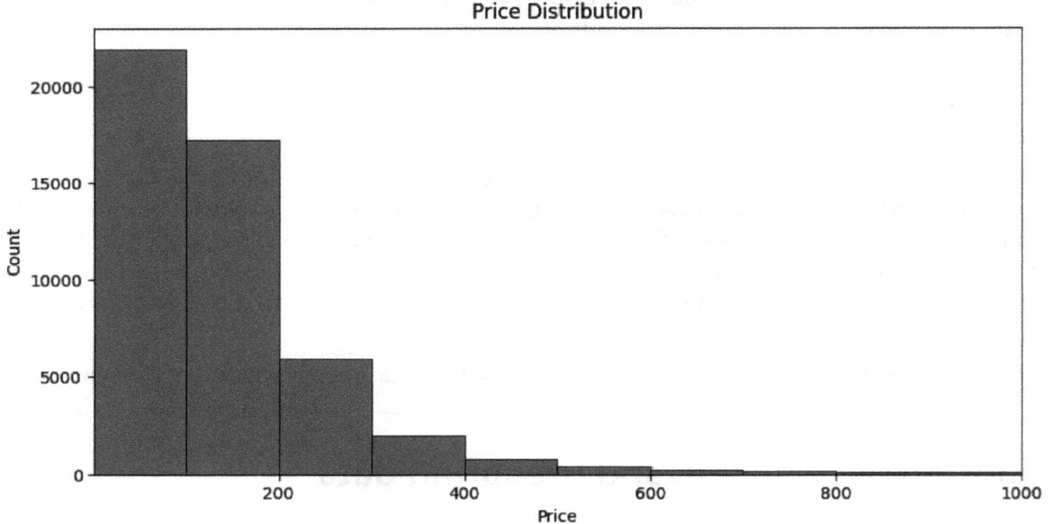


Figure 6.3: Distribution of prices in the dataset

As we can see, the majority of the accommodation options cost less than $200 per night, but there are some data points (although not many) between $600 and $1,000 per night. Those are either very expensive accommodation options or they could be potential outliers/errors in the data. You can see additional data visualization graphs in the notebook.

Regarding data cleanup activities, to clean up potential pricing outliers, for example, we use the following piece of code to set a limit of $800 (although still high) and remove any listings above that nightly rate:

```
price_range = (data_cleaned['price'] >= 10) & (
    data_cleaned['price'] <= 800)
data_cleaned = data_cleaned.loc[price_range]
```

To remove features that are not likely to be useful for training a regression model to predict accommodation prices, we use the following piece of code:

```
columns_to_drop = ['id', 'name', 'host_name', 'last_review',
    'reviews_per_month']
data_cleaned = data.drop(columns=columns_to_drop)
```

These are just a couple of examples of the data preparation steps we perform in the notebook.

When you have completed executing all of the activities in the notebook, we will move on to see how we can turn those activities into an automated pipeline in production. We will start by implementing a batch data pipeline.

# Batch data pipelines

Now that we've used our Jupyter notebook to explore our data and figure out what kinds of transformations we want to perform on our dataset, let's imagine that we want to turn this into a production workload that can run automatically on very large files, without needing any further human effort. As I mentioned earlier, this is essential in any company that implements large-scale data analytics and AI/ML workloads. It's simply not feasible to get somebody to manually perform those transformations every time, and for very large data volumes, the transformations could not be performed on a notebook instance. For example, imagine we get thousands of new postings every day, and we want to automatically prepare that data for an ML model. We can do this by creating an automated pipeline to perform data transformations every night (or however often we wish).

## Batch data pipeline concepts and tools

Before we start diving in and building our batch data processing pipeline, let's first cover some important concepts and tools in this domain.

### *Apache Spark*

Apache Spark is a highly popular development and execution framework that can be used to implement very large-scale data processing workloads and other types of large-scale computing use cases such as ML. Its power lies both in its in-memory processing capabilities and its ability to implement multiple large computing and data processing tasks in parallel.

While Spark can be used for both batch and streaming (or micro-batching) data processing workloads, we're going to use it to perform our batch data transformations in this chapter.

### *Google Cloud Dataproc*

As we discussed in *Chapter 3*, Google Cloud Dataproc is a fully managed, fast, and easy-to-use service for running Apache Spark and Apache Hadoop clusters on GCP. In this chapter, we will use it to execute our Spark processing jobs.

### *Apache Airflow*

Apache Airflow is an open source platform used for orchestrating complex data workflows. It was created by Airbnb and later contributed to the **Apache Software Foundation** (**ASF**). Airflow is designed to help developers and data engineers create, schedule, monitor, and manage workflows, making it easier to handle tasks that depend on one another. It is commonly used in data engineering and data science projects for tasks such as ETL, ML pipelines, and data analytics, and it is widely used by organizations across various industries, making it a popular choice for managing complex data workflows.

#### Directed acyclic graphs

Airflow represents workflows as **directed acyclic graphs** (**DAGs**), which consist of tasks and their dependencies. Each task in the workflow is represented as a node, and the dependencies between tasks are represented as directed edges. This structure ensures that tasks are executed in a specific order without creating loops, as depicted in *Figure 6.4*:

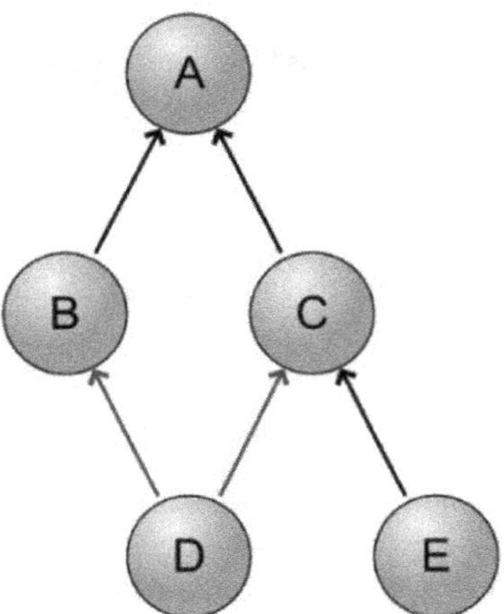


Figure 6.4: A simple DAG
(excerpted from source: https://www.flickr.com/photos/dullhunk/4647369097)

In *Figure 6.4*, we can see that tasks *b*, *c*, *d*, and *e* all depend on task *a*. Similarly, task *d* also depends on tasks *b* and *c*, and task *e* also depends on tasks *c* and *d*.

In Airflow, a DAG is defined in a Python script, which represents tasks and their dependencies as code.

### *Google Cloud Composer*

**Google Cloud Composer** (GCC) is a fully managed workflow orchestration service built on Apache Airflow. It allows you to create, schedule, and monitor data workflows across various Google Cloud services, as well as on-premises or multi-cloud environments.

Cloud Composer simplifies the process of setting up and managing Apache Airflow by providing an easy-to-use interface and automating the infrastructure management. This allows you to focus on creating and maintaining your workflows while Google takes care of the underlying infrastructure, scaling, and updates.

Now that we've covered the important concepts for implementing batch data pipelines, let's start building our pipeline.

# Building our batch data pipeline

In this section, we will create our Spark job and run it on Google Cloud Dataproc and will use GCC to orchestrate our job. What this means is that we can get GCC to automatically run our job every day. Each time it runs our job, it will create a Dataproc cluster, execute our Spark job, and then delete the Dataproc cluster when our job completes. This is a standard best practice that companies use to save money because you should not have computing resources running when you are not using them. The architecture of our pipeline on Google Cloud is shown in *Figure 6.5*:

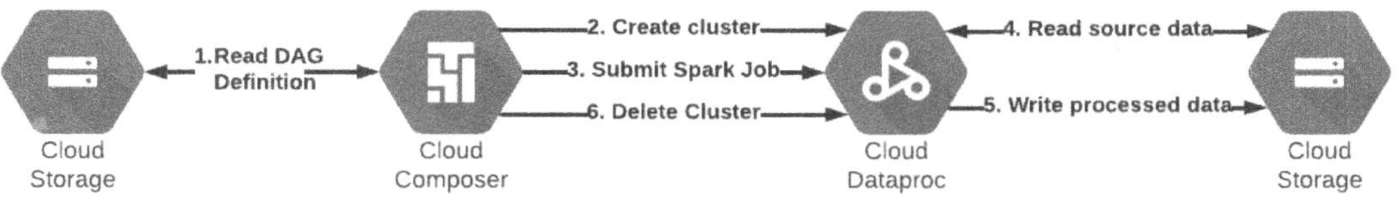


Figure 6.5: Batch data pipeline architecture

Let's begin by setting up Cloud Composer.

## Cloud Composer

In this section, we will set up Cloud Composer to schedule and run our batch data processing pipeline.

### *Cloud Composer environment*

Everything we do in Cloud Composer happens within a Cloud Composer environment. To set up our Cloud Composer environment, perform the following steps:

1. In the Google Cloud console, navigate to the **Google Cloud services** menu → **Composer** → **Environments**.
2. Select **Create Environment**.
3. If prompted, select **Composer 2**.
4. On the screen that appears, enter a name for your Composer environment. See *Figure 6.6* for reference:

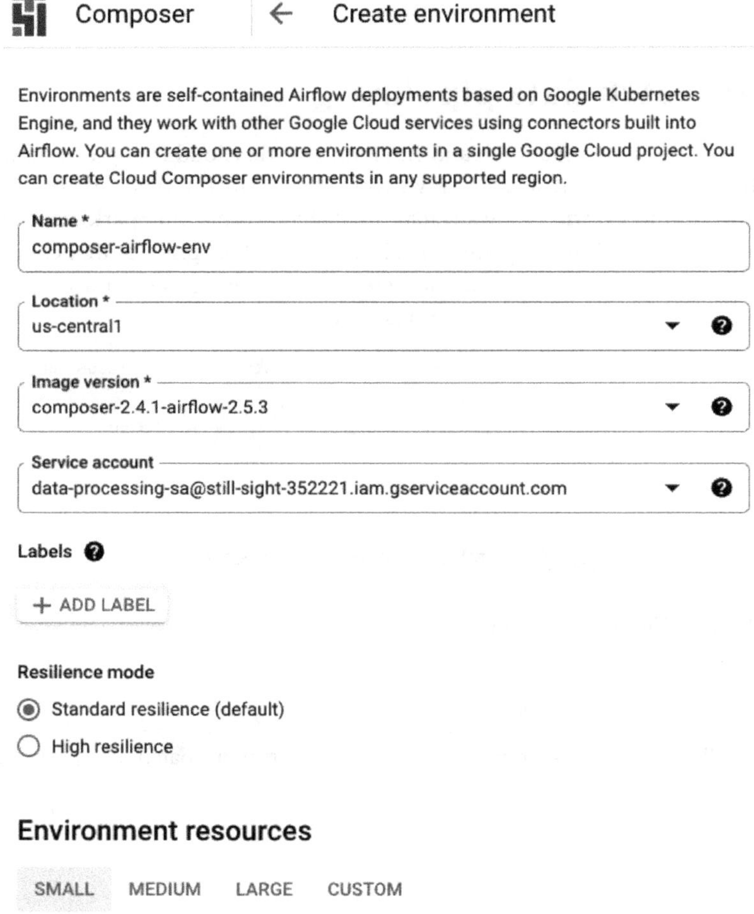


Figure 6.6: Creating a Composer environment

5. Select your preferred region. (Remember that it's better if you use the same region for each activity throughout this book, if possible.)
6. Select the latest image version. See *Figure 6.6* for reference.
7. *IMPORTANT*: Select the service account you created earlier in this chapter (if you used the suggested name, then it will include `data-processing-sa` in the name).
8. In the **Environment resources** section, select a small environment.
9. Leave all other options at their default values and select **Create**.
10. The environment will take up to 25 minutes to spin up.

While waiting for the environment to spin up, let's move on to the next section.

### Cloud Composer Python code

In this section, we will review and prepare the Python code for our Cloud Composer Spark workload. There are two code resources that we will use with Cloud Composer, which can be found in our GitHub repository (`https://github.com/PacktPublishing/Google-Machine-Learning-for-Solutions-Architects/tree/main/Chapter-06`):

- `composer-dag.py`, which contains the Python code that defines our Cloud Composer Airflow DAG
- `chapter-6-pyspark.py`, which contains the PySpark code that defines our Spark job

Perform the following steps to start preparing those files for use with Cloud Composer:

1. Locate those files in the clone of our GitHub repository that you created on your local machine, and open them for editing.
2. In the `chapter-6-pyspark.py` file, you just need to update the storage locations for the source and destination datasets. To do that, search for the `GCS-BUCKET-NAME` string in the file and replace it with your own GCS bucket name that you created earlier.

> IMPORTANT
>
> The `GCS-BUCKET-NAME` string exists in two locations in the file (once near the beginning, and again near the end). One location specifies the source dataset, and the other location specifies the destination where our Spark job will save the processed data. Replace both occurrences of the string with your own GCS bucket name.

3. In the `composer-dag.py` file, you will see the following block of variables near the beginning of the file:

```
PROJECT_ID = "YOUR PROJECT ID"
REGION = "us-central1"
```

```
ZONE = "us-central1-a"
SERVICE_ACCOUNT_EMAIL = "data-processing-sa@YOUR-PROJECT-ID.iam.
gserviceaccount.com"
PYSPARK_URI = "gs://GCS-BUCKET-NAME/code/chapter-6-pyspark.py"
```

All of those variables need to be updated with specific values from your GCP project. The comments in the file provide additional details on the replacements.

4. In addition to making the aforementioned changes, review the contents of the code to get an understanding of how Cloud Composer will execute our jobs. The code contains comments to describe what it's doing in each section so that you can understand how it works.
5. When you have completed the preceding steps, we're ready to upload the resources to be used by Cloud Composer.
6. Cloud Composer requires the preceding code resources to be stored in GCS, but in two separate locations:
   - For `chapter-6-pyspark.py`: Upload this file in the `code` folder you created earlier in GCS. To do that, navigate into the `code` folder you created, select **Upload files**, then select the `chapter-6-pyspark.py` file from the clone of our GitHub repository that you created on your local machine (that is, the file you edited in the previous steps).
   - `composer-dag.py`: This file will be uploaded to a special folder that Cloud Composer will create. When your Cloud Composer environment is fully created, click on the name of your newly created environment in the Cloud Composer console, and your environment details screen will open. Near the top of the screen, select **OPEN DAGS FOLDER**. Then, you can click **Upload files** and select the `composer-dag.py` file from the clone of our GitHub repository that you created on your local machine (that is, the file you edited in the previous steps).

That's it! As soon as the `composer-dag.py` file is uploaded, Cloud Composer will completely automate everything for you. It will take a few minutes, but Cloud Composer will create your DAG, and you will see it appearing in the Cloud Composer console. It will then execute the DAG, meaning that it will create a Dataproc cluster, execute a Spark job to perform the data transformations we specified, and delete the Dataproc cluster when the job completes.

You can see the various tasks happening in the Composer and Dataproc consoles (give it some time in each case), and the final test will be to verify that the processed data appears in the GCS destination you specified in the PySpark code.

When you have completed all of the aforementioned steps and you no longer need your Composer environment, you can delete the environment.

### *Deleting the Cloud Composer environment*

Perform the following steps to delete the Composer environment:

1. In the Google Cloud console, navigate to the **Google Cloud services** menu → **Composer** → **Environments**.
2. Select the checkbox next to the name of your environment.
3. Select **DELETE** at the top of the screen. See *Figure 6.7* for reference:

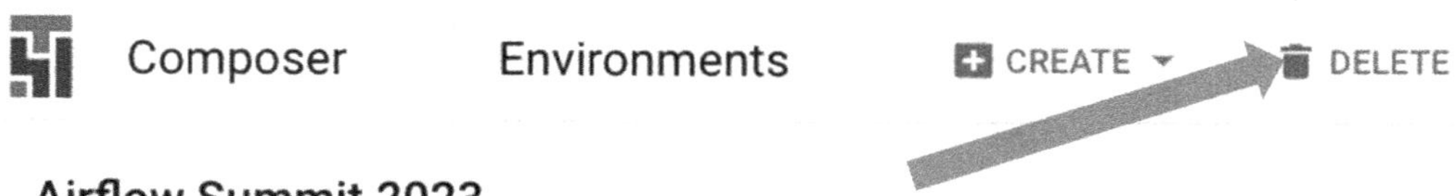


Figure 6.7: Deleting the Composer environment

4. Select **DELETE** in the confirmation screen that appears.

It will take a few minutes for the environment to delete, after which it will disappear from your list of environments.

Awesome work! You have officially created your first data processing pipeline on Google Cloud!

Note that the method we used in this chapter is an example of one (very popular) pattern for using multiple Google Cloud services to implement a data pipeline. Google Cloud also provides other products that could be used to implement similar outcomes, such as Google Cloud Data Fusion, which enables you to create pipelines using a visual user interface. We will explore other services in later chapters of this book, and one other important offering that was launched by Google Cloud in 2023 is Serverless Spark, which we'll briefly discuss next.

## Google Cloud Serverless Spark

Google Cloud Serverless Spark is a fully managed, serverless Apache Spark service that makes it easy to run Spark jobs without having to provision or manage any infrastructure. It also automatically scales your jobs up or down based on demand, so you only pay for the resources that you use, which makes it a cost-effective way to run Spark jobs, even for short-lived or intermittent workloads.

It's also integrated with other Google Cloud services, such as BigQuery, Dataflow, Dataproc, and Vertex AI, and popular open-source tools such as Zeppelin and Jupyter Notebook, making it easy to explore and analyze data and build and run **end-to-end** (E2E) data pipelines directly with those services.

With Serverless Spark on Dataproc, for example, you can select from pre-made templates to easily perform common tasks such as moving and transforming data between **Java Database Connectivity** (**JDBC**) or Apache Hive data stores and GCS or BigQuery, or you can build your own Docker containers that Serverless Spark will run in order to implement custom data processing workloads. For more information on how to develop such custom containers, see the following Google Cloud documentation: `https://cloud.google.com/dataproc-serverless/docs/guides/custom-containers`.

Now that we've learned how to build a batch data processing pipeline, we will move on to implementing streaming data pipelines.

# Streaming data pipelines

In this section, we will deal with a different kind of data source and will learn about the differences in processing data in real time versus the batch-oriented methods we used in the previous sections.

## Streaming data pipeline concepts and tools

Again, before we start to build a streaming data processing pipeline, there are some important concepts and tools that we need to introduce and understand.

### *Apache Beam*

Apache Beam is an open source, unified programming model for processing and analyzing large-scale data in batch and streaming modes. It was initially developed by Google as a part of its internal data processing tools, and later, it was donated to the ASF. Beam provides a unified way to write data processing pipelines that can be executed on various distributed processing backends such as Apache Flink, Apache Samza, Apache Spark, Google Cloud Dataflow, and others. It supports multiple programming languages, including Java, Python, and Go, and it allows developers to write both batch and streaming data processing pipelines using a single API, which simplifies the development process and enables seamless switching between batch and streaming modes. It also provides a rich set of built-in I/O connectors for various data sources and sinks, including Kafka, Hadoop, Google Cloud

Pub/Sub, BigQuery, and others. Additionally, developers can build their own custom connectors if needed. In this chapter, we will use Apache Beam on Google Cloud Dataflow to create a pipeline to process data in real time.

### Apache Beam concepts

In this section, we discuss some basic concepts of the Apache Beam programming model, which include the following:

- **Pipelines**: Just like the concept of a pipeline in Apache Airflow, which we used earlier in this chapter, an Apache Beam pipeline is a DAG representing a sequence of data processing steps or the overall data processing workflow in Apache Beam workloads.
- **PCollections**: A **Parallel Collection** (**PCollection**) is an immutable distributed dataset representing a collection of data elements. It is the primary data structure used in Apache Beam pipelines to hold and manipulate data.
- **PTransforms**: A **Parallel Transform** (**PTransform**) is a user-defined operation that takes one or more PCollections as input, processes the data, and produces one or more PCollections as output. PTransforms are the building blocks of a pipeline and define the data processing logic.
- **Windowing**: Windowing is a mechanism that allows grouping data elements in a PCollection based on timestamps or other criteria. This concept is particularly useful for processing unbounded datasets in streaming applications, where data elements need to be processed in finite windows.
- **Watermarks**: Watermarks are a way to estimate the progress of time in a streaming pipeline, and they help to determine when it's safe to emit results for a particular window.
- **Triggers**: A trigger determines when to aggregate the results of each window, based on factors such as the arrival of a certain number of data elements, the passage of a certain amount of time, or the advancement of watermarks.
- **Runners**: Runners are the components responsible for executing a Beam pipeline on a specific execution engine or distributed processing platform.

The Beam model decouples the pipeline definition from the underlying execution engine, allowing users to choose the most suitable platform for their use case.

#### Google Cloud Dataflow

We introduced Dataflow in *Chapter 3*, and now we'll dive deeper into it. Dataflow is a Google Cloud service on which you can run Apache Beam. In other words, it provides one of the execution environments or runners on which you can run Apache Beam workloads.

Dataflow provides multiple features for various types of use cases, and in this section, we'll briefly discuss the various options and their applications.

### Dataflow Data Pipelines and Dataflow Jobs

People often get confused about the difference between Dataflow Data Pipelines and Dataflow Jobs. The best way to look at this is that Dataflow Data Pipelines refer to the definition of a data pipeline, which could be executed on a recurring basis, whereas Dataflow Jobs refer to a single execution of a data pipeline. The reason this can be confusing is that you can create a new pipeline definition either in the Dataflow Jobs console or in the Dataflow Data Pipelines console.

In either case, when creating a pipeline definition, we have the option to use predefined templates that cover common types of tasks that people often want to do with Dataflow, such as transferring data from BigQuery to Bigtable, or from Cloud Spanner to Pub/Sub, and there are lots of different templates to choose from, covering a wide variety of data sources and destinations. These templates make it very easy for us to implement a data transfer workload without requiring much or any development effort on our part. Alternatively, if we have more complex data processing needs that are not included in one of the standard templates, then we can create our own custom pipeline definitions. We will look at both options later in this chapter.

### Dataflow Workbench notebooks

One way in which we can develop custom data processing pipelines is by using Dataflow Workbench notebooks. This may sound somewhat familiar, because you may remember creating and using a Vertex AI Workbench notebook in *Chapter 5*. In the Dataflow Workbench console, we can create notebooks that come with Apache Beam already installed. We will create a notebook in the Dataflow Workbench console later in this chapter.

### Dataflow snapshots

Dataflow snapshots save the state of a streaming pipeline, which allows you to start a new version of your Dataflow job without losing state. This is useful for backup and recovery, testing, and rolling back updates to streaming pipelines.

### SQL Workspace

The Dataflow console also includes a built-in SQL Workspace that enables you to run SQL queries directly from the console and send the results to BigQuery or Pub/Sub. This can be useful for ad hoc use cases in which you want to simply run a SQL query to fetch information from a given source and store the results in one of the supported destinations.

Now that we've covered the important concepts for implementing streaming data pipelines, let's start building our pipeline.

# Building our streaming data pipeline

Sticking with the theme of planning our trip to New York City, the activities in the previous sections of this chapter gave us some good insights into what kinds of accommodation options are available to us, and now we want to assess our transportation options; specifically, how much it's likely to cost us to travel around in taxis while we're there.

Our streaming data pipeline will take input data from Google Cloud Pub/Sub, perform some processing in Dataflow, and place the outputs into BigQuery for analysis. The architecture of our pipeline on Google Cloud is shown in *Figure 6.8*:

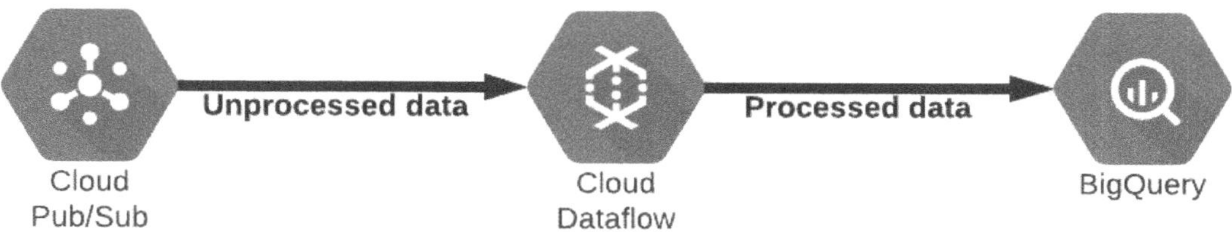


Figure 6.8: Streaming data pipeline

Google Cloud provides a public stream of data that can be used to test these kinds of stream processing workloads, which contains information relating to New York City taxi rides, and which we will use in our example in this section. Let's start by creating a destination for our streamed data, also referred to as a **sink** for our pipeline.

## Creating a BigQuery dataset

We will use Google Cloud BigQuery as a storage system for our data. To get started, we first need to define a dataset in BigQuery that we will use as our pipeline destination. To do this, perform the following steps:

1. In the Google Cloud console, navigate to the **Google Cloud services** menu → **BigQuery**.
2. In the top-left corner of the screen, you will see your project name. Click on the symbol of three vertical dots to the right of your project name (see *Figure 6.9* for reference):

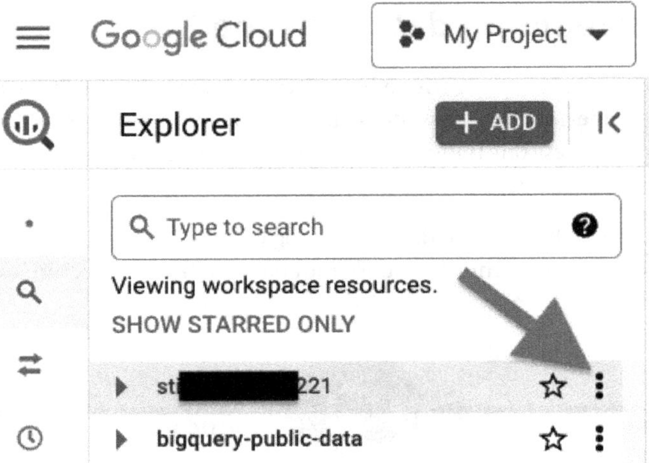


Figure 6.9: BigQuery project menu

3. In the menu that gets displayed, select **Create dataset**.
4. Give your dataset a name: `taxirides`.
5. Select your preferred region, and select **Create dataset**.

Now that we've created our dataset, we will need to create a table within that dataset.

## Creating a BigQuery table

A BigQuery dataset generally contains one or more tables that contain the actual data. Let's create a table into which our data will be streamed. To do this, perform the following steps:

1. In the BigQuery editor, click on the `taxirides` dataset you just created and select **Create table**.
2. In the **Table** field, enter `realtime` as the table name.
3. In the **Schema** section, click the plus sign (+) to add a new field. Our first field has the following properties (leave all other options for each field at their default values):

| **Field name** | **Type** | **Mode** |
|---|---|---|
| `ride_id` | `STRING` | `NULLABLE` |

Table 6.1: Properties for the first field in the BigQuery table

4. Repeat *step 3* to add more fields, until the schema looks like that shown in *Figure 6.10*:

| Field name | Type | Mode |
|---|---|---|
| ride_id | STRING | NULLABLE |
| point_idx | INTEGER | NULLABLE |
| latitude | FLOAT | NULLABLE |
| longitude | FLOAT | NULLABLE |
| timestamp | TIMESTAMP | NULLABLE |
| meter_reading | FLOAT | NULLABLE |
| meter_increment | FLOAT | NULLABLE |
| ride_status | STRING | NULLABLE |
| passenger_count | INTEGER | NULLABLE |

Figure 6.10: Table schema

5. Select **Create table**.

Now our table is ready for us to stream data into it, let's move on to creating our data streaming pipeline.

## Creating a Dataflow job from a template

We're going to use a Dataflow template to create our first data streaming pipeline. To do this, perform the following steps:

1. In the Google Cloud console, navigate to the **Google Cloud services** menu → **Dataflow** → **Jobs**.
2. Select **Create job from template**.
3. For **Job name**, enter `taxi-data-raw`.
4. Select your preferred region.
5. In the drop-down menu for **Dataflow template**, select the **Pub/Sub to BigQuery** template.

> **Note**
>
> Pub/Sub has also released a direct integration with BigQuery, but considering that we want to illustrate how to use a Dataflow template, we're using the Dataflow connection method in this chapter.

6. Next, in the **Input Pub/Sub topic** field, select **Enter topic manually**.
7. Enter the following topic:

```
projects/pubsub-public-data/topics/taxirides-realtime
```

8. In the **BigQuery output table** field, click **BROWSE**, and then select the `realtime` table you created in the previous section.
9. Click **SELECT** at the bottom of the screen.
10. In the **Temporary location** field, enter your desired storage location path in the following format (replace [BUCKET-NAME] with your bucket name): gs://[BUCKET-NAME]/dataflow.
11. Expand the **Optional Parameters** section and scroll down until you find the **Service account email** field.
12. In that field, select the service account you created earlier in this chapter (it will contain `data-processing-sa` in the name if you used the suggested name).
13. Leave all other options at their default values.
14. Select **RUN JOB**.
15. After a few minutes, you will see the job details appearing, and a graph that looks similar to the one shown in *Figure 6.11*:

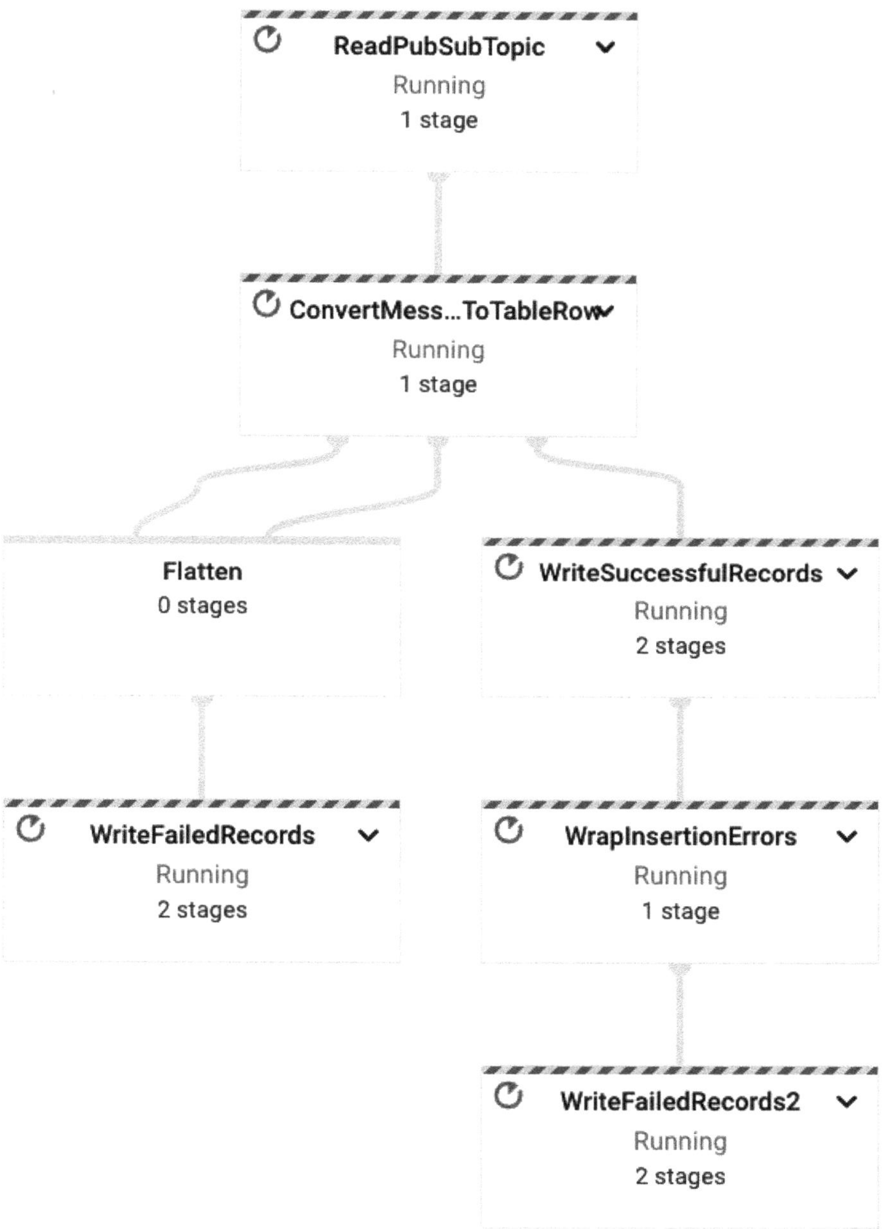


Figure 6.11: Dataflow execution graph

16. Try clicking around on each of the steps and sub-steps in the graph to get a better understanding of what each step is doing.

After some time, you can head over to the BigQuery console and verify that the data is streaming into the table. Move on to the next section in order to do that. However, don't close the Dataflow console yet because you will come back here after verifying the data in BigQuery.

## Verifying data in BigQuery

To verify the data in BigQuery, perform the following steps:

1. In the Google Cloud console, navigate to the **Google Cloud services** menu → **BigQuery**.
2. In the top-left corner of the screen, you will see your project name. Click on the arrow symbol to the left of your project name to expand it (see *Figure 6.12* for reference).
3. Then, click on the arrow symbol to the left of your dataset name (`taxirides`) to expand it.
4. Select your `realtime` table.
5. Select the **PREVIEW** tab.
6. You should then see a screen that looks like the one shown in *Figure 6.12*:

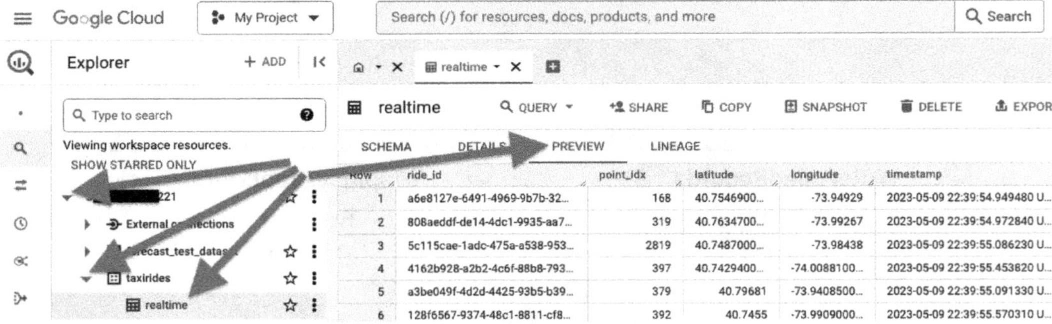


Figure 6.12: BigQuery data

7. When you have verified the data in BigQuery, you can go back to the Dataflow console and stop the Dataflow job by clicking **STOP** at the top of the screen.

Now that you've seen how easy it is to set up a Dataflow job from a template, let's move on to more complex Dataflow use cases.

## Creating a Dataflow notebook

Because we want to use a customized Apache Beam notebook, we will create a notebook in the Dataflow Workbench console. Perform the following steps to create the notebook:

1. In the Google Cloud console, navigate to the **Google Cloud services** menu → **Dataflow** → **Workbench**.
2. At the top of the screen, select the **Instances** tab.
3. Now, at the top of the screen , select **Create New**.
4. In the screen that appears (see *Figure 6.13* for reference), you can either accept the default notebook name or create a name of your preference:

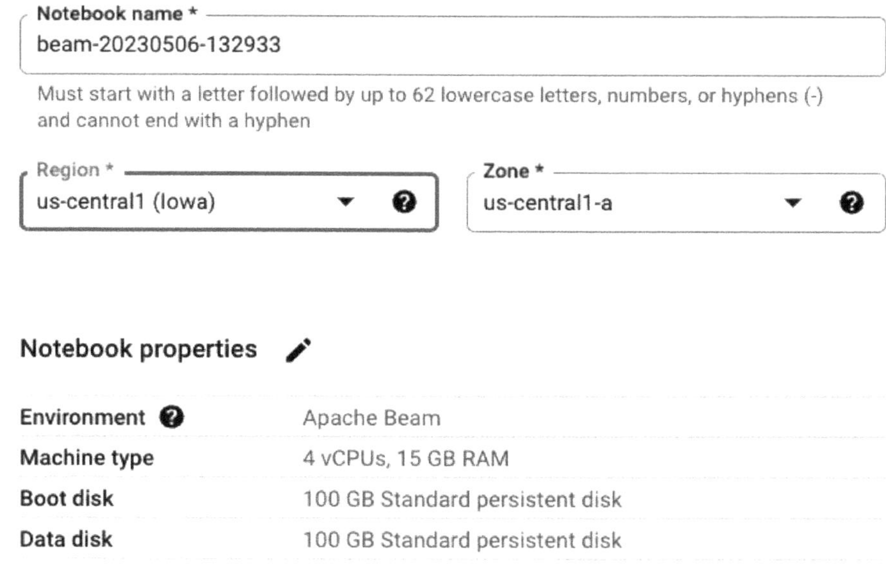


Figure 6.13: Creating a user-managed notebook

5. Select your preferred region and zone. The zone selection doesn't really matter in this case, but I recommend selecting the same region you have been using in previous activities in this book.
6. Select **Continue**, and then **Continue** again.
7. Select E2 as the compute instance type.
8. You can also configure an idle timeout period after which the machine will automatically shut down if it is idle for that amount of time. This helps to save costs.
9. Select **Continue** multiple times until you reach the **IAM** and **security** screen

10. In the **IAM and security** screen, select the **Single user** option. See *Figure 6.14* for reference:

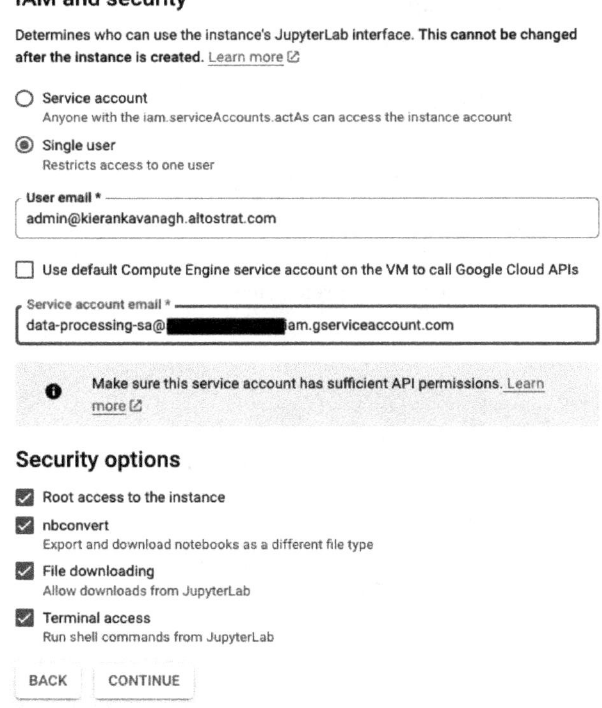


Figure 6.14: Dataflow notebook – IAM and security

> **Note**
>
> You may remember performing a similar step (that is, selecting the **Single user authentication** option) when we created our managed notebook in *Chapter 5*. This allows us to directly use the notebook without needing to authenticate it as a service account beforehand.

11. The **User email** box that appears should automatically be populated with your login email address. If not, enter the email address that you use for logging in to the Google Cloud console.
12. Also, we want our notebook instance to use the service account we created earlier in this chapter, so **uncheck** the option that says **Use default Compute Engine service account on the VM to call Google Cloud APIs**.
13. In the **Service account email** box that appears, start typing `data`, and you should see the name of the service account you created earlier in this chapter appearing in the list of available service accounts (assuming that you named it `data-processing-sa`, as recommended in that section).

14. Select that service account in the list.
15. Also, take note of the **Pricing summary** option in the top-right corner of the screen. This is an estimate of how much it would cost if you were to leave the notebook running all month. Fortunately, you will only use it for a short time in this chapter. If you did not configure an idle shutdown period when creating the notebook, remember to shut it down when you're finished using it.
16. You can leave all other options at their default values and select **Create** at the bottom of the screen.
17. It will take a few minutes for the notebook instance to be created. When the instance creation has completed, you will see it in the list of user-managed notebooks, and an option to open JupyterLab will appear.
18. Select **Open Jupyterlab**.
19. When the JupyterLab screen opens, it's time to clone our repository into your notebook. This is similar to the process you performed in *Chapter 5*, but you have created a separate notebook instance in this chapter, so we need to clone the repository into this instance. The steps, again, to clone the repository, are as follows.
20. Click on the **Git** symbol in the menu on the left of the screen. The symbol will look like the one shown in *Figure 6.15*:

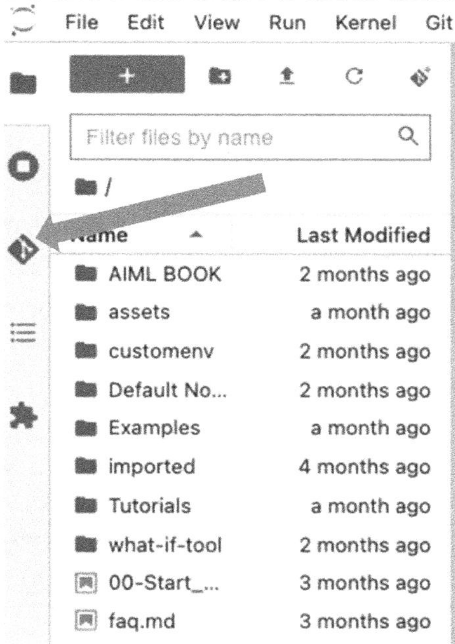


Figure 6.15: Git symbol

21. Select **Clone Repository**.
22. Enter our repository URL: `https://github.com/PacktPublishing/Google-Machine-Learning-for-Solutions-Architects`.
23. If any options are displayed, leave them at their default values.
24. Select **Clone**.
25. You should see a new folder appear in your notebook, named `Google-Machine-Learning-for-Solutions-Architects`.
26. Double-click on that folder, double-click on the `Chapter-06` folder within it, and then double-click on the `Streaming_NYC_Taxi_Data.ipynb` file to open it.
27. In the **Select Kernel** screen that appears, select the latest version of Apache Beam. At the time of writing this, the latest available option in the launcher is Apache Beam 2.4.6 (see *Figure 6.16* for reference):

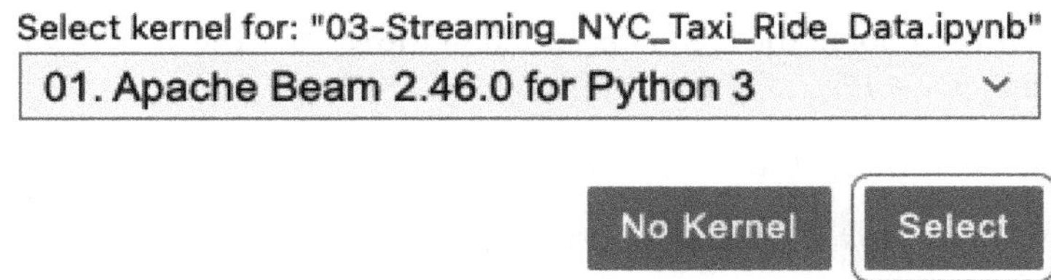


Figure 6.16: Selecting a notebook kernel

28. The notebook we have opened contains a lot of Apache Beam Python code that we can use to process the data that's streaming in from the public Pub/Sub topic.
29. Run each of the cells in the notebook, and read the explanations in the markdown and comments to understand what we're doing. We're using Apache Beam to define a streaming data processing pipeline that we will run in Google Cloud Dataflow.

When you have completed running the cells in the notebook, you can go to the Dataflow Jobs console, and you will see your new pipeline running there. Give it a few minutes for the pipeline to start up and for data to flow through the pipeline. Just like before, I recommend that you click on various parts of the pipeline execution graph in order to get a better understanding of how the pipeline is structured. Next, let's verify this new pipeline data in BigQuery, but again, don't close the Dataflow console yet, because you will come back here after verifying the data in BigQuery.

## Verifying data in BigQuery

To verify the data in BigQuery, open the BigQuery console, and under the `taxirides` dataset, you will see a new table that has been created by our custom Dataflow pipeline, called `run_rates`. Click on the **PREVIEW** tab again to see the data that is streaming in. As you will see, in this case, we just have a single column in our database, which contains the aggregated `run_rates` values that were computed by our pipeline.

When you have verified the data in BigQuery, you can go back to the Dataflow console and stop the Dataflow job by clicking **STOP** at the top of the screen.

Now that we have completed the activities in this section, you can shut down your user-managed notebook by performing the following steps:

1. In the Google Cloud console, navigate to the **Google Cloud services** menu → **Dataflow** → **Workbench**.
2. At the top of the screen, select the **User-managed notebooks** tab.
3. Select the checkbox next to your notebook name, and click **STOP** at the top of the screen (just above the **User-managed notebooks** tab).

The notebook will shut down after a few minutes.

If everything worked as expected, you have now successfully created a custom pipeline that processes and transforms streaming data in flight!

# Summary

In this chapter, you learned how to ingest data into Google Cloud from various sources, and you discovered important concepts on how to process data in Google Cloud.

You then learned about exploring and visualizing data using Vertex AI and BigQuery. Next, you learned how to clean and prepare data for ML workloads using Jupyter notebooks, and then how to create an automated data pipeline to perform the same transformations at a production scale in a batch method using Apache Spark on Google Cloud Dataproc, as well as how to automatically orchestrate that entire process using Apache Airflow in GCC.

We then covered important concepts and tools related to processing streaming data, and you finally built your own streaming data processing pipelines using Apache Beam on Google Cloud Dataflow.

In the next chapter, we will spend additional time on data processing and preparation, with a specific focus on the concept of feature engineering.

# 7
# Feature Engineering and Dimensionality Reduction

In this chapter, we will dive progressively deeper into the kinds of data processing steps that are common in many data science projects, and how to perform those steps using Vertex AI in Google Cloud. We'll begin this chapter by taking a more detailed look at how features are used in machine learning workloads, and what kinds of challenges often arise concerning how features are used.

We will then transition our discussion to focus on how to address those challenges, and how to use our machine learning features effectively in Google Cloud.

This chapter covers the following topics:

- Fundamental concepts related to dimensions or features in machine learning
- An introduction to the curse of dimensionality
- Dimensionality reduction
- Feature engineering
- Vertex AI Feature Store

## Fundamental concepts in this chapter

In this section, we'll briefly cover concepts that provide additional context for this chapter's learning activities.

## Dimensions and features

We introduced the concept of features in *Chapter 1* and described examples of features using the King County housing sales dataset for illustration. To briefly recap, features are individual, measurable properties or characteristics of the observations in our dataset. They are the aspects of our dataset from which a machine learning algorithm learns to create a model. In other words, a model can be seen as a representation of patterns learned by the algorithm from the features in our dataset.

The features of a house, for example, include information such as how many rooms it contains, the year it was constructed, where it is located, and other factors that describe the house, as depicted in *Table 7.1*:

| price | bedrooms | bathrooms | sqft_living | sqft_lot | floors | waterfront | view | condition | grade | yr_built | yr_renovated |
|---|---|---|---|---|---|---|---|---|---|---|---|
| 221900 | 3 | 1 | 1180 | 5650 | 1 | 0 | 0 | 3 | 7 | 1955 | 0 |
| 538000 | 3 | 2.25 | 2570 | 7242 | 2 | 1 | 0 | 3 | 7 | 1951 | 1991 |
| 257500 | 3 | 2.25 | 1715 | 6819 | 2 | 0 | 0 | 3 | 7 | 1995 | 0 |
| 291850 | 3 | 1.5 | 1060 | 9711 | 1 | 0 | 0 | 3 | 7 | 1963 | 0 |
| 229500 | 3 | 1 | 1780 | 7470 | 1 | 0 | 0 | 3 | 7 | 1960 | 0 |
| 323000 | 3 | 2.5 | 1890 | 6560 | 2 | 0 | 0 | 3 | 7 | 2003 | 0 |
| 650000 | 4 | 3 | 2950 | 5000 | 2 | 0 | 3 | 3 | 9 | 1979 | 0 |

Table 7.1: King County house sales features

When we're dealing with tabular data, features are generally represented as columns in our dataset, and each row represents an individual data point or observation, sometimes referred to as an **instance**.

Features are also referred to as variables, attributes, or dimensions. So, when we talk about the dimensionality of a dataset, it relates to how many features or dimensions our dataset has, and how that affects our machine learning workloads.

## Overfitting, underfitting, and regularization

We briefly discussed the concepts of overfitting and underfitting in *Chapter 2*, and we will continue to revisit these topics in more detail throughout this book since they are so fundamentally important to the process of machine learning. In this section, we'll discuss how the number of features in our dataset can affect how our algorithms learn from our data. A key concept to keep in mind is that overfitting and underfitting can be strongly influenced by how many observations we have in our dataset, and how many features we have for each observation.

We usually need to find the right balance between these two aspects of our dataset. For example, if we have very few observations and a lot of features for each observation, our model is likely to overfit the dataset because it learns very specific patterns for those observations and their features, but it cannot generalize well to new observations. Conversely, if we have many observations, but very few pieces of information (that is, features) for each observation, then our model may not be able to learn any valuable patterns, meaning it will underfit our dataset. Because of this, reducing the number of features can help reduce overfitting, but only to an extent – removing too many features may result in underfitting. Also, we don't want to remove features that contain useful information for our models to learn, so another way that we can address overfitting while keeping many of our features is to use a mechanism called regularization. We also briefly mentioned this in *Chapter 2*, and it's something we will discuss in more detail here.

### *Regularization*

To begin our discussion on regularization, we need to bring up the concept of the loss function in machine learning once more, something we introduced in *Chapter 1*. Remember that many machine learning algorithms work by trying to find the best coefficients (or weights) for each feature that result in the closest approximation of the target feature. So, overfitting is influenced by the mathematical relationship between the features and their coefficients. If we find that a model is overfitting to specific features, we can use regularization to reduce the influence of those features and their coefficients on the model.

Because overfitting usually happens when a model is too complex, such as having too many features relative to the number of observations, regularization addresses this issue by adding a penalty to the loss function, which discourages the model from assigning too much importance to any feature. This helps improve the generalizability of the model.

There are several ways to implement regularization in machine learning, but I'll explain two of the most common types here – that is, **L1** and **L2** regularization – as well as a combination of both approaches, referred to as **elastic net**.

#### L1 regularization

This type of regularization is also known as **Lasso** regularization, and it works by adding a penalty equivalent to the L1 norm (or the absolute value) of the coefficients or weights in the cost function by using the following formula:

$$Cost\ Function + \lambda * |weights|$$

Here, λ is the regularization parameter, which controls the strength of the penalty and can be considered a hyperparameter whose optimal value can vary depending on the problem. Note that if the penalty is too strong, it can result in underfitting, so it's important to find the right balance in this regard.

The effect of L1 regularization is to shrink some of the model's coefficients to exactly zero, effectively excluding the corresponding feature from the model, which makes L1 regularization useful for feature selection (something we'll cover in more detail shortly) when dealing with high-dimensional data.

### L2 regularization

This type of regularization is also known as **Ridge** regularization. This method adds a penalty equivalent to the L2 norm (or the square) of the coefficients in the cost function by using the following formula:

$$Cost\ Function + \lambda * (weights \wedge 2)$$

Unlike L1 regularization, L2 regularization doesn't result in the exclusion of features but rather pushes the coefficients close to zero, distributing the weights evenly among the features. This can be beneficial when we're dealing with correlated features as it allows the model to keep all of them under consideration.

### Elastic net

Elastic net, as a combination of L1 and L2 regularization, was designed to provide a compromise between these two methods, incorporating the strengths of both. Like L1 and L2 regularization, elastic net adds a penalty to the loss function, but instead of adding an L1 penalty or an L2 penalty, it adds a weighted sum of both by using the following formula:

$$Cost\ Function + \lambda 1 * |weights| + \lambda 2 * (weights \wedge 2)$$

Here, λ1 and λ2 are hyperparameters that control the strength of the L1 and L2 penalties, respectively. If λ1 is zero, elastic net reduces to Ridge regression, and if λ2 is zero, it reduces to Lasso regression.

Elastic net has the feature selection capability of L1 regularization (since it can shrink coefficients to zero) and the regularization strength of L2 regularization (since it can distribute weights evenly among correlated features). The trade-off with elastic net is that it has two hyperparameters to tune, rather than just one in the case of Lasso or Ridge, which can make the model more complex and computationally intensive to train.

Now that we've covered the important topic of regularization in more detail, let's dive into feature selection and feature engineering.

## Feature selection and feature engineering

We briefly discussed feature engineering in previous chapters of this book, but we will explore these concepts in more detail here. In *Chapter 2*, we used the example of creating a new feature named `price-per-square-foot` in our housing data by dividing the total cost of each house by the total area of that house in square feet. We will explore many additional examples of feature engineering in this chapter.

However, creating new features from existing ones is not the only type of activity we need to perform on our features when preparing our dataset for training a machine learning model. We also need to select which features we think could be most important for achieving the task that we want our model to achieve, such as predicting the price of a house. As we saw in *Chapter 6*, we may also need to perform transformations on our features, such as ensuring that they are all represented on a common, standardized scale.

The goal in selecting and engineering features is to provide our model with the most relevant information, in the most digestible format, so that it can most effectively learn from the data.

## The curse of dimensionality

High-dimensional datasets contain many dimensions or features for each observation in the dataset. It would be possible to presume that, the more features we include for each data point, the more information our model will learn, and therefore, the more accurate our model will be. However, note that not all features are equally useful. Some may contain little to no useful information for our model, others may contain redundant information, and some may even be harmful to the model's ability to learn. Part of the art and science of machine learning is figuring out which features to use and how to prepare them in a way that allows the model to perform its best.

Also, bear in mind that, the more information we have in our dataset, the more information our model has to process. This directly translates to additional computing resources being required for our machine learning algorithm to process our dataset, which, in turn, directly translates to longer model training times, and increased monetary cost. Too much irrelevant data or **noise** in the dataset can also make it harder for the algorithm to identify (that is, to learn) patterns in the data. The ideal scenario, then, is to find the minimum number of features that provide the maximum amount of useful information to our model. If we can achieve the same results with three features or ten features, for example, it's generally better to go with the option of using three features. The "maximum amount of useful information" can be measured in terms of **variance**, whereby features with high relative variance are those that influence our model's outcomes most prominently, and features with little relative variance are often not as useful in training our model to identify patterns.

The "curse of dimensionality" is a term that's used in the data science industry to describe the challenges that arise when dealing with datasets that contain higher numbers of dimensions. Let's take a look at what some of these challenges are. In subsequent sections, we will discuss mechanisms to address these challenges.

### Data exploration challenges

This is an important point at which we can introduce the link between "dimensions" in our dataset, and the dimensions of physical space. As we all know, and as mentioned earlier in this book, humans can only perceive our physical world in up to a maximum of three dimensions (width, height, and depth), with "time" considered as the fourth dimension of our physical reality. For datasets that have two or three dimensions, we can easily create visualizations representing various aspects of those datasets, but we can't create graphs or other visual representations of higher-dimensional datasets. In such cases, it helps if we can try to find other ways of visually interpreting those datasets, such as by projecting them down into lower-dimensional representations, something we will explore in more detail shortly.

### Feature sparsity

In high-dimensional space, points (that is, instances or samples) in our dataset tend to be far away from each other, leading to sparsity. Generally, as the number of features increases relative to the number of observations in our dataset, the feature space becomes increasingly sparse, and this sparsity makes it more difficult for algorithms to learn from the data since they have fewer examples to learn from in the vicinity of any given point.

As an example, let's imagine that our dataset consists of information regarding a company's customers, in which case each instance in the dataset represents a person, and each feature represents some characteristic of a person. If our dataset stores hundreds or even thousands of characteristics, then it's unlikely that all characteristics for every person will be populated. As such, the overall feature space for our dataset will be sparsely populated. On the other hand, if we have fewer features, this is less likely to occur.

### Distance measurements

In previous chapters, we talked about how Euclidean distance is used in many machine learning algorithms to find potential relationships or differences between data points in our dataset. One of the most direct examples of this concept that we've explored already is the K-means clustering algorithm. In high-dimensional spaces, distances can sometimes become less meaningful, because the difference between the maximum and minimum possible distances becomes increasingly smaller. This means that traditional distance metrics, such as Euclidean distance, also become less meaningful, which can particularly affect algorithms that rely on distance, such as **k-nearest neighbors** (**kNN**) or clustering algorithms.

### Overfitting and increased data requirements

Referring back to our discussion of how overfitting or underfitting are influenced by the ratio of observations to features in our dataset, high-dimensional datasets are more prone to overfitting unless we have enormous amounts of observations to help our model generalize. Bear in mind what we said earlier in this chapter, regarding the relationship between the amount of data our algorithms need to process and the cost of training our models. Datasets with lots of observations and lots of features will be more expensive to train and manage.

### *Interpretability and explainability*

Interpretability and explainability refer to our ability to understand and explain how our machine learning models work. This is important for numerous reasons, all of which we will discuss briefly here. Firstly, a lack of understanding of how our models work inhibits our ability to improve those models. However, more importantly, we need to ensure that our models are as fair and unbiased as possible, and this is where explainability plays a crucial role. If we cannot explain why our models are producing specific results, then we cannot adequately assess their fairness. Generally, the higher the dimensionality of our dataset, the more complex our models tend to be, which can directly affect (that is, reduce) interpretability and explainability.

Now that we've reviewed some of the common challenges associated with high-dimensional datasets, let's take a look at some mechanisms to address those challenges.

## Dimensionality reduction

As you might imagine, one of the first ways to address the challenge of having too many dimensions is to reduce the number of dimensions, and there are two main types of techniques we can use for this purpose: feature selection and feature projection.

### *Feature selection*

This involves selecting a subset of the original features. There are several strategies for feature selection, including the following:

- Filter methods, which rank features based on statistical measures and select a subset of features with the highest ranking
- Wrapper methods, which evaluate multiple models using different subsets of input features and select the subset that results in the highest model performance
- Embedded methods, which use machine learning algorithms that have built-in feature selection methods (such as Lasso regularization)

It's also important to understand that the feature projection methods we will discuss can be used to help select the most important subset of features from our dataset, so there is some overlap between these concepts.

### *Feature projection*

With feature projection, we use mathematical transformations to project our features down into a lower-dimensional space. In this section, we'll introduce three popular feature projection techniques: **Principal Component Analysis** (**PCA**), **Linear Discriminant Analysis** (**LDA**), and **t-distributed Stochastic Neighbor Embedding** (**t-SNE**).

## PCA

PCA is an **unsupervised** algorithm that aims to reduce the dimensionality of our feature space while maintaining as much of the original data's variance as possible. In the high-dimensional data space, PCA identifies the axes (principal components) along which the variation in the data is maximized. These principal components are orthogonal, meaning they're at right angles to each other in this multi-dimensional space. The **first principal component** (**PC1**) captures the direction of the greatest variance in the data. The **second principal component** (**PC2**) captures the maximum amount of remaining variance while being orthogonal to PC1, and so on. The process of PCA generally involves the following steps:

1. Standardize the data if the features have different scales. This is important because PCA is sensitive to feature scales, whereby features with larger scales could mistakenly be perceived as more dominant.
2. Calculate a covariance matrix to understand how different features vary together. The covariance matrix is a square matrix that contains the covariances between each pair of features. The covariance between two features measures how those features vary together: a positive covariance indicates that the features increase or decrease together, while a negative covariance indicates that one feature increases while the other decreases.
3. Calculate the **eigenvalues** and **eigenvectors** of the covariance matrix. The eigenvectors represent the directions or components of the new space, and the eigenvalues represent the magnitude or explained variance for each component. The eigenvectors are often called the principal components of the data, and they form a basis for the new feature space.
4. Sort the eigenvalues and their corresponding eigenvectors. After computing the eigenvalues and their associated eigenvectors, the next step is to sort the eigenvalues in descending order. The eigenvector with the highest corresponding eigenvalue is PC1. The eigenvector with the second highest corresponding eigenvalue is PC2, and so on. The reason for this ordering is that the significance of each eigenvector is given by the magnitude of its eigenvalue.
5. Select a subset of the principal components. PCA creates as many principal components as there are variables in the original dataset. However, since the goal of PCA is dimensionality reduction, we usually select a subset of the principal components, referred to as the **top k** principal components, which capture the most variance in the data. This step is what reduces dimensionality because the smaller eigenvalues and their vectors are dropped.
6. Transform the original data. The final step in PCA is to transform the original data into the reduced subspace defined by the selected principal components, which is done by multiplying the original data matrix by the matrix of the top $k$ eigenvectors.

The transformed data is now ready to be used for further analysis and visualization (as depicted in *Figure 7.1*), or used as input to a machine learning algorithm. Importantly, the reduced dataset retains as much of the variance in the original data as possible (given the reduced number of dimensions):

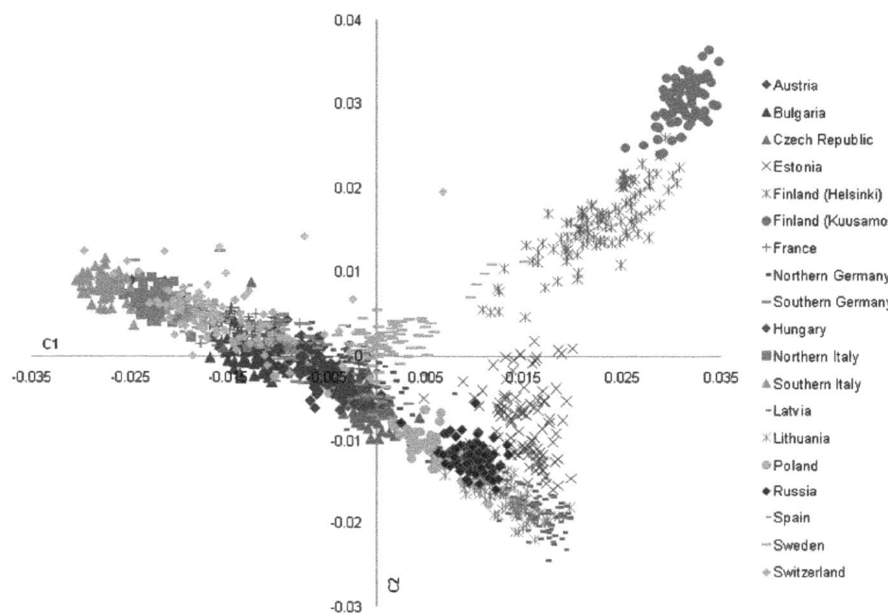


Figure 7.1: PCA visualization of European genetic structure (source: https://commons.wikimedia.org/wiki/File:PCA_plot_of_European_individuals.png)

PCA is a powerful technique with many uses, but it also has limitations. For example, it assumes that the principal components are a linear combination of the original features. If this is not the case (that is, if the underlying structure in the data is non-linear), then PCA may not be the best dimensionality reduction technique to use. It's also worth noting that the principal components are less interpretable than the original features – they don't have an intuitive meaning in terms of the original features.

## LDA

LDA is a **supervised** algorithm that aims to find a linear combination of features that best separates classes of objects. The resulting combination can then be used for dimensionality reduction. It involves the following steps:

1. Compute the class means. For each class in the dataset, calculate the mean vector, which is simply the average of all vectors in that class.
2. Compute the within-class covariance matrix, which measures how the individual classes are dispersed around their respective means.
3. Compute the between-class covariance matrix, which measures how the class means are dispersed around the overall mean of the data.

4. Compute the linear discriminants, which are the directions in the feature space along which the classes are best separated.
5. Sort the linear discriminants. Just like in PCA, the eigenvectors are sorted by their corresponding eigenvalues in descending order. The eigenvalues represent the amount of the data's variance that is accounted for by each discriminant. The first few linear discriminants, corresponding to the largest eigenvalues, are the ones that account for the most variance.
6. Finally, the data is projected onto the space spanned by the first few linear discriminants.

This results in a lower-dimensional representation of the data where the classes are maximally separated. We can visualize this as follows:

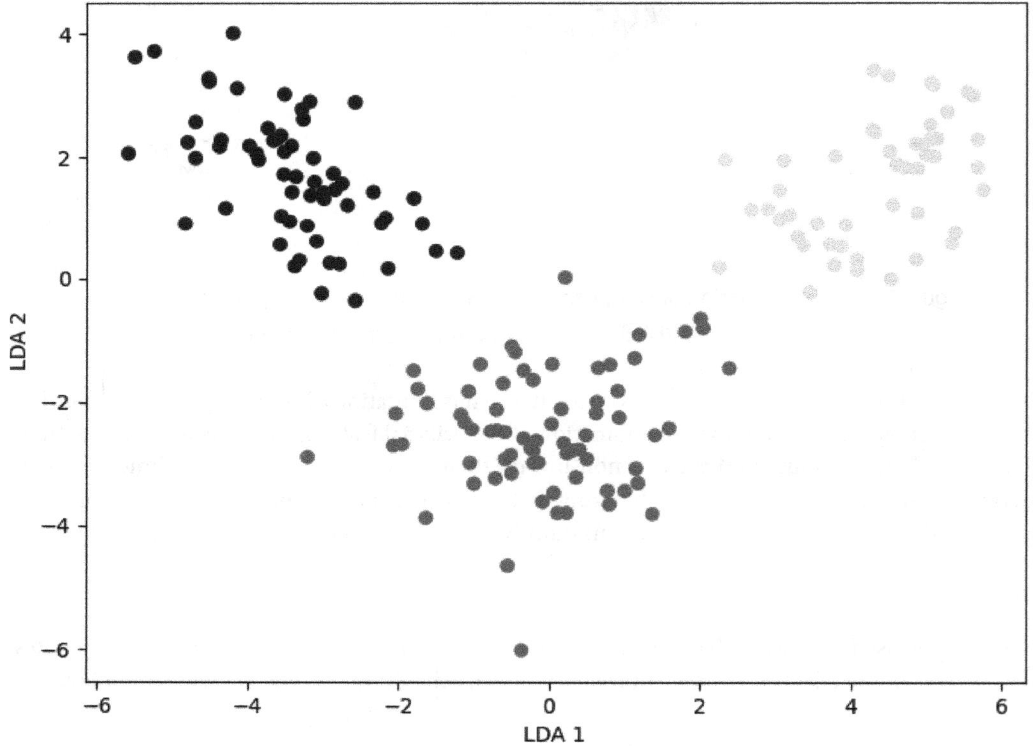


Figure 7.2: LDA plots of wine varieties

It's important to note that the main assumption of LDA is that the classes have identical covariance matrices. If this assumption is not met, LDA might not perform well.

## t-SNE

This method has perhaps the coolest name of them all. It's an **unsupervised**, non-linear dimensionality reduction algorithm that is particularly well-suited for embedding high-dimensional data into a space of two or three dimensions while aiming to keep similar instances close and dissimilar instances apart. It does this by mapping the high-dimensional data to a lower-dimensional space in a way that retains much of the relative distances between points. It involves the following steps:

1. **Compute similarities in the high-dimensional space**: t-SNE begins by calculating the probability that pairs of data points in the high-dimensional space are similar. Points that are close to each other have a higher probability of being picked, while points that are far away have a lower probability.
2. **Computing similarities in low-dimensional space**: t-SNE then calculates the probabilities of similarity for pairs of points in the low-dimensional representation.
3. **Optimization**: Finally, t-SNE uses gradient descent to minimize the difference between the probabilities in the high-dimensional and low-dimensional spaces. The goal is to have similar objects modeled by nearby points and dissimilar objects modeled by distant points in the low-dimensional space.

The result of t-SNE is a map that reveals the structure of the high-dimensional data in a way that it's easier for humans to comprehend.

It should be noted that while t-SNE is excellent for visualization and can reveal clusters and structure in your data (as depicted in *Figure 7.3*), it doesn't provide explicit information about the importance or meaning of the features in your data like PCA does. It's more of an exploratory tool that can help in dimensionality reduction, rather than a formal dimensionality reduction technique:

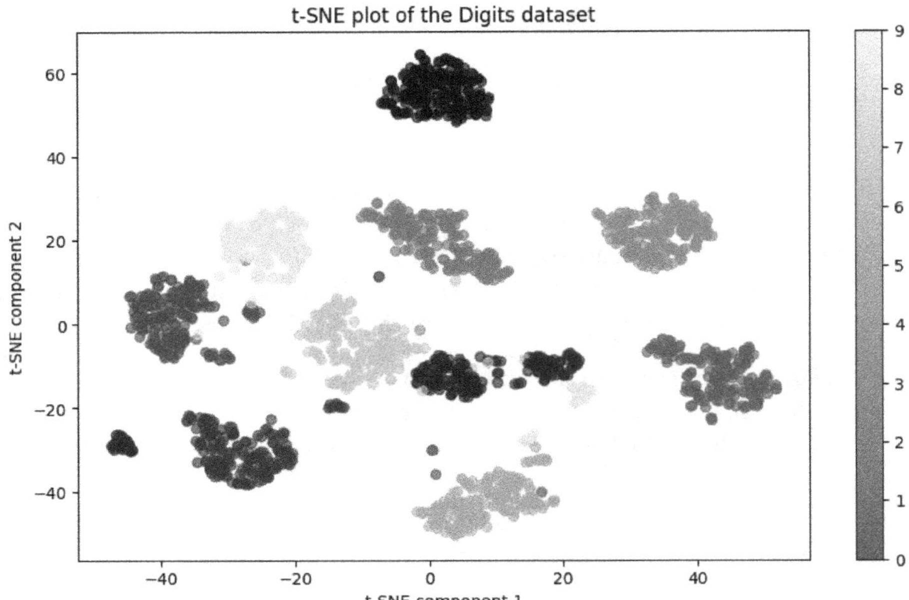


Figure 7.3: t-SNE visualization of digits dataset

Now that we've covered some of the most popular feature projection techniques, let's discuss one more set of important concepts regarding how the features in our datasets influence model training.

# Using PCA and LDA for dimensionality reduction

We'll start our hands-on activities in this chapter with dimensionality reduction using PCA and LDA. We can use the wine dataset within scikit-learn as an example. I always wish I could impress my friends by being a wine expert, but I can barely tell a $10 bottle from a $500 bottle, so instead, I'll use data science to develop impressive knowledge.

The wine dataset is an example of a multivariate dataset that contains the results of a chemical analysis of wines grown in the same region in Italy but derived from three different types of grapes (referred to as `cultivars`). The analysis focused on quantifying 13 constituents found in each of the three types of wines.

Using PCA on this dataset will help us to understand the important features. By looking at the weights of the original features in the principal components, we can see which features contribute most to the variability in the wine dataset.

Again, we can use the same Vertex AI Workbench Notebook Instance that we created in *Chapter 5* for this purpose. Please open JupyterLab on that notebook instance and perform the following steps:

1. In the directory explorer on the left-hand side of the screen, navigate to the `Chapter-07` directory and open the `dimensionality-reduction.ipynb` notebook.
2. Choose **Python (Local)** as the kernel.
3. Run each cell in the notebook by selecting the cell and pressing *Shift* + *Enter* on your keyboard.

The code in the notebook performs the following activities:

1. First, it imports the necessary libraries.
2. Then, it loads the dataset.
3. Next, it standardizes the features of the wine dataset. It applies PCA to reduce the dimensionality to two dimensions (that is, the first two principal components). This is done using the `fit_transform()` method, which fits the PCA model to the data and then transforms the data.
4. Finally, it visualizes the data in the space of those two principal components, coloring the points according to the type of wine:

```
from sklearn.datasets import load_wine
from sklearn.preprocessing import StandardScaler
from sklearn.decomposition import PCA
import matplotlib.pyplot as plt
import pandas as pd

# Load dataset
data = load_wine()
df = pd.DataFrame(data.data, columns=data.feature_names)

# Standardize the features
scaler = StandardScaler()
df = scaler.fit_transform(df)

# Apply PCA
pca = PCA(n_components=2)
principalComponents = pca.fit_transform(df)
principalDf = pd.DataFrame(data = principalComponents,
    columns = ['principal component 1',
        'principal component 2'])

# Visualize 2D Projection
plt.figure(figsize=(8,6))
plt.scatter(principalDf['principal component 1'],
principalDf['principal component 2'], c=data.target)
```

```
plt.xlabel('Principal Component 1')
plt.ylabel('Principal Component 2')
plt.show()
```

The resulting visualization should look similar to what's shown in *Figure 7.4*:

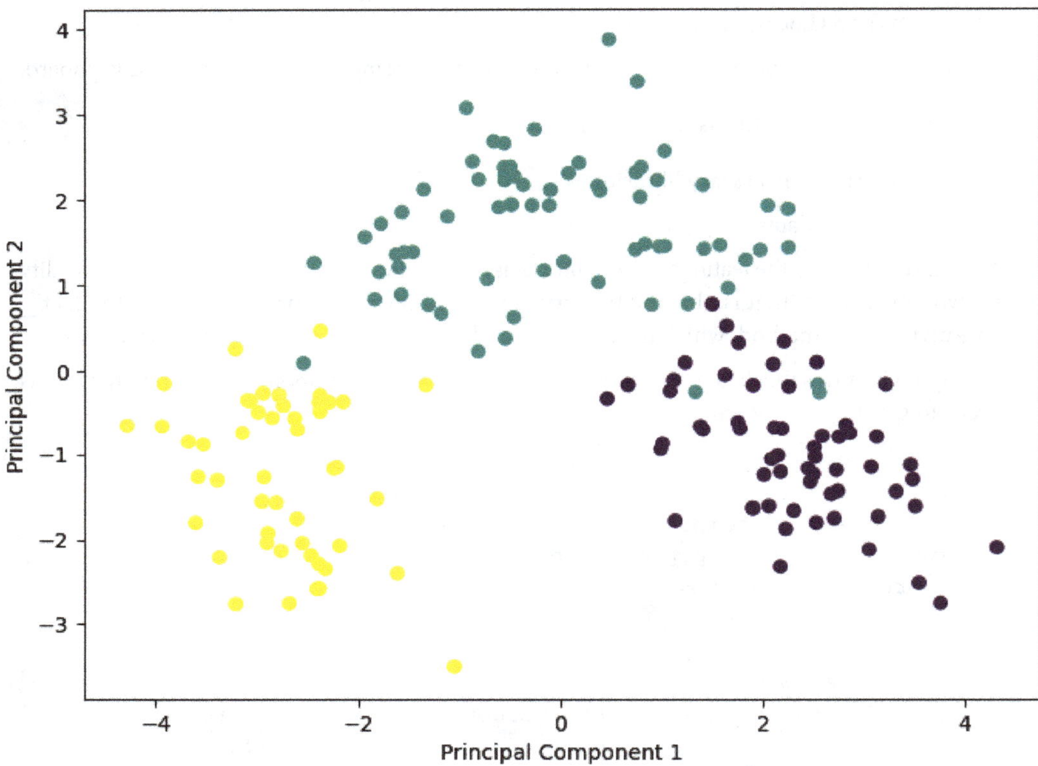


Figure 7.4: PCA scatter plot for the wine dataset

The scatter plot should show a clear separation between the different types of wine, which suggests that the type of wine is closely related to its chemical constituents. Moreover, the PCA object (named `pca` in our code) stores the `components_` attribute, which contains the mappings of each feature concerning the principal components. By examining these, we can find out which features are the most important in distinguishing between the types of wine.

Each data point in the visualization represents a single sample from our dataset, but instead of being plotted in the original, high-dimensional feature space, it's plotted in the lower-dimensional space defined by the principal components.

In the context of the wine dataset, each data point in the PCA visualization represents a single wine sample. Because we have mapped our original 13 features to two PCA dimensions, the position of each point on the *X* and *Y* axes corresponds to the values of the first and second principal components for that wine sample.

The color of each point represents the true class of the wine sample (derived from three different cultivars). By coloring the points according to their true class, you can see how well the PCA transformation separates the different classes in the reduced dimensional space.

Remember that each principal component is a linear combination of the original features, so the position of each point is still determined by the values of its original features. In this way, PCA enables us to visualize the high-dimensional data, and it highlights the dimensions of greatest variance, which are often the most informative.

We can then access the `components_` attribute of the fitted PCA object to view the individual components. This attribute returns a matrix where each row corresponds to a principal component and each column corresponds to an original feature.

So, the following code will print a table, where the values in the table represent the weights of each feature in each component:

```
components_df = pd.DataFrame(pca.components_,
    columns=data.feature_names, index=['Component 1', 'Component 2'])
print(components_df)
```

The results should look similar to what's shown in *Table 7.2*:

```
              alcohol  malic_acid       ash  alcalinity_of_ash  magnesium  \
Component 1  0.144329   -0.245188 -0.002051          -0.239320   0.141992
Component 2 -0.483652   -0.224931 -0.316069           0.010591  -0.299634

             total_phenols  flavanoids  nonflavanoid_phenols  proanthocyanins  \
Component 1       0.394661    0.422934             -0.298533         0.313429
Component 2      -0.065040    0.003360             -0.028779        -0.039302

             color_intensity       hue  od280/od315_of_diluted_wines   proline
Component 1        -0.088617  0.296715                      0.376167  0.286752
Component 2        -0.529996  0.279235                      0.164496 -0.364903
```

Table 7.2: PCA components and features for the wine dataset

By looking at the absolute values of these weights, we can determine which features are most important for each principal component. Large absolute values correspond to features that play a significant role in the variation captured by that principal component, and the sign (positive or negative) of the weight can tell us about the direction of the relationship between the feature and the principal component.

Next, let's see how we could use LDA to identify the constituents that account for the most variance between the types of wine. Again, we'll first import the necessary libraries and standardize the data. We'll then apply LDA to the standardized features, specifying `n_components=2` to get a two-dimensional projection, and then fit the LDA model to the data and transform the data for the first two LDA components. Finally, we'll visualize the transformed data:

```
from sklearn.discriminant_analysis import LinearDiscriminantAnalysis
as LDA

# Apply LDA
lda = LDA(n_components=2)
lda_components = lda.fit_transform(df, data.target)
lda_df = pd.DataFrame(data = lda_components,
     columns = ['LDA 1', 'LDA 2'])

# Visualize 2D Projection
plt.figure(figsize=(8,6))
plt.scatter(lda_df['LDA 1'], lda_df['LDA 2'], c=data.target)
plt.xlabel('LDA 1')
plt.ylabel('LDA 2')
plt.show()
```

The resulting visualization should look similar to what's shown in *Figure 7.5*:

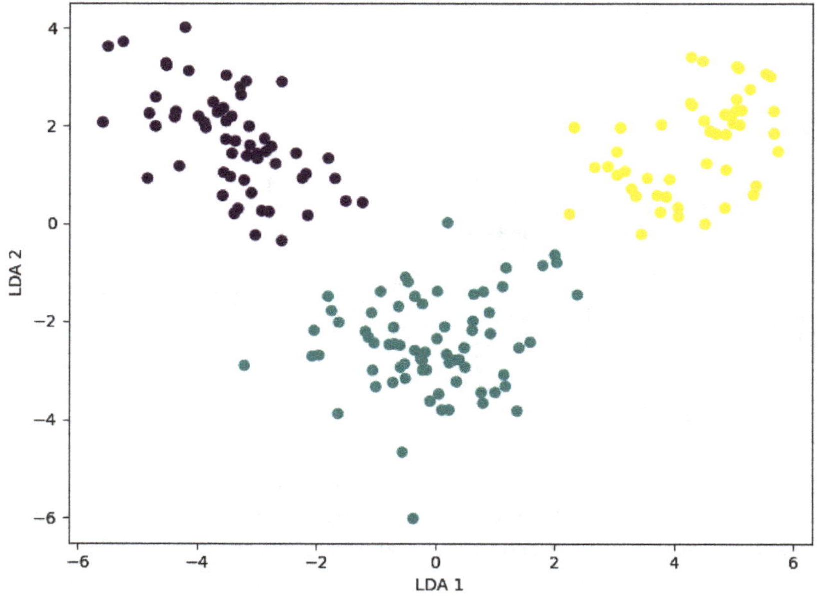


Figure 7.5: LDA scatter plot for the wine dataset

In this case, the scatter plot shows the data points in the space of the first two LDA components, and the points are again colored according to the type of wine. We should also see a clear separation between the different types of wine, indicating that they have different distributions of their various chemical constituents.

As we did with the `components_` attribute of the fitted PCA object, we can inspect the `coef_` attribute of the fitted LDA object to view the most discriminative features, as shown in the following code and its respective output:

```
# Create and print a DataFrame with the LDA coefficients and feature
names
coef_df = pd.DataFrame(lda.coef_, columns=data.feature_names,
    index=['Class 1 vs Rest', 'Class 2 vs Rest', 'Class 3 vs Rest'])
print(coef_df)
```

The results should look similar to what's shown in *Table 7.3*:

```
                  alcohol  malic_acid       ash  alcalinity_of_ash  magnesium  \
Class 1 vs Rest  2.311586   -0.054563  1.431213          -2.588949   0.094308
Class 2 vs Rest -1.719050   -0.855860 -1.578800           1.164253   0.018753
Class 3 vs Rest -0.298563    1.333026  0.576110           1.460125  -0.143660

                 total_phenols  flavanoids  nonflavanoid_phenols  \
Class 1 vs Rest      -1.354133    4.833933              0.292931
Class 2 vs Rest       0.018958    1.343652              0.515282
Class 3 vs Rest       1.636414   -7.929195             -1.122249

                 proanthocyanins  color_intensity       hue  \
Class 1 vs Rest        -0.558434        -1.818874  0.053736
Class 2 vs Rest         0.427280        -1.512956  0.869060
Class 3 vs Rest         0.054389         4.473614 -1.351536

                 od280/od315_of_diluted_wines   proline
Class 1 vs Rest                      2.866183  4.407904
Class 2 vs Rest                     -0.024265 -2.147873
Class 3 vs Rest                     -3.487126 -2.240987
```

Table 7.3: LDA classes and features for the wine dataset

In the resulting table, each row corresponds to a class (compared to the rest), and each column corresponds to an original feature. So, the values in the table represent the coefficients of each feature in the context of the linear discriminants. Similar to our PCA assessment, large absolute values indicate features that contribute significantly to separating the classes.

Now that we've looked at how to reduce the dimensionality of our datasets, let's assume that we've identified our required features and begin to explore how we could further engineer features to ensure we have the best possible set of features to train our models.

# Feature engineering

Feature engineering can constitute a large portion of a data scientist's activities, and it can be just as important to their success, or sometimes even more important, than choosing the right machine learning algorithm. In this section, we will dive deeper into feature engineering, which can be considered both an art and a science.

We will use the *Titanic* dataset available on OpenML (`https://www.openml.org/search?type=data&sort=runs&id=40945`) for our examples in this section. This dataset contains information about passengers aboard the Titanic, including demographic data, ticket class, fare, and whether they survived the sinking of the ship.

In the `Chapter-07` directory in JupyterLab on your Vertex AI Workbench Notebook Instance, open the `feature-eng-titanic.ipynb` notebook and choose **Python (Local)** as the kernel. Again, run each cell in the notebook by selecting the cell and pressing *Shift* + *Enter* on your keyboard.

In this notebook, the code performs the following steps:

1. First, it imports the necessary libraries.
2. Then, it loads the dataset.
3. After, it performs some initial exploration to see what our dataset looks like.
4. Finally, it engineers new features.

Let's take a look at each step in more detail, starting with importing the required libraries, and loading and exploring the dataset. We use the following code to perform those tasks:

```
import pandas as pd
import numpy as np
from sklearn.impute import SimpleImputer
from sklearn.preprocessing import OneHotEncoder
# Load the data
titanic_raw = pd.read_csv('./data/titanic_train.csv')
titanic_raw.head()
```

The resulting output from the `head()` method should look similar to what's shown in *Table 7.4*:

| | pclass | survived | name | sex | age | sibsp | parch | ticket | fare | cabin | embarked | boat | body | home.dest |
|---|---|---|---|---|---|---|---|---|---|---|---|---|---|---|
| 0 | 1 | 1 | Allen, Miss. Elisabeth Walton | female | 29.0000 | 0 | 0 | 24160 | 211.3375 | B5 | S | 2 | NaN | St Louis, MO |
| 1 | 1 | 1 | Allison, Master. Hudson Trevor | male | 0.9167 | 1 | 2 | 113781 | 151.5500 | C22 C26 | S | 11 | NaN | Montreal, PQ / Chesterville, ON |
| 2 | 1 | 0 | Allison, Miss. Helen Loraine | female | 2.0000 | 1 | 2 | 113781 | 151.5500 | C22 C26 | S | NaN | NaN | Montreal, PQ / Chesterville, ON |
| 3 | 1 | 0 | Allison, Mr. Hudson Joshua Creighton | male | 30.0000 | 1 | 2 | 113781 | 151.5500 | C22 C26 | S | NaN | 135.0 | Montreal, PQ / Chesterville, ON |
| 4 | 1 | 0 | Allison, Mrs. Hudson J C (Bessie Waldo Daniels) | female | 25.0000 | 1 | 2 | 113781 | 151.5500 | C22 C26 | S | NaN | NaN | Montreal, PQ / Chesterville, ON |

Table 7.4: Titanic dataset head() output

The fields in the dataset are as follows:

- `survived`: Indicates if a passenger survived or not. It's a binary feature where 1 stands for survived and 0 stands for not survived.
- `pclass` (passenger class): Indicates the class of the passenger's ticket. It has three categories: 1 for first class, 2 for second class, and 3 for third class. This could also indicate the socio-economic status of the passengers.
- `name`: The name of the passenger.
- `sex`: The gender of the passenger; male or female.
- `age`: The age of the passenger. Some ages are fractional, and for passengers less than 1 year old, the age is estimated as a fraction.
- `sibsp`: The total number of the passengers' siblings and spouses aboard the Titanic.
- `parch`: The total number of the passengers' parents and children aboard the Titanic.
- `ticket`: The ticket number of the passenger.
- `fare`: The passenger fare – that is, how much the ticket cost.
- `cabin`: The cabin number where the passenger was staying. Some entries are NaN, indicating that the cabin number is missing from the data.
- `embarked`: The port where the passenger boarded the Titanic. C is for Cherbourg; Q is for Queenstown; S is for Southampton.
- `boat`: Which lifeboat the passenger was assigned to (if the passenger survived).
- `body`: Body number (if the passenger did not survive and their body was recovered).
- `home.dest`: The passenger's home and destination.

Assuming that we want to use this dataset to build a model that could predict the likelihood of a passenger surviving based on the information recorded about them, let's see if we can use some domain knowledge to assess which features are likely to be the biggest contributors to the outcome of surviving or not surviving, and whether we could use any data manipulation techniques to engineer more useful features.

`Passenger Class` stands out as a potentially important feature to begin with because the first-class and second-class passengers had cabins on higher levels of the ship, which were closer to the lifeboats.

It's unlikely that the passenger's name would affect the outcome, nor their ticket number or port of embarkation. However, we could engineer a new feature named `Title` that's extracted from the passengers' names and could provide valuable information related to social status, occupation, marital status, and age, which might not be immediately apparent from the other features. We could also clean up this new feature by merging similar titles such as `Miss` and `Ms` and identifying elevated titles as `Distinguished`. The code to do that would be as follows:

```
# We first define a function to extract titles from passenger names
def get_title(name):
    if '.' in name:
        return name.split(',')[1].split('.')[0].strip()
    else:
        return 'Unknown'

# Create a new "Title" feature
titanic['Title'] = titanic['Name'].apply(get_title)

# Simplify the titles, merge less common titles into the same category
titanic['Title'] = titanic['Title'].replace(['Lady', 'Countess',
     'Capt', 'Col', 'Don', 'Dr', 'Major', 'Rev', 'Sir', 'Jonkheer',
     'Dona'], 'Distinguished')
titanic['Title'] = titanic['Title'].replace('Mlle', 'Miss')
titanic['Title'] = titanic['Title'].replace('Ms', 'Miss')
titanic['Title'] = titanic['Title'].replace('Mme', 'Mrs')
```

Next, let's consider the `Fare` and `Cabin` features. These could be somewhat correlated with class, but we will dive into these features in more detail. For the `Cabin` feature, we could extract another feature named `CabinClass`, which more clearly represents the class associated with each entry. We could do this, for example, by extracting the first letter from the cabin number, using it to represent the cabin class (for example, A, B, C, and so on), and storing it in the new `CabinClass` feature. The code to do that would be as follows:

```
# Create "CabinClass" feature
titanic['CabinClass'] = titanic['Cabin'].apply(lambda x: x[0])
```

Let's also ensure that we represent the fare as accurately as possible by considering that people may have purchased fares as families traveling together. To do this, we can create a new feature named `FamilySize` as a combination of the `SibSp` and `Parch` features (adding an additional "1" to account for the current passenger), and then compute `FarePerPerson` by dividing the `Fare` feature by the `FamilySize` feature by using the following code:

```
titanic['FamilySize'] = titanic['SibSp'] + titanic['Parch'] + 1
# Create "FarePerPerson" feature
titanic['FarePerPerson'] = titanic['Fare'] / titanic['FamilySize']
```

Whether somebody is traveling alone or with their family could also affect their chances of survival. For example, family members could help each other when trying to get to the lifeboats. So, let's create a feature from the `FamilySize` feature that identifies whether a passenger was traveling alone:

```
# Create new feature "IsAlone" from "FamilySize"
titanic['IsAlone'] = 0
titanic.loc[titanic['FamilySize'] == 1, 'IsAlone'] = 1
```

Next, let's consider how age affects the likelihood of survival. People who are very young, or elderly, may, unfortunately, have less likelihood of surviving unless they have people to help them. However, we may not need yearly and fractional-yearly granularity when considering age in this context, and perhaps grouping people into age groups may be more effective. In that case, we can use the following code to create a new feature named `AgeGroup` that will group passengers by decades such as 0-9, 10-19, 20-29, and so on:

```
# Create "AgeGroup" feature
bins = [0, 10, 20, 30, 40, 50, 60, 70, np.inf]
labels = ['0-9', '10-19', '20-29', '30-39', '40-49', '50-59', '60-69',
    '70+']
titanic['AgeGroup'] = pd.cut(titanic['Age'], bins=bins, labels=labels)
```

We also want to convert the categorical features into numerical values using one-hot encoding since machine learning models typically require numeric values. We could do this as follows (we need to do this for all of our categorical features):

```
# Convert "Title" into numerical values using one-hot encoding
one_hot = OneHotEncoder()
title_encoded = one_hot.fit_transform(titanic[['Title']]).toarray()
title_encoded_df = pd.DataFrame(title_encoded,
    columns=one_hot.get_feature_names_out(['Title']))
titanic = pd.concat([titanic, title_encoded_df], axis=1)
```

Now, let's drop the features that we know will not be valuable for predicting the likelihood of survival (as well as the original features that we encoded, because only their encoded versions are needed):

```
titanic = titanic.drop(['name', 'ticket', 'Title', 'cabin', 'sex',
    'embarked', 'AgeGroup', 'CabinClass', 'home.dest'], axis=1)
```

Then, we can take a quick peek at what our updated dataset looks like:

```
titanic.head()
```

The resulting output from the `head()` method should look similar to what's shown in *Table 7.5*:

| | pclass | survived | age | sibsp | parch | fare | boat | body | FamilySize | IsAlone | ... | CabinClass_D | CabinClass_E | CabinClass_F | CabinClass_G | CabinClass_T | sex_female |
|---|---|---|---|---|---|---|---|---|---|---|---|---|---|---|---|---|---|
| 0 | 1 | 1 | 29.0000 | 0 | 0 | 211.3375 | 2 | 155.0 | 1 | 1 | ... | 0.0 | 0.0 | 0.0 | 0.0 | 0.0 | 1.0 |
| 1 | 1 | 1 | 0.9167 | 1 | 2 | 151.5500 | 11 | 155.0 | 4 | 0 | ... | 0.0 | 0.0 | 0.0 | 0.0 | 0.0 | 0.0 |
| 2 | 1 | 0 | 2.0000 | 1 | 2 | 151.5500 | 13 | 155.0 | 4 | 0 | ... | 0.0 | 0.0 | 0.0 | 0.0 | 0.0 | 1.0 |
| 3 | 1 | 0 | 30.0000 | 1 | 2 | 151.5500 | 13 | 135.0 | 4 | 0 | ... | 0.0 | 0.0 | 0.0 | 0.0 | 0.0 | 0.0 |
| 4 | 1 | 0 | 25.0000 | 1 | 2 | 151.5500 | 13 | 155.0 | 4 | 0 | ... | 0.0 | 0.0 | 0.0 | 0.0 | 0.0 | 1.0 |

Table 7.5: Output from the head() method for the updated dataset

At this point, we have an augmented dataset with engineered features that can be used to train a model. It's important to bear in mind that any feature engineering steps we perform on our source dataset also need to be taken into account when we use our model to make predictions. This common need in machine learning is what gave rise to the requirement for Google Cloud to develop a service named Feature Store. We'll explore this in the next section.

# Vertex AI Feature Store

We've done a lot of feature engineering work in this chapter. Bear in mind that we performed data transformations and engineered new features because we had reason to believe that the raw data was insufficient to train a machine learning model to suit our business case. This means that the raw data our model will see in the real world would usually not contain the enhancements we performed on the data during training. After all of that work, we would generally want to save the updated features we've engineered so that our model can reference them when it needs to make predictions. Vertex AI Feature Store was created for this purpose. We briefly mentioned Vertex AI Feature Store in *Chapter 3*, and in this section, we will dive into more detail regarding what it is and how we can use it to store and serve features for both training and inference.

## Introduction to Vertex AI Feature Store

Here's the official definition from the Google Cloud documentation:

> *Vertex AI Feature Store is a managed, cloud-native feature store service that's integral to Vertex AI. It streamlines your machine learning feature management and online serving processes by letting you manage your feature data in a BigQuery table or view. You can then serve features online directly from the BigQuery data source. Vertex AI Feature Store provisions resources that let you set up online serving by specifying your feature data sources. It then acts as a metadata layer interfacing with the BigQuery data sources and serves the latest feature values directly from BigQuery for online predictions at low latencies.*

In addition to storing and serving our features, Vertex AI Feature Store also integrates with Google Cloud Dataplex to provide feature governance capabilities, including the ability to track feature metadata such as feature labels and versions. In *Chapter 13*, we will dive into the importance of data governance and discuss how Dataplex can be used as an important component in building a robust governance framework.

At this point, it's important to highlight that an entirely new version of Vertex AI Feature Store was launched in 2023. As a result, there are now two different variants of the Feature Store service we can choose in Google Cloud, whereby the prior version is referred to as Vertex AI Feature Store (Legacy), and the new version is simply referred to as Vertex AI Feature Store. We will discuss both variants in this chapter, as well as some of the main distinctions between them. To provide context for the content in subsequent sections, I will briefly describe the topic of online versus offline feature serving.

## Online versus offline feature serving

Simply put, **online serving** refers to a scenario in which the interaction is happening in real time – that is, the requesting entity or client sends a request and synchronously waits for a response. In this scenario, latency needs to be reduced as much as possible. On the other side of the coin is **offline serving**, which refers to a scenario in which the requesting entity or client does not synchronously wait for a response, and the operation is allowed to happen over a longer period. In this case, latency is generally not a primary concern. This concept relates closely to the topic of online and offline inference, which we will cover in detail in *Chapter 10*.

In the case of offline feature serving, Vertex AI Feature Store allows us to store and serve features directly in a Google Cloud BigQuery dataset. This is quite a convenient option as many Google Cloud customers already use BigQuery to store and analyze large amounts of their data.

In the case of online feature serving, there are now two ways in which we can serve our features in Vertex AI Feature Store. The first option uses Google Cloud Bigtable to serve our features. Google Cloud Bigtable is a powerful service that is designed for serving large data volumes (terabytes of data).

The second option for online feature serving, which is referred to as **optimized online serving**, and was added as part of the new version of Vertex AI Feature Store, allows us to create an online store that is optimized specifically for serving feature data at ultra-low latencies.

Choosing one option or the other depends on the needs of your use case, specifically whether you need to handle very large volumes of data or whether you need to serve your features with ultra-low latency. Cost is also a consideration in this decision, bearing in mind that the Bigtable solution generally costs less than the optimized online serving solution.

We will focus primarily on the optimized online serving approach in this chapter and the accompanying Jupyter Notebook. The following section dives deeper into the process of setting up online feature serving in Vertex AI Feature Store.

### Online feature serving

At a high level, the following steps are required to set up online serving using Vertex AI Feature Store. We will elaborate on these steps in subsequent sections:

1. Prepare data sources in BigQuery.

   Optional: Register data sources in the feature registry by creating feature groups and features.

2. Set up the online store and feature view resources to present the feature data sources.
3. Serve the latest online feature values from a feature view.

Let's take a look at these concepts in more detail.

#### Feature registry

When using the optimized online serving approach, we can perform an optional step to register our features in the Vertex AI feature registry, which has also been added as a component of the new version of Vertex AI Feature Store.

This involves the process of creating resources referred to as **feature groups**, which represent logical groupings of feature columns and are associated with specific BigQuery source tables or views. In turn, feature groups contain resources referred to as **features**, which represent specific columns containing feature values within the data source represented by the feature group.

We can still serve features online even if we don't add our BigQuery data sources to the feature registry, but note that the feature registry provides additional functionality, such as storing historical time series data associated with your features. As a result, we will use the feature registry in the practical exercises accompanying this chapter. Now, let's take a look at the process of setting up online feature serving in more detail.

#### Online feature store and feature views

After we have set up our feature data in BigQuery, and optionally registered feature groups and features in the feature registry, there are two main types of resources that we need to set up to enable online feature serving:

- An **online serving cluster instance**, which is referred to as the **online store**. Remember that we can either use Bigtable for online feature serving or the newly released optimized online feature serving option.
- One or more **feature view instances**, where each feature view is associated with a feature data source, such as a feature group in our feature registry (if we have chosen the option of registering our features in the feature registry), or a BigQuery source table or view.

After we create a feature view, we can configure synchronization settings to ensure that our feature data in BigQuery is synchronized with our feature view. We can trigger a synchronization manually, but if our source data is expected to be updated over time, then we can also configure a schedule to periodically refresh the contents of our feature view from the data source.

Now that we've covered many of the important concepts related to Vertex AI Feature Store, it's time to dive in and build our very own feature store!

## Building our feature store

In this section, we will perform practical exercises that implement the concepts we learned about in the previous sections.

### *Use our Vertex AI notebook to build the feature store*

In the Vertex AI Notebook Instance we created in Chapter 5, we can perform the following steps to build the feature store:

1. Select **Open JupyterLab** for the Vertex AI Notebook Instance we created in Chapter 5.
2. When the JupyterLab opens, you should see a folder in your notebook called `Google-Machine-Learning-for-Solutions-Architects`.
3. Double-click on that folder, then double-click on the `Chapter-07` folder within it, and then double-click on the `feature-store.ipynb` file to open it.
4. On the **Select Kernel** screen that appears, select **Python (Local).**
5. Press *Shift* + *Enter* to run each of the cells in the notebook and read the explanations in the Markdown and comments to understand what we're doing.

Now that you've followed the steps in the notebook to perform feature selection and engineering, have built a feature store, and used some of the features to train a model, let's take a look at how those features could be used during inference time. In later chapters, you will learn how to deploy models for online inference and send inference requests to those models, but for now, I'll explain the process at a conceptual level.

## How features are used during online inference

In this section, I'll use the taxi fare prediction model use case that we built in the accompanying Jupyter Notebook as an example to explain how we can use features from our feature store during inference. We'll take a look at each step in the process.

### *Combining real-time and precomputed features*

Features such as the current pickup time (`pickup_datetime`), `pickup_location`, and `passenger_count` can be obtained in real time as each taxi journey begins.

Our feature store also contains precomputed features, such as historical trip distances, fares per mile, pickup times, and locations. These features can be selected based on the current journey's context from the available real-time features.

To get the precomputed features, the application handling the taxi journey can send a request to our feature store, passing identifiers such as the current time and location, after which the feature store can return the relevant feature values for these identifiers.

### *Data assembly for prediction*

At this point, we can assemble the real-time data and fetched features into a feature vector matching the format expected by the model, and then pass the assembled feature vector to the model. The model then processes this vector and outputs a fare prediction, which can then be displayed in the app.

Well done! You have successfully built a feature store on Google Cloud. Let's summarize everything we've discussed in this chapter.

## Summary

In this chapter, we discussed how the quality of our features, and the ratio of features to observations in our dataset, influences how our algorithms learn from our data. We discussed challenges that can occur when our dataset contains many features and how to address those challenges by using mechanisms such as dimensionality reduction. We dived into details on dimensionality reduction techniques such as feature selection and feature projection, including algorithms such as PCA, LDA, and t-SNE, and we looked at examples of how to use some of these algorithms using hands-on activities.

Next, we dived into feature engineering techniques in which we augmented a source dataset to create new features that contained information that was not readily available in the original dataset. Finally, we dived into Vertex AI Feature Store to learn about how we can use that service to store and serve our engineered feature sets.

In the next chapter, we will shift our focus away from the datasets and parameters that our models learn from and discuss different types of parameters that influence how our models learn. There, we'll explore the concepts of hyperparameters and hyperparameter optimization.

# 8
# Hyperparameters and Optimization

We introduced the concepts of hyperparameters and hyperparameter optimization (or tuning) in *Chapter 2*. In this chapter, we will dive into these concepts in more detail, and we will use Google Cloud products such as Vertex AI Vizier to define and run hyperparameter tuning jobs.

Following our established pattern, we will begin by covering some prerequisites that are required for the hands-on activities in this chapter. Then, we cover some important basic concepts that relate to the content covered in this chapter, and finally, we perform hands-on activities that teach you how to apply those concepts in real-world scenarios.

This chapter covers the following topics:

- Prerequisites and basic concepts
- What are hyperparameters?
- Hyperparameter optimization
- Hands-on: performing hyperparameter tuning in Vertex AI

Let's begin by reviewing the prerequisites for this chapter.

## Prerequisites

The steps in this section need to be completed before we can perform the primary activities in this chapter.

## Enabling the Artifact Registry API

We're going to create Docker images in order to run our custom code in conjunction with the Google Cloud Vertex AI Vizier service. The Google Cloud Artifact Registry is a fully managed artifact repository that we can use for storing our container images. It can be seen as the next generation of the **Google Cloud Container Registry** (**GCR**) that can be used to store artifacts such as Java JAR files, Node.js modules, Python wheels, Go modules, Maven artifacts, and npm packages (in addition to Docker images, which were already supported in GCR).

To enable the Artifact Registry API, perform the following steps:

1. In the Google Cloud console, navigate to **Google Cloud services menu** → **APIs & Services** → **Library**
2. Search for `Artifact Registry` in the search box.
3. Select the API in the list of results.
4. On the page that displays information about the API, click **Enable**.

Next, let's set up the required permissions for the steps in this chapter.

## Creating an AI/ML service account

In *Chapter 6*, we created a service account for using Google Cloud's data processing services. In this chapter, we will create a service account that will be used in our hyperparameter tuning job for administering resources in Google Cloud Vertex AI.

Perform the following steps to create the required service account:

1. In the Google Cloud console, navigate to **Google Cloud services menu** → **IAM & Admin** → **Service accounts**.
2. Select **Create service account**.
3. For the service account name, enter `ai-ml-sa`.
4. Click **Create and Continue**.
5. In the section titled **Grant this service account access to project**, add the roles shown in *Figure 8.1*.

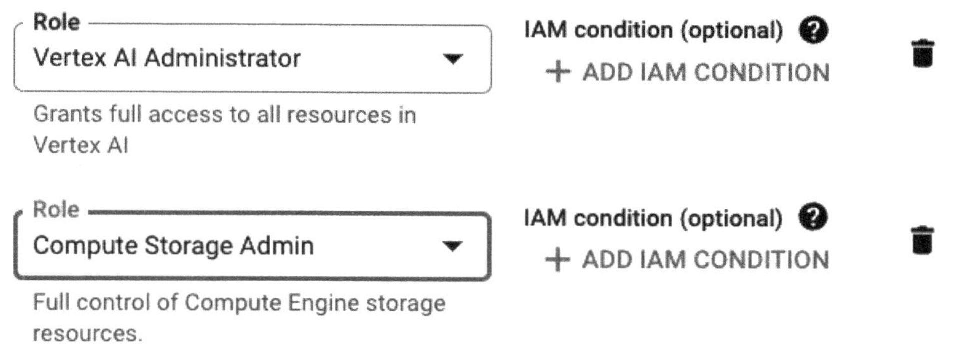


Figure 8.1: AI/ML service account permissions

6. Select **Done**.

Our service account is now ready to be used later in this chapter.

Now that we've covered the prerequisites, let's discuss some concepts that we need to understand before performing the hands-on activities in this chapter.

# Concepts

This section describes concepts that underpin the practical activities we will cover in this chapter.

## Model evaluation metrics used in this chapter

We've already discussed the topic of model evaluation metrics in previous chapters. We first introduced the concept in *Chapter 1*, where we briefly discussed metrics such as the **mean squared error** (**MSE**) for regression use cases and accuracy for classification use cases. In *Chapter 5*, we used functions in scikit-learn to calculate some of these metrics for the models we created, and we suggested looking up additional metrics as a supplemental learning activity at the end of that chapter.

In this chapter, we will train models for a classification use case, and we will introduce some additional metrics to evaluate our models. The main metric we will use is something called **AUC ROC**, which stands for **area under the receiver operating characteristic curve**. That sounds like a lot, but don't worry, we will explain this metric in more detail in this section. In order to do so, we need to first introduce some concepts and simpler metrics that are used in calculating the AUC ROC.

Note that in a binary classification use case, the model needs to predict one of two possible outcomes for each data point in the dataset, true or false, also referred to as **positive** or **negative**. They are usually represented by 1 and 0.

Models are rarely perfect, so they will sometimes make mistakes. Let's take a look at the possible outcomes of a binary classification model's predictions.

### True positives, false positives, true negatives, and false negatives

A binary classification model's predictions generally have four possible outcomes:

- When our model predicts something to be true (or positive) and it really is true (or positive), we call this a **true positive** (**TP**)
- When our model predicts something to be true (or positive) but really it is false (or negative), we call this a **false positive** (**FP**)
- When our model predicts something to be false (or negative) and it really is false (or negative), we call this a **true negative** (**TN**)
- When our model predicts something to be false (or negative) but it is really true (or positive), we call this a **false negative** (**FN**)

Let's take a look at how these outcomes are related to each other in more detail.

### Confusion matrix

The preceding concepts can be represented visually in something called a confusion matrix, which is demonstrated in *Table 8.1*.

| | **Predicted negative** | **Predicted positive** |
|---|---|---|
| Actual negative | TN | FP |
| Actual positive | FN | TP |

Table 8.1: Confusion matrix

We can use the preceding concepts to calculate metrics that measure how well our models are performing when trying to accurately identify positive or negative data points in our dataset. We define these metrics next.

### True positive rate

The **true positive rate** (**TPR**) represents a count of all of the data points in our dataset that our model correctly predicted as positive compared against all of the data points in the dataset that are positive, including any data points that our model erroneously predicted to be negative (i.e., our model said they were negative, even though they were positive, which means it was a false negative).

The formula to calculate TPR is TPR = TP / (TP + FN)

TPR is also referred to as **recall** or **sensitivity**.

### False positive rate

The **false positive rate** (**FPR**) is the ratio of false positives (the number of negative instances incorrectly predicted as positive by our model) to the sum of false positives and true negatives (the number of negative instances correctly predicted by the model). In other words, it's the proportion of actual negatives that are incorrectly identified as positive.

The formula to calculate FPR is FPR = FP / (FP + TN)

TPR is also referred to as **fall out**.

### True negative rate

The **true negative rate** (**TNR**) is described very similarly to how we described the TPR, just with positive and negative switched around. That is, the TNR represents a count of all of the data points in our dataset that our model correctly predicted as negative compared against all of the data points in the dataset that are negative, including any data points that our model erroneously predicted to be positive (i.e., our model said they were positive, even though they were negative, which means it was a false positive).

The formula to calculate TNR is TNR = TN / (TN + FP)

TPR is also referred to as **specificity**.

### False negative rate

The **false negative rate** (**FNR**) is described very similarly to how we described the FPR, just with positive and negative switched around. It is the ratio of wrongly predicted negative observations to the actual positives. In other words, it's the proportion of actual positives that are incorrectly identified as negative.

The formula to calculate FNR is FNR = FN / (FN + TP)

### Precision

Precision is the ratio of correctly predicted positive observations to the total predicted positives. In other words, out of all the instances the model predicted as positive, how many were actually positive?

The formula to calculate FNR is P = TP / (TP + FP)

When I first started learning all of this stuff, I wondered why there were so many different metrics for measuring slightly different aspects of binary classification model performance.

There are at least a couple of reasons for this:

- **Statistical basis**: These metrics are natural outcomes from statistical analysis in binary classification use cases
- **Trial and error**: Each metric may be more important than the others, based on the desired outcomes of the use case

Consider the following example for the second point. If you're trying to predict credit card fraud, you may want to maximize the sensitivity of your model, which would reduce the number of false negatives as much as possible, even if that ends up causing more false positives. In other words, it's better to accidentally flag a transaction as fraudulent even if it's not fraudulent than to accidentally allow a fraudulent transaction to occur.

On the other hand, if you're creating a spam filter, you would probably prefer allowing a few spam emails to accidentally reach your inbox (false negatives) than having valid emails flagged as spam (false positives).

The preceding metrics are often too simple to use independently, and you will therefore usually want to find a more complex combination of metrics that provides a more balanced outcome. Even in the credit card fraud use case, too many false positives would be disruptive and frustrating for credit card customers. The balance depends on the threshold you specify (between 0 and 1) for determining whether something is positive or negative. For example, a low threshold results in more positives, while a higher threshold results in more negatives. This brings us to more advanced metrics such as F1 score and AUC ROC, which we describe next.

### F1 score

F1 score is defined as the **harmonic mean** of precision and recall, which is calculated using the following formula:

F1 = 2 * (precision * recall) / (precision + recall)

The F1 score is particularly useful when you care more about the positive class and you want to balance precision and recall.

### AUC ROC

To understand AUC ROC, let's first break down its name. The **receiver operating characteristic** (**ROC**) is a pretty fancy name for a curve that is generated by plotting the TPR against the FPR at various classification threshold settings, as depicted in *Figure 8.2*.

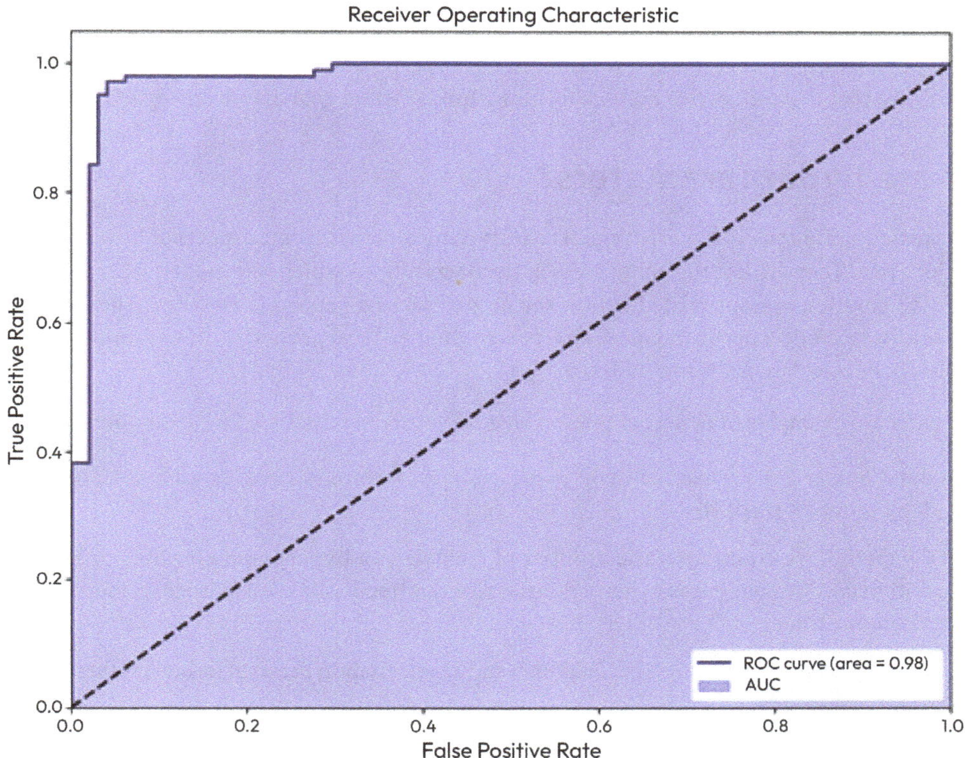


Figure 8.2: AUC ROC

The **area under the curve** (**AUC**) is a measure of the entire two-dimensional area underneath the ROC curve from (0, 0) to (1, 1), as represented by the blue area in *Figure 8.2*. The AUC provides an aggregate measure of performance across all possible classification thresholds, and the objective is to maximize the area under the curve, so in the best possible scenario, the curve would stretch up into the top-left corner, filling up the entire graph.

Let's take a look at how to interpret the AUC ROC score values in more detail:

- An AUC ROC score of 1.0 means that the model is able to perfectly distinguish between all the positive and the negative data points correctly, in which case it has no false negatives and no false positives (i.e., no mistakes).
- An AUC ROC score of 0.5 means that the model is not able to accurately distinguish between positive and negative data points and performs no better than random guessing.
- An AUC ROC score of less than 0.5 means that the model is performing worse than random guessing, predicting negatives as positives and positives as negatives.

Understanding these metrics is important because they are generally what our ML algorithms are trying to optimize. In the next section, we will discuss hyperparameters and hyperparameter tuning and we will see that these objective metrics form the fundamental goal of our tuning jobs.

# What are hyperparameters?

As we discussed in *Chapter 2*, hyperparameters are parameters that define aspects of how our model training jobs run. They are not the parameters in the dataset from which our models learn but rather external configuration options related to how the model training process is executed. They influence how the resulting models perform and they represent higher-level properties of the model, such as its complexity or how quickly it should learn.

The following are examples of hyperparameters that we've already discussed in this book:

- In our *Chapter 2* discussion of hyperparameters, we covered examples such as learning rate and the number of epochs
- In *Chapter 5*, we configured the number of clusters as a hyperparameter for our K-means algorithm and we configured hyperparameters for our tree-based models, such as the maximum depth of our trees
- We talked about regularization in *Chapter 7*, and regularization parameters are another example of hyperparameters.

There are many more types of hyperparameters for different kinds of algorithms, and we will encounter more as we progress through this book.

# Hyperparameter optimization

How do we know what kinds of hyperparameters to use and what their values should be? Hyperparameters can be chosen based on domain knowledge, experience, or trial and error, but to most efficiently choose the best hyperparameters, we can use a process called hyperparameter optimization, or hyperparameter tuning, which is a systematic process that can be implemented via different mechanisms that we will discuss next. Ultimately, the goal of hyperparameter optimization is to tune the hyperparameters of a model to achieve the best performance as measured by running it against a validation set, which is a subset of our source dataset.

## Methods for optimizing hyperparameter values

In *Chapter 2*, we described hyperparameter tuning mechanisms such as grid search, random search, and Bayesian optimization, summarized here as a quick refresher:

- **Grid search**: This is an exhaustive search of the entire hyperparameter space (i.e., it tries out every possible combination of all hyperparameter values). This is usually impractical and unnecessarily computationally expensive.

- **Random search**: The random search approach uses a subsampling technique in which hyperparameter values are selected at random for each training job experiment. This will not result in all possible values of every hyperparameter being tested, but it can often be quite an efficient method for finding an effective set of hyperparameter values.
- **Bayesian optimization**: This uses an optimization algorithm, and it is something that is provided as a managed service in Google Cloud Vertex AI.

The following are some additional hyperparameter tuning mechanisms that exist in the industry:

- **Gradient-based optimization**: This method uses Gradient Descent, which we've already covered in depth earlier in this book. These methods are often used when training neural networks. We provide a separate section later in this book that describes how to train neural networks in detail.
- **Evolutionary algorithms**: These are **population-based** optimization algorithms loosely modeled on the process of evolutionary natural selection. The term "population-based" refers to the practice of building a pool (or population) of potential candidates. In this context, each candidate in the population represents a different set of hyperparameters, and candidates are evaluated based on their validation performance. The best-performing ones are then selected to produce "offspring" for the next generation. These algorithms are also more likely to be used for advanced use cases such as neural networks, where the hyperparameter search space can be large and complex, and it can be expensive to evaluate the performance of individual solutions.
- **Automated machine learning** (**AutoML**) **systems**: We discussed the process of AutoML in previous chapters. It can be used to automate the entire ML lifecycle, including hyperparameter tuning.

In any case, the general tuning process works as follows:

1. Split your source dataset into three subsets:

   A. **Training dataset**: Used to train the model

   B. **Validation dataset**: Used to evaluate each combination of hyperparameters during the tuning process

   C. **Test dataset**: Used to test the final model

2. Select which type of machine learning model we want to create (e.g., linear regression, decision tree, neural network). This determines which specific hyperparameters can be tuned.
3. Set an initial range or grid of hyperparameters and values. This can be based on domain knowledge or research or we could just start with a random broad range and refine it over time.
4. Choose a method for searching through the model's hyperparameter space (e.g., random search, Bayesian optimization).
5. For each combination of hyperparameters, fit the model to the training data and evaluate its performance by testing it against the validation data and measuring the appropriate objective metrics for the chosen type of model (e.g., MSE for regression, AUC ROC for binary classification).

6. Once all combinations have been evaluated, choose the combination of hyperparameter values that resulted in the best model performance.
7. Train a final model using those hyperparameters and test the resulting model using the `test` dataset to confirm the model's ability to generalize to unseen data.

Note that finding the best set of hyperparameters and values could require iterating through the outlined steps hundreds or even thousands of times, which would be extremely time-consuming or potentially impossible to perform manually. This is why hyperparameter tuning jobs, which automate the steps, are often required.

Now that we've covered many of the important theoretical concepts related to hyperparameter tuning, it's time for us to shift our focus to the practical implementation of these concepts.

# Hands-on: performing hyperparameter tuning in Vertex AI

Considering that Google Cloud Vertex AI provides tools that make it easy for us to implement every step in the data science project lifecycle, this gives us the perfect environment to put our knowledge into practice and start implementing hyperparameter tuning jobs. In fact, as we mentioned previously, Vertex AI provides a tool called Vizier that is specialized for the purpose of automating hyperparameter tuning jobs, which we will dive into in more detail next.

## Vertex AI Vizier

Vertex AI Vizier is a service in Google Cloud that automates the hyperparameter tuning process that we outlined in the previous section of this chapter. In this section, we discuss some terminology used by the Vertex AI Vizier service and we describe some details on how it works. Then, we will actually use it in our hands-on activities to implement hyperparameter tuning jobs.

### *Vertex AI Vizier terminology*

Google Cloud uses some terminology that is specific to the Vertex AI Vizier service. We will briefly describe some important terms here and relate them back to the generic concepts we covered earlier in this chapter.

#### Studies, study configurations, and trials

In Vertex AI Vizier, a **study** represents the overall objective we are trying to achieve and all of the steps and other details involved in working towards that objective. For example, if we look at the general tuning process steps we outlined in the *Methods for optimizing hyperparameter values* section of this chapter, a study encapsulates all of those steps. A **study configuration** is the actual configuration object that contains all of the details of our study, such as the objective metric that we want the study to optimize, what parameters to test, and what kind of parameter search method to use.

A **trial** is an individual experiment in our study, or a single iteration in the tuning process (i.e., a single training and evaluation job that uses a specific set of hyperparameter values). A study will run many trials when working toward our specified objective.

After you create a study, Vertex AI Vizier will start running trials on its own. In each test, a different set of hyperparameters will be used. Vertex AI Vizier will keep track of the results of each run and use this knowledge to choose the best set of hyperparameters (it will automatically stop running trials when it has found the best set of hyperparameters). Vizier will also summarize all of the trials and rank them according to which ones performed best with regard to the objective metric. Then, we can train our ML model with the hyperparameters from the top-ranking trial.

Now that we've gotten the terminology covered, let's dive into the hands-on activities!

### Use case and dataset

In this section, we will develop an XGBoost model to detect credit card fraud using the `Credit Card Fraud Detection` dataset available on Kaggle (`https://www.kaggle.com/datasets/mlg-ulb/creditcardfraud`).

### Implementation

We're going to use Jupyter Notebook for the hands-on activities in this chapter, and we're going to customize the contents of the notebook, so we will use a `user-managed` notebook instance. We can use the same Vertex AI Workbench user-managed notebook instance that we created in *Chapter 7*. Please open JupyterLab on that notebook instance. In the directory explorer on the left side of the screen, navigate to the `Chapter-08` directory and open the `vizier-hpo.ipynb` notebook. You can choose `Python (Local)` as the kernel. Again, you can run each cell in the notebook by selecting the cell and pressing *Shift* + *Enter* on your keyboard. In addition to the relevant code, the notebook contains markdown text that describes what the code is doing.

### How our hyperparameter tuning job works

Using Vertex AI Vizier for hyperparameter tuning involves several steps that we implement in the model. Let's take a look at the salient steps in the process:

1. First, we create a training application, which consists of a Python script that trains our model with the given hyperparameters. This script must also track and report the performance of the model when testing it on the validation set so that Vertex AI Vizier can use those performance metrics to determine the best hyperparameters. For this reason, we use the `cloudml-hypertune` Python library in our code to periodically report the hyperparameter tuning metric back to Vertex AI.

2. Next, we create a configuration object for the hyperparameter tuning job, which specifies the hyperparameters to tune and the range of their possible values to try, as well as the objective metric we want to optimize (in our case, we're using AUC ROC, referred to simply as `auc` in the code). One important thing to note at this point is that the more hyperparameters we include, the more combinations of trials will need to be run. This could result in additional time and computing resources (and therefore cost) being needed for our tuning job. For this reason, it is best to use domain knowledge wherever possible to determine which hyperparameters we want the tuning job to focus on. We can also use the `maxTrials` variable in the configuration for the hyperparameter tuning job to control the number of trials.

   Understandably, it's not always possible to use domain knowledge to narrow the parameter search space, and we often will need to find a trade-off between the quality of our hyperparameter tuning job outputs and the time and costs required to run them. For example, running the tuning job for a very long time may get us as close as possible to finding the perfect set of hyperparameter values, but running it for a shorter time may get us results that are just good enough, depending on the needs of our use case.

3. The final stage in our hyperparameter tuning implementation is to use the Vertex AI Vizier client library to submit the hyperparameter tuning job to Vertex AI, which then runs our training application with different sets of hyperparameter values and finds the best ones.

### *Using the results of our hyperparameter tuning job*

Of course, we're not just running hyperparameter tuning jobs for fun (although it is also fun)! When our tuning job finds the best set of hyperparameters, we will want to access and review them, and usually, we will want to then use them to train the final version of our model.

## Accessing the results via the Google Cloud console

When we run the hyperparameter tuning job in the notebook, the output from our code will display a link that will enable us to view the status of the tuning job in the Google Cloud console. The most important thing to view at that link is the list of trials performed by our tuning job, which will look similar to *Figure 8.3*:

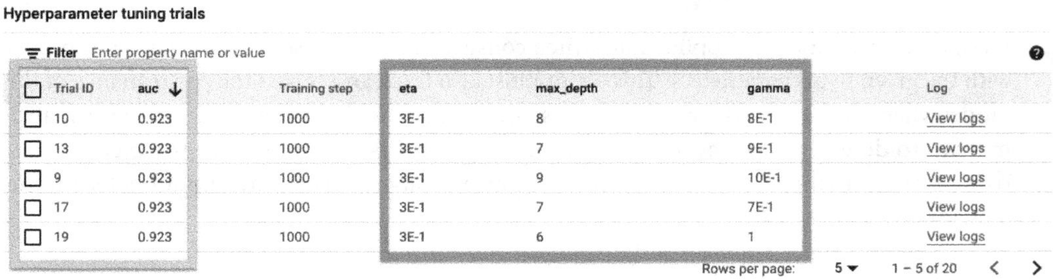

**Hyperparameter tuning trials**

Filter Enter property name or value

| Trial ID | auc ↓ | Training step | eta | max_depth | gamma | Log |
|---|---|---|---|---|---|---|
| 10 | 0.923 | 1000 | 3E-1 | 8 | 8E-1 | View logs |
| 13 | 0.923 | 1000 | 3E-1 | 7 | 9E-1 | View logs |
| 9 | 0.923 | 1000 | 3E-1 | 9 | 10E-1 | View logs |
| 17 | 0.923 | 1000 | 3E-1 | 7 | 7E-1 | View logs |
| 19 | 0.923 | 1000 | 3E-1 | 6 | 1 | View logs |

Rows per page: 5 1 – 5 of 20

Figure 8.3: Hyperparameter tuning trials

In the list of our hyperparameter tuning trials in the Google Cloud console, we can see the AUC metric for each trial, as well as the related trial ID (depicted within the box on the left in *Figure 8.3*), and we can also see the hyperparameter values that were used for each trial (depicted within the box on the right in *Figure 8.3*). We can click the arrow symbol in the header of the `auc` column to sort that column by ascending or descending AUC score. In our case, we want to sort it in descending order because we want the maximum score to appear at the top. This then tells us which trial had the hyperparameters that resulted in the best-performing model. In *Figure 8.3*, you may notice that at least the top five trials all have the same AUC score. This is common because there may be multiple different combinations of hyperparameter values that can result in the same metric score. You can use the arrow in the bottom-right of the screen to peruse through the pages of additional trials, and you will see other trials that resulted in lower AUC scores.

### Accessing the results programmatically

While it's useful and interesting to view the results of our hyperparameter tuning jobs in the Google Cloud console, we likely won't want to have to manually copy and paste them into the final training job to create our resulting model.

Fortunately, we can access all of those details programmatically via the Vertex API, and we can use the Vertex client library to do that from our development environment. In our notebook, after the tuning job finishes, we can continue with the additional activities in that notebook, which will show you how to access and use the best set of hyperparameter values produced by our tuning job. We then use those hyperparameter values to train a new model in our notebook, and then we finally test that model against the test dataset and calculate and display the resulting final AUC score. Note that when I ran this, I got a ROC-AUC score of 0.9188, which is pretty good!

Great job, you have now learned quite a lot about the topic of hyperparameter tuning and you should be ready to start applying what you've learned to other types of ML problems. Let's summarize what we've learned in this chapter.

# Summary

In this chapter, we dived deeper into the important concept of objective metrics in machine learning. We covered, in detail, many of the most popular metrics that are used for evaluating binary classification models, such as precision, recall, F1 score, and ROC AUC. We then moved on to discuss hyperparameter optimization, including some of the important theoretical information in this area, such as the different types of methods that can be used to search for the optimal set of hyperparameters and associated values. This also provided some insight into why it can be very difficult or even impossible to efficiently perform hyperparameter tuning manually due to the large number of trials that can be required.

Next, we dived into the Google Cloud Vertex AI Vizier service, which can be used to automate the hyperparameter tuning process for us. We then performed hands-on activities in Jupyter Notebook on Vertex AI, and we used Vizier to automatically find the best set of hyperparameters for training a credit card fraud detection model using XGBoost.

Next, we used the outputs of our hyperparameter tuning job to train a final version of our model, and we then evaluated that model against our test dataset.

In the next chapter, begin exploring beyond the simpler machine-learning algorithms such as linear regression and decision trees, and delve into the realm of Artificial Neural Networks (ANNs). Let's move on and discover this fascinating category of concepts and technologies.

# 9
# Neural Networks and Deep Learning

In this chapter, we will discuss **neural networks** (**NNs**) in **machine learning** (**ML**), often referred to as **artificial NNs** or **ANNs**. We will introduce many important topics in this field of science, including fundamental concepts that led to the development of ANNs, and relevant use cases for their application. At this point, it's important to note that the term **deep learning** (**DL**) refers to ML that is implemented with the use of **deep NNs** (**DNNs**). We will explain the term "DNN" later in this chapter.

We will also introduce some tools and frameworks that make it easier for us to create NNs, such as TensorFlow and Keras, and we will use those tools to build an NN in our hands-on activities later in this chapter. Finally, we will expand the discussion to include different types of NN architectures, common challenges in NN implementations, and some practices for optimizing our NN architectures.

As a side note, I first started learning about ANNs when I was in college, and I remember being absolutely fascinated by the concept because I also had an avid interest in understanding how the human brain works. Although the concept of NNs is indeed loosely based on the theorized workings of the human brain, in this chapter, we will separate hype from reality and will focus on the practical, mathematical descriptions of this technology. Let's start by covering some important concepts in this space.

This chapter covers the following topics:

- NN and DL concepts
- Libraries
- Implementing a **multilayer perceptron** (**MLP**) in TensorFlow
- NN architectures, challenges, and optimization

# NN and DL concepts

In this section, we discuss concepts that are important to understand in the context of NNs and DL. We begin by discussing how ANNs are linked to our understanding of the human brain.

## Neurons and the perceptron

While the link between artificial neurons and biological neurons (as found in the human brain) is often over-emphasized, there is a conceptual link between them that helps us form a mental model of how they work. Biological neurons generally consist of three main parts, as depicted in *Figure 9.1*:

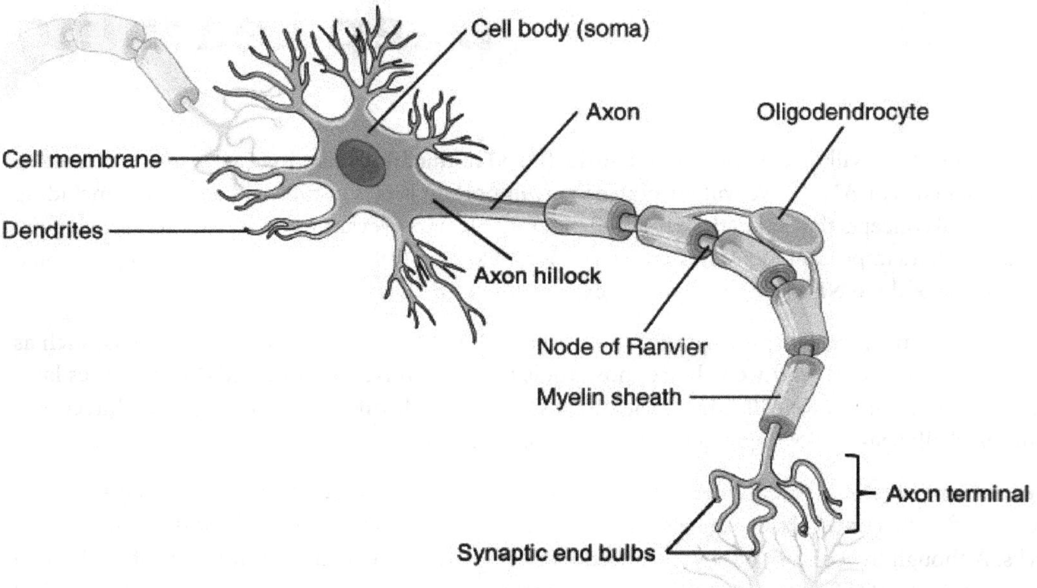


Figure 9.1: Neuron (source: https://www.flickr.com/photos/187096960@N06/51173238594)

The cell body is the core part of the neuron that contains the nucleus and other important components. The dendrites (coming from the Greek word "dendron," meaning "tree") are structures that branch out from the cell body. They receive information from other neurons and transmit this information to the cell body. Finally, the axons are long, tube-like structures that extend from the cell body and send information out to the dendrites of other neurons (across an interface called a synapse). That's about as deep as we're going to get with regard to the biology of a neuron; I've simplified it quite a bit here because we just need high-level context for the comparison we're going to make with ANNs, but the next thing we need to understand is how they transmit information, and this works at a high level as follows (again, simplified for relevant context).

When a neuron receives a signal from another neuron, this causes a change in what's called the **electrical potential** across the neuron's cell membrane (the difference in voltage between the inside and outside of the neuron), triggering what's called an **action potential**, which is an electrical impulse that travels down the axon. When it reaches the end of the axon, it triggers the release of neurotransmitters, which are chemical messengers. These neurotransmitters cross something called the **synaptic gap** (the tiny space between neurons) and bind to receptors on the dendrites of the next neuron, and this binding can then either trigger or inhibit a new action potential in the second neuron.

Okay – we've just introduced a lot of biological terminology in the past two paragraphs, but those concepts are important to understand when we want to draw a comparison with ANNs. With this in mind, let's move on and discuss how ANNs are constructed, beginning with their most basic concept, the perceptron, which I briefly mentioned in *Chapter 1* when I summarized various milestones in the evolution of ML.

A perceptron can be seen as one of the simplest types of ANNs and as a building block for larger, more complex networks. It was developed by Frank Rosenblatt in the late 1950s, and it's basically a binary classifier that maps its input X (a vector) to an output value f(x) (a single binary value) using a set of weights that are applied to the input features.

To understand this process in more detail, let's dive deeper into how a perceptron works, which can be summarized using the following set of steps:

1. The perceptron receives input values, which could be features or attributes from a dataset.
2. Each input has an associated **weight** that represents its importance. The weights are often just given random values at the beginning of training, and the values are then refined during the training process. We will discuss this in more detail shortly.
3. A **bias unit** is also added to the perceptron model to increase the model's flexibility. This is not related to the topic of bias that we will discuss in the context of fairness in later chapters of this book; it's simply a mathematical trick that provides an additional control mechanism for refining our model's performance when attempting to produce the desired output.
4. Next, each input is multiplied by its corresponding weight, and all of the results of those multiplications are added together (as well as the bias), which results in a **weighted sum**, so this is simply a **linear transformation** of the inputs, based on the weights and bias values.
5. This weighted sum is then passed through an **activation function** that produces a binary output. We will explain activation functions in more detail shortly, but at a high level, a non-linear transformation is performed on the weighted sum, and the results of that transformation are produced as an output from the perceptron. In the case of a single perceptron, a simple example would be that if the weighted sum of the inputs is greater than a threshold value, the perceptron would output a value of 1, or if the weighted sum is less than or equal to the threshold, it would output a value of 0, so this is basically an implementation of the process referred to as **logistic regression**.

Mathematically, this can be written as the following:

- If Σ (weights * inputs) + bias > 0, output 1
- If Σ (weights * inputs) + bias ≤ 0, output 0

*Figure 9.2* provides a visual representation of how a perceptron works:

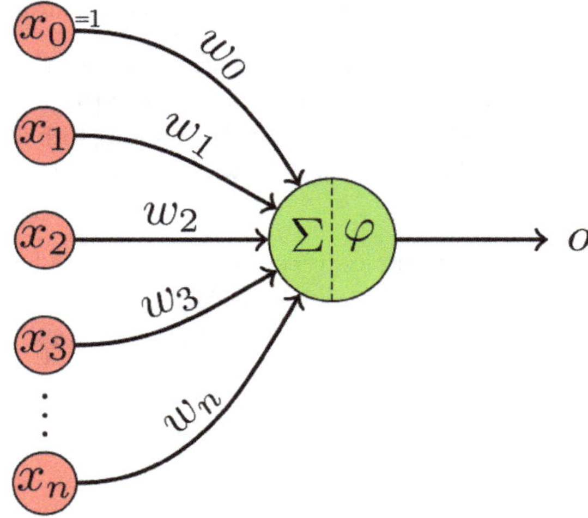


Figure 9.2: Perceptron (source: https://commons.wikimedia.org/wiki/File:Perceptron-unit.svg#file)

In *Figure 9.2*, the *x* values on the far left of the diagram represent inputs to the perceptron. The $x_0$ input is the bias, and $x_1$ through $x_n$ represent the input features from our dataset. The *w* values represent the weights, the Greek characters (sigma and phi) within the green circle represent the activation function, and the Greek character (omicron) on the far right represents the output.

The important concept to understand is that the values of the weights and bias are what our perceptron model is trying to learn. That is, our model is trying to figure out what the best weights are for each feature, which results in a pattern that gets us as close as possible to the target outcome after we perform the linear and non-linear transformations we described (in combination with the bias). If we think back to the more traditional ML models we created in earlier chapters, such as linear regression, you may remember that our models were trying to figure out the optimal values of the coefficients for each of our features that would result in the desired outcomes. In perceptrons, and in ANNs in general, the weights (and the bias) are the coefficients that our models are trying to optimize.

With this understanding of how perceptrons work, let's discuss how they can be used to build more complex NNs.

### *MLPs and NNs*

While the perceptron is a simple and powerful algorithm, it can only model linearly separable functions. This means that if our data isn't linearly separable (that is, we can't draw a straight line to separate the classes), the perceptron won't be able to accurately distinguish between the classes in the dataset. To overcome this limitation, multiple perceptrons can be combined in layers to form an MLP, which has the potential to solve non-linear problems. An MLP is a form of NN, so basically, when we combine multiple perceptrons in a sequential manner (that is, where the outputs of some perceptrons become the inputs for other perceptrons), we form a type of ANN.

Considering that the perceptron can be considered a type of artificial neuron, we will use the terms "perceptron" and "artificial neuron" (or sometimes just "neuron") interchangeably from this point onward.

#### Summarizing the comparison to biological neural activity

As we've already discussed, a neuron receives inputs from sensory organs or from other neurons, and, depending on the resulting electrical potential, an action potential causes the neuron to fire (or not fire) a message to other neurons.

Similarly, a perceptron (or artificial neuron) receives inputs from our dataset, or from other artificial neurons if we are chaining the perceptrons together in an NN. Then, depending on the linear combination of these inputs and their weights and biases, the activation function will influence the output of the perceptron, which can be used as an input to another perceptron in the network.

Next, let's dive into more detail on how NNs are typically structured, and introduce the important concept of **layers** in NNs.

### Layers in an NN

When artificial neurons are combined together to form an NN, they are not just connected randomly but instead are connected in a structured manner using the concept of layers, as depicted in *Figure 9.3*:

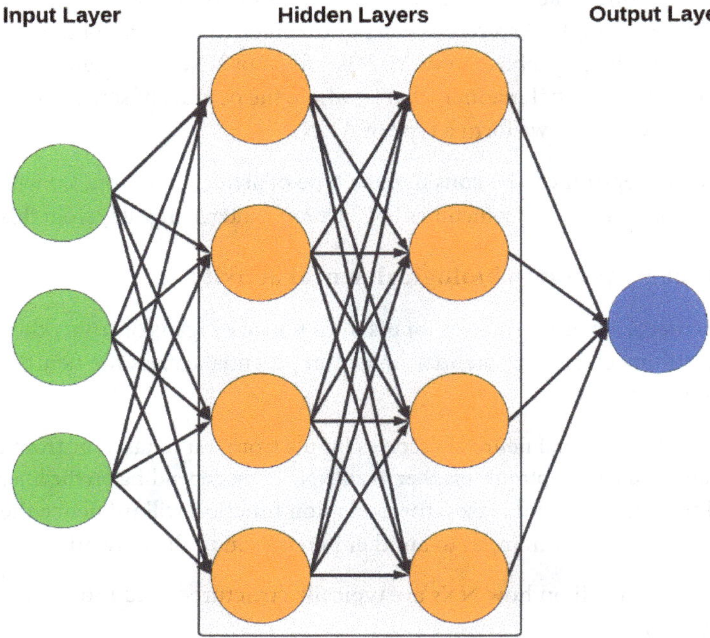


Figure 9.3: NN layers

As shown in *Figure 9.3*, the layers of an NN are generally categorized into three different types:

1. **The input layer**, which, as the name suggests, is how our inputs enter our NN. It is, of course, the first layer in the network.
2. **Hidden layers**, which sit between the input and output layers. Their job is to transform the inputs into something the output layer can use. The term "hidden" just means that they do not interface with the outside world (they are neither the input nor the output of the network). We usually do not have control over or direct interaction with them; they learn to represent the data on their own. The number of hidden layers and the number of neurons in each hidden layer define the complexity and structure of the NN.
3. **The output layer**, which, again suggested by the name, presents the output of our NN, which is usually some kind of prediction.

In addition to the number of hidden layers and the number of neurons in each hidden layer, exactly how the layers are connected together depends on the architecture of the NN. We will discuss different types of common NN architectures later in this chapter, but what generally happens is that after our input data is fed into the NN's input layer, each neuron in the subsequent layers of the network behaves similarly to how we described the perceptron earlier in this chapter.

For example, each input is assigned a weight that represents its importance. These weights are typically initialized with random values and then refined during the learning process. The inputs are multiplied by their corresponding weights and the results are summed together, in addition to a bias value. The summed result is then used as the input to the activation function in a neuron in the subsequent layer (that is, beginning with the first hidden layer) of the NN. Generally, this process is performed by every neuron in the subsequent layers. The important thing to remember is that the weights and biases will be different for each neuron. Therefore, even though the exact same input data is seen by each neuron in the first hidden layer, how each neuron reacts to the data will vary because the various weights and biases will influence each neuron's activation function differently.

The next thing to understand is that the outputs from the activation functions in each layer serve as inputs to the same process that will be performed again in subsequent layers of the network. So, the same process that we just described will be performed in each subsequent layer, but rather than our original dataset being used as an input in every layer, each subsequent layer will use the activation values from the previous layer as the input. This means that multiple transformations are being implemented as information travels through our network, and this is what makes NNs so powerful.

Let's just be sure we have a clear understanding of this process. If we take the second hidden layer as an example, the process works as follows:

Each activation function output from the first layer is assigned a weight that represents its importance. These weights are typically initialized with random values and then refined during the learning process. The activation function outputs are multiplied by their corresponding weights and the results are summed together, in addition to the bias values. The summed result is then used as the input to the activation function in the neurons in the next layer of the NN. This process is repeated in each layer until we get to the final, output layer of the network.

**Note**

Different NN architectures can use slightly different ways of propagating information through the network. We will discuss some common types of NN architectures in more detail in this chapter, but what we've described so far can be considered the building blocks of how ANNs work.

You may also hear the term "DNN" being used. Traditionally, any NN with at least two hidden layers is considered a DNN.

The fact that activations in neurons in one layer of the network influence the activations of neurons in the next layer brings us back to the loose analogy with the human brain, in which certain neurons firing can cause other neurons to fire, resulting in varied combinations of interactions that can produce much more complex higher-level functions. We must take this analogy with a pinch of salt, however, because even the most complex ANNs contain thousands of neurons, whereas the human brain has billions of neurons, and each neuron is capable of much more complex functionality than the relatively simple mathematical transformations performed by artificial neurons.

Now that we've discussed how information travels through an ANN, let's dive into more detail on how an ANN learns, and to do that, we must introduce the concept of **backpropagation**.

## Backpropagation

What we described in the previous section can be referred to as **forward propagation**, whereby information is being propagated forward from one layer to another in our NN. In order to discuss backpropagation, let's think back on what we learned earlier in this book with regard to how **supervised ML (SML)** algorithms work. Remember that we use labels to describe what each data point is in our dataset. When we train a model, the model tries to learn patterns between the features in our dataset that will help it accurately predict the label for each data point. We then use a loss function to calculate how far off our model's predictions were from the correct label; the primary purpose of the model training activity is to minimize the loss function (that is, to minimize the errors produced by our model), and we can use techniques such as gradient descent to minimize the loss function.

Let's take a basic linear regression model trained on tabular data as an example. You may remember that in such a case, each row in the table of our dataset represents a data point or observation, and each column in the table is a feature. The linear regression model tries to guess which coefficients it could use for each feature, such that multiplying each feature by its coefficient and adding all of the results together would result in something as close as possible to the target label.

In the case of linear regression, each time the model predicted incorrectly, we would use the loss function to calculate the error, then calculate the gradients of the loss function with respect to each coefficient, and then use gradient descent to determine how to adjust the coefficients accordingly, and the process would repeat many times until the model improved or was stopped for some reason.

This is where NNs become more complex than the simple regression models we implemented earlier in this book. In the case of NNs, we don't have a one-to-one mapping of input features to coefficients. Instead, our data propagates through a complex network consisting of multiple layers, which each contain multiple neurons, and each neuron in each layer has a different set of weights (and bias values). Therefore, when our model makes a prediction and we use a loss function to calculate the error, it's no longer simply a case of calculating the loss function's gradient with respect to the coefficient of each input feature and then updating each feature's coefficient and trying again. Instead, we must perform that process for all of the weights in each layer of the NN. One way to do this is referred to as "backward propagation of errors" or "backpropagation."

With backpropagation, we update the weights in each layer, starting with the last layer (that is, the one closest to our output layer, which represents our model's prediction), and then moving back through our network, layer by layer.

Since the loss function for our NN is composed of several nested functions (due to the layers in the network), the calculation of gradients in the backpropagation step uses something called the **chain rule**, which is a technique in calculus to compute the derivatives of the loss function with respect to the weights in each layer. These results are then used to determine how to update the weights in each pass through the network.

We will come back to the topic of backpropagation and the chain rule later in this chapter, but first, let's dive into more detail on what kinds of algorithms we can use to optimize our cost function in each training pass.

## *Cost function optimization algorithms*

We've already discussed using mechanisms such as gradient descent for optimizing our cost function during model training. In this section, we will briefly discuss some other common types of optimization algorithms that we can use.

### Momentum

This can be considered an upgrade to the basic gradient descent algorithm. When we discuss gradient descent in cost function optimization, we often use the analogy of walking downhill in a mountainous landscape until we reach the bottom of the mountain (or at least the bottom of a valley, which could be a "local minimum"). In the case of **stochastic gradient descent** (**SGD**), the analogy is more akin to jumping around somewhat randomly, in which case we may sometimes jump a little bit uphill (that is, in the incorrect direction), but overall, we usually end up going downhill (that is, in the correct direction). This case of jumping around in slightly different directions is referred to as **oscillating**. The Momentum algorithm accelerates SGD by navigating more prominently in the correct directions and reducing oscillations in incorrect directions. It does this by averaging the gradients of the updates in each step, which results in a smoother descent down the error gradient and often leads to reaching the bottom more quickly. The analogy used in this case is that of a ball rolling down the hill, in which case the movements are smoother than jumping around sporadically. Note that the ball can also gain momentum, which can help it to perform better even when the slope of the gradient is small (small gradient slopes result in slower learning for traditional gradient descent). In practice, Momentum almost always outperforms basic gradient descent.

### Adaptive Gradient Algorithm (Adagrad)

Adagrad, as the name suggests, is an adaptive optimization algorithm. That is, in each optimization cycle, Adagrad adapts the learning rate to each individual parameter. It performs smaller updates for parameters that have large gradients and larger updates for parameters with small gradients, which makes it particularly useful for dealing with sparse data and DL models with millions of parameters. Although it can be a useful algorithm, it can cause the learning rate to get too small too quickly, and effectively stop learning. This can be a problem in long training scenarios (such as in DL) where the learning process could prematurely stop. More recent variants such as **Root Mean Square Propagation** (**RMSProp**) and **Adaptive Moment Estimation** (**Adam**) were developed to tackle this issue, and we will discuss those next.

### RMSProp

RMSProp resolves Adagrad's rapidly diminishing learning rates by dividing the learning rate over a **moving average** (**MA**) of the squared gradients. Basically, it's faster and often better than Adagrad.

### Adam

Adam combines the benefits of Momentum and RMSProp. It averages the gradients (like Momentum) and uses squared gradients (like RMSProp). It's often the best choice of optimizer, especially for DL.

There are many more optimization algorithms in addition to the ones we have covered in this section, and we may use other optimizers in later chapters, but we will use Adam in this chapter, so for now, we're just introducing the popular algorithms that Adam builds upon.

We will dive deeper into one more important concept before we begin our hands-on activities, and that is the concept of activation functions.

## Activation functions

So far, we've touched upon the topic of activation functions and how they work at a high level. In this section, we dive into more details on this topic and discuss some common types of activation functions that we can use in our NNs.

### *Linear (identity) activation function*

This activation function simply returns whatever input we provide to it, unchanged, and it's typically used for simple tasks, or it's often used as the output layer in regression use cases.

It is represented mathematically as $f(x) = x$.

This is generally not something we would use for the hidden layers in our network because it doesn't allow for any kind of complex relationships to be learned.

> **Note**
>
> It's important to specifically call out the significance of non-linear transformations in NNs because the main power of NNs is their ability to combine multiple non-linear transformations in order to learn complex relationships in the data. Non-linear activation functions are, therefore, an important ingredient in complex NNs.
>
> Even if we combined many layers together in our network, if all of them just implemented a linear transformation, our whole network would just perform one big linear transformation, which we could implement without the need for NNs.

### *Sigmoid activation function*

The sigmoid function is an implementation of logistic regression, which maps any input into a range between 0 and 1.

It is represented mathematically as f(x) = 1 / (1 + exp(-x)).

See *Figure 9.4* for a visual representation of this function:

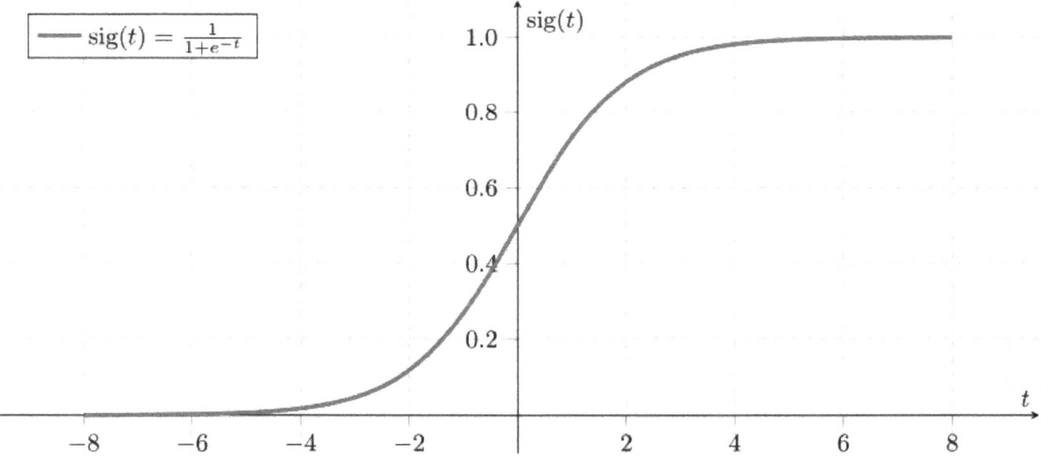


Figure 9.4: Sigmoid function (source: https://commons.wikimedia.org/wiki/File:Sigmoid-function-2.svg)

The sigmoid function is one of the simpler activation functions, and it has generally been superseded by newer functions that we will discuss next. One of its limitations is that it is susceptible to a problem known as the vanishing gradient problem, which we will describe later in this chapter. However, it can still be useful in the context of output neurons in binary classification problems, where we interpret the output as the probability of the input being in one class or the other.

### Hyperbolic tangent (tanh) activation function

The tanh function is like the sigmoid function but it maps any input to a value in the range between -1 and 1.

It is represented mathematically as f(x) = (exp(x) - exp(-x)) / (exp(x) + exp(-x)).

See *Figure 9.5* for a visual representation of this function:

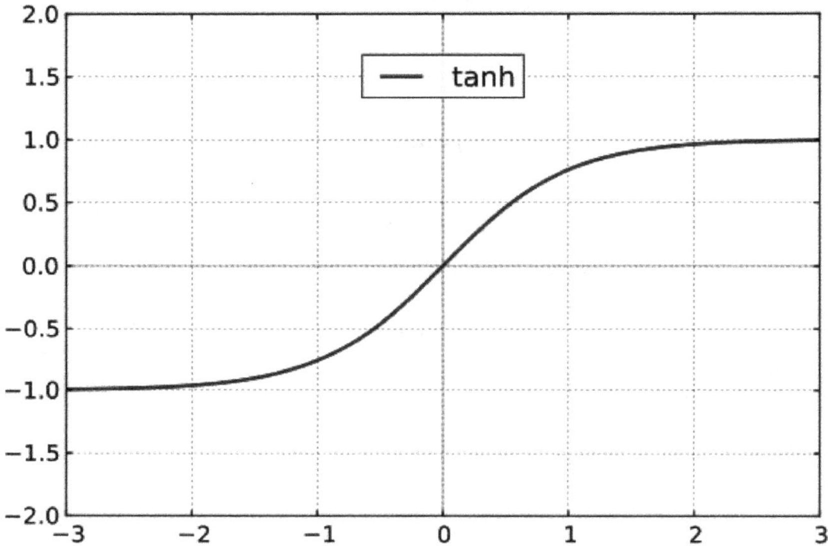


Figure 9.5: tanh function (source: https://commons.wikimedia.org/wiki/File:Mplwp_tanh.svg)

Like the sigmoid function, tanh also suffers from the vanishing gradient problem, but it can still be useful in practice.

### Rectified Linear Unit (ReLU) activation function

The ReLU function has become very popular in the last few years. Quite simply, it maps any positive number to itself and any negative number to zero.

It is represented mathematically as f(x) = max(0, x).

See *Figure 9.6* for a visual representation of this function:

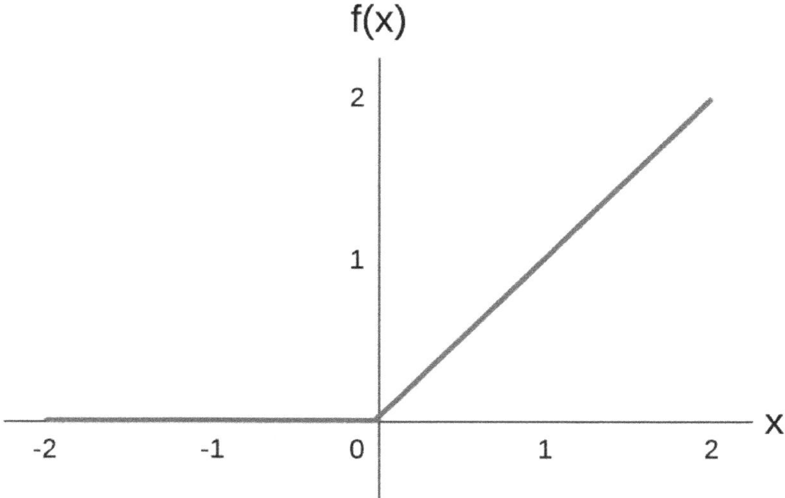


Figure 9.6: ReLU function

A major benefit of ReLU is that it is computationally efficient because the function is essentially just checking whether the input is greater than zero and simply returning the input directly if it is greater than zero, or returning zero if it isn't. This is an easy mathematical computation to perform, and this simplicity leads to much faster training.

Another major benefit is that it doesn't suffer from the vanishing gradient problem. However, it suffers from another problem known as the dying ReLU problem, which is a phenomenon where neurons effectively become useless due to consistently outputting zero. This happens because when the inputs are zero, or negative, then the gradient of the function becomes zero. This means that during backpropagation, when the weights get updated, the weights of that neuron will not be adjusted. This situation leads to the neuron becoming "stuck" and continually outputting zero – effectively causing the neuron to "die" and play no role in discriminating the input.

### *Leaky ReLU activation function*

This function attempts to solve the dying ReLU problem by outputting small negative values when the input is less than zero.

It is represented mathematically as $f(x) = \max(0.01x, x)$.

In this case, the value of 0.01 represents a small, nonzero gradient for x, and it's a hyperparameter that can be changed.

See *Figure 9.7* for a visual representation of this function:

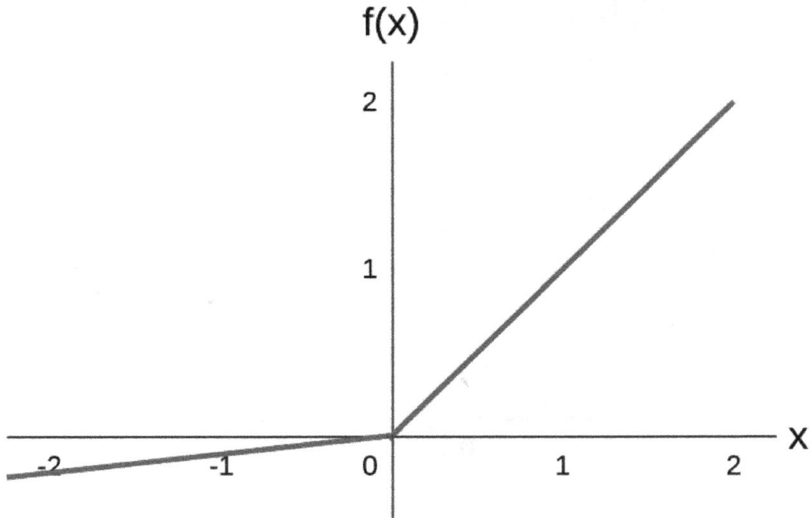


Figure 9.7: Leaky ReLU

As we can see in *Figure 9.7*, Leaky ReLU avoids the problem of the outputs becoming zero, even when the input is negative. There is also an extension of Leaky ReLU called **Parametric ReLU** (**PReLU**), which allows the small, nonzero gradient for x to be learned by backpropagation during learning, rather than specifying it as a static number via a hyperparameter.

### Softmax activation function

The softmax function is often used for the output layer of an NN in multiclass classification use cases, where it transforms the raw outputs of the network into a vector of probabilities (that is, creating a probability distribution for the classes).

It is represented mathematically as f(xi) = exp(xi) / Σ(exp(xj)), where j runs over the set of neurons in the output layer.

It's an extension of the sigmoid function; while the sigmoid function can be used to provide the probability that the input is a member of one class or another (in a selection between two classes), the softmax function can provide the range of probabilities of the input being a member of multiple classes. For example, if our network tries to identify images of numbers between 1 and 10, then there are 10 possible classes to choose from. If we provide an image of the number 1 and use softmax in the output layer of our network, it would hopefully determine that the number in the image has a high probability of being 1 and lower probabilities for each of the other potential classes (that is, 2 through 10).

There are more activation functions in addition to the ones we have covered in this section, but the ones we've covered here are among the most well known and widely used. Our choice of activation function can depend on the specific use case and other factors.

Now that we've covered many of the important theoretical concepts in the field of DL, let's put what we've learned into practice by building our first NN. To do that, we're going to introduce some important libraries.

# Libraries

In this section, we describe the libraries we will use in this chapter, such as TensorFlow and Keras.

## TensorFlow

TensorFlow is an **open source software** (**OSS**) library developed by the Google Brain team for ML and DNN research. However, it is also possible to run it on a wide range of systems, from mobile devices to multi-GPU setups, and it can have many applications beyond just ML. In this section, we will discuss some of its important aspects, such as the following:

- Tensors, which are a generalization of vectors and matrices in multiple dimensions (we can think of them as multi-dimensional arrays or lists). They are the basic building blocks in TensorFlow.
- **Data-flow graphs** (**DFGs**), in which nodes in the graph represent mathematical operations and edges represent the data (tensors) transmitted between these nodes. This approach enables parallel computation across multiple devices, making TensorFlow suitable for training large NNs.
- Multiple options for model deployment, such as TensorFlow Serving for server-side deployments or TensorFlow Lite for mobile and IoT devices.
- Backpropagation optimization by automatically computing the gradient of the loss with respect to the model weights.

Anyone interested in ML and DL should be familiar with TensorFlow because it is widely used in the industry.

## Keras

Keras is a high-level NN API written in Python that can run on top of lower-level frameworks such as TensorFlow, Theano, and **Cognitive Toolkit** (**CNTK**). It was developed to enable fast experimentation and has become the official high-level API of TensorFlow (as of TensorFlow 2.0). Some of its main features include the following:

- **User-friendliness**: It has a simple and consistent interface that has been optimized for common use cases, and it provides clear error messages, as well as useful documentation and developer guides.

- **Modularity**: A Keras model is assembled by connecting configurable building blocks together. For example, we can easily construct an NN by stacking multiple layers together.
- **Extensibility**: We can write custom building blocks to express new ideas for research, and create new layers, loss functions, and models.

The core data structure of Keras is a model, which is a way to organize layers. The main type of model is the Sequential model, which is a linear stack of layers, but for more complex architectures, we can use the Keras functional API, which allows us to build our own custom graphs of layers. A layer in Keras is a class that implements common NN operations, and Keras includes a wide range of pre-defined layers that we can use to build our models. It also allows us to specify the loss function and the metrics we want to evaluate during the training phase, and it provides many pre-defined loss functions such as `mean_squared_error` and metrics such as `accuracy`. Keras also includes many optimization algorithms, such as SGD. Overall, it includes lots of useful tools that make it easy for us to create NNs.

Now that we've introduced the relevant libraries, let's dive in and build our first NN!

## Implementing an MLP in TensorFlow

In this section, we will build an MLP using TensorFlow. We will use Keras as the high-level API for interacting with TensorFlow. We can use the same Vertex AI Workbench notebook we created in *Chapter 5* for this purpose. In that notebook, perform the following steps:

1. Navigate into the folder named `Google-Machine-Learning-for-Solutions-Architects`.
2. Double-click on the `Chapter-09` folder within it and then double-click on the `Chapter-9-TF-Keras.ipynb` file to open it.
3. When prompted to select a kernel, select **TensorFlow**.
4. The notebook we have opened contains some Python code that creates and tests an MLP using Keras and TensorFlow.
5. Run each of the cells in the notebook by clicking on each cell and pressing *Shift* + *Enter* on your keyboard. If you see any errors related to CUDA, you can ignore them because we are not using GPUs in this notebook.

Our code in the first cell of the notebook imports the necessary libraries and modules, loads a dataset using the `make_moons` function from `sklearn.datasets`, and then visualizes the data using `matplotlib`.

In this case, we're using the `moons` dataset, which is a mathematically generated dataset for binary classification that is often used as a simple test case for ML algorithms, especially those designed to handle non-linear data (such as NNs). The dataset consists of a two-dimensional array of two features (usually visualized on an X-Y plane) and a binary label (0 or 1) for each sample, and the samples are generated in such a way that they form two crescent moon-like shapes when plotted (hence the name "moons"), with each "moon" corresponding to one class. See *Figure 9.8* for reference:

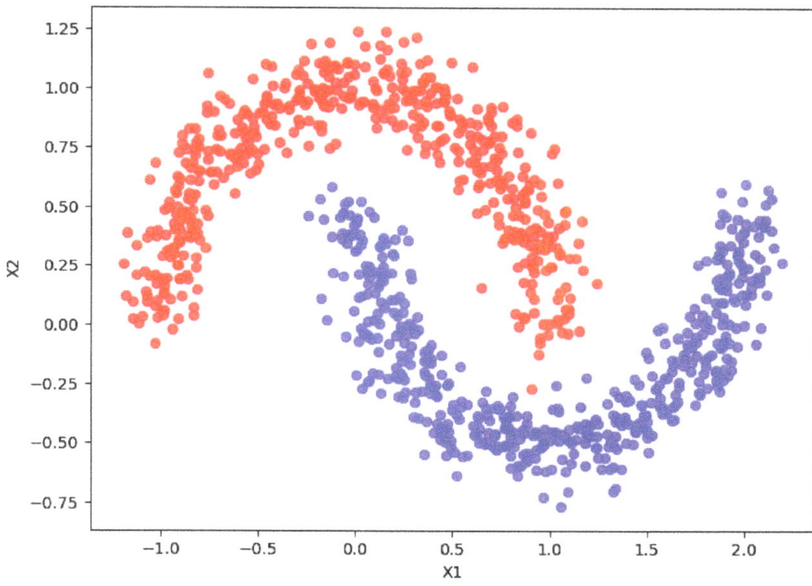


Figure 9.8: The moons dataset

Note that the main characteristic of the dataset is its non-linearity (that is, the decision boundary separating the two classes is not a straight line).

The code in the second cell of our Jupyter notebook then does the following:

- Splits the dataset into a training set and a test set
- Defines a Sequential model (which means that the layers are stacked on top of each other)
- Adds an input layer and the first hidden layer with 32 neurons and the `relu` activation function
- Adds a second hidden layer with 32 neurons and the `relu` activation function
- Adds an output layer with one neuron (for binary classification) and the `sigmoid` activation function

- Compiles the model with the `adam` optimizer and the `binary_crossentropy` loss function (suitable for binary classification)
- Trains the model for 50 epochs

When the code is running, you should see outputs from each epoch, as depicted in *Figure 9.10*:

```
loss: 0.6145 - accuracy: 0.6012 - val_loss: 0.5268 - val_accuracy: 0.8600

loss: 0.4672 - accuracy: 0.8600 - val_loss: 0.4178 - val_accuracy: 0.8700
```

Figure 9.9: Training epoch outputs

Note that `loss` and `val_loss` (validation loss) should decrease as the training progresses, and `accuracy` and `val_accuracy` (validation accuracy) should increase as the training progresses.

Next, our code in the third cell does the following:

- Evaluates our model using the `model.evaluate` method, which returns the loss value and metric values (in this case, `accuracy`) for the model, in test mode.
- Gets some predictions from our model using the `model.predict` method, which outputs the probabilities that each input sample belongs to the positive class.
- In order to treat this as a binary classification use case, our code then converts these probabilities into binary class labels based on a threshold of 0.5 (that is, anything with a probability of more than 0.5 is deemed to be a member of the positive class).
- Finally, we print out the first 10 predictions for a quick check. These outputs will be in the form of 0s and 1s, denoting the predicted class labels.

That's it! You've just created an NN! It might not be quite as smart as a human, but this is a basic example of an MLP in TensorFlow with Keras. We could extend this to more advanced DL use cases by adjusting things such as the number of layers, neurons, types of activation functions, types of optimizers, loss functions, and training configurations (such as the number of epochs, batch size, and so on). Great job!

Next, we will dive into additional DL concepts, such as different types of NN architectures, challenges in applications of DNN, and optimization considerations.

## NN architectures, challenges, and optimization

We've primarily covered the basics of NNs so far in this chapter, and in this section, we will expand our discussion to include different types of NN architectures that can be used for different types of real-world use cases, as well as some challenges that are often encountered when training them. Finally, we will discuss how to optimize our NNs to address those challenges.

### Common NN architectures

The "architecture" of an NN refers to its structure in terms of the number of layers it contains and the number of neurons in each layer, as well as any special characteristics that influence how information is propagated through the network. The NN architectures we've described so far in this chapter are the simplest forms of ANNs, which are referred to as **feed-forward NNs** (**FFNNs**). Information in these networks travels in one direction only, from the input layer, through the hidden layers, to the output layer. Next, let's take a look at some other commonly used NN architectures. We will introduce them at a high level here, and we will dive into much more detail in later chapters.

> Note
>
> When we talk about information traveling through the network in this section, we are not referring to the backpropagation step, because that is a separate step implemented during the iterative learning process. We are simply referring to how data is processed through our network in each training pass, or at inference time.

#### *Convolutional NNs (CNNs)*

CNNs are commonly used in **computer vision** (**CV**) for use cases such as object recognition and picture categorization. Consider the scenario where we wish to teach our model to recognize images of cats, keeping in mind that these images might take many different forms, such as being captured at various distances and perspectives. Our model would need to establish a kind of visual understanding of what a cat is, such as the shape of its face, ears, body, and tail, in order to correctly identify cats in the images.

CNNs do this by breaking pictures down into smaller components or "features" and learning each one separately. In this way, the network learns to detect small details in a larger image, such as an edge or a curve, and then combines those into larger features such as a single whisker or a portion of a cat's ear, regardless of where they appear in the image, and then combining those characteristics to identify a cat. The concepts of **convolutional layers**, **pooling layers**, and **fully connected (FC) layers** are used by CNNs to do this. In later chapters of this book, we will dive further into those ideas and how they function.

### *Recurrent NNs (RNNs) and long short-term memory (LSTM) networks*

RNNs are designed to find patterns in sequential data such as language or time series. In RNNs, the network contains loops, which enable information to be persisted from one step to the next, and this is what allows RNNs to create a kind of memory, unlike basic NNs that assume all inputs (and outputs) are independent of each other. In this way, the network blends current data with inputs from earlier phases in each step, which is important for activities such as language comprehension, in which the model needs to understand each word in relation to the other words in the input (that is, the words are not entirely independent).

However, one of the problems with RNNs is that, for reasons we'll cover later, they "forget" prior inputs when dealing with long sequences. To address these issues, variations such as LSTM networks and **gated recurrent unit** (**GRU**) networks have been invented, which use gates and other techniques to maintain memory.

### *Autoencoders (AEs)*

AEs are used to learn efficient encodings of unlabeled data, usually to reduce the dimensionality of the data. An AE's basic idea is pretty straightforward: it is trained to try to duplicate its input to its output. Although it might seem like a simple (and redundant) operation, the restrictions we place on the network force it to discover interesting aspects of the data. Most commonly, we limit the number of nodes in the hidden layers, forcing the network to learn a compressed view of the data.

AEs consist of an encoder, which encodes the input data as a compressed representation in a reduced-dimensional space, and a decoder, which attempts to reconstruct the input from the reduced-dimensional representation (that is, it "decodes" the compressed representation). The goal during training is to create a reconstructed output that is as close as possible to the input so that the network can learn to rebuild the input from the compressed representation.

In the real world, AEs are used for applications such as anomaly detection and recommendation systems. One particularly popular application of AEs is in **generative AI** (**GenAI**) models. In this case, after the AE has been trained, the decoder can generate new data that mimics the training data.

### *Generative adversarial networks (GANs)*

The basic objective of GANs, which are made up of two competing NNs called a generator and a discriminator, is to create new fake data that closely matches "real" data (as determined by the training data). The two networks are trained in tandem using a **minimax** game, which is a type of game where the generator tries to trick the discriminator, and the discriminator tries to reliably distinguish real data from the generated data. During training, the generator becomes ever more accurate at providing data that appears real, while the discriminator becomes increasingly skillful at identifying forgeries.

### *Transformer networks*

Transformer networks are a type of model architecture invented by Google, in which the breakthrough innovation is the use of **self-attention** mechanisms, or **multi-head attention**, which enables the model to consider the relative significance of various words in a phrase while generating an output. For example, in a sentence such as "the dog tried to jump over the pond, but it was too wide," the word "it" could refer to either the pond or the dog. To humans, it seems pretty obvious that it refers to the pond, but that's because we're using contextual awareness to intuit what makes the most sense. However, this is not inherently obvious to an ML model, and self-attention is the mechanism that allows the model to get a better understanding of the contextual meaning of each word in the sentence. The Transformer architecture also includes the concept of an encoder and a decoder, which are both made up of a stack of identical layers. And, because the self-attention mechanism alone does not account for the location of words in the input sequence, the Transformer design additionally contains a **positional encoding** system to track the positions of the words. We will dive into all of these components in much greater detail in a later chapter.

Another advantage provided by Transformer models is that they handle all inputs in parallel, as opposed to sequence-based models such as RNNs or LSTM networks, and this can significantly speed up training and inference.

Transformers have proven to be very useful for **natural language processing** (**NLP**) tasks including **sentiment analysis** (**SA**), text summarization, and machine translation, and the Transformer architecture serves as the foundation for models such as **Generative Pre-trained Transformer** (**GPT**), **Bidirectional Encoder Representations from Transformers** (**BERT**), and T5. If you're interested in learning more about this ground-breaking technology, I recommend reading the pivotal research paper that first introduced the concept of Transformers (Vaswani, A. et al., 2017), which can be found at the following URL: `https://arxiv.org/abs/1706.03762`.

There are more types of NN architectures in addition to the ones we have covered in this section, but the ones we've covered here are among the most well known and widely used. Researchers are constantly creating and experimenting with new NN configurations, and the choice of network configuration relies on the issue we're seeking to resolve because each network type has advantages and disadvantages.

Next, let's discuss some common challenges that people run into when training and using NNs.

## Common NN challenges

In addition to all of the challenges we covered in earlier chapters with regard to traditional ML implementations, DNNs introduce their own set of challenges, such as interpretability, cost, and vanishing or exploding gradients.

### *Interpretability*

Interpretability refers to how easily we can understand the inner workings of our models, and the reasons behind the decisions they produce. Models such as linear regression are generally quite easy to understand and explain because their outputs are just a straightforward mathematical transformation of whatever input is provided to the model.

However, DNNs can be extremely complex, with thousands of neurons and billions of parameters that influence their outputs. Also, their outputs are usually not just linear transformations of their inputs but instead are non-linear in nature.

In a later chapter, we will discuss the importance of interpretability in much more detail and will introduce mechanisms that can help us to better understand how our models work.

### *Cost*

DNNs can require a lot of computing resources to train and host, which can result in monetary expenses. If we consider the example of highly complex models with billions of parameters, those models can take weeks or even months to train, using large numbers of very powerful servers with the latest generation of cutting-edge GPUs. Those kinds of resources are not cheap, so we need to ensure that our models are optimized to use computing resources as efficiently as possible.

### *Vanishing gradient problem*

We briefly touched on this topic in earlier sections of this chapter, so let's take a look at this concept in more detail. The vanishing gradient problem develops when the gradients of the loss function are so tiny that they essentially vanish, leading to sluggish weight updates in the network's first layers. Remember that backpropagation uses the chain rule of differentiation, which involves multiplying a sequence of derivatives (or gradients). This means that if the values of the derivatives are less than 1, we are essentially dividing them in an exponential manner into smaller and smaller values as we propagate back through the network. Because of this, early layers learn much more slowly than later layers.

When utilizing activation functions such as the sigmoid or tanh functions, which condense their input into a small range, this problem becomes more prominent. The sigmoid function, for example, compresses input values into a range between zero and one, which means that even when the inputs are large in size (either positive or negative), the output of the sigmoid function is between zero and one, and the result is that the gradients reduce to tiny values, causing backpropagation-based learning to slow down significantly.

### *Exploding gradient problem*

On the other hand, the exploding gradient problem happens when the gradient becomes too large, causing the weights in the network to be updated by excessive increments. This can result in the network's performance oscillating wildly, causing the model training process to fail.

The exploding gradient problem is more common in RNNs, especially with long sequences, but it can occur in any type of network.

### *Optimizations to help prevent vanishing or exploding gradients*

Now that we have a better understanding of how vanishing and exploding gradient issues take place, the following considerations can help reduce their likelihood of occurring.

#### Weight initialization

Effective weight initialization can help mitigate problems of vanishing and exploding gradients. For example, techniques such as Xavier (Glorot) initialization and He initialization can help set the initial weights to values that prevent the gradients from becoming too small or too large early in training.

#### Choice of activation functions

As we discussed, using activation functions that squash their inputs, such as sigmoid or tanh, can increase the likelihood of encountering vanishing and exploding gradient issues. Therefore, it's best to avoid those activation functions in cases that are prone to those issues. We can instead use activation functions such as Leaky ReLU or PReLU, as these functions do not squash their inputs.

#### Batch normalization

Using this technique, we can normalize the output of a layer to stabilize the means and variances of each layer's inputs. This helps control the scale of gradients, mitigating both vanishing and exploding gradient problems.

#### Gradient clipping

This technique puts a pre-defined limit or threshold on the gradients to prevent them from getting too large, which is especially useful for tackling the exploding gradient problem.

#### Architectural methods

When we discussed RNNs earlier in this section, we mentioned that they "forget" prior inputs when dealing with long sequences. The vanishing gradient problem is one factor that causes this to happen, and this is one of the reasons why certain architectures such as LSTM and GRU were designed to address these issues in the context of RNNs by using a form of gating in their structure. Therefore, sometimes the choice of NN architecture can reduce the likelihood of encountering vanishing and exploding gradient problems.

These problems and their solutions are major factors in understanding how DL models work and how to effectively train DNNs.

We've covered a lot of new concepts and terminology in this chapter. Let's take a moment to summarize everything we've learned.

# Summary

In this chapter, we started by exploring the comparison between artificial neurons (in the form of the perceptron) and biological neurons in the human brain. We then extended this idea to describe the activity of multiple neurons in an NN, both in terms of combining multiple perceptrons together and in terms of how the tiny neurons in our brain work together to produce extremely complex higher-level functions.

We then dived deeper into the inner workings and components of ANNs, including concepts such as activation functions and backpropagation. We discussed many different types of activation functions, including how they work and what use cases are most appropriate for them.

In the context of backpropagation, we learned about various types of commonly used cost function optimization algorithms, such as Momentum and Adam, and then we introduced two very important libraries for DL: TensorFlow and Keras.

Next, we built our first NN using those libraries, and we tested that network by successfully getting predictions based on the `moons` dataset, which we also explored in some detail.

After building our first simple NN, we expanded our discussion to cover more advanced kinds of NN architectures and their use cases, and we explored common challenges that people often run into when training and using NNs, as well as some approaches we can use to optimize our networks in order to reduce the likelihood of running into those issues.

These are some of the more advanced concepts in ML, so if you have understood the content we've covered in this chapter, then you have built an important foundation for the deeper dives we will perform on these concepts in later chapters.

In the next chapter, let's explore how we can bring trained models into production to host them for serving real-world use cases.

# 10
# Deploying, Monitoring, and Scaling in Production

Some people may read this book from beginning to end to gain an overall understanding of as many concepts as possible in the realm of AI/ML on Google Cloud, while others may use it as a reference, whereby they pick it up and read certain chapters on specific topics whenever they need to work with those topics as part of a project or client engagement. If you've been reading this book from the beginning, then you have come a long way, and we have journeyed together through the majority of the **ML model development life cycle** (**MDLC**). While model training is what often gets the most attention in the press – and that is where a lot of the magic happens – you know by now that training is just one piece of the overall life cycle.

When we've trained and tested our models, and we believe they're ready to be exposed to our clients, we need to find a way to host them so that they can be used accordingly. In this chapter, we will dive into that part of the process in more detail, including some of the challenges that exist when it comes to hosting and managing models and monitoring them on an ongoing basis to ensure that they stay relevant and perform optimally in perpetuity. We'll begin by discussing how we can host our models.

This chapter covers the following topics:

- How do I make my models available to my applications?
- Fundamental concepts for serving models
- A/B testing
- Common challenges of serving models in production
- Monitoring models in production
- Optimizing for AI/ML at the edge

## How do I make my models available to my applications?

We introduced this concept in *Chapter 1*, and we talked about the various things you would need to do to host a model on your own, such as setting up all of the required infrastructure, including load balancers, routers, switches, cables, servers, and storage, among other things, and then managing all of that infrastructure on an ongoing basis. This would require a lot of your time and resources.

Luckily, all of that stuff was in the old days, and you no longer need to do any of that. This is because Google Cloud provides the Vertex AI prediction service, which enables you to host models in production within minutes, using infrastructure that is all managed for you by Google.

For completeness, I will also mention that if you would like to host your models on Google Cloud without using Vertex, numerous other Google Cloud services can be used for that purpose, such as **Google Compute Engine** (GCE), **Google Kubernetes Engine** (GKE), **Google App Engine** (GAE), Cloud Run, and Cloud Functions. We described all of these services, as well as some pointers on how to choose between them, in *Chapter 3*.

Remember that choosing the right platform to host your ML models depends on your specific use case and requirements. Factors such as scalability, latency, costs, development effort, and operational management all play a role in choosing the best solution. You may also make certain decisions based on the framework you're using to build your ML models. For example, you might want to use TensorFlow Serving if you're building models in TensorFlow, or TorchServe if you're building models in PyTorch.

In most cases, my recommendation would be to start with a service that is dedicated and optimized for the task at hand, and in the case of building and hosting ML models, Vertex AI is that service. In this chapter, we will deploy our first model using Vertex AI, but before we dive into the hands-on activities, we'll introduce some important concepts.

## Fundamental concepts for serving models

In this section, we will introduce some important topics related to how we can host our models so that our clients can interact with them.

### Online and offline model serving

In ML, there are generally two options we have for serving predictions from models: **online** serving (also known as **real-time** serving) and **offline** serving (also known as **batch** serving). The high-level use cases associated with each of these methods are **online (or real-time) inference** and **offline (or batch) inference**, respectively. Let's take a few minutes to introduce these methods and understand their use cases.

### *Online/real-time model serving*

As the name suggests, in the case of real-time model serving, the model needs to respond "in real time" to prediction requests, which usually means that a client (perhaps a customer, or some other system) needs to receive an inference response as quickly as possible, and may be waiting synchronously for a response from the model. An example of this would be a fraud detection system for credit card transactions. As you can imagine, credit card companies want to detect possible fraudulent transactions as quickly as possible – ideally during the transaction process, in which case they could prevent the transaction from processing completely, if possible. It would not be as useful for them to check for fraudulent transactions at some arbitrary point later.

Considering that online inference requests usually require a response to be returned as quickly as possible, factors such as ensuring low latency, handling high request volumes, and providing a reliable, always-available service are some of the main challenges in this area.

In this case, when we refer to low latency, we mean that the prediction response time is usually in the order of milliseconds or, at the very most, seconds. The actual response time requirements will depend on the business use case. For example, some users may accept needing to wait for a few seconds for their credit card transaction to be approved or rejected, but another example of real-time inference is the use of ML models in self-driving cars, and in that scenario, for example, a car must be able to respond to its environment in milliseconds if it needs to suddenly take some kind of action, such as avoiding an unexpected obstacle that comes into its path. *Figure 10.1* shows an example of a batch prediction workflow that we will implement in the practical exercises in this chapter. We will explain each of the components that are depicted in detail. In *Figure 10.1*, the solid lines represent steps we will explicitly perform in the practical exercises, whereas the dotted lines represent the steps that Vertex AI will perform automatically on our behalf:

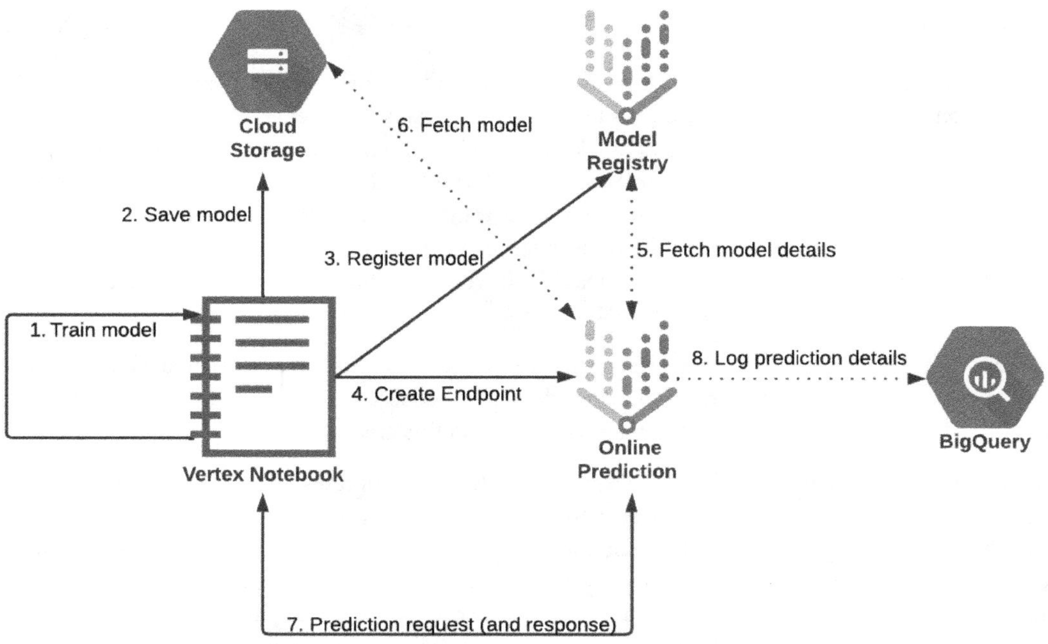


Figure 10.1: Online prediction

The steps outlined in *Figure 10.1* are as follows:

1. Train the model in our notebook.
2. Save the resulting model artifacts to Google Cloud Storage.
3. Register the model details in the Vertex AI Model Registry (explained in more detail in this chapter).
4. Create an endpoint to host and serve our model.
5. The Vertex AI Online Prediction service (explained in more detail in this chapter) fetches our model's details from the Vertex AI Model Registry.
6. The Vertex AI Online Prediction service fetches our saved model from Google Cloud Storage.
7. We send a prediction request to our model and receive a response. In this case, we are sending the request from our Vertex AI Workbench notebook, but it's important to note that when our model is hosted on an endpoint, prediction requests could be sent from any client application that can access that endpoint.
8. The Vertex AI Online Prediction service saves the prediction inputs, outputs, and other details to Google Cloud BigQuery. This is an optional feature that we can enable so that we can perform analytical queries on the inputs, outputs, and other details related to our model's predictions.

It should also be noted that in the case of online model serving, predictions are generally made on demand, and usually for a single instance or a small batch of instances. In this case, your model and its serving infrastructure need to be able to quickly react to sudden – and possibly unexpected – changes in inference traffic volume.

### Offline/batch model serving

Given our description of online model serving, it may have become obvious that offline serving means that no client is waiting for an immediate response in real time. In fact, rather than our model receiving on-demand inference requests from individual clients, we can feed many input observations into our model in large batches, which it can process over longer periods; perhaps hours or even days, depending on the business case. The predictions produced by our models can then be stored and used later, rather than being acted on immediately. Examples of batch inference use cases include predicting the next day's stock prices or sending out targeted emails to users based on their predicted preferences. *Figure 10.2* shows an example of a batch prediction workflow that we will implement in the practical exercises in this chapter. In *Figure 10.2*, the solid lines represent steps we will explicitly perform in the practical exercises, whereas the dotted lines represent the steps that Vertex AI will perform automatically on our behalf:

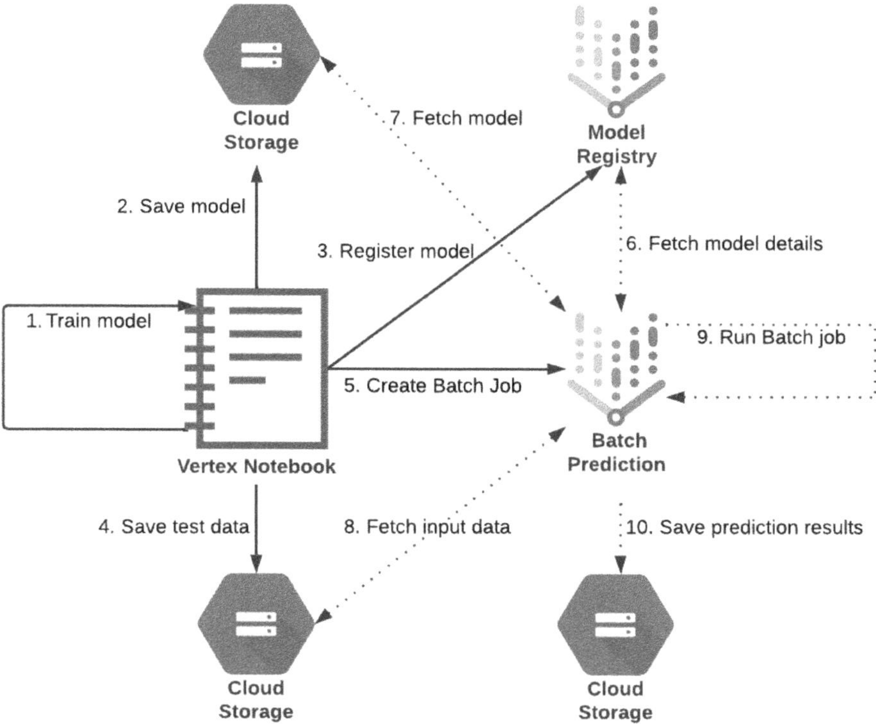


Figure 10.2: Batch prediction

The steps outlined in *Figure 10.2* are as follows:

1. Train the model in our notebook.
2. Save the resulting model artifacts to Google Cloud Storage.
3. Register the model details in the Vertex AI Model Registry.
4. Save the test data in Google Cloud Storage. This will be used as the input data in our batch prediction job later.
5. Create a batch prediction job.
6. The Vertex AI Batch Prediction service (explained in more detail in this chapter) fetches our model's details from the Vertex AI Model Registry.
7. The Vertex AI Batch Prediction service fetches our saved model from Google Cloud Storage.
8. The Vertex AI Batch Prediction service fetches our input data from Google Cloud Storage.
9. The Vertex AI Batch Prediction runs a batch prediction job, using our model and input data.
10. The Vertex AI Batch Prediction service saves the prediction outputs to Google Cloud Storage.

Rather than being optimized for low latency, batch prediction systems are generally optimized to handle a large number of instances at once (that is, high throughput), and therefore usually fall into the category of large-scale distributed computing use cases, which can benefit from parallelized execution, and can be scheduled to automatically execute periodically (for example, once a day), or can be triggered by an event (for example, when a new batch of data is available).

It's important to note that the decision between online and offline serving is not always strictly binary, and you may find that a combination of online and offline serving best meets your needs. For example, you may use offline serving to generate large-scale reports, while also using online serving to make real-time predictions in user-facing applications. In either case, both online and offline serving require the model to be deployed in a serving infrastructure. This infrastructure is responsible for loading the model, receiving prediction requests, making predictions using the model, and returning the predictions. In Google Cloud, there are various tools and platforms available to assist with this, such as TensorFlow Serving, and the Vertex AI prediction service, which we will discuss in more detail in subsequent sections in this chapter. First, however, let's introduce another tool that is important for managing our models as part of our end-to-end model development life cycle.

## Vertex AI Model Registry

Earlier in this book, we made an analogy between the traditional **software development life cycle** (**SDLC**) and the MDLC. To dive into this analogy in more depth, let's consider some important tools in the SDLC process:

- **Shared repositories**: As a part of the SDLC process, we usually want to store our code and related artifacts in registries or repositories that can be accessed by multiple contributors who need to collaborate on a specific development project. Such repositories often include metadata that helps describe certain aspects of the code assets so that people can easily understand how those assets were developed, as well as how they are used.
- **Version control**: As contributors make changes to code assets in a given project, we want to ensure that we are tracking such changes and contributions and that all contributors can easily access that information. This also enables us to roll back to previous versions if we notice issues in a newly deployed version of our software.

Experience has taught us that similar tools are required when we want to efficiently deploy and manage ML models, especially when doing so at a large scale (remember that some companies may have thousands of ML models, owned by hundreds of different teams).

Also, bear in mind that data science projects are often highly experimental, in which the data science teams may try out lots of different algorithms, datasets, and hyperparameter values, training lots of different models to see which options produce the best results. This is especially true in the early stages of a data science project, but this also often holds true on an ongoing basis, whereby the data scientists constantly strive to improve the models even after they have been deployed to a production environment. They do this to keep abreast of emerging trends in the industry and produce better results.

Google Cloud's Vertex AI platform includes a Model Registry service that allows us to manage our ML models in a centralized place, making it much easier for us to track how models are developed, even when multiple teams are contributing to the development of those models. Let's take a look at some of the Model Registry's important features:

- **Model versioning**: The Model Registry allows us to create multiple versions of a model, where each version can correspond to a different set of training parameters or a different set of training data. This helps us keep track of different experiments or deployments.
- **Model metadata**: For each model, we can record metadata such as the model's description, the input and output schemas, the labels (useful for categorization), and the metrics (useful for comparing models). For each version, we can record additional metadata such as the description, the runtime version (corresponding to the version of the Vertex AI platform services), the Python version, the machine type used for serving, and the serving settings.

- **Model artifacts**: These are the artifacts that are used to produce the model. They can be stored in Google Cloud Storage and linked to the model in the registry.
- **Access control**: We can control who can view, edit, and deploy models in the registry through Google Cloud's **Identity and Access Management** (**IAM**) system.

It's also important to understand that the Model Registry is well-integrated with other Vertex AI components. For example, we can use the Vertex AI training service to train a model and then automatically upload the trained model to the registry, after which we can deploy the model to the Google Cloud Vertex AI prediction service, which we'll describe next. We can compare model versions against each other and easily change which versions are deployed to production. We can also automate all of those steps using Vertex AI Pipelines, something we'll explore in the next chapter.

### Vertex AI prediction service

The Google Cloud Vertex AI prediction service is an offering within the Vertex AI ecosystem that makes it easy for us to host ML models and serve them to our clients, thus supporting both batch and online model-serving use cases. It's a managed service, so when we use it to host our models, we don't need to worry about managing the servers and infrastructure required to do so; the service will automatically scale the required infrastructure and computing resources up and down based on the amount of traffic being sent to our models.

As we mentioned in the previous section, it integrates with Vertex AI Model Registry so that we can easily control which versions of our models are deployed to production, and it also integrates with many other Google Cloud services, such as Vertex AI Pipelines, which allows us to automate the development and deployment of our models, and Google Cloud Operations Suite, which provides integrated logging and monitoring functionality.

We will perform hands-on activities in which we will use the Vertex AI Model Registry and the Vertex AI prediction service to store and serve models in Google Cloud shortly, but first, let's cover one more concept that's important in the context of model deployment and management: **A/B testing**.

## A/B testing

A/B testing is the practice of testing one model (model A) against another model (model B) to see which one performs better. While the term could technically apply to testing and comparing any models, the usual scenario is to test a new version of a model to improve model performance concerning the business objective.

Vertex AI allows us to deploy more than one model to a single endpoint, as well as control the amount of traffic that is served by each model by using the `traffic_split` variable, as shown in *Figure 10.3*:

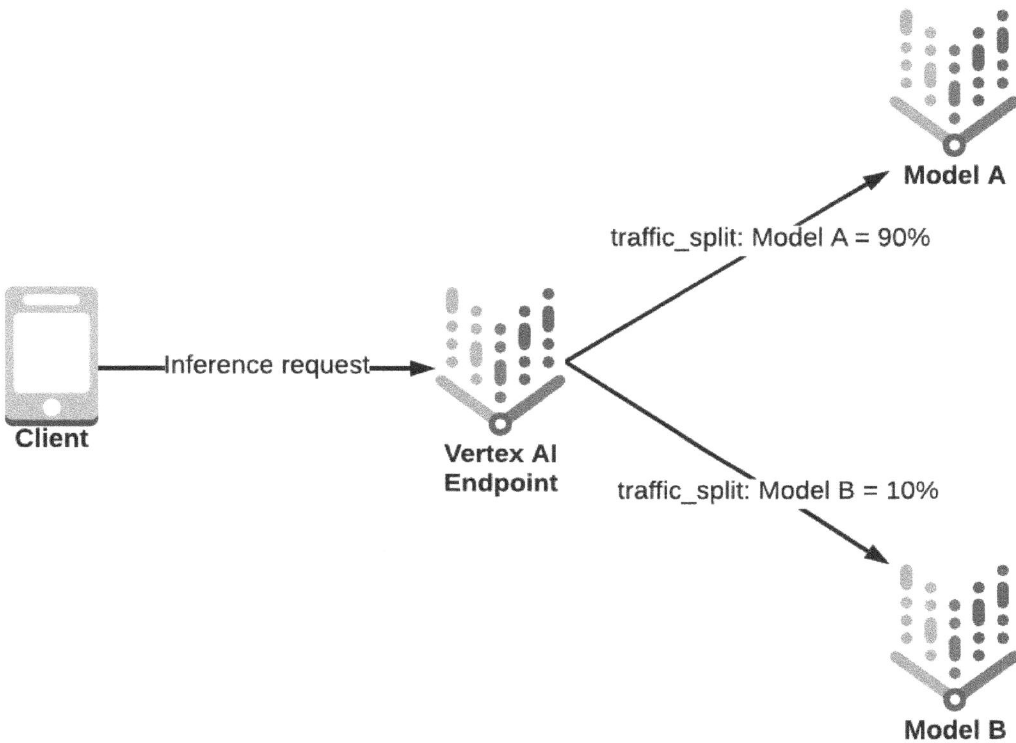


Figure 10.3: A/B configuration using Vertex AI's traffic_split

As you will see in the practical exercises in this chapter, if we don't set any value for the `traffic_split` variable, the default behavior is to keep all traffic directed to the original model that was already deployed to our endpoint. This is a safety mechanism that prevents unexpected behavior in terms of how our models serve traffic from our clients. The `traffic_split` configuration gives us very granular control over how much traffic we want to send to each deployed model or model version. For example, we could set the `traffic_split` configuration so that it suddenly starts sending all traffic to our new model by allocating 100% of the traffic to that model, which would effectively perform an in-place replacement of our model. However, we may want to test our new model version with a small subset of our production traffic before completely replacing our prior model version, which equates to the idea of **canary testing** in software development.

When we determine that our new model is behaving as intended, we can gradually (or suddenly, depending on the business requirements) change the `traffic_split` variable to send more (or all) traffic to the new model.

Now that we've covered many of the important concepts related to hosting and serving models in production, let's see how this works in the real world by deploying a model using Vertex AI.

We've prepared a Vertex AI Workbench notebook that will walk you through all of the steps required to do this. Again, we can use the same Vertex AI Workbench managed notebook instance that we created in *Chapter 5* for this purpose. Please open JupyterLab on that notebook instance. In the directory explorer on the left-hand side of the screen, navigate to the `Chapter-10` directory and open the `deployment-prediction.ipynb` notebook. You can choose TensorFlow 2 (Local) as the kernel. Again, you can run each cell in the notebook by selecting the cell and pressing *Shift* + *Enter* on your keyboard. In addition to the relevant code, the notebook contains markdown text that describes what the code is doing. I recommend only executing the model training and deployment sections, and A/B testing, and then reading through some more of the topics in this chapter before proceeding to the other activities in the notebook.

We must also enable a feature named `prediction-request-response-logging` in the notebook, which will log our models' responses for the prediction requests that are received. We can save those responses in a Google Cloud BigQuery table, which enables us to perform analysis on the prediction responses from each of our models and see how they are performing.

Once you have completed the model training and deployment sections of the notebook (or if you'd like to just keep reading for now), you can move on to the next section, where we will discuss the kinds of challenges that companies usually run into when deploying, serving, and managing models in production.

# Common challenges of serving models in production

Deploying and hosting ML models in production often comes with numerous challenges. If you're developing and serving just one model, you may encounter some of these challenges, but if you are developing tens, hundreds, or thousands of models, then you will likely run into the majority of these challenges and concerns.

## Deployment infrastructure

Choosing the right infrastructure to host ML models, setting it up, and managing it can be complex, particularly in hybrid or multi-cloud environments. Again, Google Cloud Vertex AI takes care of all of this for us automatically, but without such cloud offerings, many companies find this to be perhaps one of the most challenging aspects of any data science project.

## Model availability and scaling in production

This is an extension of deployment infrastructure management. As demand increases, our model needs to serve more predictions. The ability to scale services up and down based on demand is crucial and can be difficult to manage manually. The Vertex AI Autoscaling feature enables us to do this easily. All we have to do is specify the minimum and maximum number of machines that we want to run for each model.

For example, if we know that we always need at least three nodes running to handle our regularly expected traffic, but we sometimes get increased levels of traffic up to double the normal amount, we could specify that we want Vertex AI to always run at least three machines, and to automatically scale up to a maximum of six machines when needed.

We can specify our autoscaling preferences by configuring the `minReplicaCount` and `maxReplicaCount` variables, which are elements of the machine specification for our deployed model. We've provided steps on how to do this in the Jupyter Notebook that is associated with this chapter. Note that we can specify these details per model, not just per endpoint. This enables us to scale each of our models independently, which gives us flexibility, depending on our needs.

Vertex AI also gives us the flexibility to either deploy multiple models to the same endpoint or to create separate, dedicated endpoints for each model. If we have completely different types of models for different use cases, then we would typically deploy those models to separate endpoints. At this point, you might be wondering why we would ever want to deploy multiple models to a single endpoint. The most common reason for doing this is when we want to implement A/B testing. Again, we've covered the steps for implementing A/B testing use cases in the Jupyter Notebook that accompanies this chapter.

## Data quality

Often, the data that's used in a production environment may contain errors or may not be as clean as the data used for model training, which could lead to inaccurate predictions. As a result, we may need to implement data processing steps in production that prepare and clean up the data "on-the-fly" at prediction time. In general, any data transformation techniques that we may have applied when preparing the data for training also need to be applied at inference time.

## Model/data/concept drift

In addition to preparing and cleaning up our data, if you think back to some of the data exploration activities we performed in previous chapters of this book, you may remember that one important aspect of our data is the statistical distribution of the variables in our dataset. Examples include the mean value of each variable in the dataset, the observed range between the minimum and maximum values of each variable, or the kinds of values of each variable, such as discreet or continuous. Collectively, we can refer to these characteristics as the "shape" of our data. *Figure 10.4* shows some examples of different types of data distributions:

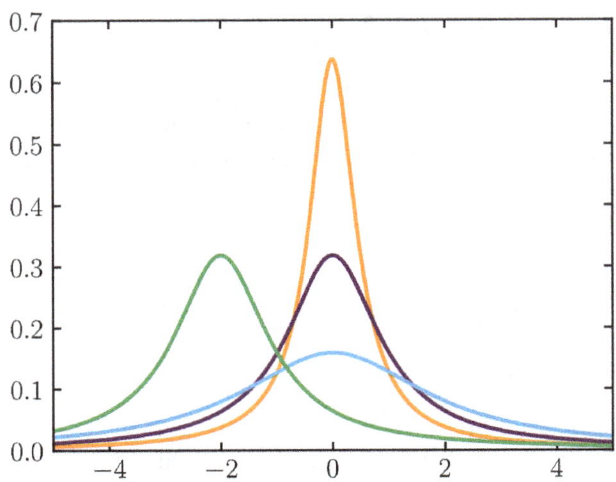


Figure 10.4: Data distributions
(source: https://commons.wikimedia.org/wiki/File:Cauchy_pdf.svg)

In an ideal world, the shape of the data that is used to train our model should be the same as the shape of the data that the model is expected to encounter in production. For this reason, we usually want to use real-world data (that is, data that was previously observed in production) to train our models. However, over time, the shape of the data that's observed in production might begin to deviate from the shape of the data that was used to train the model. This can happen suddenly, such as when a global pandemic drastically changes the purchasing behavior of consumers within a short period (for example, everybody is suddenly purchasing toilet paper and hand sanitizer, rather than fashion and cosmetic products), or seasonally, or it could occur more gradually due to natural evolutionary changes in a given business domain or market.

In any case, such a deviation is generally referred to as "drift," or more specifically **model drift**, **data drift**, or **concept drift**. Model drift refers to the overall degradation of the model's objective performance over time, and it can be caused by factors such as data drift or concept drift. Data drift is when the statistical distribution of the data in production differs from the statistical distribution of the data that was used to train our model. On the other hand, concept drift is when the relationships change between the input features and the target variable that a model is trying to predict. This means that even if the input data stays the same, the underlying relationships between the variables may change over time, which can affect the accuracy of our model. The example of consumer behavior changing during a pandemic is an instance of concept drift.

Regardless of the type of drift, it can lead to degrading model performance. As such, we need to constantly evaluate the performance of our models to ensure they consistently meet our business requirements. If we find that the performance of our models is degrading, we need to assess whether it is due to drift, and if we detect that it is, we need to take corrective measures accordingly. We will discuss such measures in more detail later.

## Security and privacy

Ensuring that the data used by our model complies with security and privacy regulations can be complex, particularly in industries dealing with sensitive data. Depending on the industry and use case, this can be considered one of the most important aspects of model development and management.

## Model interpretability

Often, models are treated as "black box" implementations, in which case we don't have much visibility into how the model is operating internally, making it difficult to understand how they're making predictions. This can cause issues, especially in regulated industries where explanations are required. We will explore this topic in a lot more detail later in this book.

## Tracking ML model metadata

As we mentioned earlier, ML model development usually involves some experimentation, in which data scientists evaluate different versions of their datasets, as well as different algorithms, hyperparameter values, and other components of the development process. Also, the development process often involves collaboration among multiple team members or multiple separate teams. Considering that some companies develop and deploy hundreds or even thousands of ML models, it's important to track all of these experiments and development iterations.

For example, if we have a model deployed in production, and we observe that the model is making inaccurate or inappropriate predictions, then we need to understand why that model is behaving as it is. To do that, we need to know all of the steps that were performed to create that specific model version, as well as all of the inputs and outputs associated with the development of that model, such as the version of the input dataset that was used to train the model, as well as the algorithm and hyperparameter values used during the training process. We refer to this information as the metadata of the model. Without this, it would be very difficult to understand why our model behaves the way it does.

This concept is somewhat linked to the topic of model interpretability but also includes additional aspects of ML model development.

### Integration with existing systems

Most companies have various software systems that were developed or procured over many years, and integrating the ML model into these systems can be complex.

### Monitoring

Setting up robust monitoring for the performance of ML models in production can be a challenge if you manage the infrastructure yourself. We need to constantly monitor the model's predictive performance to ensure it is still valid and hasn't degraded over time.

This is such an important aspect of model management that we will dedicate the next section of this chapter specifically to this topic.

## Monitoring models in production

Our work isn't over once we've developed and deployed a model – we need to track the model's performance and its overall health over time and make adjustments if we observe that the model performance deteriorates. In this section, we'll discuss some of the characteristics of our models that we typically need to monitor.

### Objective model performance

Not surprisingly, the most prominent aspect of our model that we need to monitor is how it performs concerning the objective it was created to achieve. We discussed objective metrics in previous chapters of this book, such as **Mean Squared Error** (**MSE**), Accuracy, F1 score, and AUC-ROC, among others. An example of AUC-ROC is depicted in *Figure 10.5* for reference:

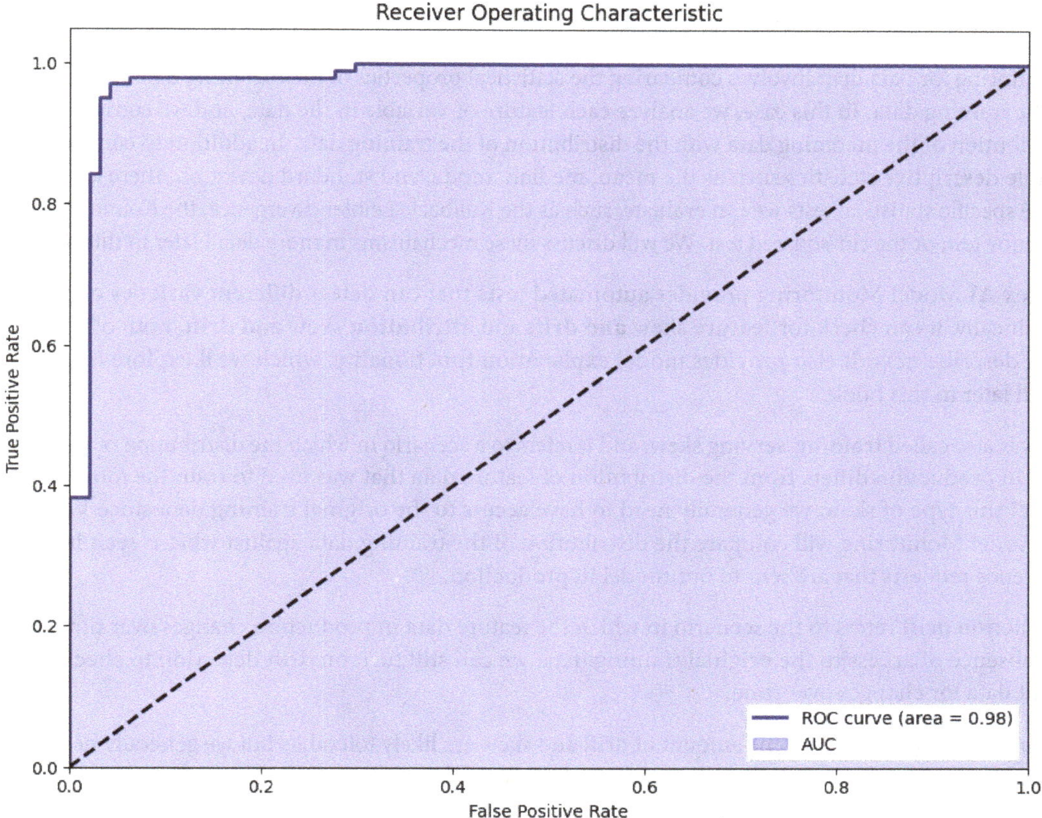


Figure 10.5: AUC-ROC

Objective metrics tell us how our model is performing in terms of the main purpose our model is intended to serve, such as predicting housing prices or identifying cats in photographs. If we notice a sustained degradation in these metrics beyond a certain threshold that we deem acceptable to the needs of the business, then we usually need to take corrective action.

A common cause of degradation of model performance is data drift, which we discussed in the previous section. If we find that data drift has occurred, a corrective action is often to retrain our model with updated data that matches the shape or distribution of the data that is currently being observed by the model in production. This may require gathering fresh data from our production systems or other appropriate sources.

## Monitoring specifically for data drift

Monitoring for data drift involves comparing the statistical properties of the incoming data with those of the training data. In this case, we analyze each feature or variable in the data, and we compare the distribution of the incoming data with the distribution of the training data. In addition to comparing simple descriptive statistics such as the mean, median, mode, and standard deviation, there are also some specific statistical tests we can evaluate, such as the Kullback-Leibler divergence, the Kolmogorov-Smirnov test, or the chi-squared test. We will discuss these mechanisms in more detail later in this book.

Vertex AI Model Monitoring provides automated tests that can detect different varieties of drift. Specifically, it can check for **feature skew and drift** and **attribution skew and drift**, both of which we'll describe next. It also provides model explanation functionality, which we'll explore in great detail later in this book.

Skew is also called **training-serving skew**, and it refers to a scenario in which the distribution of feature data in production differs from the distribution of feature data that was used to train the model. To detect this type of skew, we generally need to have access to the original training data since Vertex AI Model Monitoring will compare the distribution of the training data against what is seen in the inference requests that are sent to our model in production.

**Prediction drift** refers to the scenario in which the feature data in production changes over time. In the absence of access to the original training data, we can still turn on drift detection to check the input data for changes over time.

It's important to note that some amount of drift and skew are likely tolerable, but we generally need to determine and configure thresholds beyond which we need to take corrective actions. These thresholds depend on the needs of the business for our particular use case.

## Anomalous model behavior

Anomalies could be spikes in prediction requests, unusual values of input features, or unusual prediction response values. For example, if a model that is built for fraud detection starts flagging an unusually high number of transactions as fraudulent, it may be an anomaly and warrant further investigation. Anomalies may also occur due to data drift, or due to temporary changes in the environment.

## Resource utilization

This includes monitoring aspects such as CPU usage, memory usage, network I/O, and others to ensure that the model's serving infrastructure is operating within its capacity. These metrics are important indicators for determining when to scale resources up or down based on factors such as the amount of traffic currently being sent to our models.

## Model bias and fairness

We need to ensure that our models are making fair predictions on an ongoing basis. This can involve tracking fairness metrics and checking for bias in our models' predictions. We will explore this topic in much more detail in the next chapter.

These are some of the main aspects of our models that we need to monitor. The exact items that need to be prioritized depend on our business case.

Now that we've talked about many of the most common topics in the space of model monitoring, let's discuss what to do if we detect that our model's performance is degrading.

## Addressing model performance degradation

If we find that the values of the metrics we're monitoring for our models are showing a decline in performance, there are some actions we can take to correct the situation, and possibly prevent such a scenario from occurring in the future. This section discusses some relevant corrective actions.

Perhaps the most effective way to ensure that our models stay up to date and are performing optimally is to implement a robust MLOps pipeline. This includes **continuous integration and continuous deployment** (**CI/CD**) for ML models, and a major component of this is to regularly train our models on new data that is seen in production – this is referred to as **continuous training**. For this purpose, we need to implement a mechanism for capturing data in production, and then automatically update our models.

We could either update our models periodically (for example, every night or every month), or we could trigger retraining to occur based on outputs from a Vertex AI Model Monitoring job. For example, if a Model Monitoring job detects that drift has occurred beyond an acceptable threshold, an automated notification could be sent, and it could automatically kick off a pipeline to train and deploy a new version of our model.

The next chapter is dedicated entirely to the topic of MLOps, so we will dive into these concepts in much more detail there.

In the meantime, if you'd like to switch from theory to practical learning, now would be a good time to execute the *Model Monitoring* section of the Vertex AI Workbench notebook that accompanies this chapter. Otherwise, let's continue with the rest of the topics in this chapter, which include optimizing for AI/ML use cases at the edge.

# Optimizing for AI/ML at the edge

Serving ML models at the edge refers to running your models directly on user devices such as smartphones or IoT devices. The term "edge" is based on traditional network architecture terminology, in which the core of the network is in the network owner's data centers, and the edge of the network is where user devices connect to the network. Running models and other types of systems at the edge can provide benefits such as lower latency, increased privacy, and reduced server costs. However, edge devices usually have limited computing power, so we may need to make some changes to our models for them to run efficiently on those devices. There are several things we can do to optimize our models to run at the edge, all of which we will discuss in this section.

## Model optimization

Let's start by discussing what kinds of measures we can take to optimize our models so that they can be used at the edge.

### *Model selection*

First, we should try to choose lightweight models that can still provide good performance on our objectives. For example, decision trees and linear models often require less memory and computational power than deep learning models. Of course, our model selection process also depends on our business needs. Sometimes, we will need larger models to achieve the required objective. As such, this recommendation for optimizing AI/ML workloads at the edge is simply a first-step guideline. We will discuss more advanced strategies next.

### *Model pruning*

Pruning is a technique for reducing the size of a model by removing parameters (for example, "weight pruning") or whole neurons (for example, "neuron pruning") that contribute the least to the model's performance. An example of neuron pruning is depicted in *Figure 10.6*, in which a neuron has been removed from each hidden layer (as represented by the red X covering each of the removed neurons):

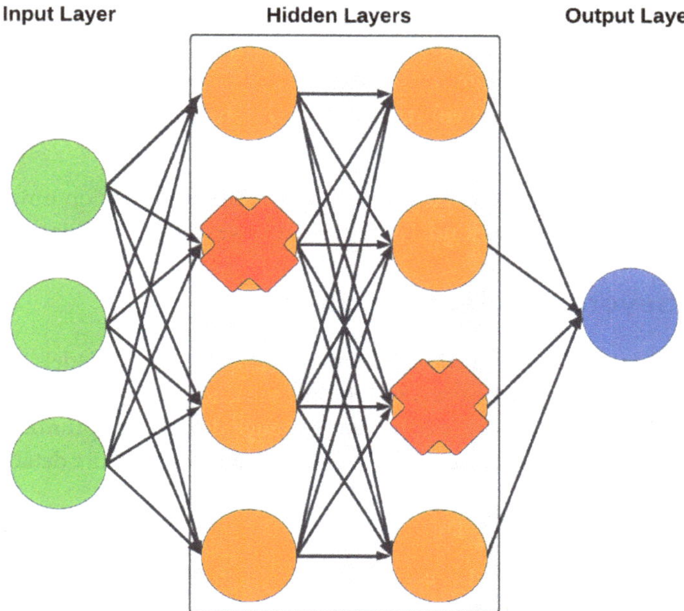


Figure 10.6: Neuron pruning

The resulting pruned model requires less memory and computational resources. If we remove too many weights or neurons, then it could affect the accuracy of our model, so the idea, of course, is to find a balance that reduces the computational resources required, with minimal impact on accuracy.

### Model quantization

Quantization is a way of reducing the numerical precision of the model's weights. For example, weights might be stored as 32-bit floating-point numbers during training, but they can often be quantized as 8-bit integers for inference without a significant reduction in performance. This reduces the memory requirements and computational cost of the model. This is particularly useful in the context of **Large Language Models** (**LLMs**), which can have hundreds of billions of weights. We will cover this in detail in the *Generative AI* section of this book.

### Knowledge distillation

This technique involves training a smaller model, sometimes referred to as a "student" model, to mimic the behavior of a larger, "teacher" model, or ensemble of models. The smaller model is trained to produce outputs as similar as possible to those of the larger model so that it can perform a somewhat similar task, with perhaps a tolerable reduction in model accuracy. Again, we need to find a balance in the trade-off between model size reduction and reduction in accuracy. Distillation is also particularly useful in the context of LLMs.

### *Making use of efficient model architectures*

Some model architectures are designed to be efficient on edge devices. For example, MobileNet and EfficientNet are efficient variants of **convolutional neural networks** (**CNNs**) that are suitable for mobile devices.

Now that we've discussed some changes that we can make to our models to optimize them for edge use cases, let's take a look at what other kinds of mechanisms we can use for this purpose.

## Optimization beyond model techniques

All of the previous optimization techniques involved making changes to our models to make them run more efficiently on edge devices. There are also additional measures we can take, such as converting trained models into other formats that are optimized for edge devices or optimizing the hardware that runs our models at the edge. Let's discuss these mechanisms in a bit more detail.

### *Hardware-specific optimization*

Depending on the specific hardware on the edge device (for example, CPU, GPU, **tensor processing unit** (**TPU**), and so on), different optimization strategies can be used. For example, some libraries provide tools for optimizing computation graphs based on specific hardware.

### *Specialized libraries*

Libraries such as TensorFlow Lite and the **Open Neural Network Exchange** (**ONNX**) Runtime can convert models into a format optimized for edge devices, further reducing the memory footprint and increasing the speed of models. We'll discuss these libraries in more detail in this section.

#### TensorFlow Lite

TensorFlow Lite is a set of tools provided by TensorFlow to help us run models on mobile, embedded, and IoT devices. It does this by converting our TensorFlow models into a more efficient format for use on such edge devices. It also includes tools for optimizing the model's size and performance, as well as for implementing hardware acceleration. We've used TensorFlow Lite to convert our model; this can be found in the Jupyter Notebook that accompanies this chapter.

#### Edge TPU Compiler

Google's Edge TPU Compiler is a tool for compiling models to run on Google's Edge TPUs, which are designed for running TensorFlow Lite models on edge devices.

### ONNX Runtime

ONNX is an open format for representing ML models that enables models to be transferred between various ML frameworks. It also provides a cross-platform inference engine called the ONNX Runtime, which includes support for various kinds of hardware accelerators and is designed to provide fast inference and lower the resource requirements of ML models.

### TensorRT

TensorRT is a deep learning model optimizer and runtime library developed by NVIDIA for deploying neural network models on GPUs, particularly on NVIDIA's embedded platform called Jetson.

### TVM

Apache TVM is an open source ML compiler stack that aims to enable efficient deployment of ML models on a variety of hardware platforms. TVM supports model inputs from various deep learning frameworks, including TensorFlow, Keras, PyTorch, ONNX, and others.

These are just some of the tools that exist for optimizing ML models so that they can run at the edge. With the constant proliferation of new types of technological devices, edge optimization is an active area of research, and new tools and mechanisms continue to be developed.

In our hands-on activities in the Jupyter Notebook that accompanies this chapter, we used TensorFlow Lite to optimize our model, after which we stored the optimized model in Google Cloud Storage. From there, we can easily deploy our model to any device that supports the TensorFlow Lite interpreter. A list of supported platforms is provided in the TensorFlow Lite documentation, which also contains a lot of additional useful information on how TensorFlow Lite works in greater detail: `https://www.tensorflow.org/lite/guide/inference#supported_platforms`.

At this point, we've covered a lot of different topics related to model deployment. Let's take some time to recap what we've learned.

## Summary

In this chapter, we discussed various Google Cloud services for hosting ML models, such as Vertex AI, Cloud Functions, GKE, and Cloud Run. We differentiated between online and offline model serving, whereby online serving is used for real-time predictions, and offline serving is used for batch predictions. Then, we explored common challenges in deploying ML models, such as data/model drift, scaling, monitoring, performance, and keeping our models up to date. We also introduced specific components of Vertex AI that make it easier for us to deploy and manage models, such as the Vertex AI Model Registry, the Vertex AI prediction service, and Vertex AI Model Monitoring.

Specifically, we dived quite deep into monitoring models in production, focusing on data drift and model drift. We discussed mechanisms to combat these drifts, such as automated continuous training.

Next, we explained A/B testing for comparing two versions of a model, and we discussed optimizing ML models for edge deployment using methods such as model pruning and quantization, as well as libraries and tools for optimizing our models, such as TensorFlow Lite.

At this point, we have covered all the major steps in the MDLC. Our next topic of focus will be how to automate the entire life cycle using MLOps. Join us in the next chapter, where you'll continue your journey of becoming an expert AI/ML solutions architect, in which you have already come a very long way and are making great progress!

# 11

# Machine Learning Engineering and MLOps with Google Cloud

It's generally estimated that almost 90% of data science projects never make it to production. Data scientists spend a lot of time training and experimenting with models in the lab, but often don't succeed in bringing those workloads out into the real world. A major reason for this is because, as we have discussed in the previous chapters of this book, there are difficult challenges at every step in the model development lifecycle. Following on from our previous chapter, we will now dive into more detail on deployment concepts and challenges, and describe the importance of **Machine Learning Operations** (**MLOps**) in addressing these challenges for large-scale production AI/ML workloads.

Specifically, this chapter will cover the following topics:

- An introduction to MLOps
- Why MLOps is needed for deploying large-scale ML workloads
- MLOps tools
- Implementing MLOps on Google Cloud using Vertex AI Pipelines

While we've already touched on some of these concepts so far in the book, we'll kick off this chapter with a more formal introduction to MLOps.

## An introduction to MLOps

MLOps is an extension of the DevOps concept from the software development industry but with a specific focus on managing and automating ML model and project lifecycles. It goes beyond the tactical steps required to create machine learning models and addresses requirements that come to light when companies need to manage the entire lifecycle of data science use cases at scale. This is a good time to reflect on the ML model lifecycle stages we outlined in previous chapters in this book, as depicted in *Figure 11.1*.

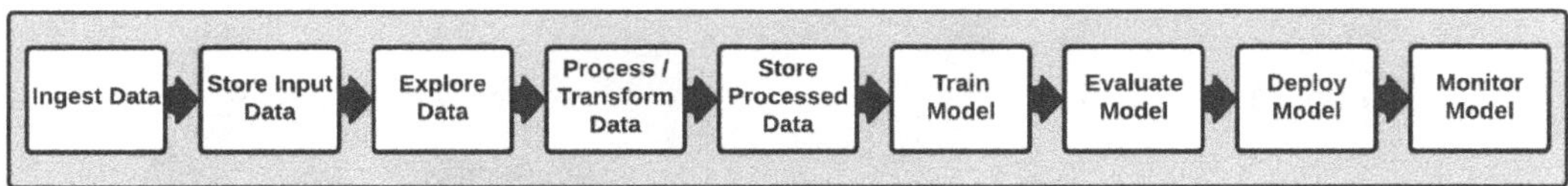


Figure 11.1: Data science lifecycle stages

At a high level, MLOps aims to automate all of the various steps in the ML model lifecycle, such as data collection, data cleaning and preprocessing, model training, model evaluation, model deployment, and model monitoring. As such, it can be seen as a type of engineering culture that aims to unify ML system development and ML system day-to-day operations. This includes a practice for collaboration and communication among all relevant stakeholders in an ML project, such as a company's data scientists, ML engineers, data engineers, business analysts, and operations staff, to help manage the overall machine learning lifecycle on an ongoing basis.

In this context, managing the ML lifecycle includes defining, implementing, testing, deploying, monitoring, and managing ML models to ensure that they work reliably and efficiently in a production environment.

In another similarity to DevOps, MLOps practices also encompass the concepts of continuous integration, continuous delivery, and continuous deployment (CI/CD), but with a specific focus on how ML models are developed. This ensures that new changes to the models are correctly integrated, thoroughly tested, and can be deployed to production in a systematic, reliable, and repeatable way. The goal in this regard is to ensure that any part of the ML lifecycle (data preprocessing, training, etc.) can be repeated at a later point with the same results. Just as with CI/CD in the context of software DevOps, this includes automating steps such as validation checks and integration tests, but in the case of ML model development, it also adds extra components such as data quality checks and model quality evaluations.

As we discussed in previous chapters, deploying a model to production is a great achievement, but the work doesn't stop there. Once ML models are deployed to production, they need to be continuously monitored to ensure their performance does not degrade over time due to changes in the underlying data or other factors. If any issues are detected, the models need to be updated or replaced with newer models. MLOps also includes tools to automate that part of the process, such as automatically retraining (for example, with fresh data), validating, and deploying a new model version.

Another important component of this is keeping track of various versions of the artifacts (e.g., data, models, and hyperparameter values) produced at various stages in our data science project. Not only does this enable easy rollback to previous versions if required, but it also provides an audit trail for compliance purposes, and to help with model explainability and transparency, as well as appropriate governance mechanisms, helping to ensure that ML models are used responsibly and ethically.

Now that we understand what MLOps is, let's dive into more detail on why it's especially important when developing, deploying, and managing models at a large scale.

> **Note**
>
> Later in this book, we will discuss generative AI/ML. There are particular types of ML models that are used in generative AI/ML that are called **Large Language Models** (**LLMs**). These models also have additional kinds of resources and artifacts associated with them, and a new term, LLMOps has emerged in the industry to refer to the operationalization of those workloads. For now, we will focus on traditional MLOps. While many of these concepts also apply to LLMOps workloads, we will discuss the additional considerations for LLMOps in later chapters.

# Why MLOps is needed for deploying large-scale ML workloads

The most important aspect of MLOps is that it helps organizations develop ML models in a faster, more efficient, and reliable manner, and it allows data science teams to experiment and innovate while also meeting operational requirements.

We know by now that ML has become an essential component of many industries and sectors, providing invaluable insights and decision-making capabilities, but that deploying ML models, especially at scale, presents many challenges. Some of these are challenges that can only be solved by MLOps, and we dive into more detail on such challenges in this section, as well as providing examples of how MLOps helps to address them.

Before we dive in, I'm going to point out that the kinds of challenges we will discuss in this section actually apply to any industry that creates products at a large scale, whether those products are cars, safety pins, toys, or machine learning models.

I'm going to take an analogy from the automobile industry in order to highlight the kinds of concepts that are required for large-scale production of any type of product. Consider that before cars were invented, a major form of transport would have been horse-drawn carts, and such carts would have been made in a kind of workshop like the one shown in *Figure 11.2*.

Figure 11.2: Forge (source: https://www.hippopx.com/en/middle-ages-forge-workshop-old-castle-wagon-wheel-297217)

Notice the characteristics of the production environment shown in *Figure 11.2*:

- It's a small environment, which could only accommodate a few people working together
- Whoever works here has lots of random tools lying around, and there doesn't seem to be much standardization implemented

With those kinds of characteristics, any kind of large-scale collaboration would not be possible, and at most, only three or four people could work together on creating a product. In such circumstances, a company could not possibly create a large number of products every month (for example). Bear in mind that this is how most companies begin to implement data science projects, in which data scientists perform various experiments and develop models on their own computers, using any random tools based on personal preferences, and trying to share learnings and artifacts with colleagues in a non-standardized and unsystematic way.

To overcome these kinds of challenges, the automotive industry invented the concept of the assembly line, and some of the first assembly lines would have looked similar to the one shown in *Figure 11.3*.

Figure 11.3: Assembly line (source: https://pxhere.com/en/photo/798929)

Notice the characteristics of the production environment shown in *Figure 11.3*:

- The environment can accommodate many people working together
- The tools and manufacturing processes have been standardized

This kind of environment enables large-scale collaboration, and it's possible to produce many more products in any given timeframe.

As automotive companies continued to evolve, they continued to apply more standardization and automation to each step in the process, leading to more efficient processes and reproducibility, until today's assembly lines look like the one shown in *Figure 11.4*.

Figure 11.4: Modern assembly line (source: https://www.rawpixel.com/image/12417375/photo-image-technology-factory-green)

As we can see in *Figure 11.4*, the production process has become highly automated and more efficient to operate at scale. With this analogy in mind, let's look at how this idea maps to the concepts and steps in the ML model development lifecycle.

## Model management and versioning

As an organization scales up its ML efforts, the number of models in operation can increase exponentially. Managing these models, keeping track of their versions, and knowing when to update or retire them can become quite difficult. Without a proper versioning and management system, ML models and their corresponding datasets can become disorganized, which leads to confusion and errors in the model development process. In fact, when we remember that many large companies have hundreds or even thousands of models in production, and are constantly developing new versions of those models to improve their performance, managing all of those models without specialized tools is basically impossible. MLOps tools provide mechanisms to facilitate easier model management and versioning, allowing teams to track and control their models in an organized and efficient way.

## Productivity and automation

Training, testing, deploying, and retraining models often involves many repetitive tasks, and as the scale of ML workloads increases, the time and effort required to manually manage these processes can become prohibitive. Considering that MLOps introduces automation at various stages of the ML lifecycle, such as data preprocessing, model training, testing, deployment, and monitoring, this frees up the data science teams to focus on more valuable tasks. Another important aspect of automation is that it reduces the likelihood of human error, which can help to avoid business-affecting mistakes, and increase productivity.

## Reproducibility

Reproducibility is arguably one of the most important factors in any industry that creates products at a large scale. Imagine if every time you tried to purchase the same product from a company, you received something slightly different. Most likely, you would lose faith in that company and would stop purchasing their products. Without well-established reproducibility in production processes, a company simply will not scale. In the context of machine learning, when we have a large number of models and datasets, and different types of development environments, ensuring the reproducibility of experiments can be a significant challenge. It can be difficult to recreate the exact conditions of a previous experiment due to the lack of version control on data, models, and code. This could lead to inconsistent results and make it difficult to build upon previous work. MLOps frameworks provide tools and practices for maintaining a record of all experiments, including the data used, the parameters set, the model architecture, and the outcomes. This allows experiments to be easily repeated, compared, and audited.

## Continuous Integration/Continuous Deployment (CI/CD)

In addition to being able to reliably reproduce a product, developing products at a large scale also requires the ability to incrementally improve your products over time. CI/CD pipelines automate the testing and deployment of models, ensuring that new changes are integrated and deployed seamlessly, efficiently, and with minimal errors. This enables data scientists to experiment and try out different kinds of updates quickly, easily, and in a controlled manner that conforms to your company's required standards. We should note that CI/CD is such an integral component of DevOps that the terms are often used almost interchangeably.

What's even more interesting is that we can also manage an MLOps pipeline within a standard DevOps CI/CD pipeline. For example, when using Vertex AI Pipelines, we can define our pipeline using code, and that code can be stored in a code repository such as Google Cloud Source Repositories and then managed via a DevOps CI/CD pipeline in Google Cloud Build. This enables us to apply all the benefits of DevOps, such as code versioning and automated testing, to the code that defines our pipelines, so that we can control how updates are made to our pipeline definitions.

## Model validation and testing

Testing machine learning models involves unique challenges due to their probabilistic nature, and traditional DevOps software testing techniques may not be sufficient. MLOps introduces practices and methodologies specifically for automating the validation and testing of ML models, ensuring they perform as expected before being deployed to production, and for their entire lifetime in production after deployment.

## Monitoring and maintenance

As we've discussed, once ML models are deployed, their performance needs to be continuously monitored to ensure they consistently provide accurate predictions. We also discussed how changes in real-world data can lead to a decrease in model performance over time, and as a result, we need tools and processes for continuous monitoring and alerting, enabling teams to react quickly to any degradation in performance and retrain models as necessary. This is an important component of any MLOps framework.

## Collaboration and communication

As ML projects scale, they typically require collaboration across various teams and roles, including data scientists, ML engineers, data engineers, business analysts, and IT operations. In many organizations, there is a gap between the data scientists who develop ML models and the IT operations teams who deploy and maintain these models in production. Without a common platform and standard practices, different team members might use inconsistent methods, tools, and data, leading to inefficient workflows and potential mistakes. MLOps fosters effective collaboration and communication among these stakeholders, enabling them to work together more efficiently and avoid misunderstandings or misalignments.

## Regulatory compliance and governance

In industries such as healthcare and finance, ML models must comply with specific regulatory requirements. MLOps provides mechanisms for ensuring transparency, interpretability, and auditability of models, therefore helping to maintain regulatory compliance. Also, without MLOps, managing multiple models across different teams and business units can become a challenge. MLOps allows for centralized model governance, making it easier to keep track of various models' performance, status, and owners.

Now that we have dived into why MLOps is needed, especially in the context of managing ML workloads at scale, let's take a look at some of the popular tools that have been developed in the industry for implementing those concepts.

# MLOps tools

So far, we've talked about the goals and benefits of MLOps, but how can we actually achieve these goals and benefits in the real world? Many tools have been developed to facilitate various aspects of MLOps, from data versioning to model deployment and monitoring. In this section, we discuss the tools that make MLOps a reality.

## Pipeline orchestration tools

When we want to standardize the steps in a process and run them automatically, we typically configure them as a set of sequential actions. This sequential set of actions is often referred to as a pipeline. However, simply configuring the order of the steps is not enough; we also need some kind of system to "orchestrate" (i.e., execute and manage) the pipeline, and we can use some popular workflow orchestration tools for this purpose, such as Kubeflow, Airflow, and **TensorFlow Extended** (**TFX**). We already covered Airflow in *Chapter 6*, but let's take a look at Kubeflow and TFX in more detail.

### *Kubeflow*

This is an open source project created by Google, and now maintained by an open source community of contributors, aimed at making it easier to run machine learning workflows on Kubernetes. Fundamental principles of Kubeflow include portability, scalability, and extensibility. Portability refers to the concept of "write once, run anywhere," which builds on the core tenet of containerization, whereby you can run your workloads across different types of computing environments in a consistent manner. Scalability refers to the ability to use Kubernetes to easily scale your model training and prediction workloads as needed, and extensibility means that you can implement customized use cases by adding popular tools and libraries that easily integrate with Kubeflow. In this case, rather than simply being a tool, Kubeflow is a framework and ecosystem that we can use to orchestrate and manage our machine learning workflows. The following are some core components of the Kubeflow ecosystem:

- **Kubeflow Pipelines** (**KFP**): For defining, deploying, and managing ML workflows
- **Katib**: For hyperparameter tuning
- **KFServing**: For serving models in a scalable way
- **TensorFlow and PyTorch Operators**: For running TensorFlow and PyTorch jobs
- **Training Operators**: For distributed training

We will particularly dive deep into Kubeflow Pipelines in the practical exercises associated with this chapter, in which we will build a pipeline and execute it in Google Cloud Vertex AI. In addition to KFP, Vertex AI also supports using TFX to build and manage ML pipelines. Let's take a look at that next.

### *TensorFlow Extended (TFX)*

TFX is an end-to-end platform, also created by Google, that manages and deploys production ML pipelines. As the name suggests, it was designed as an extension to TensorFlow, with the aim of making it easier to bring trained models to production. Just like KFP, TFX offers components for implementing each stage of the ML lifecycle, covering everything from data ingestion and validation to model training, tuning, serving, and monitoring. Also similar to KFP, TFX's core principles include scalability, and of course, extensibility. Other core principles of TFX include reproducibility and modularity. Reproducibility, as we discussed earlier in this chapter, is a critical requirement for producing pretty much anything at a large scale. Modularity, in this context, refers to the fact that the various components of TFX can be used individually or together, enabling customization. Speaking of which, the various components and responsibilities of TFX include the following:

- ExampleGen, which imports and ingests data into the TFX pipeline
- StatisticsGen, which computes statistics for the ingested data, essential for understanding the data and feature engineering
- SchemaGen, which examines dataset statistics and creates a schema for the dataset
- ExampleValidator, which identifies and analyzes anomalies and missing values in the dataset
- Transform, which can be used for feature engineering on the dataset
- Trainer, which defines and trains a model using TensorFlow
- Tuner, which can be used to optimize hyperparameters using Keras Tuner
- InfraValidator, which ensures that the model can be loaded and served in a production environment
- Evaluator, which uses the TensorFlow Model Analysis library to evaluate the metrics of the trained model
- BulkInferrer, which performs batch processing on a model.
- Blessing and deployment: If the model is validated, it can be "blessed" and then deployed using a serving infrastructure such as TensorFlow Serving.

Considering that both TFX and KFP can be used to build and run pipelines in Google Cloud Vertex AI, it may be challenging to decide which one to use. On that matter, the official Google Cloud documentation recommends the following:

*"If you use TensorFlow in an ML workflow that processes terabytes of structured data or text data, we recommend that you build your pipeline using TFX... For other use cases, we recommend that you build your pipeline using the Kubeflow Pipelines SDK."*

We're not going to focus on TFX in a lot of detail in this chapter, but if you would like to learn how to build a TFX pipeline, then there's a useful tutorial in the TensorFlow documentation at the following link: `https://www.tensorflow.org/tfx/tutorials/tfx/gcp/vertex_pipelines_simple`.

There are also many other open source tools for implementing MLOps pipelines, such as MLFlow, which provides interfaces for tracking experiments, packaging code into reproducible runs, and sharing and deploying models.

While KFP and TFX can be used to orchestrate your entire MLOps pipeline, there are also tools that focus on more specific components of the ML lifecycle, and that can integrate with KFP and TFX in order to customize your overall pipeline. Let's take a look at some of those tools next.

## Experiment and lineage tracking tools

As we discussed earlier in this chapter, and in earlier chapters, when running ML workloads at a large scale, it's important to keep track of every step in the development of every model, so that you can easily understand how any given model was created. This is required for characteristics such as reproducibility, explainability, and compliance. Imagine if you have thousands of models in production, and compliance regulators ask you to explain how a particular model was created. Without a well-established system for tracking every aspect of every model's creation, such as which versions of which datasets were used to train the model, what kinds of algorithms and hyperparameters were used, what the initial evaluation metrics were, and who deployed the model to production, this would be an impossible task. This is where experiment and lineage tracking tools come into the picture. Let's consider what kinds of tools we have at our disposal in this regard.

### *Vertex AI Experiments*

Vertex AI Experiments is a managed service that helps you track, compare, and analyze your machine learning experiments. It provides a central place to manage your experiments, and it makes it easy to compare different models and hyperparameters. Once an experiment is created (using the Vertex AI SDK or the Vertex AI user interface), you can start tracking your training runs, and Vertex AI Experiments will automatically collect metrics from your training runs, such as accuracy, loss, and training time.

Once your training runs are complete, you can view graphs and tables of the metrics, and you can use statistical tests to determine which model is the best. You can also use the Experiments dashboard to visualize your results, and you can use the Experiments API to export your results to a spreadsheet or a notebook.

#### *TensorBoard*

TensorBoard is a web-based tool that comes with TensorFlow and is designed to help visualize and understand machine learning models, as well as to debug and optimize them. It provides an interactive interface for various aspects of machine learning models and training processes and can make it easy to create metric graphs such as ROC curves, which we introduced in previous chapters.

### Model deployment and monitoring tools

In this context, we have tools such as Vertex AI Prediction and Model Monitoring, which we reviewed in detail in *Chapter 10*. There are also open source tools such as **TensorFlow Serving** (**TFServing**) for TensorFlow models, and TorchServe.

### Model interpretability and explainability tools

As we've discussed previously, model interpretability and explainability are extremely important concepts in machine learning. They relate to reproducibility, compliance, and performance. For example, if you don't have a good understanding of how your models work, then it will be harder to continuously improve them in a systematic way. These concepts also relate to fairness. That is, in order to ensure that a model is making fair and ethical predictions, you need to understand how it works in great detail. Fortunately, there are tools that can help us in this regard, which we discuss here.

Google Cloud Vertex AI provides tools such as Explainable AI and Fairness Indicators, which can help us to understand how our machine learning models work, and how they would behave under different conditions. There are also open source tools such as **SHAP** (short for **SHapley Additive exPlanations**), which uses a game theory approach to explain the output of any machine learning model, and **LIME** (short for **Local Interpretable Model-agnostic Explanations**), which is a Python library that allows us to explain the predictions of any machine learning classifier.

The next chapter in this book is dedicated to the concepts of bias, fairness, explainability, and lineage, so we will dive into these concepts and tools in a lot more detail there.

Now that we've covered many of the different kinds of tools that can be used to implement MLOps workloads, let's dive into how we can specifically do this on Google Cloud Vertex AI.

# Implementing MLOps on Google Cloud using Vertex AI Pipelines

In this section, we will cover the steps to build an MLOps pipeline on Google Cloud using Vertex AI Pipelines.

## Prerequisite: IAM permissions

In this section, we will use the same Vertex AI Workbench notebook instance that we created in *Chapter 5*. That user-managed notebook uses the default Compute Engine service account, which is granted the IAM basic Editor role by default. When we build and execute our pipeline in our notebook, we decide to let our pipeline inherit the same permissions used by our notebook. This is a decision we make by not specifying a different role for our pipeline executions. We could specify a different role if we wanted to, but for simplicity purposes, we'll take the approach of having our pipeline use the default Compute Engine service account. By default, the Editor role (used by the default Compute Engine service account) will allow us to perform all of the required activities in Vertex AI in this chapter, but our pipeline will also need to run some steps on Dataproc. For this reason, we will add the Dataproc Worker and Dataproc Editor roles to the default Compute Engine service account. To do that, perform the following steps:

1. In the Google Cloud console, navigate to **Google Cloud services** → **IAM & Admin** → **IAM**.
2. In the list of principals, click on the pencil symbol for the default Compute Engine service account (the service account name will have the format - `[PROJECT_NUMBER]-compute@developer.gserviceaccount.com`; see *Figure 11.5* for reference).

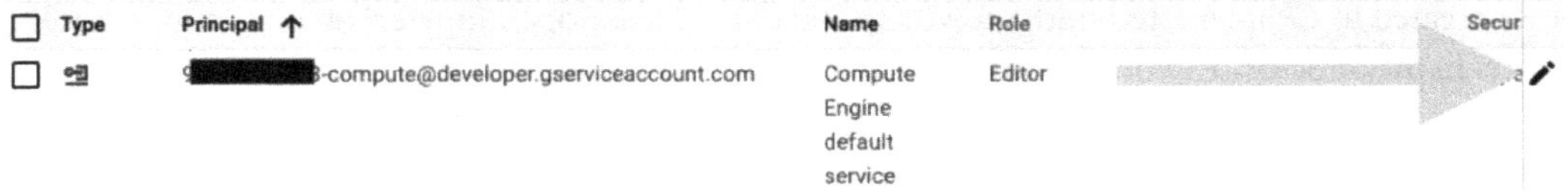


Figure 11.5: Edit service account

3. On the screen that appears, select **Add Another Role**, and in the textbox that then appears, type `Dataproc` and select **Dataproc Editor** from the list of roles.
4. Repeat step 3 and select **Dataproc Worker**.
5. The updated roles will look as shown in *Figure 11.6*.

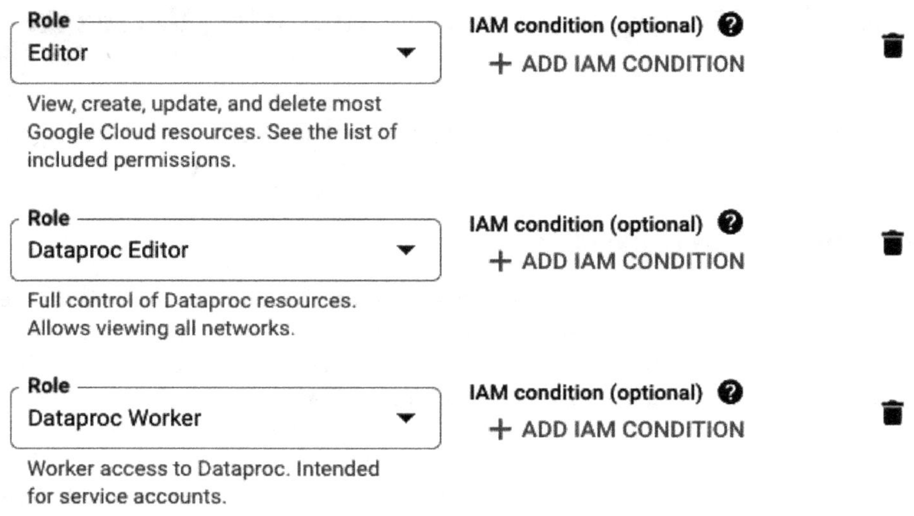


Figure 11.6: Updated roles

6. Click **Save**.

And that's it – you have successfully added the required roles.

## Implementation

As I mentioned in the previous section, we can use the same Vertex AI Workbench notebook instance that we created in *Chapter 5* to build our MLOps pipeline. Please open JupyterLab on that notebook instance. In the directory explorer on the left side of the screen, navigate to the `Chapter-11` directory and open the `mlops.ipynb` notebook. You can choose **Python (Local)** as the kernel. Again, you can run each cell in the notebook by selecting the cell and pressing *Shift* + *Enter* on your keyboard. In addition to the relevant code, the notebook contains markdown text that describes what the code is doing.

In the practical exercises, we will build and run an MLOps pipeline that will execute all of the phases of the ML model development lifecycle that we have covered so far in this book. Our pipeline is depicted in *Figure 11.5*.

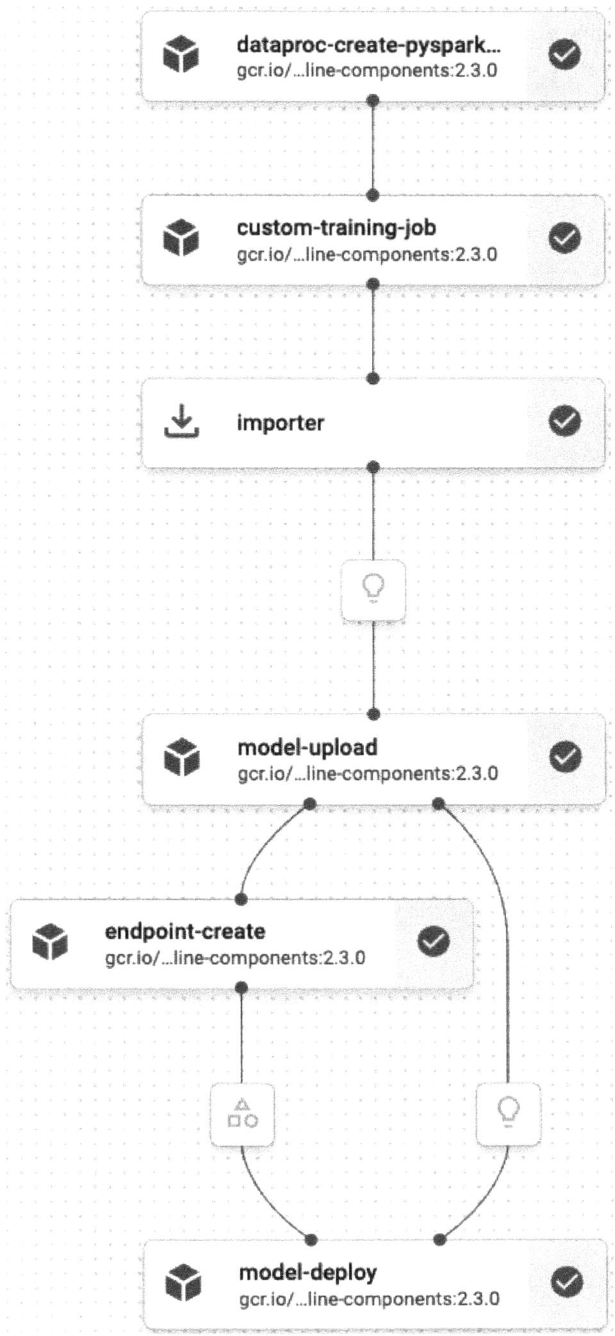


Figure 11.7: Vertex AI pipeline

At a high level, our pipeline will perform the following steps:

1. **Ingest data**: First, we need to get our data into the Google Cloud environment. We're using Google Cloud Storage to store our data, and when our pipeline kicks off a data processing and model training job, our data is read in from there.
2. **Preprocess our data**: Google Cloud offers several tools for data preprocessing, including Dataflow, Dataprep, and Dataproc, which we explored in *Chapter 6*. More recently, Google Cloud also released a service named Serverless Spark, which enables us to run Spark jobs without having to provision and manage the required underlying infrastructure. This is what we use to implement our data preprocessing job in our pipeline in the practical exercises.
3. **Develop and train our model**: Our pipeline trains a TensorFlow model in Vertex AI, using the processed data created in the previous step.
4. Register our model in the Vertex AI Model Registry.
5. **Deploy our model**: Our pipeline moves to the next step, which is to deploy our model to production. In this case, our pipeline creates a Vertex AI endpoint and hosts our model on that endpoint.

Let's take a look at what our pipeline is doing in more detail, by inspecting the solution architecture that we have just created, as depicted in *Figure 11.6*.

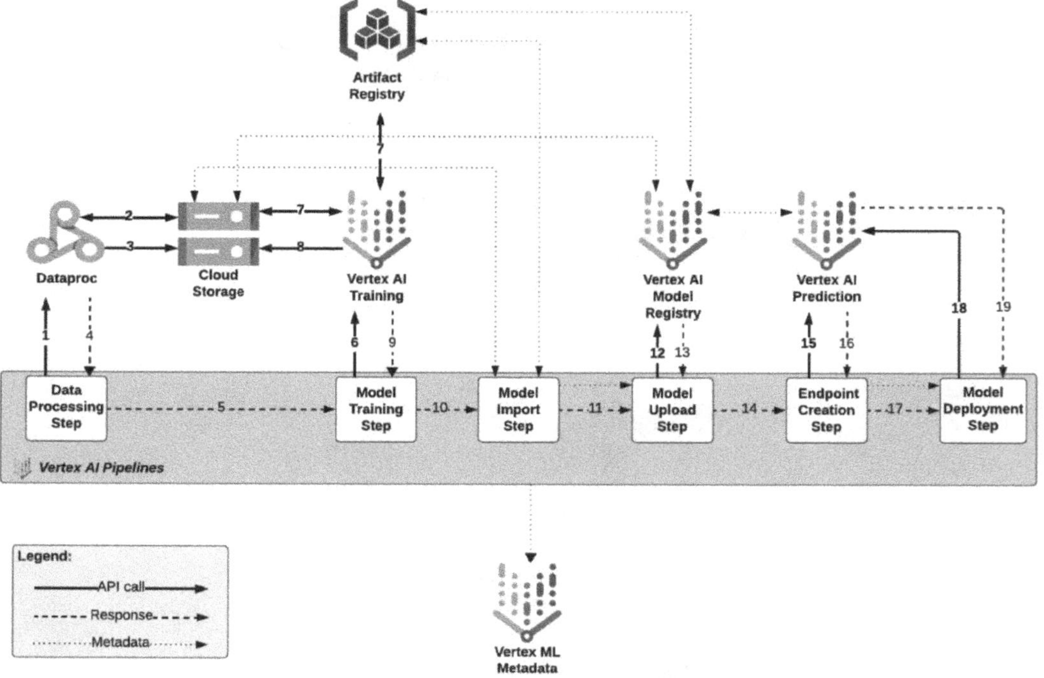


Figure 11.8: MLOps pipeline in Vertex AI

In *Figure 11.6*, the horizontal rectangular section that spans across the figure represents our pipeline that runs in Vertex AI Pipelines. Each of the steps in the process is numbered, and the numbers represent the following actions:

1. The data processing step in our pipeline submits a Spark job to Dataproc Serverless in order to execute our PySpark processing script.
2. Dataproc fetches our PySpark script and the raw data from Google Cloud Storage and executes the PySpark script to process the raw data.
3. Dataproc stores the resulting processed data in Google Cloud Storage.
4. The data processing job status is complete.
5. The next step in our pipeline — the model training step — is invoked.
6. Vertex AI Pipelines submits a model training job to the Vertex AI Training service.
7. In order to execute our custom training job, the Vertex AI Training service fetches our custom Docker container from Google Artifact Registry, and it fetches the training data from Google Cloud Storage. This is the same data that was stored in Google Cloud Storage by our data processing job (i.e., it is the processed data that was created by our data processing job).
8. When our model has been trained, the trained model artifacts are saved in Google Cloud Storage.
9. The model training job status is complete.
10. The next step in our pipeline — the model import step — is invoked. This is an intermediate step that prepares the model metadata to be referenced in later components of our pipeline. The relevant metadata in this case consists of the location of the model artifacts in Google Cloud Storage and the specification of the Docker container image in Google Artifact Registry that will be used to serve our model.
11. The next step in our pipeline — the model upload step — is invoked. This step references the metadata from the model import step.
12. The model metadata is used to register the model in Vertex AI Model Registry. This makes it easy to deploy our model for serving traffic in Vertex AI.
13. The model upload job status is complete.
14. The next step in our pipeline — the endpoint creation step — is invoked.
15. An endpoint is created in the Vertex AI Prediction service. This endpoint will be used to host our model.
16. The endpoint creation job status is complete.
17. The next step in our pipeline — the model deployment step — is invoked.

18. Our model is deployed to our endpoint in the Vertex AI Prediction service. This step references the metadata of the endpoint that has just been created by our pipeline, as well as the metadata of our model in the Vertex AI Model Registry.
19. The model deployment job status is complete.

In addition to all of the above steps that are explicitly performed by our pipeline, the Vertex AI Pipelines service also registers metadata related to our pipeline execution in the Vertex ML Metadata service.

Our model is now ready to serve inference requests! Isn't it impressive that all of these activities and API calls across multiple Google Cloud services are being orchestrated automatically by our pipeline? After our pipeline has been defined, it can run automatically whenever we wish, without any need for human interaction throughout the process, unless we deem it necessary for humans to be involved in any of the steps.

If you have completed the activities in the notebook, then you are now officially an AI/ML guru! Seriously, you have just implemented an end-to-end MLOps pipeline. That is an extremely complex and advanced task in the ML industry.

With all of that success under your belt, let's take some time to reflect on what we learned in this chapter.

## Summary

In this chapter, we started with an introduction to MLOps at a high level, which is essentially a blend of machine learning, DevOps, and data engineering, in which the main goal is to automate the ML lifecycle, resulting in improved workflows and collaborations between data scientists and engineers. We discussed how MLOps allows organizations to streamline their ML operations, increase the speed of deployment, and maintain high-quality models in production, leading to a more efficient, effective, and reliable ML workflow, and thereby maximizing the value that organizations get from their ML initiatives.

We touched upon the various pain points that MLOps addresses, including but not limited to challenges related to managing and versioning models, ensuring reproducibility and consistency, monitoring and maintaining models, and fostering collaboration between different teams.

We then dived into why MLOps is important for deploying large-scale machine learning workloads. It resolves many challenges that can crop up in machine learning systems, ranging from managing and versioning models to ensuring the reproducibility of experiments. It also facilitates continuous integration, deployment, monitoring, and validation of models and promotes better collaboration among teams.

Next, we discussed various MLOps tools such as Kubeflow Pipelines and TensorFlow Extended, among others. Each of these tools offers unique functionalities catering to different stages of the ML lifecycle, including data versioning, experiment tracking, and model deployment.

We then performed an implementation of MLOps on Google Cloud using Vertex AI Pipelines, which involved multiple steps, including managing datasets, preprocessing data, training models, and monitoring models.

Next, we will explore four important and somewhat interrelated topics in the machine learning industry, which are bias, explainability, fairness, and lineage. Let's move on to the next chapter to explore these concepts in detail.

# 12
# Bias, Explainability, Fairness, and Lineage

Now that we have learned all of the steps required to build and deploy models in Google Cloud and to automate the entire **machine learning** (**ML**) model development life cycle, it's time to dive into yet more advanced concepts that are fundamental to developing and maintaining high-quality models.

In addition to our models providing predictions that are as accurate as possible for a given use case, we need to ensure that the predictions provided by our models are as fair as possible and that they do not exhibit bias or prejudice against any individuals or demographic groups.

The topics of bias, fairness, and explainability are at the forefront of ML research today. This chapter discusses these concepts in detail and explains how to effectively incorporate these concepts into our ML workloads. Specifically, we will cover the following topics in this chapter:

- An overview of bias, fairness, and explainability in **artificial intelligence** (**AI**)/ML
- How to detect and mitigate bias in datasets
- Using explainability to understand ML models and reduce bias
- The importance of lineage tracking in ML model development

Let's begin by defining and describing the relevant concepts.

## An overview of bias, explainability, and fairness in AI/ML

While the terms "bias," "explainability," and "fairness" are not specific to ML, in this section, we will explore these terms as they apply to the development and use of ML models.

## Bias

Bias in AI/ML refers to tendencies or prejudices in data and algorithms that can lead to unfair outcomes. One of the most common sources of bias in ML model development is when biases exist in the training data; for example, when the data points in the training data do not fairly represent the reality or the population that the model's predictions will serve, which we refer to as **data bias**. For example, using a dataset in which the data points predominantly represent only one demographic group to train a model can result in poorer performance when that model is required to make predictions based on data points that represent other demographic groups. More specifically, this is an example of something called **sampling bias**. To tie this to a real-world scenario, let's imagine that we're training a model to perform facial recognition. If we train the model with images of people mainly from one specific demographic group, the model may not perform facial recognition tasks well when it is later presented with images of people from other demographic groups. There are a number of different ways in which we may encounter bias during the development and use of ML models, which we outline in the following paragraphs.

### *Collection or measurement bias*

A major cause of data bias is the manner in which the data is collected or measured. Sometimes, bias can occur due to the way we measure or frame a problem, which might not correctly represent the underlying concept that we're trying to measure. For example, let's imagine that we own a small company and we want to expand our product offerings to attract new customers. We may decide to use a survey to gather data to train ML models to predict what kinds of new products we should offer, based on how popular those new products are likely to be. There are a number of ways in which we could distribute this survey, such as via email or via physical mail. In the case of email, we might decide to send the survey to all of our current customers, because it's likely that we would already have their email addresses in our database. In the case of physical email, we may decide to send it out to everybody in the same postal code area, city, state, or country in which our company is physically located.

Unfortunately, both of those methods may unexpectedly introduce bias into the datasets used to train our models. For example, perhaps we traditionally happened to appeal to only a specific type of customer in the past. If we use our current customer base as the survey group, we may not get good data points for products that would appeal to new customer demographics. Similarly, if we send out a survey to everybody in a specific geographical area, such as a postal code, city, state, or even country, the people living in that area may not represent diverse demographic groups, which could inadvertently introduce biases into the resulting dataset. This phenomenon is sometimes referred to as **response bias**, in which the people providing the required information may be biased in some way, and may therefore provide biased responses. To make this even more complicated, the way in which we phrase questions in the survey could accidentally bias the respondents.

In this scenario, we also need to be aware of **observer bias** or **subjectivity**, in which the people running the survey may be biased in some way. In fact, our example of deciding to email our current customers, or to send physical mail to people in a specific area, is a type of observer bias, in which we decided to survey specific groups of people based on factors such as currently available data (for example, our customers' email addresses) or proximity to our company's location.

### *Pre-existing bias*

These are generally biases that already exist in society. Such biases can stem from factors such as cultural, historical, and societal norms, consisting of ingrained beliefs and practices that shape the way individuals view the world.

One of the most common ways to train a model is to use historical data that has been recorded. However, data from the past might be tainted with biases that were inherent in those times — I think we can all agree that the societal norms of the 1950s or even the 1990s are quite different from today's standards, especially in terms of fairness across different demographic groups. If AI/ML models are trained on such data without correction, they will likely perpetuate those biases. It's important to understand that unaddressed pre-existing biases can not only perpetuate but sometimes even amplify stereotypes, leading to unfair or discriminatory outcomes. For example, a job recommendation system might be biased toward recommending higher-paying jobs to men if it is trained on historical data that exhibits this bias.

Another common type of bias in this context is **confirmation bias**, in which people may subconsciously tend to select and interpret data in ways that confirm their pre-existing beliefs.

If data collection processes do not check for pre-existing biases, they can lead to data that's not truly representative of the reality or population it's supposed to depict. Models trained on such data may learn and replicate those biases in their predictions or actions.

### *Algorithmic bias*

This refers to biases that are introduced by the design of the algorithms themselves, not just the data they are trained on. This is an increasingly important topic, considering that algorithms are used to implement an ever-expanding plethora of important decisions in modern society, such as credit approvals, recruitment processes, and medical diagnoses.

Algorithmic bias can be more subtle and difficult to detect. For example, such bias could simply stem from the algorithm developers themselves. If the team developing the algorithm lacks diversity, it might not foresee or recognize potential biases in how the algorithm is implemented. We also need to recognize the possibility of accidentally developing biased feedback loops in ML systems. For example, in systems where algorithms are continuously trained with new data, biased outcomes can reinforce the input biases, and over time, this can lead to increasingly skewed results.

Just like the other types of biases discussed previously in this section, unaddressed algorithmic bias can lead to the perpetuation of stereotypes, misinformation, and unjust practices.

It's important to note that in this section, we have discussed only some of the most common types of biases that can affect ML model development. This is an active area of research, and additional types of bias exist that we may not be explicitly aware of as we develop and use ML models.

Figuratively speaking, the concept of bias represents one side of a coin, and on the other side of that coin is the concept of fairness, which we explore in more detail next.

# Fairness

Fairness in AI and ML refers to the practice of developing algorithms in a way that prevents discrimination and promotes equity. While fairness is a rather straightforward concept to define, such as treating all people equally, in practice, it can be difficult to monitor and uphold. A little later in this chapter, we will look at mechanisms to monitor and enhance the fairness of ML models, but let's first describe the concept in a bit more detail in the context of ML. Just as in the previous section related to bias, we will discuss a few different ways in which we can define and measure fairness with regard to ML models, which we outline in the following subsections.

## Representational fairness

This is the first line of defense against bias in ML model development, especially with regard to data bias. Representational fairness aims to ensure that the data used in model training and validation contains a fair representation of each of the various demographic groups that the resulting model's predictions will affect; for example, ensuring that genders and ethnic groups are represented fairly in the dataset.

## Procedural fairness

This involves ensuring that the processes for data collection, data handling, and algorithm development are equitable and do not favor any particular group. In *Chapter 13*, we will discuss the concept of governance in great detail, especially in relation to data governance, and the important role it plays in helping to ensure that potential biases are addressed or mitigated in datasets. Many enterprises employ entire teams or organizations dedicated to outlining and upholding governance requirements. In the case of procedural fairness, the focus is not so much on the contents of the data, but rather on the fairness of the processes that lead to the development, deployment, and operation of ML models. Some key components of procedural fairness are represented by the **TAIC** framework, which consists of the following:

- **Transparency**, such as clear documentation of the methods, data sources, and decisions made throughout the model's life cycle
- **Accountability**, meaning that there should be clarity on who is responsible for various stages of model development, deployment, and monitoring

- **Impartiality**, meaning that procedures should not favor or prejudice any individuals or groups
- **Consistency**, meaning that the procedures used to develop a model should be reproducible in a consistent manner

Google Cloud Vertex AI provides mechanisms that help to audit procedural fairness, which we will explore in more detail in later sections of this chapter, as well as in the practical activities that accompany this chapter.

### *Outcome fairness*

While representational fairness mainly focuses on the inputs used to train ML models (that is, the contents of the training data) and procedural fairness focuses on the procedures used to develop and deploy ML models, outcome fairness, as the name suggests, focuses on the results produced by ML models. Specifically, it aims to ensure that the outcomes produced by a model are equitable and do not disproportionately benefit or harm any group. Of course, one way to measure this is to monitor the predictions made by a model to determine whether any bias appears to be present. In this chapter, we will explore some mechanisms and metrics that can be used for this purpose.

As I mentioned when discussing bias in the previous section, the concepts of bias and fairness in ML are still very active and evolving areas of research. This is why fairness can be a complex topic in practice, and it's important to understand that achieving one type of fairness can sometimes lead to a violation of another type. Moreover, what is considered "fair" can be context-dependent and might vary between communities, as well as over time.

Now that we've introduced the concepts that define the two-sided coin of bias and fairness, the next step will be to introduce a topic that is inherently linked to those two concepts, which is referred to as **explainability**.

## Explainability

Explainability in ML focuses on the ability of a human to understand and explain how the outputs of an ML model were produced. This is often also referred to as the **interpretability** of a model. These concepts continue to grow in importance as ML models become increasingly complex over time. For example, it's pretty easy to interpret and explain how the outputs of a simple linear regression model were produced, because we have well-defined mathematical formulae that describe that process, such as $y = a + bx$, where $y$ (the output) is a linear transformation of $x$ (the input). Similarly, for decision tree models, we can logically trace the decision path through the tree. However, when we're dealing with large **neural networks** (**NNs**) containing millions or even billions of parameters, and incorporating different types of architectures and activation functions, we need some additional tools to help us understand how those models are making their decisions, which we will explore in this chapter.

When we discussed procedural fairness in the previous section, we talked about the importance of transparency. Explainability links closely to the concept of transparency, as it seeks to ensure that people can understand the process by which predictions are produced from a given model. For example, if we own a bank and one of our models is responsible for granting or declining credit applications, we want to thoroughly understand how it makes those decisions.

We can talk about explainability in terms of **global explainability**, in which we try to understand the general logic the model applies to make predictions across all input instances, or **local explainability**, which involves explaining why a model made a specific decision for a particular instance (for example, understanding why a specific customer's credit application was declined).

Overall, explainability is critical for building trust in AI systems, complying with legal requirements, and ensuring that humans can intervene effectively in decision-making processes.

Now that we've introduced and explained the overall concepts, let's start diving in deeper. Our first deep dive in this chapter will be on the topic of bias in datasets.

# How to detect and mitigate bias in datasets

In this section, we explore how to detect bias in our datasets, and there are various tools and methods we can use for this purpose. In fact, we've already covered some of them in previous chapters of this book, such as data exploration and visualization.

## Data exploration and visualization

When we explore our datasets using visualization, for example, charts such as histograms and scatter plots can help visualize disparities in data distribution for different demographic groups. Similarly, we've already explored descriptive statistics such as mean, median, mode, and variance to understand the contents of our datasets. If there are significant disparities in these statistics between subgroups, it may suggest the presence of bias in the dataset.

## Specific tools for detecting dependencies between features

We also want to test for potential correlations or dependence between features in our dataset in order to understand whether the values of some features are significantly influenced by the values of others. While this is something that we generally want to do as part of our regular data exploration and feature engineering anyway (that is, we need to understand underlying patterns in our data as much as possible), it becomes even more important in the context of bias and fairness, especially with regard to potential links between the target variable and protected attributes such as gender or ethnicity.

There are specific tools for detecting dependencies among features, such as **Pearson's Correlation Coefficient**, which measures the linear relationship between numeric variables, or the **Chi-Squared Test for Independence**, which can be used to determine if there is a significant association between two categorical variables.

## Mechanisms incorporating model prediction results

In addition to the aforementioned methods, we can use mechanisms that go beyond just examining the dataset by also taking a model's outputs into account, which can then help us to link any observed biases back to the training data and process. For example, we can use **disparate impact analysis** (**DIA**) to compare the ratio of favorable outcomes for different groups and measure whether any specific groups tend to get more favorable outcomes than any others. Let's take a look at DIA in more detail.

### *DIA*

In the case of DIA, we generally identify what is referred to as a **privileged group** and an **unprivileged group** (also referred to as a **protected group**), based on some protected characteristic such as gender or race, and we compare the outputs of a given model for both of those groups to try to determine whether the model appears to bias in either direction (positively or negatively) with regard to those groups. We measure this disparity using a metric referred to as the **disparate impact ratio** (**DIR**), which is defined by the following formula:

$$DIR = \frac{P(FUP)}{P(FP)}$$

Here, $P(FUP)$ represents the probability of a favorable outcome for the unprivileged group, and $P(FP)$ represents the probability of a favorable outcome for the privileged group.

The DIR value can be interpreted in the following way:

- When DIR is equal to 1, it indicates perfect fairness, where both groups receive favorable outcomes at the same rate
- When DIR is greater than 1, it indicates that the unprivileged group is more likely to receive favorable outcomes
- When DIR is less than 1, it indicates that the unprivileged group is less likely to receive favorable outcomes

A commonly accepted threshold is a DIR value of between 0.8 and 1.25; values outside this range often indicate potential **disparate impact** (**DI**).

Note that DIA can be performed directly on a dataset by using the target feature, or it can use a combination of the input features and a model's predictions.

In the next section, we will dive into the concept of explainability in ML in more detail, and explore how we can use explainability frameworks to detect and address bias. We will also expand on the concepts covered in this section, and discuss how they relate to explainability and fairness.

# Using explainability to understand ML models and reduce bias

We introduced the concept of explainability at a high level in the previous section. This section dives further into this topic, introducing tools that can be used to gain insights into how ML models are working at inference time.

## Explainability techniques, methods, and tools

Let's begin by exploring some popular techniques, methods, and tools that we can use for implementing explainability in ML, which we describe in the following subsections.

### *Performing data exploration*

By now, it should hopefully be clear that understanding the data used to train our models is one of the first steps in explaining how the model makes decisions, and it is also one of the first lines of defense to identify and combat potential biases.

In the practical activities associated with this chapter, we explore the "Adult Census Income" dataset (`https://archive.ics.uci.edu/dataset/2/adult`), which is known to contain imbalances with regard to race and gender. The dataset comprises information pertaining to people, such as their race, gender, education they've received, and their current annual income, expressed as either "<=50K" (less than or equal to $50,000 per year) or ">50K" (more than $50,000 per year), which creates a binary classification use case when the income is used as the target variable.

When exploring this data, we can ask questions such as the following:

1. Are feature values, such as race and gender, represented somewhat evenly or unevenly?
2. Are there any correlations between some feature values and the income earned by that person?

We can easily see imbalances in the dataset by using data visualization techniques. Let's take a look at some examples.

#### Income distribution by gender

The following code will show us the distributions of each gender in the dataset, with regard to the two different income categories (that is, people who earn less than or equal to $50,000 per year, or people who earn more than $50,000 per year). You can open the Jupyter notebook that accompanies this chapter if you'd like to follow along with the code examples that we will review. We can again use the same Vertex AI Workbench-Notebook Instance that we created in *Chapter 5* for this purpose. Please open JupyterLab on that notebook instance. In the directory explorer on the left side of the screen, navigate to the `Chapter-12` directory and open the `bias-explainability.ipynb` notebook. You can choose **Python (Local)** as the kernel. Again, you can run each cell in the notebook by selecting the cell and pressing *Shift* + *Enter* on your keyboard. In addition to the relevant code, the notebook contains markdown text that describes what the code is doing:

```
plt.figure(figsize=(10, 6))
sns.countplot(x='income', hue='sex', data=adult_data)
plt.title('Income Distribution by Gender')
plt.show()
```

This will display a graph like the one shown in *Figure 12.1*:

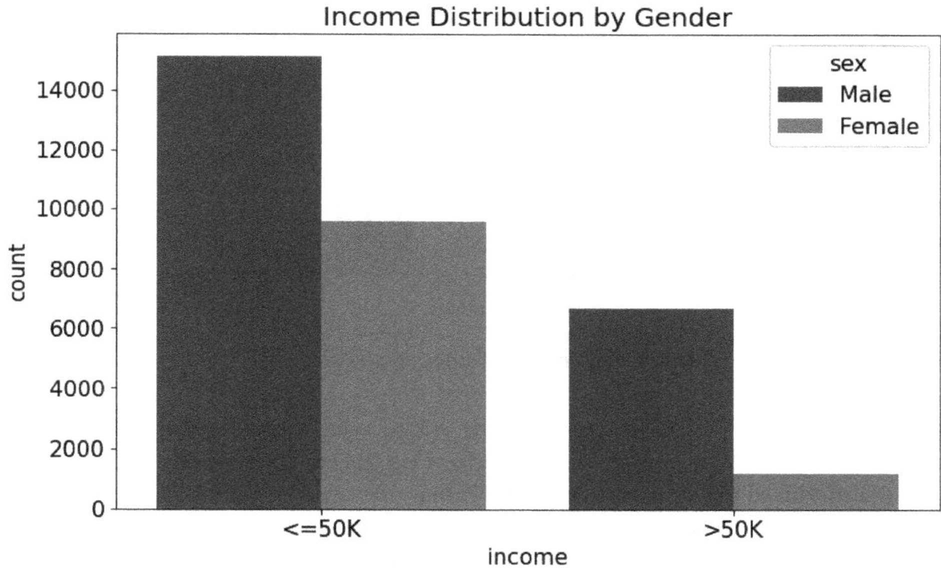


Figure 12.1: Income distribution by gender

In *Figure 12.1*, we can see that, overall, more people in the dataset earn more than $50,000 per year than those who do not. We can also see that the number of males in each group far exceeds the number of females in each group. This also tells us the entire dataset consists of more data points related to men than women.

### Income distribution by race

The following code will show us the distributions of each race in the dataset, with regard to income:

```
plt.figure(figsize=(15, 8))
sns.countplot(x='income', hue='race', data=adult_data)
plt.title('Income Distribution by Race')
plt.show()
```

This will display a graph like the one shown in *Figure 12.2*:

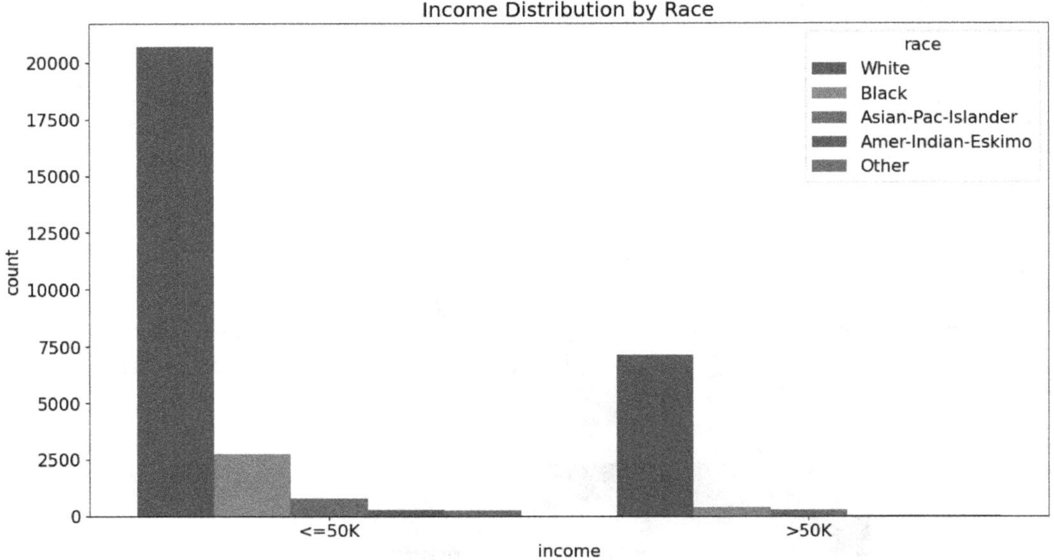


Figure 12.2: Income distribution by race

In *Figure 12.2*, for both the "<=50K" category and the ">50K" category, we see that the data contains many more data points for White people than any other race. This can be seen as a type of bias in the dataset. As we introduced earlier in this chapter, this bias may exist for multiple potential reasons, such as bias in the collection of the data, or it may happen due to other factors such as geographic location. This particular dataset represents the population of a specific area, which may somewhat explain its apparent bias toward a particular race. For example, if the data were collected in Asia, then it would contain many more data points for Asian people than any other race, or if it were collected in central Africa, then it would contain many more data points for Black people than any other race. It's important to note any imbalances in the data and determine how they may affect the training of an ML model and who that ML model is intended to serve. Generally, if features in the dataset have much higher numbers of instances of a specific value, then an ML model's predictions will likely reflect that in some way.

In the Jupyter notebook that accompanies this chapter, we also assess other types of distributions in the data, such as occupational distribution by gender and educational distribution by race. I encourage you to use the Jupyter notebook to explore the data in more detail. For now, let's move on and look at implementing DIA in more detail.

### *Implementing DIA*

The following is the code we use to implement DIA in our Jupyter notebook:

```
pivot_gender_income = adult_data.pivot_table(index='sex',
    columns='income', values='age', aggfunc='count')
pivot_gender_income['rate'] = pivot_gender_income['>50K'] / (
    pivot_gender_income['>50K'] + pivot_gender_income['<=50K'])
DI = pivot_gender_income.loc['Female', 'rate'] / (
    pivot_gender_income.loc['Male', 'rate'])
```

The code first creates a pivot table of the `adult_data` DataFrame, grouped by gender and income, with the count of people in each group as the value. Then, it adds a new column to the pivot table called `rate`, which is the proportion of people in each group who earn more than $50,000. Finally, it calculates the DI by dividing the rate for females by the rate for males.

This is a very simple example of implementing DIA on a dataset that is known to contain gender imbalances. DIA can be much more complex and may require some domain expertise to implement effectively, depending on the contents of the dataset and the intended function of an ML model trained on that data.

Next, let's discuss the topic of feature importance.

## Feature importance

Feature importance evaluates the impact that each feature has on the predictions made by a model. To explore this concept, let's imagine we have a dataset that contains information about people, and the features in the dataset include height, age, eye color, and whether or not they like coffee. We want to use this data to train a model to predict the likelihood of each person being a successful basketball player. Do you think any of the input features in our dataset may be more important than any others in terms of influencing the outcome? Is it likely that height would be more important or less important than eye color in determining whether a person is likely to be a successful basketball player? Is age an important factor? In this case, we're describing the concept of feature importance in a simple example. In reality, we may be dealing with datasets that contain thousands of features, and those features may not always represent easily interpretable concepts such as height, age, and eye color. For this reason, we can use tooling to help us to get these kinds of insights. The following subsections describe the tools we can use for this purpose.

### *Feature importance tools built into popular ML libraries*

In the Jupyter notebook that accompanies this chapter, we use scikit-learn to train a binary classifier model, and we use the `feature_importances_` attribute to inspect the apparent importance of each feature. This is an attribute of certain scikit-learn estimators, particularly tree-based estimators such as decision trees, random forests, and gradient boosted trees, which provide an array of importance scores for each feature, in which higher values indicate more importance. The following code snippet uses the `feature_importances_` attribute to create a graph similar to that shown in *Figure 12.3*:

```
feat_importances = pd.Series(clf.feature_importances_,
    index=X.columns)
feat_importances.nlargest(10).plot(kind='barh')
```

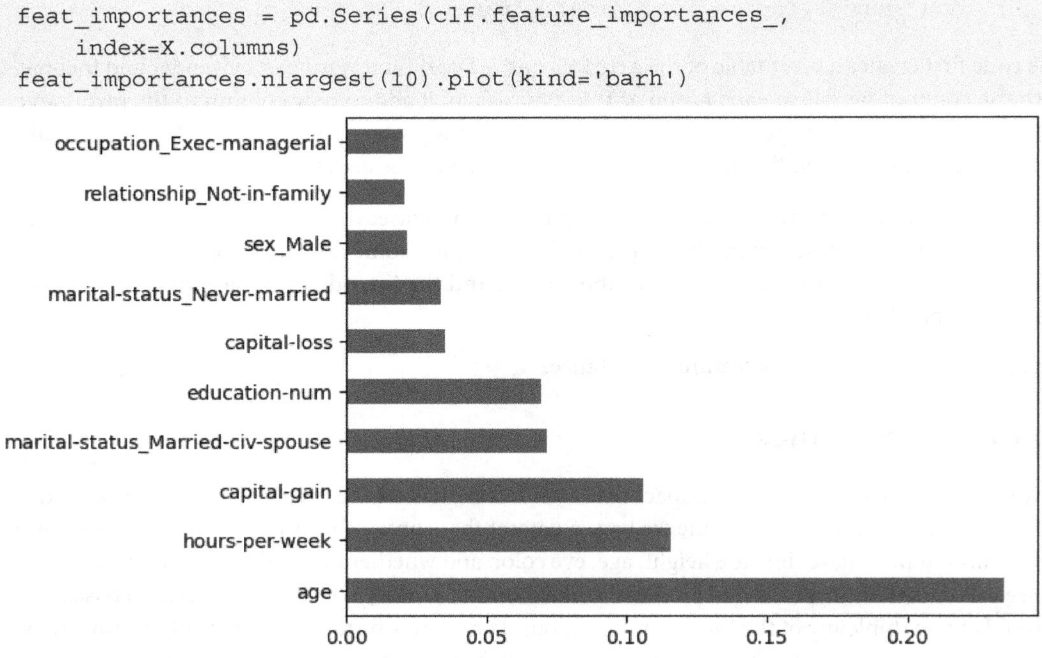


Figure 12.3: Feature importance

In *Figure 12.3*, we can see that features such as `age`, `hours-per-week`, and `capital-gain` seem to be pretty important features with regard to predicting income. Does the influence of those features seem intuitive to you?

Note that feature importance does not imply causality. Just because a feature is deemed important doesn't mean it causes the target variable to change; only that there's an association.

While we're specifically using the `feature_importances_` attribute in scikit-learn in our Jupyter notebook, other popular ML libraries also provide similar mechanisms. For example, TensorFlow's boosted trees (`tf.estimator.BoostedTreesClassifier` or `tf.estimator.BoostedTreesRegressor`) also provide `feature_importances_` as a property to get the importance of each feature. Similarly, LightGBM and CatBoost provide `feature_importances_` and `get_feature_importance()`, respectively, and XGBoost provides the `plot_importance()` function for visualization.

While `feature_importances_`, and similar mechanisms from other ML libraries, can be very useful, there are yet more advanced tools that we can use in order to assess feature importance, which I will describe next.

### Partial dependence plots

**Partial dependence plots** (**PDPs**) are graphical visualizations used to understand the relationship between an input feature (or a set of input features) and a model's predicted outcome. With PDPs, we change the value of just one feature, while keeping the values of all other features constant, in order to determine how the different values of that particular feature impact the prediction. PDPs can also be used with more than one input feature at a time, which can reveal interactions among multiple features. There's also an extended form of PDPs called **individual conditional expectation (ICE) plots**. While PDPs show the average effect of a feature on predictions, ICE plots display the effect of a feature on predictions for individual instances.

With PDPs, for every unique value of the feature of interest, the following high-level steps are performed:

1. Set the feature of interest to that value for every instance in the dataset.
2. Make predictions using the model.
3. Average the predictions.
4. Plot the averaged predictions against the unique values of the feature.

The following code uses the `PartialDependenceDisplay` attribute from scikit-learn to create and display a PDP graph:

```
features = ['age', 'hours-per-week']
PartialDependenceDisplay.from_estimator(clf, X, features)
```

This code will produce a graph similar to the one shown in *Figure 12.4*:

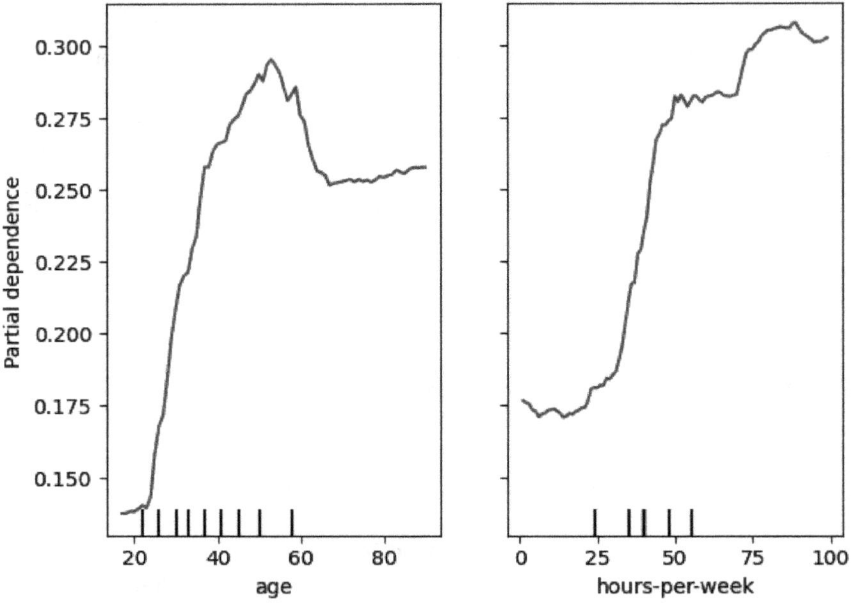


Figure 12.4: PDP

In the PDP shown in *Figure 12.4*, the model appears to predict that people tend to start increasingly earning more money from the age of about 20 onward until they reach the age of around 60, after which their income begins to decrease. This is somewhat intuitive, considering that 20 years of age is considered early adulthood, and people often tend to retire in their 60s. Similarly, the model predicts that working more hours per week may result in higher income.

Next, we move on to discuss more advanced feature importance and explanation mechanisms.

## Local Interpretable Model-agnostic Explanations

As I mentioned earlier in this chapter, some models are naturally more easily interpretable and understandable (and therefore explainable) than others. Examples provided were linear regression models and decision trees, versus large NNs with thousands, millions, or even billions of parameters (perhaps soon to be trillions!). **local interpretable model-agnostic explanations** (**LIME**) takes advantage of this fact, by training a simpler, **surrogate** model that can be more easily explained than the target model. The idea is that even if the overall model is complex and non-linear, it can be approximated well by a simpler, interpretable model.

LIME's inner workings are quite complex, and I would recommend reading the original paper (*arXiv:1602.04938*) if you want to delve into that level of detail. Such algorithmic details are generally not required for the activities of a solutions architect, and you would mainly need to understand what LIME is used for without needing to dive into the academic details of its inner workings. Moving on, then, the following code provides an example of how to use LIME:

```
explainer = lime_tabular.LimeTabularExplainer(
    X_train.values, training_labels=y_train,
    feature_names=X.columns.tolist(),
    class_names=['<=50K', '>50K'],
    mode='classification')
exp = explainer.explain_instance(
    X_test.values[0], clf.predict_proba,
    num_features=10)
exp.show_in_notebook()
```

The code will produce visualizations similar to those shown in *Figure 12.5*:

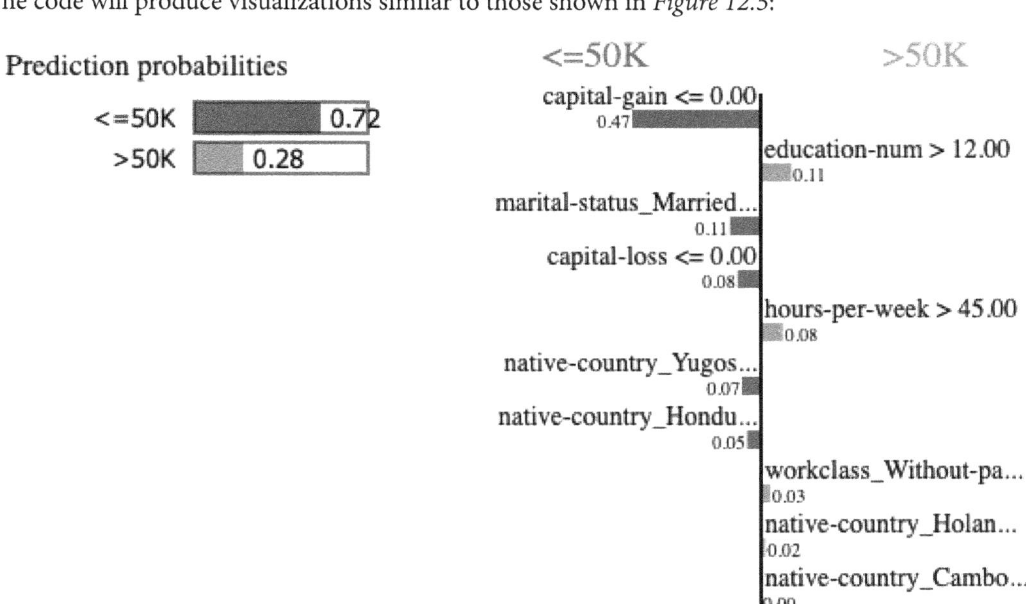


Figure 12.5: LIME outputs

At the top of *Figure 12.5*, we see the model's prediction for the instance that was used as input. In this case, since it's a binary classification problem, it shows the class that the model predicted, along with the probability score associated with that prediction. We also see a horizontal bar chart representing the influence of various features on the prediction. Each bar represents a feature and its impact. The length of the bar indicates the significance of the influence, and its direction (left or right) indicates the direction of the influence (for example, toward the "<=50K" or ">50K" class). Each bar is labeled with the feature name and a small descriptor, which indicates how that feature was quantified for the specific instance being interpreted. This provides a clear indication of how the feature value for that specific instance influenced the prediction.

It's important to highlight that LIME's explanations are local. They are specifically tailored to the instance in question and show how the model made its prediction for that one instance, not a general rule for all data. Next, we explore a mechanism that can help with both local and global model interpretations and is perhaps one of the most popular feature importance and explanation mechanisms in the industry.

### SHapley Additive exPlanations

**Shapley Additive exPlanations** (**SHAP**) helps to explain ML model predictions using a concept called **Shapley values**. These values come from a field of mathematics referred to as **game theory** or, more specifically, **cooperative game theory**, and they were originally introduced in the 1950s by Lloyd Shapley. The study of game theory looks at competitive situations (called "games") where the results of one player's choices depend on what other players do. Cooperative game theory is a sub-branch that looks at games where players can make alliances or **coalitions** with other players and work together as a team to benefit the coalition as a whole, rather than just looking out for their own interests.

Diving in further, let's imagine a game in which the winner will receive a payout of $100. People can either play the game individually, in which case an individual would simply receive $100 if they win, or they can form a team and work together to try to win the $100 payout. If three people form a team and they win the payout, in reality, it's likely they would just split the payout into three equal parts. However, what if the payout should be **fairly** divided based on the contribution of each person? This is where Shapley values come into the picture.

Shapley values represent the **average marginal contribution** of each player over all possible coalitions they could enter, as well as all possible orders in which they could enter each coalition. Again, striking a balance of how much detail to cover on this subject, I will not include the complex mathematical details here but would recommend reading the original paper if you would like to dive into those details, the formal reference for which is provided here:

*Shapley, L.S. (1953). A Value for n-Person Games. In "Contributions to the Theory of Games volume II", H.W. Kuhn and A.W. Tucker (eds.), Princeton University Press.*

Here, we will focus on how Shapley values are used in the context of ML model explainability. We begin with the concept of an **average prediction**, which represents the average of all predictions our model makes over our entire dataset. For regression models, this is simply the mean of all predicted outputs. For classification models, it is the average predicted probability for a given class over the entire dataset.

When computing Shapley values for a particular instance (that is, a row of input data), the contributions of each feature are measured with respect to how they move the prediction for that instance away from the average prediction. The Shapley value for a feature then captures its average marginal contribution over all possible combinations of features. For example, consider a binary classification model that predicts whether a bank loan will be defaulted on. If, on average, the model predicts a 5% probability of default over the entire dataset, this 5% would be the "average prediction." Now, for a particular loan application, if the model predicts a 20% probability of default, Shapley values help attribute this 15% deviation from the average to each feature in the input (for example, the applicant's income, employment status, credit score, and so on). Each feature's Shapley value will indicate how much, on average, that feature contributes to the deviation of the prediction from the average prediction.

The following code provides an example of how to get feature importance insights using the SHAP library:

```
shap_values = shap.TreeExplainer(clf).shap_values(X_test)
shap.summary_plot(shap_values[1], X_test, plot_type="bar")
```

In the code, a `TreeExplainer` object from the SHAP library is being created. This specific explainer is optimized for tree-based models, such as decision trees, random forests, and gradient boosted trees. The `clf` instance that we pass to it is a tree-based model that we've trained. Once the explainer is created, we use the `.shap_values()` method to compute the SHAP values for each sample in our `X_test` dataset. We then visualize the average impact of each feature on the model's predictions by using a bar plot (the longer the bar, the greater the feature's importance). The code will produce a graph similar to the one shown in *Figure 12.6*:

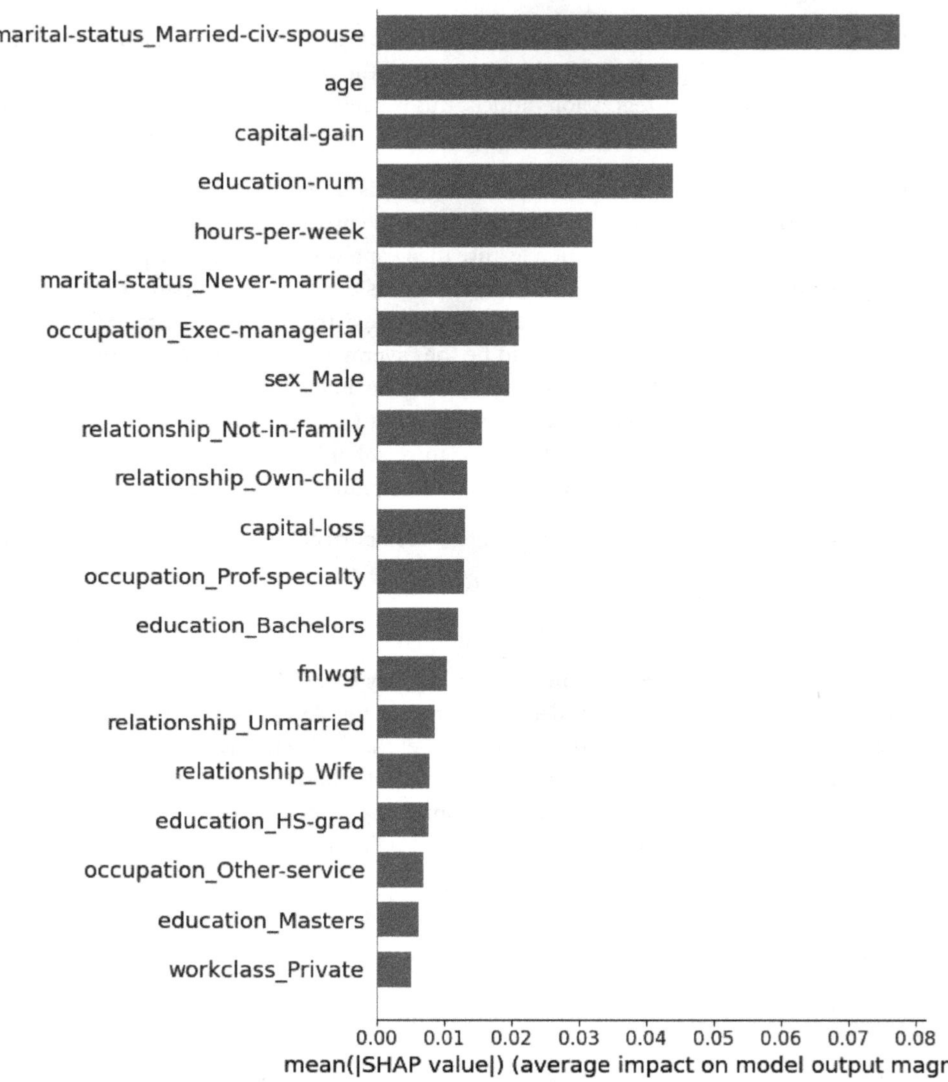


Figure 12.6: SHAP outputs

As we can see in *Figure 12.6*, perhaps surprisingly, a person's marital status appears to have the most impact on the model's output. Considering that the SHAP value for a feature is the average contribution of that feature value to every possible prediction (averaged over all instances), it takes into account intricate interactions with other features, as well as the impact of the feature itself.

While we can import tools such as SHAP into our notebooks, we can also use Vertex AI APIs to get explanations directly from models hosted in Vertex AI. The next section describes how to do this.

### *Getting explanations from a deployed model in Vertex AI*

Conveniently, Vertex AI provides APIs and an SDK that we can use to get explanations from our models. In the Jupyter notebook that accompanies this chapter, we use the `projects.locations.endpoints.explain` API to get explanations from the model that we deployed in our MLOps pipeline in the previous chapter. The following is a snippet of the code we use to do so:

```
endpoint = aiplatform.Endpoint(endpoint_name)
response = endpoint.explain(instances=[instance_dict], parameters={})
    for explanation in response.explanations:
        print(" explanation")
        # Feature attributions.
        attributions = explanation.attributions
        for attribution in attributions:
            print("  attribution")
            print("   baseline_output_value:",
                attribution.baseline_output_value)
            print("   instance_output_value:",
                attribution.instance_output_value)
            # Convert feature_attributions to a dictionary and print
            feature_attributions_dict = dict(
                attribution.feature_attributions)
            print("   feature_attributions:",
                feature_attributions_dict)
            print("   approximation_error:",
                attribution.approximation_error)
            print("   output_name:", attribution.output_name)
```

The code will produce an output similar to the following:

```
   baseline_output_value: 0.7774810791015625
   instance_output_value: 0.09333677589893341
   feature_attributions: {'dense_input': [-0.1632179390639067,
0.0, -0.2642899513244629, 0.0, 0.174240517243743, 0.0, 0.0,
-0.5113637685775757, 0.001784586161375046, 0.1180541321635246,
-0.03459173887968063, -0.004760145395994187]}
   approximation_error: 0.007384564012290227
   output_name: dense_3
```

The fields in the response can be interpreted as follows:

- `baseline_output_value`: This is the model's output value for the baseline instance. A baseline is a reference point (such as an average or neutral instance) against which the prediction for our instance of interest is compared. The difference in the model's output between the instance of interest and the baseline helps us understand the contributions of each feature.

- `instance_output_value`: This is the model's output value for the instance we passed in for explanation. In the context of a binary classifier, this can be interpreted as the probability of the instance belonging to the positive class.
- `feature_attributions_dict`:
    - `'dense_input'`: This is the name of the input tensor to the model.
    - The list of numbers represents the importance or attribution of each corresponding feature in the input for the given prediction. The length of this list matches the number of features in our model's input:
        - Each number represents the marginal contribution of that feature toward the model's prediction for the specific instance we're explaining, relative to the baseline. In other words, how much did this feature move the prediction from the average/baseline prediction?
        - Positive values indicate that the feature pushed the model's output in the positive class's direction. For binary classification, this usually means it made the model more confident in classifying the instance as the positive class.
        - Negative values indicate that the feature pushed the model's output in the negative class's direction.
        - Zero or close to zero suggests that the feature didn't have a significant impact on the prediction for this particular instance.
- `approximation_error`: This is the error in the approximation used to compute attribution values. Explanation methods often use approximations to compute attributions. The approximation error gives an idea of the confidence we can have in the attribution values (a smaller error typically indicates more reliable attributions).
- `output_name`: This is the name of the model's output tensor.

Congratulations! You have successfully retrieved an explanation for an input sent to a model hosted in Vertex AI. To dive into more detail on Vertex Explainable AI, you can reference its documentation at the following link: `https://cloud.google.com/vertex-ai/docs/explainable-ai/overview`.

So far, we've talked about getting and using explanations to understand our models. What if we found that some of our model's predictions were unfair? What kinds of actions could we take to counteract that? One type of explanation mechanism we can use for this purpose is **counterfactual explanations**, which we explore next.

### *Counterfactual explanations*

Counterfactual explanations revolve around the question, "What would need to change in my input data to alter the decision of a predictive model?" They describe a hypothetical alternative to an observed outcome that would have occurred if certain conditions were met. This could be a minimal change in input features that would change the prediction to a specified output.

Going back to our loan approval as an example, suppose an applicant, John, is denied a loan based on features such as his income, credit score, and employment history. A counterfactual explanation might tell John: "If your income was $10,000 higher, you would have been approved for the loan."

Counterfactual explanations are important for many reasons. They help individuals affected by a model's prediction to understand why a decision was made. They help data scientists understand how to augment their models based on various criteria, and they are also important for regulatory compliance reasons.

To find a counterfactual, we need to define a distance measure in the feature space. The goal is often to find the counterfactual instance that's closest to the original instance but results in a different prediction. This is often framed as an optimization problem, where the objective is to minimize the distance between the original instance and its counterfactual, while still resulting in a different prediction.

Bear in mind that for some models, especially **deep NNs** (**DNNs**), finding counterfactuals can be computationally challenging. It's also important to note that counterfactuals might suggest changes that are impossible or very hard to achieve in real life, so their real-world feasibility needs to be evaluated. In the Jupyter notebook that accompanies this chapter, we perform some simple counterfactual processing. Bear in mind that counterfactuals are a highly complex topic in a quickly evolving field.

Next, let's look at some additional mechanisms for reducing bias.

## Reducing bias and enhancing fairness

In this section, we discuss proactive steps we can use to reduce bias and enhance fairness in our datasets and ML models in each phase of the model development life cycle.

### *Starting with the data*

As with anything in the field of data science, the data is usually a good place to start. If available, we should gather more data and ensure that the data is as balanced as possible. During data preprocessing, we can also mitigate bias by using resampling techniques to adjust the representation of underrepresented groups in the training data, either by oversampling minority groups or undersampling majority groups. During feature engineering, we can create or modify features to reduce their potential discriminatory effect.

### During model training

During the model training process, we could introduce **fairness constraints** to ensure equal opportunity by making sure both protected and non-protected groups have equal positive and negative rates, for example. There are numerous fairness metrics and constraints for this purpose, such as the following:

- **Demographic parity** or **statistical parity**, which requires that the probability of a positive outcome should be the same across the different groups
- **Equal opportunity**, which mandates equality in true positive rates across different groups
- **Equalized odds**, which extends equal opportunity by requiring both true positive rates and false positive rates to be equal across groups
- **Treatment equality**, which requires the ratio of false negatives to false positives to be equal across groups

Fairness constraints can be implemented as a type of regularization during model training. There are even some fairness-aware algorithms we could use that are explicitly designed to handle fairness concerns, such as **Fair k-Means**.

### Postprocessing

There are also some steps we can take during postprocessing, such as adjusting decision thresholds for different groups to ensure fairness metrics such as equal opportunity or demographic parity, and we could also adjust model predictions to ensure they are fair across groups.

It is, of course, also important that we continuously monitor for fairness concerns in real-world predictions and retrain models as necessary. In critical decision-making scenarios, we could consider having a human in the loop to audit or override model decisions.

Note that fairness enhancement methods might lead to a trade-off with model accuracy. The challenge is often to find a balance between fairness and accuracy that's acceptable for the given application. It's also crucial to understand which fairness metric is most relevant to your specific problem, as different fairness metrics can sometimes be in conflict with each other.

To wrap up this section on explainability and fairness, let's take a brief look at some other libraries we can use for assessing and implementing these concepts.

## Additional libraries

Fortunately, there is an ever-growing list of libraries being developed for the purpose of assessing and promoting explainability and fairness. This section describes examples of such libraries.

### What-if Tool

The **What-if Tool** (**WIT**) was originally developed by Google and is an early explainability tool with a visual interface that allows the inspection of a model's predictions, comparison across different models, and examination of potential bias. It is relatively easy to use and does not require much coding, and it includes support for many of the concepts we discussed in this chapter, such as counterfactuals and PDPs.

### AI Fairness 360

IBM's **AI Fairness 360** (**AIF360**) is an open source library that includes a set of fairness metrics for datasets and models, and algorithms for mitigating bias. It can be used to provide detailed explanations to understand fairness metrics and their implications in a given context, and it enables users to visually explore bias in their datasets and models. It can also help in identifying if a trained model is producing biased outcomes and provides some tools to help mitigate bias in each phase of the model development life cycle, such as preprocessing, training, and postprocessing.

### EthicalML/XAI

This is an open source Python library designed to support interpretable ML and responsible AI. It includes tools for preprocessing, model interpretability, bias detection, and visualization, and it supports concepts we discussed in this chapter, such as feature importance, Shapley values, LIME, and DIA.

### Fairlearn

Fairlearn is another open source Python-based project that aims to help data scientists improve the fairness of AI systems. It includes algorithms for mitigating unfairness in classification and regression models, and fairness metrics for comparison. Its primary goal is to help ML practitioners reduce unfair disparities in predictions by understanding metrics and algorithms, and it provides an interactive UI experience for model assessment and comparison, which includes fairness metrics and an assessment dashboard. It again supports mitigation techniques in various phases of the ML model development life cycle, such as preprocessing, training, and postprocessing.

There are many more explainability and fairness libraries in addition to the ones mentioned here, but these ones are particularly popular libraries that are currently available in the industry. Google Cloud has recently launched more specific model evaluation mechanisms and metrics for fairness. At the time of writing this in October 2023, these mechanisms are still in preview mode and are not yet generally available. You can find out more about these features in the documentation at the following link: `https://cloud.google.com/vertex-ai/docs/evaluation/intro-evaluation-fairness`

> **A note on generative AI**
>
> *Chapters 14* through *17* in this book are dedicated to **generative AI (GenAI)**, which is a relatively new subset of AI/ML. In those chapters, we will explore the concepts of **large language models (LLMs)** and how they differ from other types of ML models that we've already covered in this book. LLMs are typically trained on extremely large datasets and therefore acquire vast amounts of knowledge that can be applied to many different kinds of use cases. In those later chapters, we will learn how LLMs can be used as auto-raters to open up new kinds of evaluation techniques for ML models, including specific evaluations focusing on bias, explainability, and fairness.

A final topic for us to explore in this chapter is the concept of lineage tracking. In the next section, we delve into this topic in detail and assess its importance in the context of explainability and fairness.

# The importance of lineage tracking in ML model development

We've touched on the concept of lineage tracking in earlier chapters, and now we will explore it in more detail. When we talk about lineage tracking, we're referring to tracking all of the steps and artifacts that were used to create a given ML model. This includes items such as the following:

- The source datasets
- All transformations that were performed on those datasets
- All intermediate datasets that were created
- Which algorithm was used to train a model on the resulting data
- Which hyperparameters and values were used during model training
- Which platform and tools were used in the training
- If a hyperparameter tuning job was used, details of that job
- Details of any evaluation steps performed on the resulting model
- If the model is being served for online inference, details of the endpoint at which the model is hosted

The preceding list is not exhaustive. We generally want to track every step that was used to create a model and all of the inputs and outputs in each step.

Why do we need to do this? Lineage tracking is important for many reasons. It complements the concept of explainability. While lineage tracking by itself will not necessarily explain why a model behaves in a particular manner, it is certainly important for researchers to understand how the model was created. It's also important for reproducibility and collaboration. We've talked about the complexity some companies encounter when they need to manage thousands of ML models created by many different teams. If a model is behaving problematically, understanding its lineage will help in troubleshooting. If one team wants to build on the work another team has already performed, such as a model they have trained or datasets they have created, understanding the lineage of those artifacts will help the consuming team to be more productive in those endeavors. Also, in order to continually enhance a model's performance, we need to know how that model was created. Lineage is also important, and sometimes required, for governance and compliance reasons.

Fortunately, Google Cloud provides tools that help us to track lineage. For example, Dataplex can be used to track data lineage, and the Vertex ML Metadata service can help us track all steps and artifacts in our ML model development life cycle. Next, we will take a look at Vertex ML Metadata in more detail; let's first start with some terminology used by the Vertex ML Metadata service.

## ML metadata service terminology

**Executions** represent the steps or operations in an ML workflow, such as data preprocessing, model training, or evaluation. **Artifacts** represent the inputs and outputs of each step, such as datasets, models, or evaluation metrics. **Events** represent the relationships between executions and artifacts, such as "Artifact X was produced by Execution Y" or "Artifact X was used as an input by Execution Y." Events help us to establish lineage data by associating artifacts and executions with each other. **Contexts** represent logical groupings that bundle related artifacts and executions together. An example of a context would be a specific pipeline run or a model version.

Collectively, all of the aforementioned resources are referred to as **metadata resources**, and they are described by a **MetadataSchema**, which describes the schema for particular types of metadata resources. In addition to the predefined metadata resources, we can also store custom metadata in the Vertex ML Metadata service. All tracked metadata is stored in a **MetadataStore**, and all of this information can be used to create a **Lineage Graph**, which is a visual representation that connects artifacts, executions, and contexts, and shows the relationships and flow between them.

Note that, as with most Google Cloud resources, access to the metadata resources can be controlled using Google Cloud **Identity and Access Management** (**IAM**), which is important for security and compliance reasons.

Now that we've covered the main terminology and concepts, let's start reviewing some metadata in Vertex AI.

## Lineage tracking in Vertex AI

To explore the lineage tracking features in Vertex AI, we will use the MLOps pipeline we built in *Chapter 11* as an example.

In the Google Cloud console, perform the following steps:

1. Navigate to **Vertex AI** > **Pipelines**.
2. Click on the name of the pipeline run that was created in *Chapter 11* (it should be the most recent pipeline run unless you have run other pipelines in this Google Cloud project in the meantime).
3. You'll see the execution graph for the pipeline run.
4. At the top of the screen, click the toggle button to the left of the words **Expand Artifacts** (see *Figure 12.7* for reference):

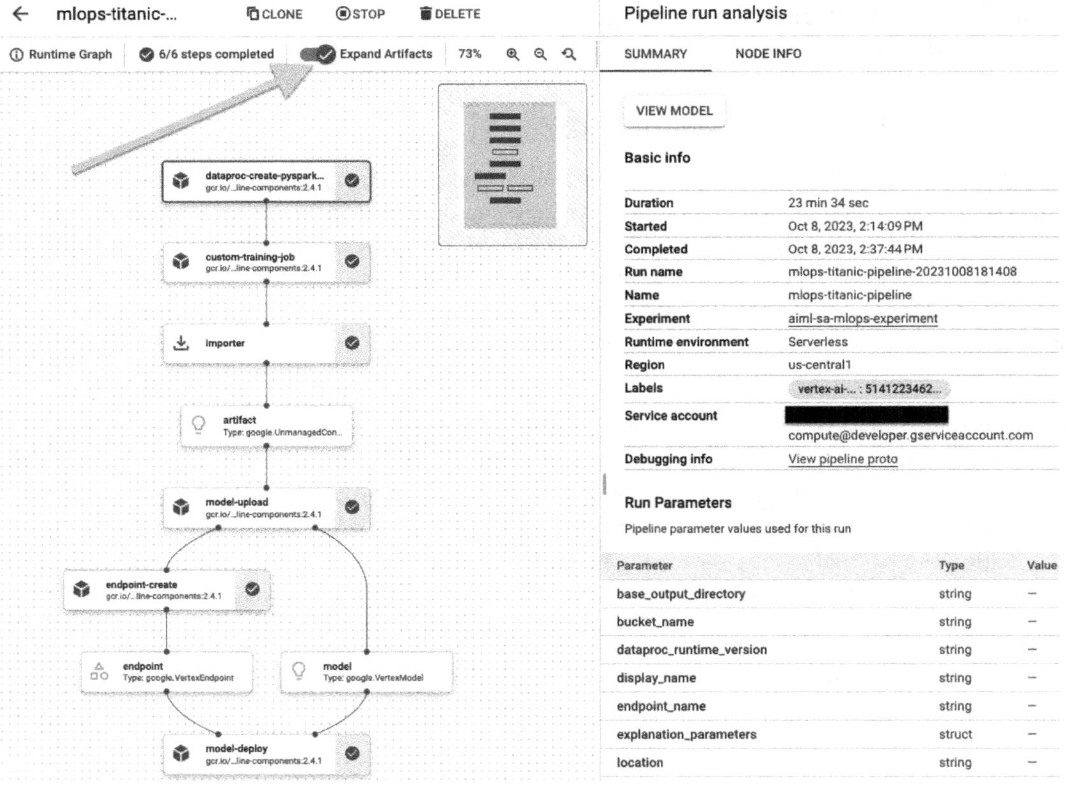


Figure 12.7: Expand Artifacts

5. We can now start exploring the metadata related to each of the steps and artifacts in our pipeline.

6. You'll also notice that the **Pipeline run analysis** section on the right side of the screen contains lots of information about this pipeline run. The **SUMMARY** tab provides information about the pipeline run itself, including the parameters that were used as inputs. Those are the parameters we defined in our pipeline definition in *Chapter 11*.
7. We can click on elements in the pipeline execution graph in order to see metadata related to that specific element. Let's start right at the beginning. We want to know which dataset was used as an initial input to our pipeline. Considering that the first step in our pipeline is the data preprocessing step, and that step fetches the dataset, click on the preprocessing step, and its metadata will be shown on the right side of the screen, as depicted in *Figure 12.8*:

Pipeline run analysis

SUMMARY NODE INFO

Execution Info Completed

VIEW JOB VIEW LOGS

| Display name | dataproc-create-pyspark-batch |
|---|---|
| Name | dataproc-create-pyspark-batch |
| Type | system.ContainerExecution |
| Duration | 6 min 18 sec |
| Started | Oct 8, 2023, 2:14:10 PM |
| Completed | Oct 8, 2023, 2:20:28 PM |

Input Parameters

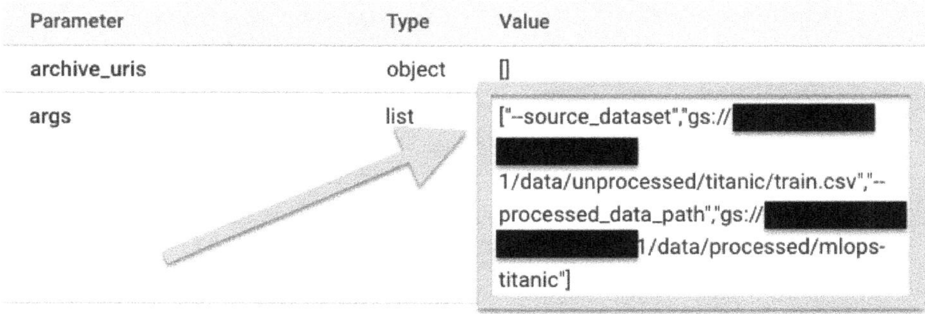


Figure 12.8: Preprocessing step details

8. The green arrow in *Figure 12.8* is pointing to the `source_dataset` input parameter, which provides the path to the source dataset (the actual details are redacted in the image in order to obscure my bucket name).
9. We can also see the `preprocessed_data_path` parameter value, which provides the path to the folder in which the preprocessing script will store the resulting processed data. If you scroll down (not shown in the screenshot), you will also see the `main_python_file_uri` parameter value, which provides the path to the PySpark script that we used in the preprocessing step in our pipeline. In fact, if we click on the **VIEW JOB** button, we can view the details of the actual Serverless Spark job that was used to execute our script in Google Cloud Dataproc, including its execution logs.
10. Now we've successfully tracked our source dataset, the script and job that performed transformations on that dataset, and the resulting processed dataset that was used to train our model, let's move to the next step in our pipeline, which is the model training step.
11. Click on the `custom-training-job` step in our pipeline execution graph. In the information panel on the right, perhaps the most important parameter is the `worker_pool_specs` parameter. As depicted in *Figure 12.9*, this parameter provides a lot of information about how our model is trained, such as the dataset that was used for training (which is the output of the previous, preprocessing step), the location at which the trained model artifacts are saved, the container image that was used to run our custom training code, the hyperparameter values used during training, as well as the machine type and number of machines that were used by the training job:

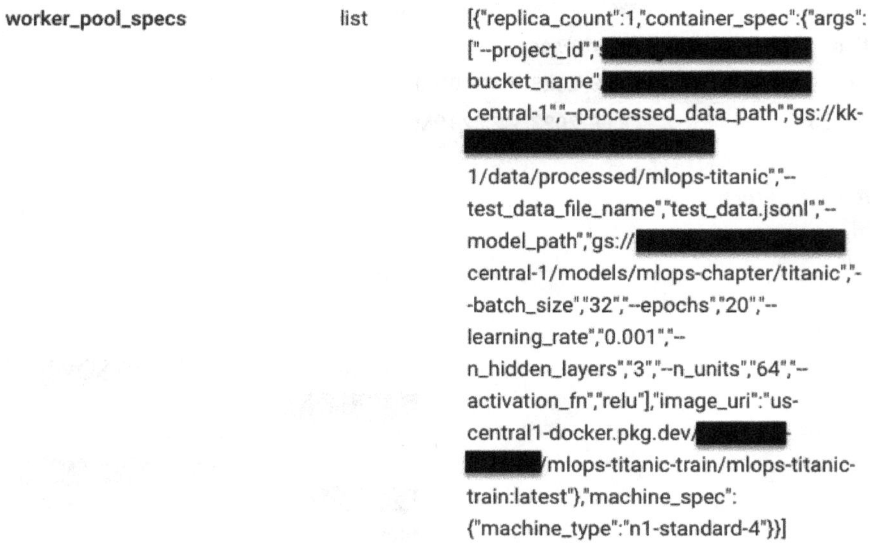

**worker_pool_specs** list

[{"replica_count":1,"container_spec":{"args":
["--project_id","
bucket_name"
central-1","--processed_data_path","gs://kk-
1/data/processed/mlops-titanic","--
test_data_file_name","test_data.jsonl","--
model_path","gs://
central-1/models/mlops-chapter/titanic","-
-batch_size","32","--epochs","20","--
learning_rate","0.001","--
n_hidden_layers","3","--n_units","64","--
activation_fn","relu"],"image_uri":"us-
central1-docker.pkg.dev/
/mlops-titanic-train/mlops-titanic-
train:latest"},"machine_spec":
{"machine_type":"n1-standard-4"}}]

Figure 12.9: worker_pool_specs

12. Again, if we click on the **VIEW JOB** button at the top of the screen, we can see the actual job that ran on the Vertex AI training service to train our model, as well as the execution logs for that job.
13. Because we used a custom script to train our model and simply saved the artifacts in Google Cloud Storage, at this point in the pipeline, our model is referred to as an **Unmanaged model**. In order to start using the model more appropriately, we needed to register it in the Vertex AI Model Registry. We use an `importer` job to import our model artifact for that purpose.
14. The `model-upload` step in the pipeline is what registers our model in the Vertex AI Model Registry. If you click on that step in the execution graph and review its metadata, in the **Output Parameters** section, you will see the URI for the resulting resource in the Vertex AI Model Registry.
15. The remaining steps, `endpoint-create` and `model-deploy`, have similar formats. As their names suggest, the `endpoint-create` step creates an endpoint in the Vertex AI prediction service, and the `model-deploy` step deploys our model to that endpoint. Their output parameters will show the URIs for the resources created by those steps.
16. I want to draw your attention to the `endpoint` and `model` artifacts in the pipeline. If you click on those, and click the **View Lineage** button that appears in the information panel on the right-hand side of the screen, it will take you directly to the console for the Vertex AI Metadata service and will show you another view of how the steps and artifacts relate to each other, as depicted in *Figure 12.10*. Again, clicking on each element in the graph will display metadata for that element:

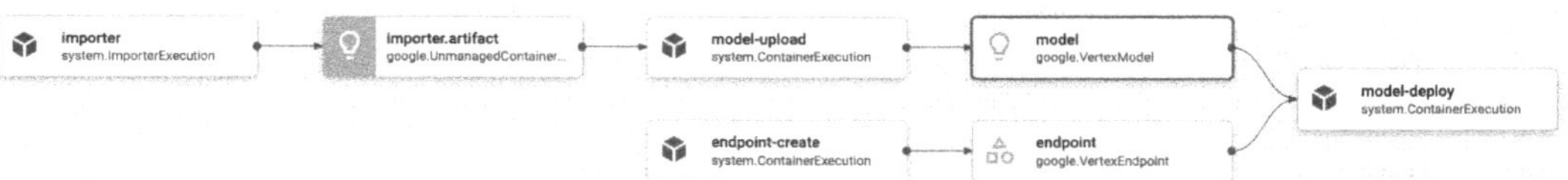


Figure 12.10: Lineage graph in Vertex AI Metadata service console

In addition to getting metadata insights via the Google Cloud console, we can also use the Vertex AI SDK and API directly to query and manage metadata programmatically. For example, the following piece of code will list all artifacts in our Google Cloud project:

```
aiplatform.Artifact.list()
```

Similarly, the following lines will list all executions and contexts in our Google Cloud project:

```
aiplatform.Execution.list()
aiplatform.Context.list()
```

We have now successfully tracked every step and artifact that was used to create our model. Next, let's explore the **Experiments** feature in Vertex AI that is closely related to lineage tracking.

### *Vertex ML Experiments*

When we created our pipeline definition in *Chapter 11*, we specified an experiment name with which to associate our pipeline runs. This provides another view of how to group the steps and artifacts related to our pipeline runs and the model versions they create. This feature can be useful for sharing and collaboration, as well as for comparing different model versions against each other. To view the experiment associated with our pipeline runs, perform the following steps:

1. In the Google Cloud console, navigate to **Vertex AI** > **Experiments**.
2. Click on the name of the experiment that we specified in our MLOps chapter (`aiml-sa-mlops-experiment`).
3. Click on the name of the most recent run, and explore the **ARTIFACTS** tab, as shown in *Figure 12.11*:

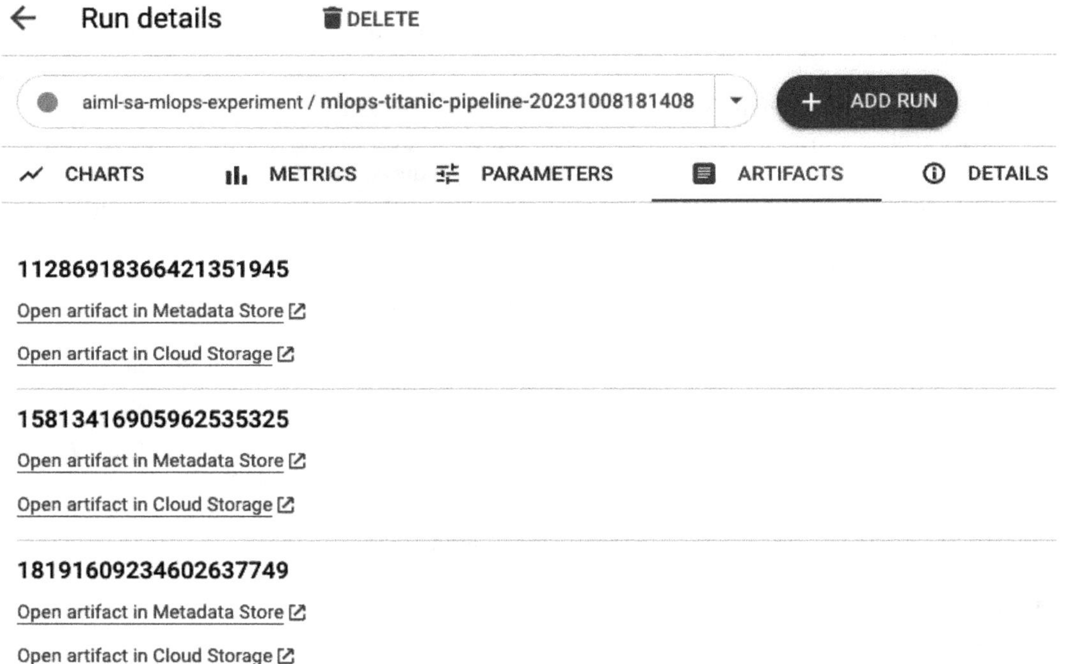


Figure 12.11: Vertex AI Experiments – Artifacts view

4. Click on the links that are shown under each artifact ID to see those artifacts, as well as their metadata in the Vertex ML Metadata service (it will bring you back to the screens we already explored in the previous section; this is just another way to access the same metadata).

We've covered a lot of information in this chapter – let's summarize what we've discussed before we move on to the next chapter.

## Summary

In this chapter, we explored the concepts of bias, explainability, fairness, and lineage. We started off by examining some of the common types of bias that can occur at various steps in the ML model development life cycle. This included sources of bias such as pre-existing bias, algorithmic bias, and collection or measurement bias, which further included sub-categories such as sampling bias, response bias, and observer bias. We talked about how to inspect for bias, using techniques such as data exploration and DIA.

Next, we dived into the use of explainability techniques to understand how our models make their decisions at inference time and to assess their fairness, particularly with regard to understanding how the input features in our dataset could influence our models' predictions. We used tools such as PDPs and SHAP for these purposes. We then looked at how to use Vertex AI to get explanations from models that were hosted on Vertex AI endpoints. Going beyond simply getting explanations, we then discussed how to proactively counteract bias by using counterfactual analysis to pose questions such as "What would need to change in my input data to alter the decision of a predictive model?"

Finally, we covered the topic of lineage tracking and its importance in terms of explainability, as well as other factors such as collaboration, troubleshooting, and compliance. We walked through an ML pipeline that we had created in a previous chapter and looked at the metadata associated with each component of the pipeline, including all of the steps and artifacts that were used to create a specific model.

While the chapters before this one focused on everything that's required to build and run an ML model, this chapter focused on more advanced topics such as ensuring that our models are explainable and fair. The next chapter in this book continues in a similar vein. We are no longer just looking at the mechanics of how to build and deploy ML models but are now incorporating broader ethical and architectural considerations. In the next chapter, we dive further into the topics of governance, compliance, and architectural best practices in the model development life cycle.

# 13
# ML Governance and the Google Cloud Architecture Framework

As technologists, we often, of course, find the technological aspects of **machine learning** (**ML**) to be the most fun and exciting parts, while legal and regulatory concepts don't always inspire us quite as much. However, these concepts are required to build robust solutions at scale in production. They are what help us make the transition from hobbyist activities to designing reliable systems that can be vital to a company's success or even affect millions of people's lives.

With that in mind, this chapter will cover the following subjects:

- ML governance
- An overview of the Google Cloud Architecture Framework
- Architecture Framework concepts about AI/ML workloads on Google Cloud

Let's go ahead and dive right into our first topic.

## ML governance

ML governance refers to everything that's required to manage ML models within an organization, throughout the entire model development life cycle. As models can play a significant role in critical decision-making processes, it's important to ensure that they are transparent, reliable, fair, and secure, and we need to implement structured frameworks to achieve these goals. These frameworks include policies and best practices that ensure responsible and ethical use of data and ML technologies.

When discussing ML governance in this chapter, I will also include data governance in the scope of the discussion, because the use of data is so inherent in the ML life cycle. Let's start there.

## Data governance

When it comes to managing data in the ML life cycle, there are a number of aspects that we need to consider, such as data quality, lineage, privacy, security, and retention. Let's take a look at each of these in more detail.

### *Data security, privacy, and access control*

Of all the aspects of a data governance strategy, data security is arguably the most important. Data security incidents tend to make headline news, and those are not the kinds of news headlines you want to be responsible for!

Fundamental to data security is the concept of data access control, which, as the name suggests, focuses on who can access the data and how they can access the data. The worst-case scenario is that somebody from outside the company gains access to sensitive data and leaks it publicly or uses it for nefarious purposes such as ransom or sabotage.

When it comes to data security, my favorite term is **defense-in-depth** (**DiD**), which alludes to the fact that a thorough data security strategy consists of using many different tools to protect the data and other resources. In the following sub-sections, I will outline steps we can take to secure our data.

#### Data categorization

Before defining security policies for your data, it's important to establish a categorization system in order to understand which elements of data need more focus in terms of protection. For instance, data that is published openly on your website, such as product descriptions and prices, is generally not considered to be top secret, whereas your customers' credit card details are highly sensitive. You can categorize your data in terms of tiers, such as the following:

- **Tier 0**: Highly-sensitive, such as customer passwords and credit card details
- **Tier 1**: Sensitive, such as customer purchase or transaction history
- **Tier 2**: Somewhat sensitive, such as customer addresses and phone numbers
- **Tier 3**: Non-sensitive, such as publicly viewable information

These are just examples; you would need to work with your organization's information security experts to determine what categories would make the most sense for your organization.

After categorizing our data, let's discuss how we can secure it.

### Network security

We can look at the DiD approach as layers in an onion. At the outermost layer, we begin with network security. If unintended users of the data do not have a network path to the data, then that's a really solid foundation in your security strategy. Network security practices include setting up devices such as firewalls to control what kinds of traffic are allowed to enter a protected network. Google Cloud provides such devices, as well as other network security constructs, such as **Virtual Private Cloud** (**VPC**), which allows you to set up your own private networks and control how to access them, and **VPC Service Controls** (**VPC-SC**), which enables you to create a security perimeter around your resources to prevent data exfiltration.

### Authentication and authorization

The next layer in the onion is authentication and authorization to grant or deny permissions to access resources. Even if somebody gets access to a protected network, the next step is to determine which resources they are allowed to access and what actions they are allowed to perform on those resources. You need to ensure that the data can be accessed only by people and systems that are authorized to do so, and authorization should be based on business criticality. In other words, a person or system should only be able to access a piece of data if they need such access to perform a required business function. At all times, you should be able to easily determine who (or what) has access to which data, and why.

Google Cloud **Identity and Access Management** (**IAM**) can be used to configure and enforce such permissions, or, for software components, additional mechanisms such as **Transport Layer Security** (**TLS**) authentication can also be used. A little later in this section, we will also cover data cataloging with Google Cloud Dataplex. Dataplex and Google Cloud BigLake can be used to make it easier for companies to manage and enforce permissions for accessing their data resources. Google Cloud BigQuery offers additional data security mechanisms such as row-level and column-level access control, meaning that not only can you grant or restrict access to tables within BigQuery, but you can more granularly grant or restrict access to specific rows and/or columns within those tables. This provides additional flexibility to protect resources from unintended access. For example, with column-level security, you could configure that only people in the finance department can view columns that contain customers' credit card details, while other employees and systems cannot. With row-level security, you could configure that sales representatives can only view details for customers in their region and not in other regions.

### Data encryption

At the innermost layer are the data resources themselves. Best practices suggest that data should be encrypted as a further security measure. In that case, even if a malicious or unintended user gets access to the data, they would need the encryption keys to unencrypt the data. It goes without saying that encryption keys should be stored in a highly secure manner in a separate system, with all of the layers of security implemented to protect them. Again, Google Cloud provides tools to implement all of those layers of security mechanisms, including encryption and key management using Google Cloud Key Management.

### Logging and auditing

In addition to all of those mechanisms, a strong data security strategy should incorporate auditing and logging implementations to monitor data access and modifications and support audits and forensic investigations to detect or investigate policy violations and data breaches. Google Cloud Logging and Audit Logs can be used for those purposes. *Figure 13.1* shows what kinds of logs can be tracked by Google Cloud Audit Logs:

LOG TYPES EXEMPTED PRINCIPALS

You can configure what types of operations are recorded in your Data Access audit logs for the selected services. There are several subtypes of Data Access audit logs:

- [ ] Admin Read
  Records operations that read metadata or configuration information.
- [ ] Data Read
  Records operations that read user-provided data.
- [ ] Data Write
  Records operations that write user-provided data.

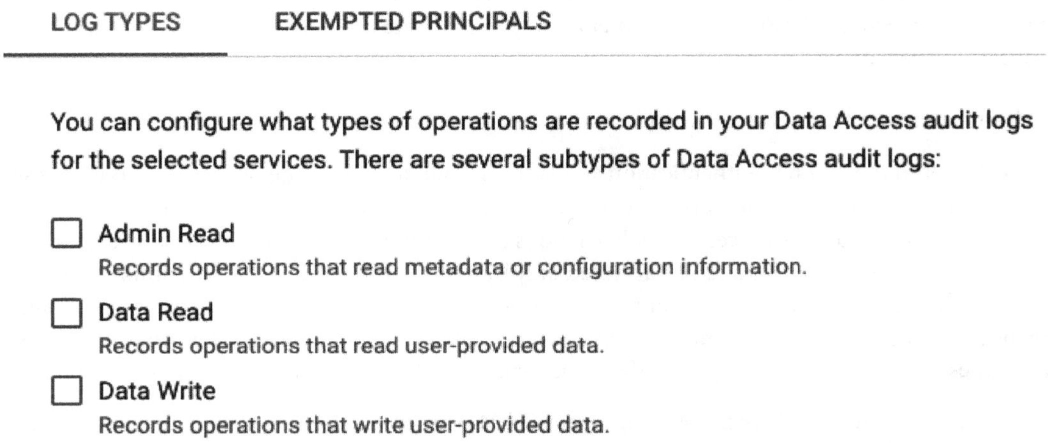

Figure 13.1: Audit log types in Google Cloud Audit Logs

As we can see in *Figure 13.1*, there are three types of logs that we can capture using Google Cloud Audit Logs.

### Data privacy

While intrinsically related to data security, data privacy more specifically focuses on the lawful, ethical, and safe handling of **personal information** (PI). It is one of the most important aspects of data security because privacy violations can severely damage a company's reputation and customer trust.

Not only that, but they can incur serious legal ramifications. There are numerous international standards and regulations that govern data privacy and protection, such as the **General Data Protection Regulation** (GDPR) in the **European Union** (EU), the **California Consumer Privacy Act** (CCPA) in the US, and many others globally. These regulations outline rules about data handling, such as collection, storage, processing, and sharing. Navigating these regulations and ensuring compliance in how your systems handle data can be quite challenging.

It can also be a challenge to keep track of sensitive data that may be dispersed throughout your datasets. Imagine a company that has petabytes of data that comes from many different sources, such as credit card readers in stores or customer-facing online forms. It's easy to identify that data being transmitted from a credit card machine needs to be protected, but this may not be as obvious for other data interfaces. For example, perhaps customers are accidentally entering their credit card details into online form fields that are not intended for that purpose (for example, if a customer accidentally pastes their credit card details into a field that was not intended for entering credit card details). Such fields may not be considered sensitive and therefore do not get any special attention from a security perspective. The Google Cloud Sensitive Data Protection service, which now incorporates the Google Cloud **Data Loss Prevention** (**DLP**) tool, can help to identify and protect sensitive information in your datasets. If sensitive data is found, you can use the Sensitive Data Protection service to implement protection mechanisms such as de-identification, masking, tokenization, and redaction.

In the context of data security, privacy, and access control, it's important to highlight the concept of shared responsibilities and shared fate on Google Cloud. Rather than risking misstating legal terminology here, I instead recommend reading Google Cloud's official policy on this topic, which can be found at the following URL:

`https://cloud.google.com/architecture/framework/security/shared-responsibility-shared-fate`

Before we move on to discuss ML model governance, we will round out this section with a discussion of data quality, cataloging, and lineage.

### *Data quality, cataloging, and lineage*

Considering that data can be such a valuable resource for a company, it's important to establish practices that ensure the accuracy, consistency, discoverability, and (depending on the use case) timeliness of data. For example, considering that data is often used to make important business decisions, inaccurate or out-of-date data could have negative impacts on a company's business. This also applies to business decisions that are automated by ML models. If we feed inaccurate data into an ML model, we will get inaccurate predictions.

An effective data governance strategy starts with clear policies, responsibilities, and standards for data quality. We then need to establish frameworks to measure and monitor data quality factors. Best practices in this regard include regular data cleaning processes for correcting data errors, de-duplication, and filling in missing values, as well as automation mechanisms such as regular, automated data quality checks.

Data discoverability is also an important factor. After all, even if you have created pristinely curated datasets, they are not very useful if nobody knows they exist. When companies have well-established data management practices, it's easier for data producers and consumers to know exactly where their data is and the current status of that data, at any given time. In such companies, a robust data catalog forms the heart of the company's data infrastructure. I've worked with many clients over the years in various consulting roles, and you'd be surprised how many companies operate without well-defined data management strategies. Given that data can be the life-blood of an organization, it's surprising to learn that many companies are not fully aware of exactly what data they own. Various systems throughout the company are producing and gathering data all day every day, but if that data is not being cataloged in some way, it may simply sit in silos in remote, disparate parts of the company, unavailable and unknown to most of the rest of the organization. Bear in mind that data can be used for all kinds of interesting business use cases. If you don't know what data you have access to, you may be missing out on significant business opportunities.

It's also important to implement data lineage tracking to understand where data comes from and how it gets transformed as it moves through various processing systems within a company. If I find a piece of data somewhere in my company, I want to know how it got there and every step it took along the way. Which systems did it pass through? What did those systems do to the data? Are there intermediate datasets that were created by those other systems in various parts of the company? This is not only important from a business operations perspective but can be required for compliance purposes. For example, if you need to comply with data sovereignty regulations, you better know where your data is at all times. If a customer decides to exercise their right to be forgotten in markets that are subject to GDPR – or other relevant regulations – you will have a really hard time complying if you do not have a good handle on your data. Similarly, if a data breach occurs, data lineage can help identify what data was compromised and understand the potential impacts and recovery steps needed.

Fortunately, Google Cloud provides numerous tools to help with each of the aforementioned activities, such as Dataproc and Dataflow for cleaning and processing data, and Dataplex for cataloging, data quality, and data lineage tracking.

Next, we will discuss ML model governance.

## ML model governance

In this section, we discuss the aspects required to ensure that the models we build and deploy are reliable, scalable, and secure and that they continue to meet those requirements on an ongoing basis. There are a number of factors that we need to incorporate in order to achieve this goal, which we discuss in the following sub-sections.

## *Model documentation*

Starting with every developer's favorite topic: documentation! While documentation is not always the most fun part of a developer's job, it is essential to building and maintaining production-grade systems. I've worked with clients and teams in various companies that have not always done a great job of developing accurate, high-quality documentation, and one thing that you can almost guarantee in the lack of such documentation is that it will make your job a lot more difficult when you need to maintain and improve your systems over time. Imagine that you join a new team and you are tasked with improving the performance of a particular application that uses ML to perform medical diagnoses, and you find that the original application and underlying model were developed years ago by people who have left the company and did not document how they implemented the system. This is not a good place to be in, and you would be surprised how commonly these kinds of scenarios exist in the industry. Perhaps most importantly in the context of this chapter, model documentation can be essential—and sometimes even legally required—for compliance purposes.

So, what does high-quality model documentation look like? Generally, our documentation should keep detailed records of factors such as model design, data inputs, transformations, algorithms, and hyperparameters. Let's take a look at each of these in more detail.

### Model design

Documentation regarding model design should clearly define the purpose of the model, such as the objectives the model intends to achieve and the context within which it needs to operate. This includes potential use cases and intended users or systems that will interact with the model. We also need to provide a detailed description of the model's architecture, such as its layers, structures, and interdependencies among different components of the model. Additionally, we should include details regarding the model's design rationale, such as the reasoning behind choosing this particular model architecture or design, including comparisons with other potential designs that were considered and an explanation of why they were not chosen.

### Data inputs

Our model documentation should describe the data collection process we used, including sources, methods of collection, and the timeframe during which data was collected. We should list all the features used by the model, including their definitions, types (for example, categorical, and numerical), any assumptions made about the data, and explanations as to why each feature is relevant to the model's predictions. Additionally, we need to document any known issues in terms of data quality, including missing values, outliers, or inconsistencies, and how these issues were handled or mitigated.

### Transformations

Another best practice is to detail the steps taken to clean and preprocess the data, such as handling missing data, normalization, encoding techniques, and feature engineering, including explanations of any methods used for feature selection or reduction (such as **principal component analysis** (**PCA**), as depicted in *Figure 13.2*), and the rationale for their use:

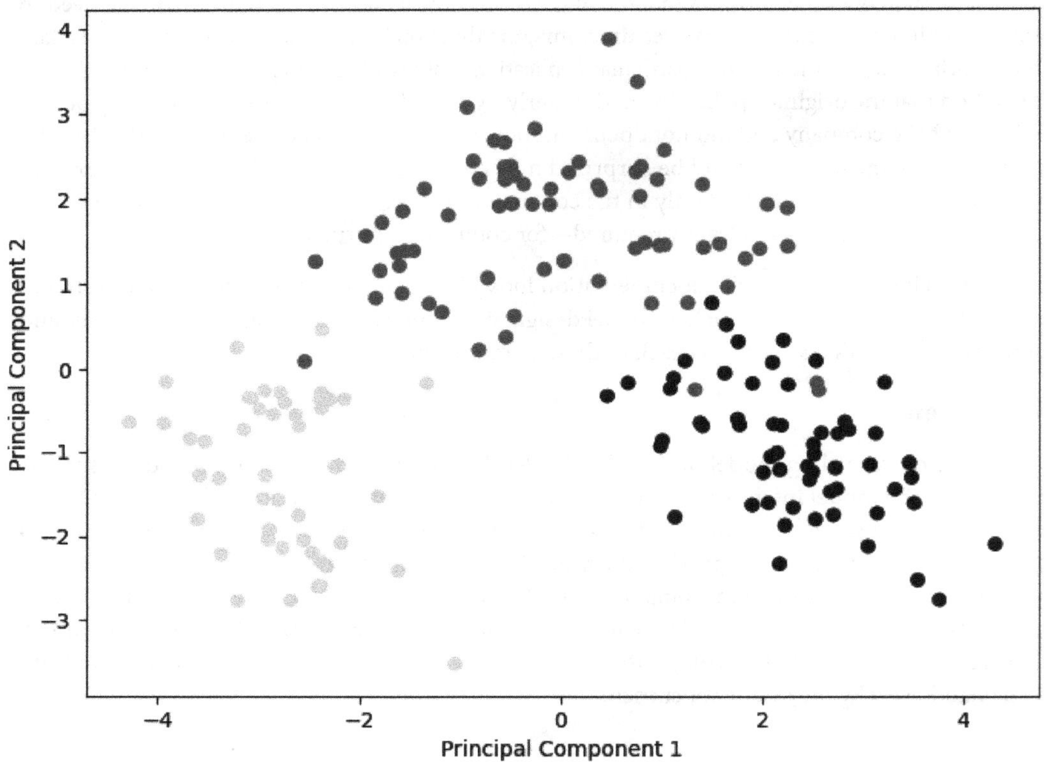


Figure 13.2: PCA

### Algorithms

In our documentation, we should discuss why a particular ML algorithm was chosen, including a comparison with other algorithms that were considered and a rationale explaining the algorithm's suitability for the problem at hand, citing relevant literature or empirical evidence, as appropriate. It's also important to detail the configuration of the algorithm, including any customizations specific to the project.

### Hyperparameters

We should include details on all hyperparameters that the model uses, providing a definition and range for each, as well as the process used for hyperparameter tuning, such as grid search, random search, or Bayesian optimization. Additionally, we should include the final values chosen for each hyperparameter and provide a rationale for why these particular values were chosen, supported by the tuning process's outcomes.

### Additional factors

Our documentation should also explain how the model's performance was evaluated, including the metrics used (**mean squared error** (MSE), **area under the ROC curve** (ROC-AUC)) and the results of these evaluations. We should document any known limitations of the model, document potential biases in the model's predictions, and describe how these biases could impact different demographic groups or individuals. We also need to detail any regulatory standards or ethical guidelines relevant to the model and discuss how compliance has been ensured. Finally, we need to outline the plan for ongoing monitoring of the model's performance and behavior in a production environment, including strategies for handling model drift, anomalies, or performance degradation.

## *Model versioning*

Just as with code versioning in traditional software development, model versioning in ML projects is essential for ensuring that teams can trace back through the evolution of models, replicate results, roll back to previous versions when necessary, and maintain a record of all changes made throughout a model's life cycle. This becomes especially important in large or collaborative environments where multiple iterations of models may be developed over time by many different people and teams. It's also important for debugging, continuous improvement, and audit and compliance purposes. *Figure 13.3* shows an example of model version metadata in the Vertex AI Model Registry:

| | |
|---|---|
| **Model ID** | 484 744 |
| **Version description** | – |
| **Created** | Oct 8, 2023, 2:25:08 PM |
| **Region** | us-central1 |
| **Encryption type** | Google-managed |
| **Dataset** | No managed dataset |
| **Objective** | Custom |
| **Source** | Custom training |
| **Container image** | us-docker.pkg.dev/vertex-ai/prediction/tf2-cpu.2-12:latest |
| **Model artifact location** | |
| **Feature attribution method** | Sampled shapley |
| **Path count** | 10 |

Figure 13.3: Model version metadata in Vertex AI Model Registry

When I speak of model versioning, I'm referring to more than just the model artifacts, such as the actual model files, weights, and architecture. We should implement version tracking for every relevant item that is used in the model development process. This includes any code associated with the model, be it for preprocessing, training, evaluation, or deployment, as well as datasets, whether they consist of raw data, preprocessed data, or feature-engineered data. Even hyperparameter values and performance evaluation metrics should be versioned so that we can easily understand which versions of these elements pertain to which versions of our models.

A well-implemented model versioning tool, such as Vertex AI Model Registry, will also enable us to add custom metadata to more comprehensively track our various model versions.

### Model monitoring

As we discussed in previous chapters, we need to continuously monitor models after we have deployed them to production. The intention is to identify any drift, anomalies, or degradation in model performance, including important markers such as fairness and explainability metrics. Again, this is not only important for **business continuity** (**BC**) purposes but also for compliance reasons. We may have certified that our model is compliant with specific regulations before and during deployment, but it could drift and lapse out of compliance if not regularly monitored on an ongoing basis. If we have set up MLOps pipelines to automate the process of continuously improving our models over time, then this monitoring should extend to all aspects of our pipeline, such as ensuring that we perform data quality monitoring during the data preprocessing steps, in addition to new model validation and other important steps in the process.

As with everything else regarding our model development and management implementations, we want our monitoring processes to be automated as much as possible. We should set thresholds beyond which some kind of corrective action is either automatically initiated or a human is notified if there is a problem. For example, if we see that performance or fairness metric values have changed by more than a specified amount, then corrective action is initiated, either by automatically training and evaluating a new model version on updated data or by paging on-call support engineers to intervene.

As we also discussed in previous chapters, Vertex AI provides built-in tools for monitoring model performance both after deployment and throughout each relevant step in the model development process.

### Auditing and compliance

Many industries have strict regulations (for example, GDPR, HIPAA, and various financial regulations) that require certain standards for factors such as data privacy, bias, transparency, and many more. Non-compliance with these regulations can lead to legal penalties and loss of customer trust. If we have workloads that are subject to regulatory standards, then we will need to establish auditing processes to help ensure our workloads remain compliant on an ongoing basis.

How we implement such processes will mainly depend on the types of regulations with which we need to comply. Such processes could consist of a regular human-review process, or, as always, it would be best if we could automate auditing processes as much as possible and notify a human only when an issue that appears not to be automatically resolvable occurs. In the case of human-review processes, this is where documentation is inherently important because good-quality documentation can greatly simplify the review process and can make it easier to determine corrective actions when issues are identified. Ideally, we would want to identify potential risks in model performance, security, or reliability before they escalate into larger issues, and establishing regular review processes can help to ensure this happens.

For some types of regulations, well-established audit checklists and **standard operating procedures (SOPs)** can be used, making the auditing task a bit easier. However, bear in mind that the regulatory landscape for ML is still evolving, and organizations must stay abreast of changes to remain compliant.

In the previous chapter, we discussed the concept of explainability. Explainability is particularly important in the context of regulatory compliance. If you can't easily or adequately explain how a given model or system works, then you will have a difficult time ensuring regulatory compliance.

Now that we've covered many of the important factors of ML governance in quite a bit of detail, let's zoom back out and focus on the bigger picture again. In the coming sections, we will discuss how to operationalize ML governance, what ML governance looks like in different industries, and how to stay abreast of the evolving ML governance landscape.

## Operationalization of ML governance

As I've alluded to in each of the previous sections, we usually would want to automate as much of our ML governance practices and processes as possible, and there are tools and platforms that can assist in achieving this goal, such as data catalogs, model management tools, and auditing tools, which I describe in the following sub-sections.

### *Data catalogs*

We briefly talked about data cataloging earlier in this chapter. Data catalogs are a kind of metadata management tool that helps companies find and manage large amounts of data spread across their organization, whether on-premises or in the cloud. You can think of a data catalog as a massive inventory of a company's data assets, designed to let users discover, organize, and understand their data sources. We've already introduced Google Cloud Dataplex, which Google describes as an "*intelligent data fabric that enables organizations to centrally discover, manage, monitor, and govern their data across data lakes, data warehouses, and data marts, with consistent controls.*" *Figure 13.4* shows an example of a catalog created by Google Cloud Dataplex:

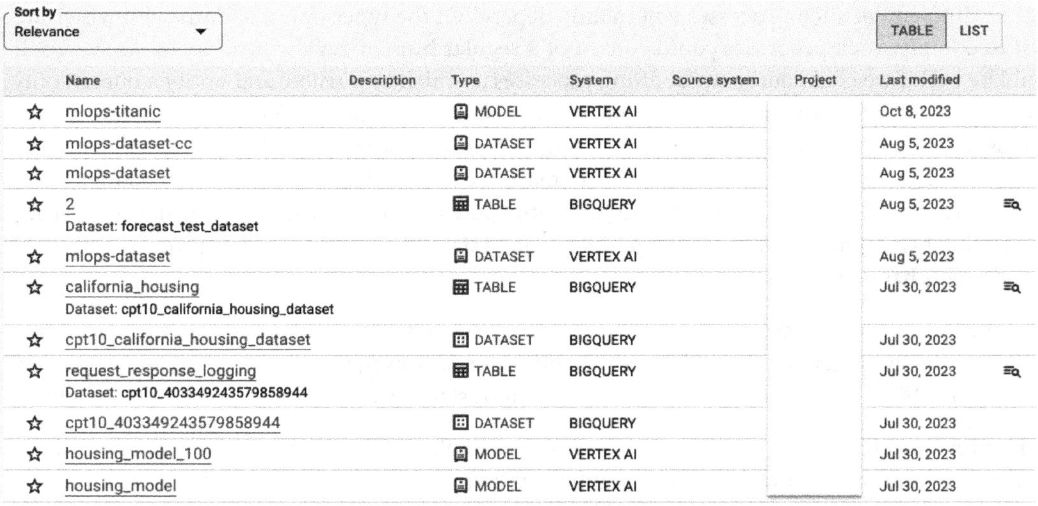

Sort by
Relevance

TABLE LIST

| Name | Description | Type | System | Source system | Project | Last modified |
|---|---|---|---|---|---|---|
| mlops-titanic | | MODEL | VERTEX AI | | | Oct 8, 2023 |
| mlops-dataset-cc | | DATASET | VERTEX AI | | | Aug 5, 2023 |
| mlops-dataset | | DATASET | VERTEX AI | | | Aug 5, 2023 |
| 2<br>Dataset: forecast_test_dataset | | TABLE | BIGQUERY | | | Aug 5, 2023 |
| mlops-dataset | | DATASET | VERTEX AI | | | Aug 5, 2023 |
| california_housing<br>Dataset: cpt10_california_housing_dataset | | TABLE | BIGQUERY | | | Jul 30, 2023 |
| cpt10_california_housing_dataset | | DATASET | BIGQUERY | | | Jul 30, 2023 |
| request_response_logging<br>Dataset: cpt10_403349243579858944 | | TABLE | BIGQUERY | | | Jul 30, 2023 |
| cpt10_403349243579858944 | | DATASET | BIGQUERY | | | Jul 30, 2023 |
| housing_model_100 | | MODEL | VERTEX AI | | | Jul 30, 2023 |
| housing_model | | MODEL | VERTEX AI | | | Jul 30, 2023 |

Figure 13.4: Dataplex catalog

This concept of consistent controls is particularly relevant in the context of governance. With Dataplex, you can use metadata to describe all of your company's data assets, and you can manage permissions in a uniform way across different Google Cloud data storage and processing tools. This is, of course, important from a governance perspective, and Dataplex also provides data quality and data lineage functionality, which are also important in the context of governance.

### *Model management platforms*

These platforms assist in the model's entire life cycle, including development, deployment, monitoring, and maintenance. They are essential for activities such as model versioning, experiment tracking, and model performance monitoring. By providing a structured environment for managing ML models, these platforms help ensure that models are reliable and reproducible and that they meet performance expectations. They also facilitate compliance by providing detailed model development and deployment process records. Of course, Vertex AI is Google Cloud's native model management ecosystem, providing all of the aforementioned features.

So far, we've focused on aspects of ML governance that are common to many industries. In the next section, let's take a look at how ML governance applies to specific industries.

## ML governance in different industries and locations

What can make regulatory compliance even more complex is that different countries, states, and industries can all have varying regulations. ML governance therefore varies significantly across different industries and geographic locations. In this section, we'll discuss governance factors for specific regions and sectors, such as healthcare and finance, as well as regional considerations.

### *Healthcare*

I currently live in the US, and when I hear the terms "regulatory compliance" and "healthcare" in the same sentence, the first thing that pops into my mind is the **Health Insurance Portability and Accountability Act** (**HIPAA**), which grants patients certain rights over their health information, including secure handling and confidentiality, and outlines significant penalties for breaches and non-compliance. If you work in healthcare in the US, you will almost certainly need to be aware of, and work within, the requirements of HIPAA. When designing and implementing ML systems in this industry, you must ensure that the data handling practices throughout the entire model development life cycle comply with those requirements. Other countries have their own regulatory requirements that you must learn and understand if you operate in those areas.

### *Finance*

In the finance industry, fraud is perhaps one of the biggest concerns or threats, and the finance industry is heavily regulated to avoid the potential of fraud occurring. If your company operates in the US, for example, then the financial operations of your company will need to abide by regulations such as the **Sarbanes-Oxley Act** (**SOX**), which is mainly intended to prevent corporate fraud and improve the reliability and accuracy of corporate disclosures to protect investors. If your systems handle credit card data in any way, then you will likely need to comply with **Payment Card Industry Data Security Standard** (**PCI DSS**) regulations, which are a set of security standards relating to the secure handling of credit card information.

### *Region-specific governance considerations*

Different regions have different regulatory requirements. For example, if you operate in the EU and the **European Economic Area** (**EEA**), then you will be subject to GDPR requirements, which protect the PI of individuals in those regions.

The state of California has the CCPA, which regulates how businesses worldwide are allowed to handle the PI of California residents.

There are also regulations that govern how data related to children must be handled, such as the **Children's Online Privacy Protection Act** (**COPPA**), which is intended to protect the privacy of children under 13 years of age.

In addition to region-specific and industry-specific regulations, we must also ensure that we remain compliant with obligations regarding fairness, transparency, explainability, accountability, and ethics. And, not only do we need to manage the complexity of regulations in different regions and industries, but that complexity is further extended by the fact that regulations may change over time. Let's discuss this topic in more detail next.

## Keeping up with the evolving landscape of ML governance

I can confidently say that the ML industry is one of the most quickly evolving industries at the moment. Companies, governments, and research institutions all over the world continue to invest heavily in this industry. As a result, compliance regulations related to this industry continue to evolve rapidly. In order to consistently remain successful and, quite frankly, ensure that you don't get into trouble, your company needs to stay current in this ever-evolving landscape. In this section, I outline important concepts and best practices in this space.

### *Ongoing education and training*

Your company's employees need ongoing education to stay current with advances in regulatory requirements. In addition to technical capabilities, your employees must understand ethical considerations and risk management strategies related to ML deployment. Well-informed and well-trained individuals are less likely to make costly errors such as violating data privacy standards. Also, since this is not a static field, new types of bias and ethical dilemmas emerge almost daily. I highly recommend implementing regular training sessions, workshops, and educational resources for employees to learn about the latest trends, tools, ethical considerations, and best practices regarding ML governance.

### *Regularly updating governance policies and practices*

Along with the technologies and ethical considerations, the legal landscape for data protection and privacy also continually evolves. Organizations must regularly update their governance policies to comply with new and changing laws and standards (GDPR, CCPA, or industry-specific regulations), and as new technologies and methodologies develop, governance policies must adapt to accommodate and manage them appropriately. What worked for a simple linear regression model might not be sufficient for a complex **deep learning** (**DL**) system. Unfortunately, the threat landscape also continues to develop as new technologies emerge, and what was secure yesterday might not be quite as secure tomorrow. As a result, we need to regularly review and update security policies and practices to protect sensitive data and ML systems against new vulnerabilities and attack strategies.

Additionally, in consideration of Google's **Site Reliability Engineering** (**SRE**) practices, while we should do everything we can to avoid a negative event occurring, if such an event does occur, we need to learn from that scenario. As such, we should conduct thorough post-mortems on any issues and use these insights to improve policies and prevent future occurrences.

Another set of concepts that are quite closely linked to SRE are encapsulated in the Google Cloud Architecture Framework, which I will discuss in the next section.

## An overview of the Google Cloud Architecture Framework

In Google's own words, "*the Google Cloud Architecture Framework provides recommendations and describes best practices to help architects, developers, administrators, and other cloud practitioners design and operate a cloud topology that's secure, efficient, resilient, high-performing, and cost-effective.*" In this section, we discuss some of the key concepts from the framework and how they can be applied to AI/ML workloads on Google Cloud, especially within the context of ML governance.

For reference, the framework documentation can be found at the following URL:

`https://cloud.google.com/architecture/framework`

Let's begin with an overview of the fundamental concepts of the framework, which are referred to as **pillars**. One way to think of it is that, by ensuring all of these pillars are implemented, we can build a solid and enduring structure (system).

The categories of the framework are the following:

- System design
- Operational excellence
- Security, privacy, and compliance
- Reliability

- Cost optimization
- Performance optimization

In the following sub-sections, I'll describe what each category represents, starting with system design.

## Pillar 1 – System design

Interestingly, the *System design* category in the Google Cloud Architecture Framework is more akin to the foundation of the overall framework than a pillar, because a well-designed system appropriately incorporates all of the pillars. The *System design* category encapsulates four core principles:

- Document everything
- Simplify your design and use fully managed services
- Decouple your architecture
- Use a stateless architecture

Let's take a look at each of these in more detail.

### Principle 1 – Document everything

We touched on this principle quite a bit in this chapter's *ML model governance* section. In a broader system design context, we're not just referring to documentation related to our model, but rather related to every aspect of how our systems are implemented, including aspects such as architecture diagrams and maintenance runbooks. The question I always ask in this context is: if a new member joined our team and were required to quickly learn everything they need to know to improve and maintain a system we've built, what are all of the details they would need to review? If any of those details are not adequately documented, then that's a gap that we need to address by developing the required documentation. This also helps other teams that we need to collaborate with or that need to interact with our system in some way, and it makes everybody's job easier if regulatory compliance officers need to audit our systems.

### Principle 2 – Simplify your design and use fully managed services

Ever since I heard of the concept of Occam's razor, I have been a big fan of it. There are a number of different ways in which it can be summarized, but a fairly common one is: "If there are two possible explanations for a particular phenomenon, use the simpler one."This can be extended to say: don't make things more complicated than needed. The opposite of this is the concept of a Rube Goldberg machine, which applies an extremely complex set of mechanisms to achieve a simple goal. While Rube Goldberg machines can be fun to watch, they are generally quite impractical, and they are not what you want to implement in the design of your large-scale, low-latency, highly sensitive production system.

Keeping your system design as simple as possible has many benefits, such as making it easier to troubleshoot, maintain, and secure your systems. Highly complex systems with many different components are difficult to troubleshoot when something goes wrong. Similarly, in terms of security, highly complex systems often have a larger **attack surface**. We will cover that concept in more detail later in this chapter.

Another way in which we can make our jobs easier is by offloading responsibilities to a **cloud service provider** (**CSP**). One of the primary benefits of cloud computing is that CSPs offer platforms and systems that have been designed to address common needs in the industry. When a company runs all of its workloads in its own data centers, it either needs to build solutions and platforms completely by itself or install and manage software created by other companies. Both of those options incur a lot of staff overhead and require specific training to manage.

Let's take the example of building and maintaining a platform to enable data scientists to develop and deploy ML models. In previous chapters, we outlined the various steps that are required in the model development life cycle. If we are running all of our workloads "on-premises" (that is, not in the cloud), then we need to either design, build, and maintain a platform that supports all of those steps or, as many companies do, we could try to hack something together, using a hodge-podge of random third-party software solutions that all require specific training and don't necessarily work very well together. In the cloud, however, we can simply use the platform provided by the cloud provider and let them do all of the hard work for us so that our teams can focus on their core competencies and primary objectives rather than building and maintaining infrastructure.

Similar concepts exist across other types of workloads. For example, on-premises, we may build and maintain our own Kubernetes environments for our company's containerized workloads. We would spend a lot of time maintaining those systems, but in the cloud, we could use managed services such as **Google Kubernetes Engine** (**GKE**) or Cloud Run. In this context, we talk about services that offload more of the infrastructure management tasks to the cloud provider as "going further up the stack." In the case of GKE and Cloud Run, Cloud Run would be seen as "further up the stack" because it provides more of a fully managed experience than the basic form of GKE, although the more recent launch of GKE Autopilot also provides a very hands-off approach, in which more of the platform management tasks are implemented by Google Cloud.

### *Principle 3 – Decouple your architecture*

This ties in with *principle 2* to an extent. It refers to breaking your overall system design into smaller components. A classic example of this is to break a monolithic application into microservices. The reason is that each microservice is easier to manage than a very large and complex monolithic system architecture. Smaller components can be developed and scaled independently, which can improve the speed of development of new features in your systems.

### *Principle 4 – Use a stateless architecture*

In a stateless architecture, each transaction is independent. When processing a request, the server does not remember any prior requests or transactions, and the client must send any necessary data for a transaction in each request. In a stateful architecture, on the other hand, the server maintains the state of the client's session. The context of previous transactions is remembered, and future transactions can be affected by what happened in the past.

I would add the words "whenever possible" to the title of this principle because sometimes your system will need to maintain state, but what you would want to do, as much as possible, is minimize the amount of state that needs to be maintained and offload the state management from your application to a separate mechanism such as a caching layer.

Stateless architectures are generally easier to scale because they don't require maintaining client states, allowing requests to be processed by any available server. Stateful architectures require more complex infrastructure to ensure that the client interacts with the same server or that the state is shared, which can be challenging in large-scale environments. Stateful systems also use more resources to manage and store session data, which can further affect scalability and system complexity.

## Pillar 2 – Operational excellence

The *Operational excellence* pillar is concerned with efficiently running, managing, and monitoring systems on Google Cloud. It includes concepts such as automation, observability, and scalability.

This pillar talks about automating system deployments using **continuous integration and continuous deployment** (**CI/CD**), and managing your infrastructure using **infrastructure as code** (**IaC**). This is a very important concept because it provides all of the benefits of traditional software development, such as version tracking and incremental updates. If you manage your infrastructure updates using version tracking mechanisms, then you can maintain strong records for auditing purposes, and if issues are introduced by any updates, then you can more easily roll back to a previous version that was known to work well. This is often referred to as **GitOps**, and the opposite of this is referred to as **ClickOps**. In the case of ClickOps, infrastructure updates are made by people clicking around in a UI. If you have hundreds of people in your technology organization, and every day they are all making updates to your infrastructure by clicking around in a UI, then it can become difficult to coordinate and track these updates over time. Terraform is a popular tool for implementing IaC on Google Cloud.

The *Operational excellence* pillar also outlines best practices for incorporating testing throughout the software delivery life cycle. This includes unit tests, integration tests, system tests, and other types of tests such as performance and security testing. Rather than testing everything at the end, we should aim to include each type of test as relevant throughout each step in the development life cycle. For example, unit tests could be automated as part of our build process.

When deploying software, the *Operational excellence* pillar recommends using approaches such as immutable infrastructure updates via blue/green deployments and A/B or canary tests. Google Cloud provides CI/CD tooling that can be used to implement these strategies. The recommendation is to use small but frequent updates to your systems, which are easier to manage and roll back than large, infrequent changes.

In terms of observability, this pillar provides recommendations on effectively setting up monitoring, alerting, and logging, including common metrics to keep an eye on, and defining thresholds beyond which some kind of alert or corrective action should be invoked. It also talks about the importance of setting up audit trails to keep track of changes to your systems. For cases in which something does go wrong, it provides guidelines on establishing support and escalation procedures, as well as review processes such as post-mortem assessments to learn from any failures.

Of course, it's also important to ensure that your infrastructure is adequately scaled to handle your expected traffic volumes and that you implement plans to proactively scale accordingly for known peak events.

Finally, this pillar covers the importance of automating as many of your system management tasks as possible to minimize how much you need to rely on potentially error-prone manual processes to keep your systems running effectively.

## Pillar 3 – Security, privacy, and compliance

This is perhaps the most relevant of the pillars in the context of ML governance, and we've already touched on these topics earlier in this chapter, but here, we will take a look at how these concepts are more formally structured within the Google Cloud Architecture Framework.

In addition to the concept of DiD that we discussed earlier, Google recommends strategies to implement **security by default**. This consists of best practices for ensuring that security is built in as the default configuration in your system architecture, including concepts such as the **principle of least privilege** (**PoLP**), in which users are given the minimum permissions required to perform their job functions, and nothing more. It also refers to locking down access at the network level. For example, if you know that, in normal operating circumstances, your system should only ever be accessed from one or two other systems, then you could set up network rules that block access from any sources other than those specific systems. Google Cloud also provides an offering called **Confidential Computing** for processing sensitive data.

I want to take this opportunity to highlight that the pillars of the Google Cloud Architecture Framework are often interrelated. For example, we talked about the concept of GitOps in the context of the *Operational excellence* pillar. This concept of using IaC to manage how you deploy to your systems is a highly recommended way to establish security-by-default practices. For example, you can create Terraform modules that undergo stringent security assessment processes for setting up your infrastructure in a way that aligns with your corporate security policies and industry-wide best practices. Once those "secure-by-default" modules have been approved, anybody in your company could safely use them to set up the required infrastructure securely. This makes it much easier for your employees to abide by your security policies in terms of provisioning infrastructure. To make it easy for you to provision infrastructure resources that align with security best practices, Google Cloud provides the **security foundations blueprint**, which you can reference at the following URL:

`https://cloud.google.com/architecture/security-foundations`

As an association between this pillar and the pillar of *Operational excellence*, the CI/CD pipelines that are used to deploy your resources should have security mechanisms built in. For example, you could use automation to check for security vulnerabilities when artifacts are created. Google Cloud also provides a mechanism called Binary Authorization to validate the contents of Docker containers that are built and deployed by your CI/CD pipelines to ensure that those images contain exactly what you expect them to contain and nothing more. It can also validate that a specific build system or pipeline created a specific container image. If a security check highlights any potential problems at any point in your CI/CD pipeline, the pipeline can automatically be halted to ensure that potential security threats are not introduced into your deployments. Similarly, you can use Google Cloud's Artifact Analysis feature to scan automatically for potential vulnerabilities in containers stored in Artifact Registry and Container Registry. Even after deployment, you can continuously scan your web applications by using Google Cloud's Web Security Scanner to identify vulnerabilities in applications deployed to Compute Engine, App Engine, and GKE.

This pillar also provides recommendations on proactively identifying and cataloging risks to your company and how to mitigate common risks. Google Cloud has also recently launched the Risk Protection Program, which includes tools such as Risk Manager, to help you manage risks.

Of course, **IAM** is an important component of this pillar. Google Cloud provides many tools that help manage this aspect, such as IAM and Cloud Audit Logs, which we discussed are essential for access management and auditing.

This pillar also calls out the importance of using cloud asset management tools such as Cloud Asset Inventory to track all of your company's technology assets and monitor for deviations from your compliance policies.

In addition to all of the topics we've covered in this section, the security pillar also covers topics such as network security, data security, privacy, and regulatory compliance, which we covered earlier in this chapter. It also provides details on how to use Google Cloud Assured Workloads to help you meet your compliance obligations, how to monitor for compliance, and how to address data sovereignty, data residency, software sovereignty, and operational sovereignty.

## Pillar 4 – Reliability

The *Reliability* pillar focuses on concepts such as **high availability** (**HA**), scalability, automated change management, and **disaster recovery** (**DR**). It covers topics that some of you may know from Google's SRE practices, such as defining **service-level indicators** (**SLIs**), **service-level objectives** (**SLOs**), **service-level agreements** (**SLAs**), and error budgets. As was the case with the *Operational excellence* pillar, the *Reliability* pillar includes observability as a major component. It also reiterates some other concepts from the *Operational excellence* pillar, such as automating deployments and incremental updates using CI/CD pipelines, and the importance of setting up appropriate observability and alerting mechanisms, **incident management** (**IM**), and post-mortem practices.

This pillar talks about ways of creating redundancy for higher availability in your system architectures, including using multiple Google Cloud zones and regions to mitigate any potential issues that may occur in a particular location. In addition to these kinds of proactive mitigation techniques, it also outlines practices for establishing DR strategies, such as synchronizing data to other regions and establishing playbooks for failing over to those regions if needed.

Finally, it goes into much detail regarding best practices for specific Google Cloud products. This pillar contains a wealth of knowledge and much more detail on many specific Google Cloud products than would be appropriate to include here.

## Pillar 5 – Cost optimization

You can be guaranteed this is very important to almost every customer. In fact, cost optimization is often one of the main factors that entice companies to move to the cloud in the first place. When companies run their workloads in their own data centers, they often have to purchase and install enough infrastructure (and more) to cater for their highest peak events that may only happen once or twice per year. For the rest of the year, that infrastructure is highly under-utilized, which amounts to a lot of wasted money. In the cloud, however, companies can scale their infrastructure up and down based on what they actually need and therefore do not need to waste money on over-provisioned infrastructure. Also, as discussed earlier in this chapter, offloading infrastructure management to a cloud provider enables companies to invest their time in innovation and developing features that support their core business.

The first major focus of this pillar is on the concept of financial operations or **FinOps**, which is a cultural paradigm that includes a set of technical processes and business best practices to help organizations optimize and manage their cloud investments more effectively. In this context, it's important to provide each technology team in the organization with visibility into their cloud spend and for each team to take accountability for that spend. To learn more about FinOps, I recommend reading the Google Cloud FinOps whitepaper, which can be found at the following URL:

`https://cloud.google.com/resources/cloud-finops-whitepaper`

Remember that we generally cannot optimize or improve something without monitoring it. As such, the *Cost optimization* pillar provides recommendations regarding monitoring costs, analyzing trends, and forecasting future costs. If you forecast that you will spend a certain amount of money in the next year or the next 3 years, you can purchase **committed use discounts** (**CUDs**) to save money on those workloads. You can also use labels to categorize your expenses in billing reports, such as attributing resource expenses to specific workloads and environments.

The *Cost optimization* pillar also provides best practices on **optimizing** resource usage to reduce costs, such as ensuring that you provision your infrastructure based on your current and projected needs (including some buffer where appropriate), and do not over-provision. This is referred to as **right-sizing**, and Google Cloud even provides a right-sizing recommender that can highlight opportunities for improving your sizing by identifying resources that appear to be under-utilized (and therefore over-provisioned). You should also use auto-scaling, which, in addition to ensuring that you have enough resources to serve your required traffic volumes, can scale resources down when they're not needed, thus saving money.

When implementing cost optimization mechanisms, it's important to set up budgets, alerts, and quotas to control your spending. For example, you can specify a certain spending budget and get alerted when you are close to reaching that budget. You can also use quotas to set hard limits on resource usage and can set API caps to limit API usage after a certain threshold is reached.

As with the *Reliability* pillar, the *Cost optimization* pillar provides detailed best practices for many specific Google Cloud products, such as optimizing storage tiers in **Google Cloud Storage** (**GCS**) or optimizing partitions in BigQuery.

## Pillar 6 – Performance optimization

Performance optimization can be linked to cost optimization, so there is some overlap in terms of these concepts. For example, if your systems are performing optimally, then they may be less costly to run. A well-implemented auto-scaling strategy is a prime example of this. The *Performance optimization* pillar provides recommendations on how to define performance requirements, how to monitor and analyze performance, and, of course, how to optimize performance.

In terms of monitoring and analyzing performance, this refers back to the concept of observability, in which we need to implement and monitor performance metrics such as latency and resource utilization.

As with the *Reliability* and *Cost optimization* pillars, the *Performance optimization* pillar also provides many in-depth recommendations for specific Google Cloud products, which is a level of detail beyond what would be appropriate to include here.

Now that you understand what the Google Cloud Architecture Framework is and what it consists of, let's look at how we can apply its concepts in the context of AI/ML on Google Cloud.

> **Note**
>
> Across all of the categories in the Google Cloud Architecture Framework, the theme that is reiterated the most is automation. The idea is to automate everything as much as possible. Vetted, repeatable processes that can run automatically tend to make our jobs easier across all pillars.

# Architecture Framework concepts about AI/ML workloads on Google Cloud

In this section, we will assess how the Google Cloud Architecture Framework can be used with regard to AI/ML workloads on Google Cloud. We will use the steps in the model development life cycle to frame our discussion. As a reminder, the steps in the model development life cycle are summarized at a high level in *Figure 13.5*:

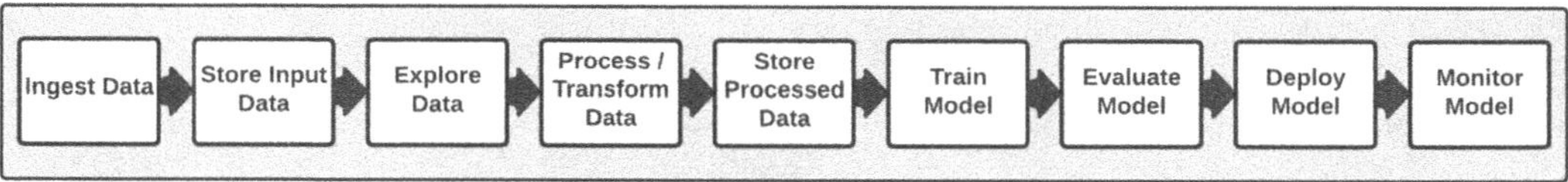


Figure 13.5: The ML model development life cycle

Let's begin with the data collection and preparation activities in the model development life cycle, which include gathering, ingesting, storing, and processing data.

> **Spoiler alert!**
>
> You will notice that we have already been using many of these practices throughout this book. Here, we are calling them out explicitly so that you can understand how they apply to workloads in general.

## Data collection and preparation

Data management is perhaps the most important of all the topics related to ML governance. As we know by now, the quality of your data directly impacts the quality of your models. Also, in terms of security, when malicious actors try to access your systems, they are usually after your data, because data is such a valuable resource, and data breaches can have catastrophic effects for your company. Let's look at how we can apply recommendations in the Google Cloud Architecture Framework regarding data handling. In this case, we will not discuss system design as a separate pillar, because an effective system design encapsulates all of the other pillars.

## *Operational excellence in data collection and preparation*

Remember that operational excellence focuses on concepts such as automation, observability, and availability. The following sub-sections explore these concepts in the context of data collection and preparation.

### Process automation and integration

In earlier chapters of this book, I talked about the importance of building data pipelines to automate our data processing steps. Essentially, we want to establish repeatable processes and then put in place mechanisms to run those processes automatically, either based on a schedule (such as daily, weekly, or monthly) or in reaction to some event, such as new data becoming available. Hence, the concept of implementing data processing pipelines is an application of the automation recommendations outlined in the *Operational excellence* pillar of the Google Cloud Architecture Framework. Google Cloud provides many services that we can integrate together to set up data processing pipelines, such as Dataproc, Dataflow, GCS, Pub/Sub, and BigQuery.

### Consistency and standardization

In the MLOps chapter, I shared experiences I've had with various organizations at different levels of maturity in terms of how they implemented their ML workload operations. When companies do not have well-established processes in place, their teams tend to use lots of random different tools that each have their own learning curves, and they maintain their artifacts in silos. These practices are not conducive to company-wide collaboration, and they hinder the scalability and efficiency of a company's ML operations. I then talked about the importance of standardizing tools and processes throughout the company in order to overcome those limitations. This all relates to the operational excellence pillar of the Google Cloud Architecture Framework. Consistency in tools, libraries, and processes across teams and projects reduces complexity and learning curves, making it easier to manage and scale operations. Perhaps the most relevant example of this, in relation to ML operations, is to use Vertex AI for all of our model development and deployment needs, since it provides a standard set of tools for every step in the model development and management life cycle.

### Observability

All of the regular system monitoring and logging requirements also apply to AI/ML workloads, and AI/ML workloads also have additional requirements regarding monitoring the quality of model prediction outputs on an ongoing basis to ensure that they do not drift over time. In earlier chapters, we discussed ML-specific metrics to monitor, such as MSE for linear regression use cases or AUC-ROC scores for classification use cases, as well as fairness metrics. Google Cloud Logging and Google Cloud Monitoring can be used for observability purposes at all points in the ML model development life cycle on Google Cloud, from evaluating training validation metrics to tracking the latency of responses from a model deployed on a Vertex AI prediction endpoint, and Vertex AI Model Monitoring can be used to watch for drift.

### *Security, privacy, and compliance in data collection and preparation*

By now, we should thoroughly understand that data security and privacy are paramount in almost every company. One sure way to quickly lose customers and damage your company's reputation is to become a victim of a data breach. My opinion is that managing sensitive data securely is the most important thing you can possibly do; it takes priority over all other considerations in this chapter, this book, and your company's business.

In our overview of this pillar earlier in this chapter, we talked about how data and systems should be protected through multiple layers of defenses (DiD), incorporating factors such as access management, encryption, and network security. The following sub-sections explore these concepts in the context of data collection and preparation.

#### Data access control

When collecting and preparing data in Google Cloud, we can use IAM to ensure that only authorized individuals and services can access the data, and we can control who can view, modify, or interact with the data based on roles and responsibilities. For example, a data scientist might have permission to read and analyze data but not delete it, or finance data may only be accessible to the finance department. This is made easier if we use Google Cloud Dataplex to build a data catalog because Dataplex allows us to centrally manage permissions for our data assets across multiple different GCS and processing services.

#### Data protection and encryption

Sensitive data should always be encrypted, both when it's stored (at rest) and when it's being transferred between services or locations (in transit). Google Cloud provides automatic encryption for data stored in its services, and we can use TLS to protect data in transit. For highly sensitive data, we can even encrypt it during processing by using Google Cloud Confidential Computing. Also, during the data preparation phase, sensitive data elements can be masked or tokenized to hide their actual values, therefore enhancing privacy while still allowing the data to be used for analysis.

#### Data classification and discovery

We can use the Google Cloud Sensitive Data Protection service to discover, classify, and mask sensitive elements in datasets. When collecting data, this service can help automatically identify information such as **personally identifiable information** (**PII**) or financial data so that we can treat it with higher levels of protection. This is another area in which Google Cloud Dataplex can help because tracking all of our data assets in a data catalog makes classification and discovery easy.

### Auditing and monitoring

We introduced Cloud Audit Logs earlier. We can use Cloud Audit Logs to keep a detailed log of who accesses the data and what operations they perform, which is important for accountability and traceability. This is especially relevant in ML workloads, where understanding who introduced what data and when can be required for explainability and troubleshooting. And, guess what?! Google Cloud Dataplex integrates with Cloud Audit Logs to generate audit logs for actions that are performed in the data catalog.

### Data retention and deletion

When using Google Cloud, we can establish policies about how long data should be retained based on its nature and relevance. When using GCS, for example, a retention policy can be specified to prevent an object from being deleted within the timeframe specified by the retention policy. This can be important for regulatory purposes or to comply with legal holds. Conversely, Object Lifecycle Management can be used to automatically delete data after a certain period (as long as it does not conflict with a data retention policy). For sensitive data that you want to delete permanently, Google Cloud provides mechanisms to ensure that the deletion is performed securely and is irrecoverable.

### Compliance frameworks and certifications

Google Cloud provides tools and documentation to help businesses comply with standards such as GDPR and HIPAA (as well as many more), and it undergoes independent third-party audits to ensure its services comply with common regulatory standards.

### Resilience against threats

Services such as Cloud Security Command Center and Event Threat Detection allow for continuous data and environment monitoring for potential threats, offering insights and actionable recommendations. Regularly scanning and assessing the systems involved in data collection and preparation for vulnerabilities can help ensure that data isn't exposed to potential breaches. You can also use VPC network security and VPC-SC to control access to your data storage and processing systems and to prevent data exfiltration.

All of the items we discussed in this section are important to ensure that data is handled securely and that user privacy is protected. Ethical considerations also come into play to ensure that data is collected and used in ways that are fair, transparent, and don't propagate biases, especially when it'll be used to train ML models that might impact individuals' lives.

## *Reliability in data collection and preparation*

Remember that the *Reliability* pillar in the Google Cloud Architecture Framework focuses on ensuring that services and applications perform consistently and meet the expected SLOs, even in the case of unexpected disturbances or increased demands. The following sub-sections discuss how we can apply concepts from the *Reliability* pillar in the data collection and preparation phases of the ML model development life cycle.

### Automated data ingestion and processing

Relying on manual processes for data collection can be prone to errors, while automated data ingestion helps to ensure that data can be collected consistently. We can also automate data validation steps to ensure that incoming data adheres to the expected formats and value ranges, which can prevent corrupted or malformed data from propagating through our data processing and ML pipelines. For data transformation scripts and configurations, we should use version control to ensure that if changes introduce errors, we can easily revert to a previous, stable version.

### Infrastructure resilience

Many of Google Cloud's data storage and processing services are either designed for HA by default or provide mechanisms to help you build resilience into your architecture, such as by using multiple machines across multiple zones and regions.

If we are designing systems ourselves, we should ensure that data storage and processing infrastructure have redundant components. In the event of a failure, backup systems can take over, ensuring uninterrupted data collection and preparation. We should also implement backup and restore mechanisms to regularly back up raw and processed data. We could store data across multiple zones or regions to safeguard against potential issues in any particular location. This not only protects against data loss but also allows for restoring to a previous state if data becomes corrupted or if there's a need to revisit earlier data versions. We could also implement load balancing for data ingestion services for high-velocity data streams to help ensure an even distribution of data loads and prevent system overloads. We should also design our infrastructure to scale (up or out) based on demand to ensure reliable performance under varying loads, and we could implement queueing mechanisms to manage data spikes.

### Continuous monitoring and alerts

As is the case with the *Operational excellence* pillar, observability is a key component of this pillar. We should regularly check the health of systems involved in data collection and preparation and implement alerting mechanisms that notify relevant teams when anomalies or failures are detected.

### Cost optimization in data collection and preparation

Managing costs is important during data collection and preparation due to potentially vast volumes of data, complex preprocessing tasks, and varying infrastructure needs. The following sub-sections discuss how we can apply concepts from the *Cost optimization* pillar in the data collection and preparation phases of the ML model development life cycle.

#### Efficient data storage

Storage systems such as GCS and BigQuery provide various classes of storage that are priced differently. From a cost optimization perspective, it's important that we use the appropriate storage class for our data. For example, frequently accessed data can be stored in Standard storage, while infrequently accessed data could be moved to Nearline or Coldline storage. To make this easier for us to manage, we can implement policies to automatically transition data to cheaper storage classes or delete it once it's no longer needed. We could also reduce the amount of data stored (and, therefore, reduce our costs) by removing duplicates and compressing our data. In terms of feature storage, we should evaluate the necessity of every feature during the data preparation stage. Removing redundant or low-importance features can significantly reduce storage and computation costs.

#### Optimized data processing

I'm a huge fan of using serverless solutions wherever possible. Not only do we offload the headaches of managing infrastructure to the cloud provider, but with serverless solutions such as BigQuery and Dataflow, we generally only pay for what we use, and we don't have to worry about over-provisioning (and therefore overpaying for) infrastructure. We can also opt for scalable services such as GKE or Cloud Dataflow that can handle spikes in data processing loads but scale down in low-demand periods, and for non-critical, fault-tolerant data processing tasks, we can use preemptible **virtual machines** (**VMs**), which are generally cheaper than regular instances.

It's also important to consider the location of our data, and we should generally aim to process data as near as possible to where it resides for a number of reasons, including cost and latency. For example, storing our data in the `us-central1` region while our processing infrastructure is located in the `us-east4` region would be sub-optimal from a latency perspective and would incur additional network egress costs as the data is transmitted across regions. This also applies in the cases of hybrid cloud infrastructures, in which some of your resources are in the cloud while others are located on your own premises. In such cases, consider the location at which your on-premises resources are connected to the cloud, as well as the data storage location and data processing location. We discussed the various methods of connecting your on-premises resources to Google Cloud (such as VPNs and "interconnects") in *Chapter 3*, and you can further bolster the security of such hybrid configurations by using VPC-SC to establish a trusted perimeter within which your data is transmitted and processed.

Also, if we are using VMs and those VMs need to work together to process our data, we can use **Google Compute Engine** (**GCE**) placement policies (specifically, the "compact placement policy") to specify that our VMs should be located close to each other, which can be particularly important for **high-performance computing** (**HPC**) workloads.

Finally, where real-time processing isn't necessary, we can accumulate data and process it in batches, which is often more cost-effective than streaming.

#### Cost monitoring and analytics

We can use tools such as Cost Explorer or custom dashboards in Cloud Monitoring to get insights into our spending patterns and set up billing alerts to notify us of unexpected spikes in costs so that we can intervene accordingly in a timely manner. Additionally, we should regularly analyze our billing reports to identify areas where costs can be trimmed, such as by looking for under-utilized resources or services.

#### Cost governance

A good practice is to set budgets for projects or departments and implement quotas for specific services to prevent unintentional overspending. It's also important to establish a resource organization and cost attribution strategy. We can use Google Cloud projects, folders, and labels to organize and attribute costs, which makes it easier to track and optimize expenses for specific tasks or teams, and we should promote a culture in which teams are aware of the costs associated with their data handling and processing activities, and encourage cost-saving practices.

#### Regular reviews

We should regularly review the architecture of our data collection and preparation systems because newer and more cost-effective solutions might emerge over time. Similarly, we should periodically evaluate the relevance of the data we're collecting because some data might become irrelevant over time, and the costs associated with its collection, storage, and processing could be eliminated.

### *Performance optimization in data collection and preparation*

As we discussed in the overview sections earlier in this chapter, there are some links between performance optimization and cost optimization because a system that performs optimally will often use resources more efficiently. The following sub-sections discuss how we can apply concepts from the *Performance optimization* pillar in the data collection and preparation phases of the ML model development life cycle.

#### High-performance data collection

To optimize our data ingestion processes for real-time data ingestion, we can use services such as Cloud Pub/Sub or Cloud Dataflow, which can help to achieve minimal latency and efficient data streaming. We can also use parallel processing in our data collection strategies by using distributed systems to fetch data from multiple sources concurrently, thus making our data collection more efficient.

#### Efficient data storage

It's important to use appropriate data structures such as columnar formats (for example, Parquet) for analytics workloads, which can lead to faster querying. In the case of high-performance storage use cases, we can use storage solutions such as Cloud Bigtable for low-latency, high-throughput workloads, which can help to ensure quick data access during the preparation phase. How we index our datasets can also improve the speed of retrieval and querying, which is especially important for large datasets during the data exploration phase.

#### Accelerated data processing

We can use platforms such as Cloud Dataflow and Cloud Dataproc, which provide managed Beam, Spark, and Hadoop clusters, to distribute data processing tasks across multiple nodes. For workloads such as feature engineering or data augmentation tasks in ML, using hardware accelerators such as GPU/TPU acceleration can drastically improve performance. Also, in platforms such as BigQuery, we can write optimized SQL queries to minimize computational overhead and improve processing speed.

#### Network optimization

If we're transferring large amounts of data from on-premises systems to Google Cloud, dedicated interconnects provide a high-speed, low-latency connection. For collecting data from global sources, **content delivery networks** (**CDNs**) ensure optimal data transfer speeds, and we can also use tools such as Traffic Director to manage and optimize network traffic, ensuring efficient data flow between services.

#### Resource allocation and auto-scaling

As we discussed earlier, it's important to ensure that services automatically scale resources based on demand. For example, Cloud Dataflow can auto-scale worker instances based on the data processing load. We should also tailor VM types and configurations (in terms of memory and CPU resources) to the specific needs of the data collection and preparation tasks.

Next, let's discuss how the Google Cloud Architecture Framework applies to the model building and training steps in our model development life cycle.

## Model building and training

As we did in the previous section regarding data collection and preparation, we will discuss the concepts of each pillar in the context of this phase in the model development life cycle.

### *Operational excellence in model building and training*

Let's begin with operational excellence, and how it applies to model building and training.

#### Standardized and automated workflows

The key components here are MLOps pipelines, version control, and CI/CD tooling. We can use Vertex AI Pipelines to create standardized, end-to-end ML pipelines that automate our model training and evaluation steps, including hyperparameter optimization. We can use Google Cloud's source control tooling to manage our pipeline definition code, and Google Cloud's CI/CD tooling, such as Cloud Build and Cloud Deploy, to build and deploy our pipeline definitions to Vertex AI Pipelines.

#### Observability

There are two main types of monitoring and logging that we need to implement during the model training and building phase. First, we need to track model performance metrics during model training, such as loss, accuracy, and validation scores. We can do this by using Vertex AI and tools such as TensorBoard. The second is related to system resource monitoring, for which we can use Google Cloud Monitoring to keep an eye on the resource consumption of VMs, TPUs, or GPUs during model training, which can help to achieve optimal resource utilization and timely detection of any potential bottlenecks that might occur.

#### Managed infrastructure

We should use managed infrastructure for our model training and building steps. By using managed infrastructure such as Vertex AI, we automatically use the recommendations outlined by the *Operational excellence* pillar in the Google Cloud Architecture Framework.

### Security, privacy, and compliance in model building and training

Data security will still be a major focus in this section, considering that the training process involves data handling. The following sub-sections discuss how we can apply concepts from the *Security, privacy, and compliance* pillar in the model building and training phases of the ML model development life cycle.

#### Data security

The mechanisms here are the same as we discussed in the *Data collection and preparation* section. We should ensure that data used for model building and training is encrypted both in transit and at rest. When using sensitive data for training, we can use data masking and tokenization to mask or tokenize specific fields to prevent exposure of PII or other sensitive data points. Additionally, we can use services such as VPC-SC to restrict the services and resources that can access our data, thus creating a secure perimeter around the data used for training.

#### Environment security

We can set up secure training environments to ensure that VMs and containers are securely configured, patched, and hardened, or use managed environments such as Vertex AI, which take care of a lot of these activities for us, and we can use VPC and firewall rules to secure the network traffic related to model training.

#### Compliance monitoring

We can use tools such as Cloud Security Command Center to continuously monitor and ensure that the training environment adheres to compliance standards, and we should also regularly audit data sources for training to ensure compliance with data usage policies, especially if sourcing data from third parties.

#### Privacy

If working with sensitive datasets, we can use techniques such as differential privacy to introduce noise into the data, ensuring individual data points are not identifiable. We can also use data de-identification to remove PI so that it cannot be associated with specific individuals.

In addition to all of the aforementioned, we can use IAM to control access to the training environment and artifacts.

### *Reliability in model building and training*

The following sub-sections discuss how we can apply concepts from the *Reliability* pillar in the model-building and training phases of the ML model development life cycle.

#### Data reliability

As we've done in earlier chapters of this book, we can implement validation checks for incoming data to ensure consistency, quality, and completeness. We should also regularly back up training data to prevent data loss and use data versioning for reproducibility.

#### Training infrastructure reliability

We can provision redundant resources in regional or multi-regional deployments to ensure training can continue even if one data center faces issues. In terms of infrastructure scalability, Vertex AI can automatically scale resources based on the training workload. Of course, it's important to use monitoring tools to keep an eye on resource utilization and health.

#### Model training resilience

We can use checkpointing to save model states at regular intervals during training. In case of interruptions, training could resume from the latest checkpoint rather than starting from scratch. For transient failures in any stage of the model building process, we should implement retry policies to automatically attempt the task again before raising an error.

#### Dependency management

Vertex AI lets us use containerization to ensure consistent software and library versions across training runs, preventing "it works on my machine" issues. This also brings with it all of the other established benefits of containerization, such as standardization and scalability. Think about how we used containers during the practical exercises in our MLOps chapter. We packaged our custom data processing and training code into containers, and then we could use them seamlessly in later stages of the model development process, simply by pointing to the container locations during various steps being implemented on systems such as Vertex AI and Dataproc. This kind of packaging, which facilitates repeatable execution results, is essential for automating steps in an MLOps pipeline, as well as automatically scaling our training and inference workloads based on their varying resource requirements. Such automation is a core benefit of MLOps practices. Furthermore, by using discrete packages of code for each step in the MLOps life cycle in this way, we can scale each step independently, providing flexibility in accordance with the best practices outlined in the *Operational excellence* and *Reliability* pillars of the Google Cloud Architecture Framework.

To further mitigate potential dependency issues, if we have dependencies on external systems such as data providers, we should ensure they have uptime guarantees and fallback mechanisms.

#### DR

It's important to regularly back up model architectures, configurations, trained weights, and other essential components to allow for quick recovery in case of data corruption or loss. We should establish clear protocols for restoring from backups, ensuring minimal downtime and a quick return to operational status in case of disruptions. The following point cannot be emphasized enough: we must periodically test our recovery procedures. Companies often focus only on backup mechanisms and not on testing the recovery processes. We want to ensure that our recovery processes are effective (that is, they actually work) and efficient (that is, they work as quickly as possible).

### *Cost optimization in model building and training*

The following sub-sections discuss how we can apply concepts from the *Cost optimization* pillar in the model building and training phases of the ML model development life cycle.

#### Resource efficiency

To optimize costs during model building and training, we should ensure that VMs, GPUs, or TPUs used for training are appropriately sized for the workload. Initially, this may take some experimentation to find the best configuration of resources, but by the time we have standardized our training process into an MLOps pipeline, we should have a good idea of the required resources. Using Vertex AI and serverless services can help us optimize costs because those services can scale our resources based on demand. We can also utilize CUDs to save on computing costs.

For training jobs that can handle interruptions, we can use preemptible VMs, which can offer substantial savings.

It should also be noted that simpler model architectures can be easier, quicker, and cheaper to train than complex model architectures. It's also important to shut down all resources when not in use so that we don't pay when they are idle.

#### Review and optimize regularly

We can use tools such as Google Cloud Cost Management to regularly review and analyze infrastructure costs, and identify opportunities for optimization. As always, we can use budgets, quotas, and billing alerts to help keep our costs under control, and we should periodically review our ML infrastructure, data storage, and associated processes to identify and eliminate inefficiencies.

### *Performance optimization in model building and training*

The following sub-sections discuss how we can apply concepts from the *Performance optimization* pillar in the model building and training phases of the ML model development life cycle.

#### Compute optimization

To optimize performance, we can use hardware acceleration and specialized hardware such as GPUs and TPUs, which can significantly accelerate the training process.

#### Distributed training

We can distribute the training process across multiple nodes to run in parallel and reduce training time. Also, for hyperparameter tuning, we can use services such as Vertex AI Vizier to perform concurrent trials, significantly reducing the time required to find optimal model parameters.

#### Data I/O optimization

We should use high-throughput data sources and systems for performant workloads so that the data coming into the training process isn't a bottleneck.

As mentioned in other sections of this chapter, it's important to continuously track performance metrics such as processing speed, memory usage, and I/O throughput using tools such as Google Cloud Monitoring, and then adjusting resources or configurations as needed. We can also use profiling to analyze ML training code, identify performance bottlenecks, and then optimize the most time-consuming segments.

Next, let's discuss how the Google Cloud Architecture Framework applies the model evaluation and deployment steps in our model development life cycle.

## Model evaluation and deployment

In this section, we will discuss the concepts of each pillar in the context of model evaluation and deployment in the model development life cycle. Note that, in some phases, the same concepts that we've already discussed in previous phases of the model development life cycle still apply. In the remaining sections of this chapter, I will briefly call out when the same concepts apply again.

### *Operational excellence in model evaluation and deployment*

Let's begin with operational excellence, and how it applies to model evaluation and deployment.

#### Automation, observability, and scalability

In this phase of the ML model development life cycle, the same concepts from the *Operational excellence* pillar, such as automated workflows, observability, and scalability, which we've already discussed in the context of the model building and training phase, apply again here. Basically, we can set up MLOps pipelines that automate our model evaluation and deployment steps using Vertex AI Pipelines, and we can use Google Cloud Monitoring and logging tools to track metrics related to our model evaluations and deployed model performance. We can also use load balancers and Vertex AI auto-scaling infrastructure to ensure that our models can handle varying levels of demand.

#### A/B testing and canary deployments

When deploying new model versions, we can use A/B testing to gradually shift traffic and compare performance against previous versions. Of course, we want to ensure that the newer versions being deployed perform better than the previous versions and that they don't negatively impact user experience. Using canary deployments, we can deploy new model versions to a small subset of users first, closely monitor performance, and then gradually expand to a broader user base. We should also use model versioning to allow for quick rollbacks if newer versions result in unexpected behaviors or errors.

### *Security, privacy, and compliance in model evaluation and deployment*

Again, the same concepts regarding data security and privacy apply here also, as well as access control, compliance regulations, and auditing. In addition to all of that, we can use network security controls and VPC-SC to protect the endpoints on which our models are hosted.

### *Reliability in model evaluation and deployment*

In this case, the same concepts of infrastructure resilience, such as deploying resources in multiple zones or regions, also apply here, as well as health checks, load balancing, auto-scaling, DR, monitoring, alerting, and dependency management.

### *Cost optimization in model evaluation and deployment*

When discussing cost optimization in the context of model evaluation and deployment, some concepts from our previous phases apply again, such as right-sizing resources, shutting down idle resources, using CUDs, and setting budgets and alerts. It's also important to note that smaller, simpler models require fewer resources and are therefore cheaper to run than larger, more complex models.

### *Performance optimization in model evaluation and deployment*

You won't be surprised to hear the terms *auto-scaling* and *load balancing* being used in the context of performance optimization for model evaluation and deployment, as well as optimizing compute and storage resources, and hardware acceleration.

We can also use caching mechanisms to improve response times. For example, we can cache frequent prediction results so that repeated requests can be served without invoking the model again, and we can store frequently accessed data or intermediate model evaluation results in memory for quicker access.

By now, you have become an expert in the Google Cloud Architecture Framework and how it specifically applies to the ML model development life cycle. Let's take a moment to summarize everything we covered in this chapter.

# Summary

This chapter discussed various aspects of ML model governance, including documentation, versioning, monitoring, auditing, compliance, operationalization, and continuous improvement. We then explored some industry-specific and region-specific regulations, such as HIPAA for healthcare, SOX for finance, GDPR (EU), and CCPA (California).

Next, we focused on the Google Cloud Architecture Framework and how to apply its pillars—*Operational excellence*, *Security, privacy and compliance*, *Reliability*, *Cost optimization*, and *Performance efficiency*—to the various stages of the ML life cycle. We dived deep into each pillar, detailing its relevance across different phases, from data collection and preparation to model evaluation and deployment. This included important concepts, such as cost-efficient model deployment, enhancing security throughout the model life cycle, and maintaining high reliability and performance standards. Overall, this chapter covered many factors related to deploying and managing ML workloads on Google Cloud, with best practices and optimizations in mind.

In the next chapter, we'll take a look at using some other popular tools and frameworks in the industry—such as Spark MLlib and PyTorch—on Google Cloud.

# 14
# Additional AI/ML Tools, Frameworks, and Considerations

At this point, we have covered all of the major steps and considerations in a typical **machine learning** (**ML**) project. Considering that AI/ML is one of the fastest-developing areas of research in the technology industry, new tools, methodologies, and frameworks emerge every day.

In this chapter, we will discuss additional tools and frameworks that are popular in the data science industry that we haven't covered so far. This includes important topics such as **BigQuery ML** (**BQML**), various types of hardware that we can use for AI/ML workloads, and the use of open source libraries and frameworks such as PyTorch, Ray, and Spark MLlib. We will also discuss some tips on how to implement large-scale distributed training on Google Cloud.

At the end of this chapter, I will provide some additional context to help transition the focus of the remainder of this book to Generative AI. This will include diving a bit deeper into some of the commonly used neural network architectures that I described at a high level earlier in this book.

For example, in *Chapter 9*, we covered the basics of neural networks and introduced common types of neural network architectures, such as **convolutional neural networks** (**CNNs**), **recurrent neural networks** (**RNNs**), and transformers. In this chapter, we will dive into those use cases in more detail to build some foundational knowledge for discussing Generative AI in the remaining chapters. Specifically, this chapter includes the following main topics:

- Custom Jupyter kernels
- BQML
- Hardware considerations for AI/ML workloads
- Additional popular open source tools and frameworks – Spark MLlib, Ray, and PyTorch on Google Cloud

- Large-scale distributed model training
- Transitioning to Generative AI

The first topic in the list is also associated with some prerequisite steps that we need to cover to set up our environment for the practical activities in this chapter. These will be described in the next section.

# Prerequisite topics and steps

This section describes the prerequisite topics and steps for setting up our Vertex AI Workbench environment.

## *Custom Jupyter kernels and package dependency management*

When we run our code in a Jupyter Notebook (such as a Vertex AI Workbench Notebook), the environment in which our code executes is referred to as a kernel. Vertex AI Workbench instances come with various kernels already installed for popular tools and frameworks such as TensorFlow and PyTorch, which we will cover in more depth in this chapter.

However, we can also create custom kernels if we want to define isolated environments with specific packages installed. This is a good practice to follow when using packages that are in preview mode, for example, as they may have very specific dependency requirements. We will use one such library, called `bigframes`, which I will describe in detail in this chapter. As a prerequisite, I will outline how to create a custom Jupyter kernel and explain some important concepts related to that process. Let's begin with the concept of virtual environments.

### Virtual environments

When our code executes, it runs within an environment, and this environment contains all of the dependencies that our code requires, which are usually other software packages. Managing the dependencies for various packages can be complicated, especially if we have two or more pieces of software that depend on different versions of a particular package. For example, imagine the following scenario:

- Software package X depends on version 1.0.3 of software package A
- Software package Y depends on version 2.2.1 of software package A

If we installed software package Y and all of its dependencies in our environment, then our environment would contain version 2.2.1 of software package A.

Now, if we try to run software package X in our environment, it may fail because it specifically requires a different version (that is, version 1.0.3) of software package A to be installed in the environment. This problem is referred to as a **dependency conflict**.

Virtual environments can help us avoid this problem because, as the name suggests, they provide a virtual execution environment in which we can run our code. When we create a virtual environment, it's almost like creating a dedicated machine on which to execute our code because that environment, and everything in it, is isolated from other execution environments, but the isolation is virtual because it is simply a logical separation from other environments that can run on the same machine.

When we use Vertex AI Notebook instances, there are two main types of virtual environments that we can create:

- **Python virtual environments**, which we can create by using the Python `venv` module, and which are therefore specific to Python packages. This option uses `pip` for package management.
- **Conda environments**, which use the Conda package and environment management system that goes beyond Python and can also manage packages for various other languages, such as R, Ruby, and others.

When determining which option to use, bear in mind that Python virtual environments are simpler and more lightweight, but Conda offers more features and handles more complex scenarios (as a result, Conda environments can be larger and slower to set up than Python virtual environments). We will use Conda for our use case. I will describe this next.

### Creating a Conda virtual environment and a custom Jupyter kernel

Perform the following steps to create the Python virtual environment and custom Jupyter kernel that we will use later in this chapter:

1. Open JupyterLab on the Vertex AI Notebook instance you created in Chapter 5.
2. Select **File** | **New** | **Terminal** and perform the following steps on the terminal screen.
3. Create a Conda environment:

   ```
   conda create --name bigframes-env -y
   ```

4. Activate the environment:

   ```
   conda activate bigframes-env
   ```

5. Install `bigframes` (we could also install any other packages we want, but we'll keep it simple for now):

   ```
   conda install bigframes -y
   ```

6. Install JupyterLab in the new environment (this will make our new Conda environment available as a kernel in JupyterLab via JupyterLab's autodiscovery feature):

   ```
   conda install jupyterlab -y
   ```

7. Change the display name of the kernel (it can take a few minutes for this update to appear in the Vertex AI Workbench JupyterLab interface):

```
sed -i 's/"display_name": "Python 3 (ipykernel)"/"display_name":
"Python 3 (bigframes)"/' /opt/conda/envs/bigframes-env/share/
jupyter/kernels/python3/kernel.json
```

Now, our Conda environment and custom Jupyter kernel are ready to be used in the hands-on exercises in this chapter. We just have one more prerequisite step to perform before we dive into the remaining topics in this chapter, at that is to stage the required files for our serverless Spark MLlib activities.

## Staging files for serverless Spark MLlib activities

In this section, we will stage some files in Google Cloud Storage to be used in the serverless Spark MLlib activities later in this chapter. To do this, perform the following steps:

1. Go to the Google Cloud console and open Cloud Shell by clicking the **Cloud Shell** icon, as shown in *Figure 14*. This icon looks like a "greater than" symbol, followed by an underscore – that is, >_:

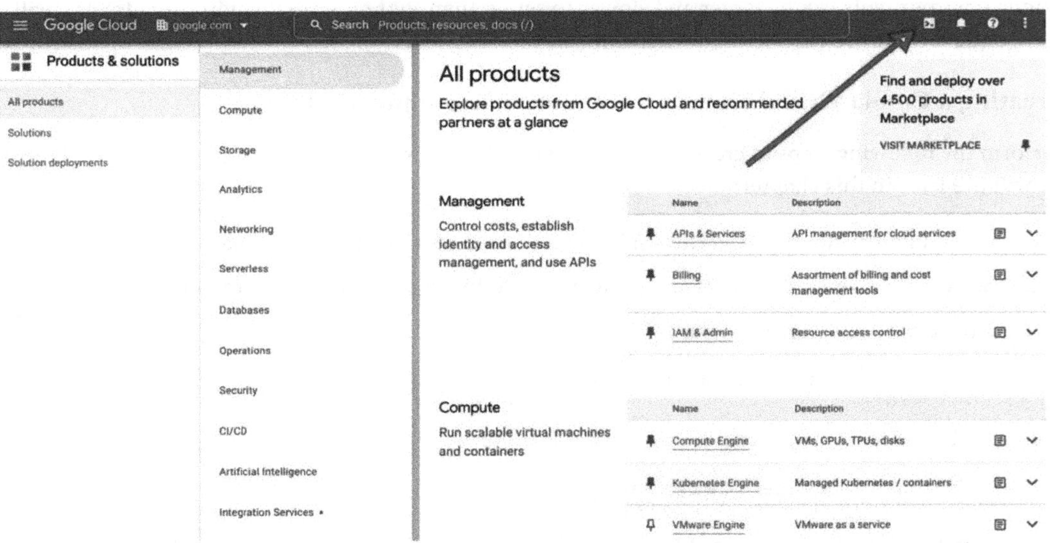


Figure 14.1: Activating Cloud Shell

2. Run the following commands in Cloud Shell to download the required files from our GitHub repository:

```
wget https://raw.githubusercontent.com/PacktPublishing/Google-
Machine-Learning-for-Solutions-Architects/main/Chapter-14/data/
data_processed_titanic_part.snappy.parquet
```

```
wget https://raw.githubusercontent.com/PacktPublishing/Google-
Machine-Learning-for-Solutions-Architects/main/Chapter-14/
pyspark-ml.py
```

3. Run the following commands in Cloud Shell to upload the required files to Google Cloud Storage (**important: replace [YOUR-BUCKET-NAME] with the name of a bucket you created in earlier chapters**):

```
gsutil cp pyspark-ml.py gs://[YOUR-BUCKET-NAME] /code/
additional-use-cases-chapter/pyspark-ml.py
gsutil cp *.parquet gs://[YOUR-BUCKET-NAME]/data/processed/
mlops-titanic
```

Now, everything is ready for our hands-on exercises later in this chapter. Next, we'll dive into the important topic of BQML.

# BQML

We first introduced BigQuery in *Chapter 3*, and we've used it in various chapters of this book for its data management and processing functionality. However, given the close relationship between large-scale data processing and ML, Google Cloud has built ML functionality directly into BigQuery, as well as native integrations with Google Cloud Vertex AI. This functionality is referred to as BQML, and this section explores this service in detail.

BQML enables us to create and execute ML models using standard SQL queries in BigQuery. Considering that many companies already store large amounts of data in BigQuery, BQML makes it easy for data scientists in those companies to train models on large datasets and make predictions directly in the database system, without needing to move the data around between different storage systems.

It supports a variety of ML algorithms, including linear regression, logistic regression, k-means clustering, and deep neural networks, as well as forecasting use cases based on time series data stored in BigQuery.

In addition to training and prediction, we can use BQML to perform many of the steps in the model development life cycle, such as feature engineering, model performance evaluation, and hyperparameter tuning. Considering that all of this can be done using standard SQL, data scientists can easily start using BQML without needing to go through a steep learning curve in terms of tooling. This is referred to as lowering the barrier of entry, where the barrier of entry represents how easy or difficult it is for people to start performing an activity. For instance, an activity with a high barrier of entry presents a lot of initial difficulty. An example of this would be a system that requires extensive training or complex prerequisites, such as provisioning infrastructure, before somebody could start using that system. Some technology systems require months of training and effort before somebody can start using them effectively. BQML, on the other hand, can easily be used by anybody who understands standard SQL syntax, and it does not require any complex infrastructure provisioning. Just like other managed services provided by Google Cloud, BigQuery (and, by extension, BQML) manages the infrastructure for us, and it can automatically scale up and down based on demand.

With this in mind, let's take a look at how we can use BQML to develop and use ML models.

## Using BQML

While the Jupyter Notebook that accompanies this chapter provides hands-on instructions on how to use BQML for various ML tasks, I will summarize some of the main features here. For context, let's recall our ML model development life cycle, as depicted in *Figure 14.3*:

Figure 14.2: ML model development life cycle

The following subsections dive into each step in more detail.

### *Data preparation*

We will combine the following steps shown in *Figure 14.3* into this section:

- Ingest data
- Store input data
- Explore data
- Process/transform data
- Store processed data

Starting with data ingestion and storage, there are many ways in which we can get our data into BigQuery. For example, we can directly upload files via the BigQuery console UI or the CLI. For larger datasets, we can stage them in Google Cloud Storage and import them to BigQuery from there. We can also use the BigQuery Data Transfer Service to automate data movement into BigQuery, either as a one-off transfer or on a scheduled cadence. We can stream data into BigQuery by integrating with other services, such as Google Cloud Pub/Sub and Dataflow. There are also third-party ETL tools that can be used to transfer data to BigQuery.

Once we have stored our data in BigQuery, we can use various tools to explore and transform the data, in addition to the BigQuery console and standard SQL. We introduced and used the `pandas` library in previous chapters, and we know that it's a very widely used library in data science, particularly for data exploration and manipulation. For this reason, many data scientists like to use pandas directly with their data stored in BigQuery. Fortunately, there are additional libraries that make it easy to do this:

- The BigQuery DataFrames Python API
- The `pandas_gbq` library

Let's take a look at each of these in more detail.

### The BigQuery DataFrames Python API

The BigQuery DataFrames Python API enables us to use Python to analyze and manipulate data in BigQuery and perform various ML tasks. It's a relatively new, open source option that was launched and is maintained by Google Cloud for using DataFrames to interact with BigQuery (at the time of writing this in December 2023, it is currently in Preview status). We can access it by using the `bigframes` Python library, which consists of two main parts:

- `bigframes.pandas`, which implements a pandas-like API on top of BigQuery
- `bigframes.ml`, which implements a `scikit-learn`-like API on top of BigQuery ML

The Jupyter Notebook that accompanies this chapter provides instructions on how to use `bigframes.pandas` in more detail. We will dive into those steps shortly, but first, I'll briefly cover `pandas_gbq`.

### pandas_gbq

If I were writing this chapter a few months ago, the `pandas_gbq` library would have been the main or only option that I would include in this section because, at the time of writing, it has been the primary option available for using pandas to interact with BigQuery. It's an open source library that is maintained by PyData and volunteer contributors and has been around for quite some time (since 2017), so it has become broadly used in the industry.

Essentially, it's a thin wrapper around the BigQuery client library (`google-cloud-bigquery`) that provides a simple interface for running SQL queries and uploading pandas DataFrames to BigQuery. The results from these queries are parsed into a `pandas.DataFrame` object in which the shape and data types are derived from the source table.

The Jupyter Notebook that accompanies this chapter also provides instructions on how to use this library in more detail. Now would be a good time to dive into those steps. Open JupyterLab on the Vertex AI Workbench instance you created in Chapter 5 and perform the following steps:

1. In the navigation panel on the left-hand side of the screen, navigate to the `Chapter-14` directory within the `Google-Machine-Learning-for-Solutions-Architects` folder.
2. Double-click on the `pandas-gbq.ipynb` notebook file to open it. When prompted to select a kernel, you can use the default **Python 3 (ipykernel)** kernel.
3. Double-click on the `bigframes.ipynb` notebook file to open it. When prompted to select a kernel, you can use the default **Python 3 (bigframes)** kernel that we created in the *Prerequisite topics and steps* section of this chapter.
4. In each of the notebooks you've opened, press *Shift* + *Enter* to execute each cell.

   The notebooks contain comments and Markdown text that describe what the code in each cell is doing.

As you can see, there are multiple options for interacting with BigQuery data in Vertex AI.

Given the broad adoption of the `pandas_gbq` library in the industry, and now that Google has launched the official BigQuery DataFrames Python API described previously, it's likely that both options will continue to be popular among data scientists. A key thing to bear in mind is that `pandas-gbq` downloads data to your local environment, whereas the BigQuery DataFrames Python API is used to run your data operations on Google Cloud distributed infrastructure.

Next, let's discuss how to create, use, and manage ML models with BQML. The Jupyter Notebook file that accompanies this chapter can be used to implement the following steps, but let's discuss them before diving into the notebook file.

### Creating ML models

We can use the `CREATE MODEL` statement in BigQuery to define and train a model. For example, the following code excerpt creates a linear regression model named `my_model_name`, which is trained on the data in `my_dataset.my_table`, using the values in `target_colum` as the labels:

```
CREATE OR REPLACE MODEL `my_dataset.my_model_name`
OPTIONS(model_type = 'linear_reg', input_label_cols=['target_column'])
AS
SELECT * FROM `my_dataset.my_table`;
```

As we've already discussed, BQML supports many types of commonly used algorithms. We can also import and export our models from/to Google Cloud Storage to interact with other tools and frameworks, and we can even perform hyperparameter tuning during the model development process.

### Evaluating ML models

Once our model has been trained, we can evaluate its performance using SQL code such as the following:

```
SELECT * FROM ML.EVALUATE(MODEL `my_dataset.my_model_name`,
TABLE `my_dataset.my_evaluation_table`);
```

This returns evaluation metrics such as accuracy, precision, recall, and more, depending on the model type.

### Generating predictions

Next, we can use the model to make predictions using SQL code such as the following:

```
SELECT * FROM ML.PREDICT(MODEL `my_dataset.my_model_name`,
TABLE `my_dataset.my_input_table`);
```

This statement will feed the data from `my_dataset.my_input_table` into our trained model to generate predictions.

Now, we can use the Jupyter Notebook file that accompanies this chapter to implement these steps. To do that, open JupyterLab on the Vertex AI Workbench instance you created in Chapter 5 and perform the following steps:

1. In the navigation panel on the left-hand side of the screen, navigate to the `Chapter-14` directory within the `Google-Machine-Learning-for-Solutions-Architects` folder.
2. Double-click on the `BQML.ipynb` notebook file to open it. When prompted, select the default **Python 3 (ipykernel)** kernel.
3. Press *Shift* + *Enter* to execute each cell.

   The notebook contains comments and Markdown text that describe what the code in each cell is doing.

Next, we'll discuss how to set up model monitoring and continuous training.

### *Model monitoring and continuous training*

To implement ongoing model monitoring and continuous training, we could set up scheduled jobs to execute the evaluation and training queries. This is a solution that we would need to architect using additional Google Cloud services, such as Google Cloud Scheduler and Google Cloud Functions, in which we could use Google Cloud Scheduler to periodically invoke Google Cloud Functions to run the training and evaluation queries. Having said that, for complex MLOps pipelines, Vertex AI offers more functionality for customization. This point brings us to our next topic, which is determining when to use BQML versus other tools for AI/ML use cases.

## When to use BQML versus other tools for AI/ML use cases

We've already covered that BQML is a great fit for data analysts and data scientists who want to use familiar SQL syntax to implement ML use cases on data stored in BigQuery, and how simplicity is one of its main benefits in that context. Bear in mind that there can often be a tradeoff between simplicity and customization. If you need to build highly advanced and customized models, then you may find that the simplicity of SQL does not provide the same level of customization that can be achieved when using specialized AI/ML frameworks such as TensorFlow, PyTorch, Ray, and Spark MLlib.

You can also have "the best of both worlds" by interacting with models hosted in Vertex AI from BigQuery by using **BigQuery remote functions**, which provide integrations with Cloud Functions and Cloud Run. With this approach, you can write code that will be executed on Cloud Functions or Cloud Run, and you can invoke that code from a query in BigQuery. The code could send an inference request to a model that has been trained and hosted in Vertex AI, and then also send the response back to BigQuery. You could even implement transformations in both the request and the response, if needed, to ensure compatibility between your query's expected data types and your model's expected data types. You can find out more about BigQuery remote functions in the Google Cloud documentation at the following link, which describes known limitations and best practices: `https://cloud.google.com/bigquery/docs/remote-functions`.

> **Note**
>
> While BigQuery is designed for large-scale analytics workloads, Google Cloud Spanner is another highly scalable database service in Google Cloud. Spanner is designed for distributed and strongly consistent transactional use cases, and it also supports SQL syntax. It recently added integrations with Vertex AI, providing somewhat similar functionality as BQML (that is, the ability to access ML models hosted on Vertex AI through a SQL interface), but intended for transactional rather than analytical workloads.

Next, I will end this section with a brief discussion of BigQuery Studio, which is a convenient way to manage all of our BigQuery tasks and workloads.

### BigQuery Studio

BigQuery Studio, as the name suggests, provides a single-pane-of-glass experience that allows us to perform many different kinds of activities in BigQuery. It includes a SQL editor interface that enables us to write and run queries directly in the Google Cloud console, with intelligent code development assistance via integration with Duet, a Google Cloud Generative AI real-time code assistant that we will cover in more detail later in this book. BigQuery Studio also enables us to schedule SQL queries to run on a periodic cadence (for example, every day) for use cases that require repeated query executions, such as generating a daily report.

Via BigQuery Studio, we can access Dataform to define and run data processing workflows within BigQuery and utilize BigQuery Studio's integrations with Dataplex for data discovery, data profiling, and data quality management. BigQuery Studio also integrates with Colab Enterprise, enabling us to use Jupyter Notebooks directly within the BigQuery console.

I encourage you to visit the BigQuery Studio interface within your Google Cloud console and explore its various capabilities and integrations.

We will revisit BigQuery in later chapters, but for now, let's discuss some hardware considerations for AI/ML workloads.

## Hardware considerations for AI/ML workloads

The majority of the topics in this book focus on the software and service-level functionalities that are available on Google Cloud. Advanced practitioners will also be interested in what kind of hardware capabilities exist. If your use cases require extreme performance, then selecting the right hardware components on which to run your workloads is an important decision. The selection and efficient usage of underlying hardware also affect the costs, which are, of course, another important factor in your solution architecture. In this section, we'll shift the discussion so that it focuses on some of the hardware considerations for running AI/ML workloads in Google Cloud, beginning with an overview of **central processing unit** (CPU), **graphics processing unit** (GPU), and **tensor processing unit** (TPU) capabilities.

### CPUs, GPUs, and TPUs

You will probably already be familiar with CPUs and GPUs, but TPUs are more specific to Google Cloud. As a brief overview, CPUs are what power the vast majority of consumer devices, such as laptops and mobile phones, as well as general-purpose servers in data centers. They are effective at multi-tasking across a broad range of tasks, but the processing they perform is somewhat sequential.

GPUs, on the other hand, are built with an architecture that optimizes parallel (rather than sequential) processing. If you can take a process and break it down into similar tasks that can run in parallel, then you'll be able to complete that process much more quickly on a GPU than on a CPU. Although, as the name suggests, GPUs were originally designed for handling graphics, which involves processing large blocks of pixels and vertices in parallel, it turns out that the kinds of matrix manipulation tasks that are inherent to many AI/ML workloads can also be sped up by the parallel architecture of GPUs.

TPUs were designed by Google to accelerate TensorFlow operations and they are optimized for both training and running models more efficiently than CPUs and GPUs for some types of workloads. Although they were created specifically for TensorFlow, they can now also be used with other frameworks, such as PyTorch, by using libraries such as PyTorch/XLA, which is a Python package that uses the **Accelerated Linear Algebra** (**XLA**) deep learning compiler to enable PyTorch to connect to TPUs and use TPU cores as devices.

Google Cloud provides many different types of hardware servers that we can choose for running our ML workloads, and these servers provide various amounts of CPU, GPU, and TPU power.

At the time of writing this in December 2023, Google Cloud has recently launched its most powerful TPU (**Cloud TPU v5p**) in conjunction with their new **AI Hypercomputer** offering, which provides highly optimized resources (that is, storage, compute, networking, and more) for AI/ML workloads.

Given the broad variety of computing options on Google Cloud, I will not list all of them here. Google Cloud is constantly launching additional options, so I encourage you to review the Google Cloud documentation to find the latest details: `https://cloud.google.com/compute/docs/machine-resource`.

Next, let's switch the discussion back to the software level and explore some popular open source tools and frameworks in the ML and data science industry that are supported in Google Cloud.

## Additional open source tools and frameworks –Spark MLlib, Ray, and PyTorch on Google Cloud

In this section, I'll introduce additional open source tools and frameworks, such as PyTorch, Ray, and Spark **Machine Learning Library** (**MLlib**), and demonstrate how they can be used to implement AI/ML workloads on Google Cloud.

## Spark MLlib

We introduced Apache Spark in previous chapters and used it for data processing to perform feature engineering in *Chapter 6*. Apache Spark MLlib is a component of Apache Spark that provides ML tools and algorithms that are optimized for parallel processing with large datasets. In addition to feature engineering, we can use the tools in MLlib to implement various stages in our ML model development life cycle, such as model training, model evaluation, hyperparameter tuning, and prediction, as well as assembling stages into a pipeline that can be executed end to end.

Just as we discussed in the context of data processing, one of the main advantages of Apache Spark (including MLlib) is its ability to execute large-scale computing workloads. While libraries such as scikit-learn work well for performing ML workloads on single machines, MLlib can distribute the computation work across many machines, which enables us to handle larger datasets more efficiently.

In the hands-on activities that accompany this chapter, we will use Spark MLlib to train a model on Google Cloud Dataproc by using the Serverless Spark features within Vertex AI. This is important to understand from a solution architecture perspective: we will perform the steps in Vertex AI, but it will run the Spark job on Dataproc in the background. There are multiple ways in which we can implement this workload, and we will cover the following two methods in our hands-on activities:

1. Using an MLOps pipeline, similar to the activities we performed in *Chapter 11*.
2. Using the Serverless Spark user interface in Vertex AI.

Let's begin with the first method: using an MLOps pipeline.

### *Serverless Spark via Kubeflow Pipelines on Vertex AI*

In *Chapter 11*, we used the `DataprocPySparkBatchOp` operator in Kubeflow Pipelines to automate the execution of a serverless Spark job in an MLOps pipeline. We will use the same operator again in this chapter, but this time, we will use it to run a model training and evaluation job using Spark MLlib. To do so, open JupyterLab on the Vertex AI Workbench instance you created in Chapter 5 and perform the following steps:

1. In the navigation panel on the left-hand side of the screen, navigate to the `Chapter-14` directory within the `Google-Machine-Learning-for-Solutions-Architects` folder.
2. Double-click on the `spark-ml.ipynb` notebook file to open it. When you're prompted to select a kernel, select the **Python 3 (ipykernel)** kernel.
3. In each of the notebooks you've opened, press *Shift* + *Enter* to execute each cell.

   The notebooks contain comments and Markdown text that describe what the code in each cell is doing.

As we can see, it's quite easy to extend the activities we performed in *Chapter 11* to use Kubeflow Pipelines to automate a Serverless Spark job for model training purposes with Spark MLlib. Let's take a look at another option for running our Serverless Spark job directly via the Serverless Spark user interface in Vertex AI.

### *The Serverless Spark user interface in Vertex AI*

To access the Serverless Spark user interface in Vertex AI, open JupyterLab on the Vertex AI Workbench instance you created in Chapter 5 and perform the following steps:

1. From the **File** menu, select **New Launcher**.
2. Scroll down to the **Dataproc Jobs and Sessions** section and select **Serverless**.
3. Select **Create Batch**.
4. On the screen that appears (see *Figure 14.4* for reference), enter the following details:

    I. **Main jar URI** (you need to select the radio button for this option to display the entry field): Enter the GCS location where you saved the `pyspark-ml.py` file. It should be in the following format **(important: replace YOUR-BUCKET-NAME with the name of the bucket you used in the prerequisites section of this chapter)**: `gs://YOUR-BUCKET-NAME/code/additional-use-cases-chapter/pyspark-ml.py`.

    II. **Arguments** (press *Enter* after specifying each of the following lines): **Important: replace YOUR-BUCKET-NAME with the name of the bucket you used in the prerequisites section of this chapter**:

    i. `--processed_data_path=gs://YOUR-BUCKET-NAME/data/processed/mlops-titanic`

    ii. `--model_path=gs://YOUR-BUCKET-NAME/models/additional-use-cases-chapter/`

5. Leave all other fields at their default values and click **Submit** at the bottom of the screen:

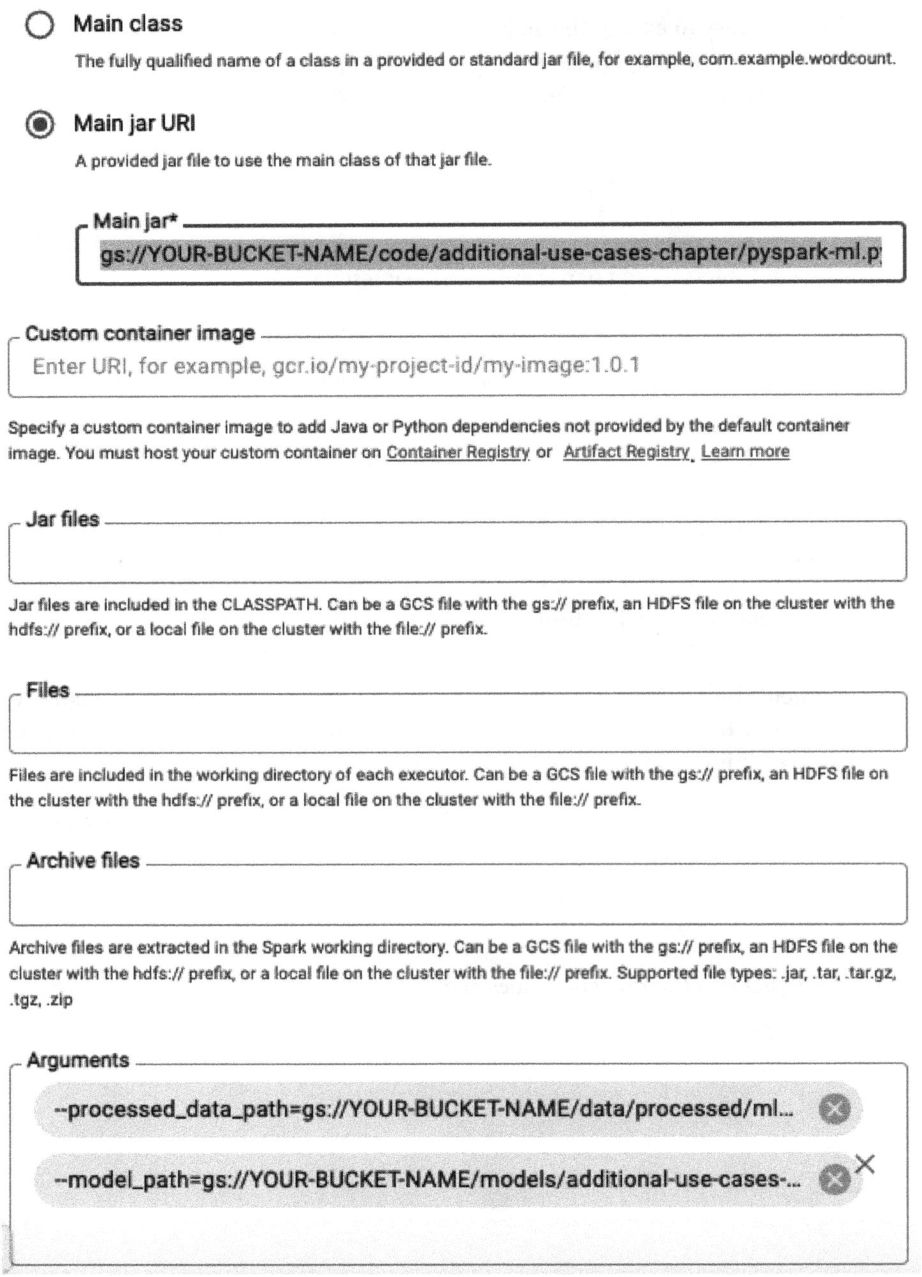

Main class

The fully qualified name of a class in a provided or standard jar file, for example, com.example.wordcount.

Main jar URI

A provided jar file to use the main class of that jar file.

Main jar*

gs://YOUR-BUCKET-NAME/code/additional-use-cases-chapter/pyspark-ml.p

Custom container image

Enter URI, for example, gcr.io/my-project-id/my-image:1.0.1

Specify a custom container image to add Java or Python dependencies not provided by the default container image. You must host your custom container on Container Registry or Artifact Registry. Learn more

Jar files

Jar files are included in the CLASSPATH. Can be a GCS file with the gs:// prefix, an HDFS file on the cluster with the hdfs:// prefix, or a local file on the cluster with the file:// prefix.

Files

Files are included in the working directory of each executor. Can be a GCS file with the gs:// prefix, an HDFS file on the cluster with the hdfs:// prefix, or a local file on the cluster with the file:// prefix.

Archive files

Archive files are extracted in the Spark working directory. Can be a GCS file with the gs:// prefix, an HDFS file on the cluster with the hdfs:// prefix, or a local file on the cluster with the file:// prefix. Supported file types: .jar, .tar, .tar.gz, .tgz, .zip

Arguments

--processed_data_path=gs://YOUR-BUCKET-NAME/data/processed/ml...

--model_path=gs://YOUR-BUCKET-NAME/models/additional-use-cases-...

Figure 14.3: The Serverless Spark user interface in Vertex AI

After clicking the **Submit** button, a screen will appear showing a list of serverless Spark jobs; your newly submitted job will appear at the top of the list (you may need to refresh your browser page to update the list). Wait until the status of your job says **Succeeded**.

6. If the job fails for any reason, click on the name of the job to display its details, and then click **View Cloud Logs** at the top of the screen.
7. It may take some time for the logs to populate. You can click the **Run query** button periodically to refresh the logs.
8. When the job has finished executing, the model artifacts will be saved in two directories named `metadata` and `stages` at the location you specified for the `--model-path` argument. Verify that those directories have been created and populated by performing the following steps in the Google Cloud console:

   i. Go to the Google Cloud services menu and choose **Google Cloud Storage**.

   ii. Browse to the path you specified for the `--model-path` argument by clicking on each successive component of the path – for example, `YOUR-BUCKET-NAME/models/additional-use-cases-chapter/`.

Great job! You have just used Spark MLlib to implement a serverless Spark workload to train a model on Dataproc via Vertex AI. Next, we'll briefly discuss another distributed computing framework that has become increasingly broadly used in the data science industry: Ray.

## Ray

I won't spend much time on Ray in this book, but I want to mention it for completeness. The developers of Ray describe it as an "*open source unified compute framework that makes it easy to scale AI and Python workloads.*" Ray is another type of distributed execution framework that has been gaining popularity in recent years, especially in AI/ML applications.

Like Spark, Ray enables parallelization of code and distribution of tasks across a cluster of machines. It also includes components that can help with specific model development steps, such as Ray Tune for hyperparameter tuning, and Ray RLlib for reinforcement learning. Google Cloud Vertex AI now directly supports Ray by enabling us to create Ray clusters.

## PyTorch

Like TensorFlow and Spark MLlib, PyTorch is an open source framework that includes a set of tools and libraries for ML and deep learning use cases. It was originally developed by Facebook's AI Research lab, and it evolved from another AI/ML framework named Torch, which is based on a programming language named Lua. PyTorch, as the name suggests, has a Pythonic interface, and it has gained popularity in recent years for its ease of use and the flexibility provided by its various components, such as TorchVision for computer vision, TorchAudio for audio processing, and TorchText for natural language processing.

PyTorch uses a dynamic (or "imperative") computational graph, referred to as a **define-by-run** approach, where the graph is built on the fly as operations are performed, which enables more intuitive and flexible model development compared to static graphs used in some other frameworks.

> **Note**
>
> TensorFlow used to only provide the option of using a static computation graph, but in recent years, it has introduced an option named "Eager mode," which offers dynamic graph functionality.

Also like TensorFlow, PyTorch supports GPU acceleration (via NVIDIA's CUDA framework) for tensor computations, which significantly speeds up model training and inference.

On the topic of speeding up model training, it's a common practice in computing to get more work done faster by performing operations in parallel across multiple devices. In the next section, we will discuss the practice of achieving this goal via distributed training.

# Large-scale distributed model training

Think back to the discussion we had in the *AI/ML and cloud computing* section of *Chapter 1*, in which I described the process of scaling our models to larger sizes, starting with small models that we could train on our laptop, progressing to larger models trained on a powerful, high-end server, and eventually getting to the scale at which a single computer (even the most powerful server on the market) couldn't handle either the size of the model or the dataset on which the model is trained. In this section, we'll look at what it means to train such large-scale models in more detail.

We've covered the model training process in great detail throughout this book, but I'll briefly summarize the process here as a knowledge refresher because these concepts are important when discussing large-scale distributed model training. For this discussion, I will focus on supervised training of neural networks.

The supervised training process, at a high level, works as follows:

1. Instances from our training dataset are fed into the model.
2. The model algorithm or network processes the training instances, and for each instance, it tries to predict the target variable. In the case of neural networks, this is referred to as forward propagation.
3. After making a prediction, the model calculates the error or loss using a loss function (for example, **mean squared error** (MSE) for regression or cross-entropy for classification), which measures the difference between the model's prediction and the actual label (the ground truth).

4. The backpropagation process then begins, which applies the chain rule from calculus to compute gradients of the loss function concerning each weight in the network.
5. An optimization algorithm such as gradient descent makes use of the gradients that were calculated during backpropagation to update the weights in the network.

Training is usually performed by looping through the preceding process multiple times in a series of epochs, where an epoch is one complete pass through the entire training dataset. However, when using large datasets, the training process can divide the overall dataset into smaller batches, and each pass through a batch is considered a training step. Also, some frameworks, such as TensorFlow, allow us to specify the number of training steps per epoch. Over time, if the model is learning effectively, its predictions will become more accurate after each epoch, and the loss will decrease (hopefully to an acceptable threshold).

**Note**

The term **processor** in these discussions refers to CPUs, GPUs, and TPUs. Also, the concepts we will discuss in this section can apply to training workloads that are distributed across multiple processors in a single machine, or across multiple processors on multiple machines.

In distributed training use cases, some communication and data sharing is required among the processors. When using a single machine, the processors usually share the same system memory (**RAM**), which simplifies data sharing, but when using multiple machines, communication needs to happen over a network (including routers, switches, and more), which can introduce latency and additional complexity.

We've implemented the training process many times previously in this book. Next, we'll discuss how this works when we're using multiple processors.

## Data parallelism and model parallelism

There are generally two main reasons we would need to implement distributed training workloads:

- We want to use a very large dataset
- We want to train a very large model

Note that both scenarios may exist at the same time (that is, we may need to train a very large model on a very large dataset). In this section, I'll explain each of these scenarios in detail, starting with the case of using large datasets.

## Data parallelism

Sometimes, we need to train our models on huge datasets that would take a long time to process on just one processor. To make the training process more efficient, we can split the dataset into smaller subsets or batches and process them in parallel across multiple processors. In this case, we can have a copy of our model running on each processor, which works through the subset of data that is loaded onto that processor. A simple example would be if we had a dataset that contained 10,000 data points and we ran 10 processors that each worked on 1,000 data points from our dataset. This approach is referred to as **data parallelism**, and it is depicted in *Figure 14.5*:

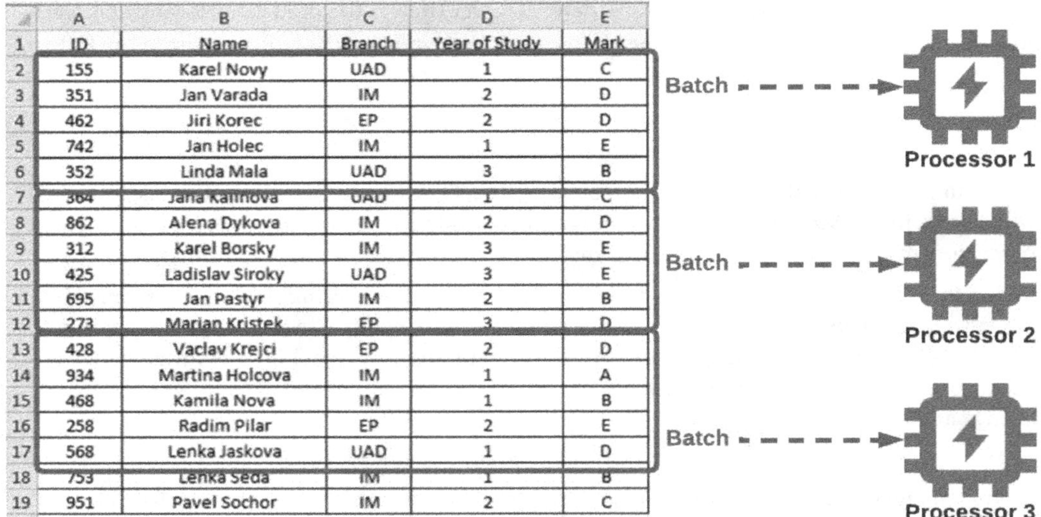

| | A | B | C | D | E |
|---|---|---|---|---|---|
| 1 | ID | Name | Branch | Year of Study | Mark |
| 2 | 155 | Karel Novy | UAD | 1 | C |
| 3 | 351 | Jan Varada | IM | 2 | D |
| 4 | 462 | Jiri Korec | EP | 2 | D |
| 5 | 742 | Jan Holec | IM | 1 | E |
| 6 | 352 | Linda Mala | UAD | 3 | B |
| 7 | 364 | Jana Kalinova | UAD | 1 | C |
| 8 | 862 | Alena Dykova | IM | 2 | D |
| 9 | 312 | Karel Borsky | IM | 3 | E |
| 10 | 425 | Ladislav Siroky | UAD | 3 | E |
| 11 | 695 | Jan Pastyr | IM | 2 | B |
| 12 | 273 | Marian Kristek | EP | 3 | D |
| 13 | 428 | Vaclav Krejci | EP | 2 | D |
| 14 | 934 | Martina Holcova | IM | 1 | A |
| 15 | 468 | Kamila Nova | IM | 1 | B |
| 16 | 258 | Radim Pilar | EP | 2 | E |
| 17 | 568 | Lenka Jaskova | UAD | 1 | D |
| 18 | 753 | Lenka Seda | IM | 1 | B |
| 19 | 951 | Pavel Sochor | IM | 2 | C |



Figure 14.4: Data parallelism

In *Figure 14.5*, each purple box represents a batch of our data, and each batch is sent to a separate processor in parallel. Each batch in the diagram contains only a few data points, but in reality, the batches would be much larger (this diagram intends to illustrate the concept at a high level). Next, we'll discuss the other main scenario in which we may need to implement a distributed training workload. This is referred to as model parallelism.

## Model parallelism

Sometimes, the model itself is so large that it cannot fit in memory on a single processor. In these cases, we need to spread the model across multiple processors, where each processor handles a different portion or segment of the model. A segment could be a layer in the network, or pieces of layers in the case of some very complex models. *Figure 14.6* shows a simple example of model parallelism, in which each layer of our neural network runs on a separate processor:

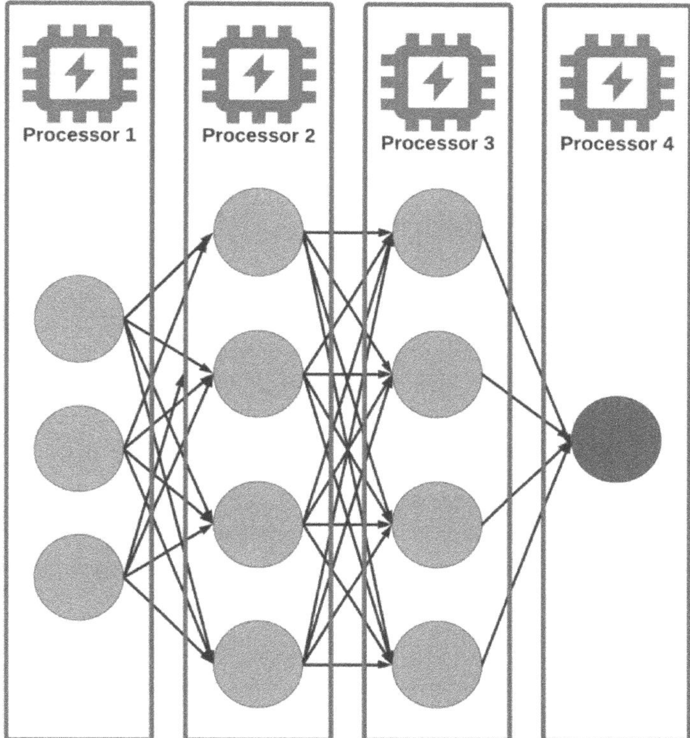


Figure 14.5: Model parallelism

There are different methods we can use to break our model down into segments, and we generally want to choose an approach that minimizes communication overhead between the processors. This is because that part of the process can introduce the most latency and complexity, especially when we're using multiple machines where communication needs to happen over a physical network.

In addition to optimizing communication between the processors, we also need to design our training procedure in a way that maximizes the utilization of each processor. Remember that the outputs from one layer in our neural network become inputs to other layers in our network, so there are **sequential dependencies** among the layers. When our layers are distributed across different processors, some processors could be idle while they wait for inputs to be propagated from the previous layer in the network. This, of course, would be an inefficient use of our processing resources, and we can use an approach called **pipelining** to improve the training efficiency by breaking the input data up into micro-batches. In that case, each segment or stage of the model processes its micro-batch and passes the outputs to the next stage. Then, while the next stage begins to process the outputs, the previous stage can start processing the next micro-batch, and so on. In this way, data flows through the network in a more streamlined manner, rather than processors for later layers being idle while waiting for an earlier layer to process an entire large dataset.

So far, I've been mainly talking about how to parallelize the training procedure in terms of how the models process the inputs (that is, the forward pass in the training process). Remember that the training procedure also includes a feedback loop to check the outputs of the model, compute the loss against the ground truth, then compute the gradients of the loss and update the weights. Next, we will dive into how those steps can be implemented in a distributed manner.

## Distributed training update process

In the case of model parallelism, data is still processed through the network sequentially – that is, each layer (or segment) in the network produces outputs, and those outputs become the inputs for the next layer (even though the layers or segments are being processed on different processors). Because of this, the backpropagation process can also happen sequentially, where gradients are computed in each segment starting from the last to the first, similar to how it's done in a non-distributed setting. Each segment independently updates its portion of the model weights based on the computed gradients.

In the case of data parallelism, however, subsets of the training data are processed in parallel across different processors, so the overall workflow is not sequential. Each processor has an identical copy of the model, but each copy works on a different subset of the data. For this reason, the individual replicas of the model on each processor are unaware of what's being done on the other processors. Therefore, some additional coordination is needed when computing the loss, the gradients, and the weights.

We can implement this coordination in various ways, but we should generally pick one of the following two options:

- Use a centralized **parameter server** to do the coordination
- Implement a protocol that allows the individual training servers to act as a community (I will refer to this as "collaborative computing")

Let's discuss each of these approaches in more detail, starting with the parameter server approach.

### *The parameter server approach*

In the parameter server approach, one or more nodes in the distributed system are designated as parameter servers, and these nodes hold the global model parameters. This means that they have a global view of the gradients and parameters for all of the model replicas across all nodes in the distributed training system or cluster.

In this case, the individual training nodes (let's call them "worker nodes") process different subsets of data and compute their respective gradients, and then each worker node sends its gradients to the centralized parameter server(s). The parameter server(s) then update the global model parameters based on these gradients and send the updated parameters back to the worker nodes. A simplified architecture diagram for this approach is shown in *Figure 14.7*:

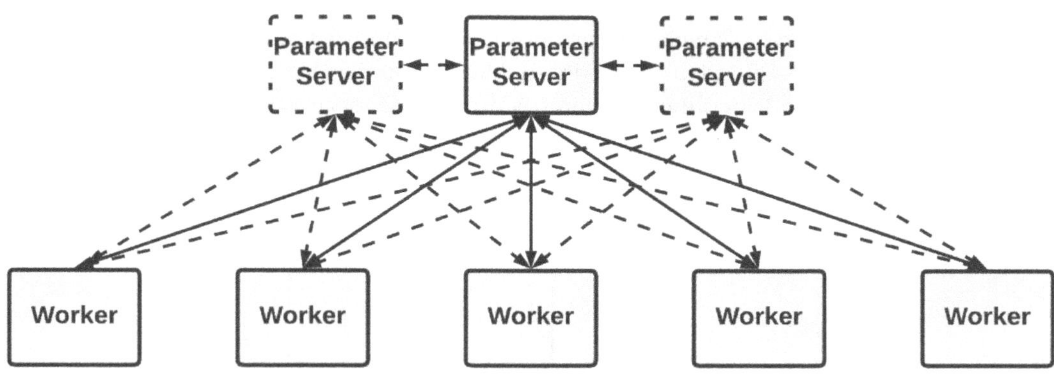


Figure 14.6: The parameter server approach

In *Figure 14.7*, the arrows represent the interchange of gradients and weight updates between the workers and the parameters server(s). Dashed lines and borders represent that there could be one or more parameter servers because a single parameter server could become a bottleneck in the solution. There also needs to be a coordinating entity in this architecture to manage all of the resources and their communication. This could be co-located in the parameter server or could be implemented as a separate server.

Next, we'll discuss the other option for implementing model training with data parallelism where, instead of using centralized servers for performing the coordination steps, all of the workers collaborate directly with each other.

### *The collaborative computing approach using Ring All-Reduce*

In the collaborative computing approach, there is no central server or set of servers that are dedicated to the purpose of performing coordination among the independent worker nodes in the training cluster. Instead, a distributed computing protocol is used for coordination among the nodes. This means that the nodes get updates (in this case, the computed gradients) from each other. Numerous communication protocols exist for coordination across distributed computing systems, and a pretty popular one is known as **All-Reduce**. These kinds of protocols are generally designed to aggregate data across all nodes in a distributed system and then share the result back to all nodes. This means that, even though each node may process a different subset of our dataset, and the model instance on each node may therefore compute different gradients and weights, each node eventually ends up with an aggregated representation of the gradients and weights that is consistent across all nodes.

When using All-Reduce, we can connect the processors in various ways, such as by implementing a fully connected mesh, as depicted in *14.8*, a ring architecture, as depicted in *Figure 14.9*, or other approaches, such as tree and butterfly architectures, which we will not cover here:

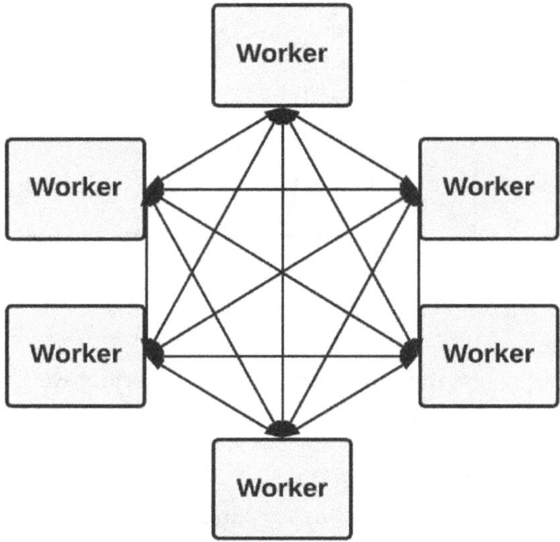


Figure 14.7: Fully connected mesh

As shown in *Figure 14.8*, the fully connected mesh architecture, in which all nodes are connected to all other nodes, would require a lot of complex coordination among all of the nodes. For this reason, the simpler and more effective ring architecture is often chosen:

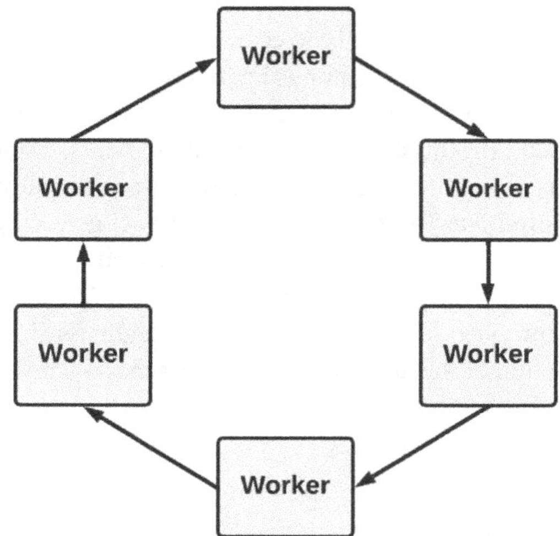


Figure 14.8: Ring architecture

In the ring architecture, as the name implies, the nodes are logically arranged in a ring. One of the important aspects of this architecture, referred to as **Ring All-Reduce**, is that each node only needs to send its computed gradients to one of its neighbors. The neighbor then combines (aggregates) the received gradients with its own gradients, passes the aggregated gradients onto the next node in the ring, and so on. This aggregation is often as simple as adding the gradients together. After *N-1* steps (where *N* is the number of nodes), each node will have a copy of the aggregated gradients for all nodes. They can then simply perform a division operation to calculate the average gradients, and use those values in the optimization step.

> **Note**
>
> To be more specific, the Ring All-Reduce algorithm is usually broken down into two phases, referred to as the "reduce-scatter" phase and the "all-gather" phase, but that's a level of detail that's not necessary for understanding how the process works at a high level, as described in this section.

At this point, you might be wondering which approach is better – that is, using a centralized parameter server or using a collaborative approach such as Ring All-Reduce. The answer, as is the case in many solution architecture decisions, is. "It depends." Let's discuss some additional factors that play into this decision, such as whether we want to implement synchronous versus asynchronous data parallelism.

### *Synchronous versus asynchronous data parallelism*

In the case of synchronous data parallelism, all the nodes need to update their parameters at the same time, or at least in a coordinated way. A benefit of this approach is that it provides consistency in parameter updates across all of the nodes. However, it can result in bottlenecks that slow down the overall training process, because all nodes must wait for the slowest one.

In the case of asynchronous data parallelism, on the other hand, the nodes update their parameters independently without waiting for others, which can lead to faster training. However, this can sometimes cause problems with model convergence and stability because the nodes can go out of sync if the procedure is not implemented correctly.

Asynchronous implementations can also be more fault tolerant because the nodes are not dependent on each other, so if a node goes out of service for some reason, the rest of the cluster can continue to function, so long as we have set up the environment in a fault-tolerant manner.

All-Reduce is generally implemented synchronously because the nodes need to share their gradients, whereas the parameter server approach can be implemented either synchronously or asynchronously, with asynchronous usually being the more common choice.

Next, we'll learn about how Google Cloud Vertex AI provides an optimized All-Reduce mechanism via the Vertex AI Reduction Server.

### *The Vertex AI Reduction Server*

While the Ring All-Reduce approach is pretty well established and popular in the industry, it also has some limitations. For example, in practice, it is found that latency tends to scale linearly with the number of workers in the ring (that is, the more workers we add, the more latency we get). Also, since each worker needs to wait on all other workers in the ring, a single slow worker can slow down the ring overall. Additionally, the ring architecture can have a single point of failure, whereby if one node fails, it can break the entire ring.

For these reasons, Google created a product called the Reduction Server, which provides a faster All-Reduce algorithm. In addition to the worker nodes, the Vertex AI Reduction Server also provides **reducer nodes**. The workers host and execute the model replicas, compute the gradients, and apply the optimization steps, while the reducers have a relatively simple job, which is just to aggregate the gradients from workers. *Figure 14.10* shows an example of the Reduction Server architecture:

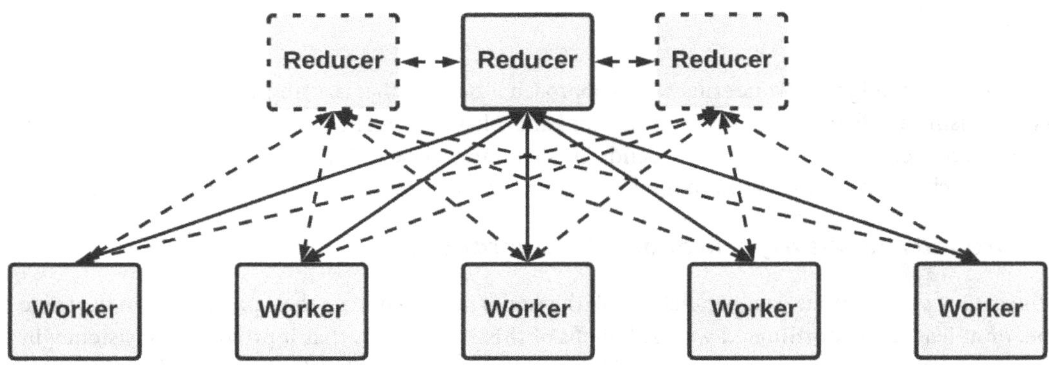


Figure 14.9: Reduction Server implementation

In *Figure 14.10*, you'll notice a similarity to the parameter server architecture, but bear in mind that the reducers perform much fewer tasks than parameter servers, and this simplicity provides some benefits. For example, the reducers are just lightweight instances that can use CPUs, which, in this case, can be significantly cheaper than GPUs. Also, unlike the ring architecture, the latency does not scale linearly as we add more reducer nodes, and there is no single point of failure that can break the overall architecture.

Based on the improvements provided by the Vertex AI Reduction Server architecture, I recommend using this approach when performing distributed training on Vertex AI.

Distributed training jobs generally involve using a lot of resources, which could incur expenses, so we will not perform a hands-on activity for this topic in this chapter. If you would like to learn more about how to implement an ML project using distributed training on Vertex AI (optionally, also including the Reduction Server), I recommend referencing the documentation at `https://cloud.google.com/vertex-ai/docs/training/distributed-training`.

Next, we'll look at some other important factors for distributed training.

### *Other important factors for distributed training*

In distributed training, the most complex and error-prone part of the process is usually managing communication among the nodes. If this is not implemented efficiently, then it can cause bottlenecks and impact training performance. We've already discussed how we generally want to minimize how much data needs to be sent between the nodes, and fortunately, there are some tricks we can use to help in this regard, such as compressing the gradients before transmitting them.

Also, considering that distributed training jobs usually need to process a lot of data, they can often run for long periods, even when the workload is being parallelized across multiple nodes. For this reason, we should put mechanisms in place to help us recover from failures that could occur during the training process. For example, if our job has been running for many hours, we wouldn't want to have to start it from the beginning again if something fails before the job finishes. A common and important mechanism to implement for this purpose is **checkpointing**, in which we regularly save the state of the model during training so that the process can continue from a recent state if failure occurs. This is just like periodically saving our work while we're working on a document just in case our laptop crashes unexpectedly.

All of the distributed training concepts we've discussed in this section so far assume that we have all of the required data stored somewhere in our environment, and we simply need to distribute that data (or the models) across multiple nodes. However, there's another popular type of distributed model training referred to as **federated learning**, which I'll briefly describe next.

## Federated learning

In previous chapters, we discussed deploying ML models to edge devices such as mobile phones or IoT devices, usually to provide low-latency inference on those devices. We also discussed how those edge devices generally have much less computing power and data storage resources than servers have. Therefore, our ability to train models on those devices is limited.

Let's imagine that the models that are deployed to the devices can be improved by being updated with data that becomes available to the devices from their local environment, such as from sensors. If we want to provide the best possible experience for users of those devices, then we will want to ensure that the models are periodically updated to account for new data that becomes available. However, in some cases, it may be inefficient or otherwise undesirable to constantly send data from the edge devices to a central location for larger model training jobs.

Federated learning is a technique that enables us to periodically update the models without needing to send data from those devices back to a centralized location. Instead, copies of the model on each device are trained locally by the data that becomes available to the device. The model training process on each device goes through the familiar steps of calculating the loss and the gradients, and then updating the model parameters or weights accordingly. Let's imagine that this kind of training is being performed on millions of devices, where each device is making updates to the model based on small pieces of data that become available to it. Given the limited training capabilities on each device, this process alone will not lead to a lot of model improvements in the long term. However, in the case of federated learning, the updated weights can be sent back to a central location to be combined to create a more powerful model that can learn from the small training processes on millions of devices. This means that all of the weights from the millions of devices can be averaged (that is, aggregated) to form a more advanced model. This updated model (or an optimized version of it) can then be sent out to each device, and the process can be repeated on an ongoing basis to keep improving the models over time.

The effective thing about this process is that only the model weights need to be communicated back to the centralized infrastructure, and none of the actual data on the devices ever needs to be transmitted. This is important for two main reasons:

- The weights are much smaller (in terms of data size) than the training data, so sending only the weights is much more efficient than sending training data from the devices to the central infrastructure
- Since only the weights are being transmitted (and no actual user data), this helps keep user data safe and preserves privacy

There are, of course, other methods for implementing distributed model training, but we've covered many of the more popular ones here.

The remainder of this book will focus on Generative AI, so this is a good point for us to revisit some important topics that we covered at a high level earlier in this book that will help us understand various developments that led to the Generative AI era.

# Transitioning to Generative AI

Before we start diving into Generative AI, I'll provide additional details on some of the important neural network architectures we discussed earlier in this book. For example, in *Chapter 9*, we briefly introduced some common neural network architectures, such as CNNs, RNNs, **long short-term memory** (**LSTM**), and transformers. In this section, we will dive deeper into how some of these notable architectures work, for two reasons:

- The practical exercises accompanying this chapter include building a CNN for a computer vision use case, so it's important to describe the inner workings of CNNs in more detail
- The rest of the neural network architectures mentioned here can be seen as milestones in the journey toward developing Generative AI technologies

Let's begin with a deeper dive into CNNs and computer vision.

## CNNs and computer vision

As we discussed in *Chapter 9*, CNNs are often used in computer vision use cases such as object recognition and visual categorization, and their trick is to break pictures down into smaller components, or "features," and learn each component separately. In this way, the network learns smaller details in the picture and then combines them to identify larger shapes or objects hierarchically.

### *The architecture of CNNs*

We could probably write an entire book on CNN architecture, but that level of detail is not required in this book. I'll cover some of the important concepts at a high level, and the Jupyter Notebook that accompanies this chapter provides code examples on how to build a relatively simple CNN for a computer vision use case.

As a brief refresher of what we discussed in *Chapter 9*, the most basic form of neural network, as depicted in *Figure 14.11*, is referred to as a **feed-forward neural network** (**FFNN**), in which the information that is being fed into the network follows a simple forward direction as it passes through the network:

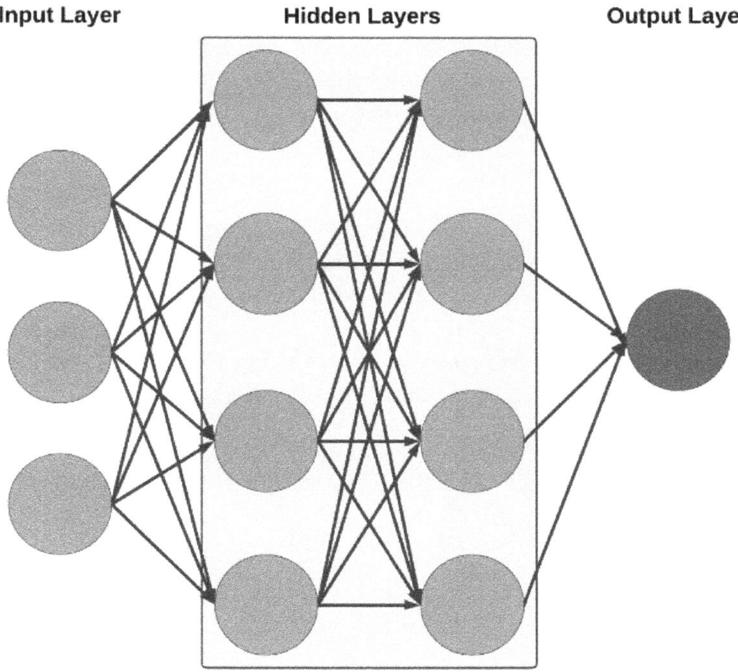


Figure 14.10: A simple neural network

A CNN, on the other hand, adds some specific types of layers into the architecture that help with implementing the hierarchical processing we mentioned in the introduction to this section. These types of layers will be explained at a high level in the following sub-sections.

### Convolutional layers

These layers perform a mathematical operation called **convolution**, which involves sliding a filter (or kernel) over the input image to produce a feature map. The filter is a small matrix that's used to detect specific features, such as edges, corners, or textures, where each filter is **convolved** across the input image, computing the dot product between the filter and input, resulting in a feature map.

The output of a convolutional layer, therefore, is a set of feature maps that represent specific features that are detected by the filters at different locations in the input image.

### Pooling layers

Pooling layers are mainly used for dimensionality reduction by downsampling the feature maps and reducing the number of parameters. We covered the concept of dimensionality reduction in detail in *Chapter 7* and discussed how it is often necessary to implement dimensionality reduction to reduce the amount of computation required during training and serving ML models, as well as to reduce the likelihood of overfitting the training data. Different pooling approaches can be used, but the most common approach is called "max pooling," which takes the maximum value from a set of pixels in a region (defined by the pool size). Another approach, called "average pooling," simply takes the average of the values from a set of pixels in a region.

### Fully connected layers

Combining the convolutional and pooling layers results in the features being extracted from the input images. After the input data is passed through the convolutional and pooling layers, the high-level reasoning in the network is then performed by using fully connected layers. These are the kinds of layers we covered when we discussed FFNNs, where the neurons in each layer have connections to all activations in the previous layer. Just as in the case of FFNNs, the fully connected layers are often used for the final classification or regression task that the network is performing.

The final output layer often uses a softmax activation function to output a probability distribution over the classes that the model is trying to identify. Alternatively, for binary classification, a sigmoid function could be used.

**Residual networks**

Residual networks, or ResNets, are a type of CNN that was developed to address the problem of vanishing and exploding gradients, something we discussed in *Chapter 9*. They do this by introducing "skip connections," which are shortcuts that "skip" over layers within the network, and directly connect the output of one layer to the input of a later layer.

That's about as much detail as we're going to cover regarding the inner workings of CNNs. Many research papers go into great detail on how CNNs work, but that's a level of detail beyond the scope of this book. Next, we'll dive in and work on some code examples to build our very first CNN for a computer vision use case!

In the Jupyter Notebooks that accompany this chapter, we will use Keras to build a simple CNN. This is where you will also start learning how to use PyTorch, which we'll then use to perform the same task.

### *Building a CNN*

To build our CNNs, open JupyterLab on the Vertex AI Workbench instance you created in Chapter 5 and perform the following steps:

1. In the navigation panel on the left-hand side of the screen, navigate to the `Chapter-14` directory within the `Google-Machine-Learning-for-Solutions-Architects` folder.
2. Double-click on the `Keras-CV.ipynb` notebook file to open it. When you're prompted to select a kernel, select the **TensorFlow** kernel.
3. Double-click on the `PyTorch-CV.ipynb` notebook file to open it. When you're prompted to select a kernel, select **PyTorch** kernel.
4. In each of the notebooks you've opened, press *Shift* + *Enter* to execute each cell.

   The notebooks contain comments and Markdown text that describe what the code in each cell is doing.

Take a moment to savor the fact that you have just used both Keras and PyTorch to build CNNs for a computer vision use case on Google Cloud Vertex AI. This is some pretty advanced stuff!

Next, we'll take a look at some types of neural network architectures that are typically useful for sequential data and use cases such as **natural language processing** (NLP).

## RNNs, LSTMs, and transformers

Again, we already introduced these concepts at a high level in *Chapter 9*. In this section, we will dive a little bit deeper into these types of architectures. We won't build these networks in this chapter, nor go into excessive detail; I'm covering these topics here mainly for historical context to pave the way for concepts and activities that I'll introduce in the Generative AI chapters in this book. This is because all of the neural network architectures in this section can be seen as stepping stones toward the development of the Generative AI models we use today, in the order shown in *Table 14.1:*

| Development | Timeframe | References |
|---|---|---|
| RNN | 1990 | Elman, J. L. (1990). *Finding structure in time*. Cognitive Science, 14(2), 179-211.<br>https://doi.org/10.1207/s15516709cog1402_1. |
| LSTM | 1997 | Hochreiter, Sepp & Schmidhuber, Jürgen. (1997). *Long Short-term Memory*. Neural computation. 9. 1735-80. 10.1162/neco.1997.9.8.1735.<br>https://doi.org/10.1162/neco.1997.9.8.1735. |
| Transformer | 2017 | Ashish Vaswani, Noam Shazeer, Niki Parmar, Jakob Uszkoreit, Llion Jones, Aidan N. Gomez, Lukasz Kaiser, and Illia Polosukhin. *Attention Is All You Need*. [2017, June 12]. arXiv preprint arXiv:1706.03762. http://arxiv.org/abs/1706.03762. |

Table 14.1: The stepping stones in sequential model development

Let's begin by diving into RNNs.

### *RNNs*

As you read each word in this sentence, you need to remember the words that came before it so that you can build an understanding of what the overall sentence is saying. This is an application of **natural language understanding** (NLU), which is a sub-field of NLP. In a more generic sense, the sentence represents sequential data because the data points (in this case, words) in the sentence relate to other words in the sentence and the order in which they are processed matters. *Figure 14.12* shows an example of sequence data being fed into a neural network:

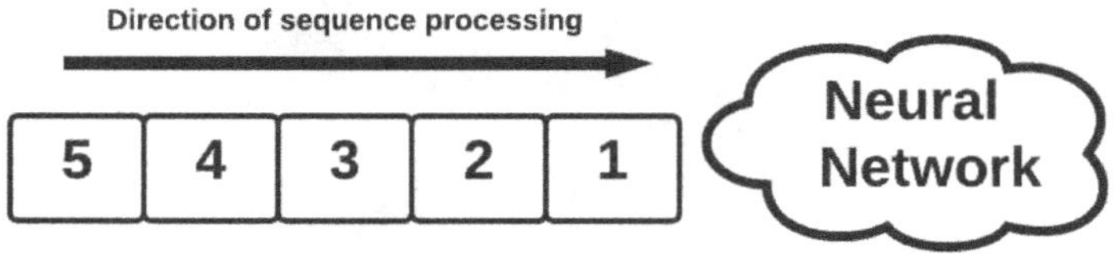


Figure 14.11: Sequence data being fed into a neural network

In *Figure 14.12*, we can see that data point 1 will be processed first by the neural network, and then data point 2 will be processed, and so on, where each data point is processed one at a time by the network.

In simple FFNNs, each data point is passed independently through the network, so the network does not associate data points with previous data points. As we briefly mentioned in *Chapter 9*, RNNs introduce a type of looping mechanism into the network architecture, and this looping mechanism is what enables the network to "remember" previous data points. Let's look at this in a bit more detail to understand what this means.

In RNNs, the concept of each data point being processed by the network is referred to as a **time step**. The important thing to note is that, with RNNs, the outputs from neurons in each time step can be saved for future reference. Let's call this the **current state** of our network; this is often referred to as the **hidden state** of the network because it is generally not exposed externally.

At each time step, neurons in the network receive two sets of inputs:

- The output from neurons in the previous layer of the network (or inputs from the input layer if we're referring to the first hidden layer in the network).
- The output from the same neuron (that is, itself) that was saved from the previous time step. This is the hidden state.

Looking at our diagram in *Figure 14.12*, in the first time step, when the first data point in the sequence (that is, data point 1) passes through the network, the process will be much like a standard FFNN, as we described in *Chapter 9*, because there is no previous state to remember.

However, when data point 2 is being processed through the network, each neuron can combine this new data with the "current state" of the network (which is the state that was created after processing data point 1). This process repeats for each new data point that passes through the network, and by doing this, the network maintains a kind of memory of the data points it processed previously.

### Backpropagation through time (BPTT)

We discussed the backpropagation process in neural network learning in *Chapter 9*. Considering that RNNs use the concept of time steps and combine the outputs of neurons over time, the backpropagation process in RNNs needs to be modified accordingly. This modified version of backpropagation is called **BPTT**.

It involves "unrolling" the RNN in time, meaning that the time steps are treated as separate layers of a deep neural network. Once the network has been unrolled, the standard backpropagation algorithm is applied – that is, the network's error is calculated at each time step, the gradients are computed, and then these gradients are used to update the weights in the network.

### Limitations of RNNs – gradients and short-term memory

In an RNN, the unrolling that is performed as part of the BPTT process can lead to very deep networks for long sequences. Remember that vanishing and exploding gradients happen because of the chain rule in calculus. As the gradients are backpropagated through the deep networks associated with long sequences, this can make the problem of vanishing and exploding gradients more pronounced.

This also affects the "memory" of the RNN. As sequences get longer, the ability for backpropagation to update the gradients of earlier time steps gets smaller and smaller (in the case of vanishing gradients). This has the effect of making the network "forget" earlier time steps in longer sequences, which means that RNNs are generally limited to short sequences.

The vanishing and exploding gradient problems and short-term memory limitations have led to the development of more advanced RNN architectures such as LSTMs, which can address these issues. We'll discuss this in more detail next.

#### *LSTMs*

LSTM is a type of RNN that was designed specifically to overcome the limitations of traditional RNNs, especially the vanishing and exploding gradient problems when processing long sequences, as we discussed in the previous section.

In addition to the hidden state mechanism we described in the *RNNs* section, LSTMs add a concept called **cell state**. This is often likened to a kind of conveyor belt that runs straight through the entire chain of an LSTM network, which allows information to be easily transported across many steps in the process. The hidden state still acts as the network's short-term memory, and the cell state acts as a longer-term memory.

LSTMs also introduce the concepts of **gates**, which can be seen as checkpoints along the conveyor belt that decide what information should be maintained, dropped, or added. In this way, the gates regulate the flow of information through the network. There are generally three types of gates:

- **Input gates**, which update the cell state with new information
- **Output gates**, which use the cell state to determine what the next hidden state should be
- **Forget gates**, which decide what information to discard from the cell state

If you'd like to learn more about how these gates are used in combination with the hidden state and the cell state, I recommend reading the original paper (Hochreiter and Schmidhuber, 1997) at `https://direct.mit.edu/neco/article-abstract/9/8/1735/6109/Long-Short-Term-Memory`.

Speaking of gates, a variation on LSTMs called **gated recurrent unit** (**GRU**) was more recently introduced (Cho et al., 2014). It combines the forget and input gates into a single **update gate** and merges the cell state and hidden state, which results in a simpler and more efficient model.

Next, we'll discuss one of the most important breakthroughs in the AI/ML industry in recent years: the transformer architecture.

### Transformers

As is the case in most fields of research, the pace of advancement generally proceeds at a somewhat linear rate, with occasional transformative breakthroughs along the timeline. The linear advancement in any field represents the step-by-step progress that occurs on an ongoing basis, and the transformative breakthroughs represent the sudden, giant leaps forward that completely change everything. When Google invented the transformer architecture (Vaswani, A., et al., 2017), this was one such giant leap forward in AI/ML research, and it has formed the basis of the Generative AI revolution that we are currently experiencing in the world.

The progression from RNNs to LSTMs is what I would consider to be a linear development. However, the progression from LSTMs to the transformer architecture was an enormous jump forward in the development of AI/ML technologies.

Firstly, unlike RNNs and LSTMs, which process sequences step by step, transformers can handle entire sequences simultaneously (in parallel), and this parallel processing allows them to manage long-range dependencies in our data much more efficiently. This also allows them to achieve larger scales than previous model architectures, resulting in what are now the largest and most powerful language models in the industry. However, parallel processing is not the most significant feature of the transformer architecture. As suggested in the title of the paper that first introduced the transformer architecture to the world – *Attention Is All You Need* (arXiv:1706.03762 [cs.CL]) – the main innovation in transformers is the self-attention mechanism, which allows the model to weigh the importance of different parts of the input data. For example, if we feed a sentence (or sequence) into a transformer model, the self-attention mechanism allows the model to understand the entire sentence and its internal relationships (for example, the relationships between the words).

Transformers run multiple self-attention **heads** in parallel on the sequence, something that's referred to as **multi-head attention** in the paper. Hopefully, you already read through the paper when we linked it in *Chapter 9*, but if not, I highly recommend reading the original paper to dive into the intricate details of how transformers work, given their pivotal importance in the AI/ML and Generative AI industry. Transformers are the basis of the current state-of-the-art Generative AI models.

Now that we've recapped the evolutionary steps that led to their development, we are ready to move on to the next chapter. But first, let's take a moment to reflect on what we've learned.

# Summary

In this chapter, we covered important AI/ML concepts that are typically used by more advanced practitioners who have specific needs or preferences in terms of how they want to implement their ML workloads, such as by using specific frameworks, or by parallelizing their workloads across multiple processors.

We started by discussing BQML, focusing on the close relationship between data processing, data analytics, and ML. We learned how to train and evaluate a model using BQML, and how to get predictions from that model, all by using SQL query syntax.

Then, we discussed different types of hardware that we can use for our AI/ML workloads, and popular tools and frameworks that we had not previously covered in this book, such as Spark MLlib, Ray, and PyTorch.

Next, we dived into CNNs and their use in computer vision before moving on to discuss neural network architectures that are particularly useful for sequential data and use cases such as NLP, such as RNNs, LSTMs, and transformers. We dived into the inner workings of these architectures, mainly to understand the evolutionary steps that have led to the development of the Generative AI technologies that we will outline in the remaining chapters of this book.

Finally, we discussed distributed model training in AI/ML, including why it's needed, and some of the mechanisms we can use to implement it, such as data parallelism and model parallelism. We also dived into different approaches for implementing data parallelism, such as using centralized parameter servers, or distributed coordination mechanisms such as Ring All-Reduce. We also explored the topic of federated learning, and how it can help to preserve privacy by avoiding the process of transmitting data from devices.

Next, we will embark on an exciting journey as we dive into the world of Generative AI. Join me in the next chapter to get started!

# Part 3: Generative AI

In this part, we begin our journey into the exciting universe of generative AI, which is currently taking the world by storm. This part of the book first introduces the fundamental concepts of generative AI, and we compare and contrast it against "traditional AI." We then move on to cover more advanced topics and learn how to implement those concepts in Google Cloud. The final chapter of this part then focuses on bringing together the main concepts that we've learned throughout the book, and lays out instructions to build AI/ML and generative AI solutions using a combination of various Google Cloud products and services.

This part contains the following chapters:

# 15
# Introduction to Generative AI

**Generative artificial intelligence** (**GenAI**) is certainly the term on everybody's lips at the moment. If you haven't already had an opportunity to "get your hands dirty" with GenAI, then you will soon learn why it is currently taking the world by storm as we dive into the kinds of amazing things we can do with this relatively new set of technologies. In this chapter, we will explore some of the concepts underpinning what GenAI is and its distinctions from other **artificial intelligence** (**AI**)/**machine learning** (**ML**) approaches. We'll also cover some of the major historical developments that have led to its meteoric rise, and examples of how it is being used today.

We'll begin the chapter by introducing some fundamental concepts before moving on to more complex topics and the evolution of various GenAI approaches, such as **autoencoders** (**AEs**), **generative adversarial networks** (**GANs**), diffusion, and **large language models** (**LLMs**). Considering that this is an introductory chapter, we will mainly lay the groundwork for the deeper dives that will follow in later chapters. Specifically, this chapter covers the following topics:

- Fundamentals of GenAI
- GenAI techniques and evolution
- LLMs

As makes logical sense, let's begin with the fundamentals!

## Fundamentals of GenAI

This section introduces the basic concepts we need to understand when discussing GenAI, starting with an overview of what GenAI is!

## What is GenAI?

I'll begin explaining this topic by focusing on what distinguishes GenAI from the other AI/ML approaches we've covered. Think about all of the various algorithms and approaches I've described so far throughout this book, and, more specifically, think about the primary goal that was being pursued in each approach. Whether we were using linear regression in `scikit-learn` to predict a numeric value for some target variable based on its features, logistic regression in XGBoost to implement a binary classifier model or using time-series data to predict the future in TensorFlow, there is one common theme, which is that we were trying to **predict** or **estimate** something.

The key concept to understand here is that the output from our various sophisticated models was generally a distinct data point that was either right or wrong, or at least as close to right as possible. Our models typically produced a single, simple answer based on relationships the model had learned (or estimated) in the data.

Even in the case of **unsupervised learning** (**UL**) approaches such as K-means clustering, the model simply finds mathematical relationships in the data and categorizes the data points into groups based on those relationships.

The huge leap forward that has been achieved with the introduction of GenAI is that the models can now go beyond simple "yes"/"no" answers or numeric estimates based on sheer mathematical number crunching and pattern recognition. With GenAI, the models can now create (or **generate**) new data. That's the big difference, and it turns out that the implications of this are pretty huge!

Taking a step back for a moment: when I was young, I thought that AI research could design machines that think like humans, and that fascinated me, so I started learning about how ML algorithms work. I was quite disappointed to learn that, although some models can do an amazing job at "understanding" consumer behavior and accurately recommending products that a given person might like to purchase, or even diagnose potential illnesses based on input data related to a patient, it was all just based on feeding large amounts of data into a mathematical algorithm that "learns" to detect a pattern in the data. There was no actual "intelligence" there, although the science is still fascinating.

The fact that GenAI models can go beyond specific mathematical answers and create new content is a significant leap forward in the pursuit of AI. Have a chat with any of the latest and greatest GenAI models out there, and I don't doubt for a moment that you will be very impressed by the kinds of mind-boggling things they can do, such as composing music, painting an imaginative scene, writing a catchy poem, or creating a web application. You'll learn how to harness GenAI for implementing many different kinds of use cases later in this book, and I'm sure you'll agree that it's a dramatic technological advancement.

It's important to understand, however, that the amazing feats performed by GenAI models are still powered by many of the AI/ML concepts we have already covered in this book. At its core, GenAI works by understanding and replicating the underlying patterns and structures in the data it has been trained on, and it still uses algorithms and **neural networks** (**NNs**) to perform these kinds of activities, although the network architectures used, and the novel ways in which they are used, are considerably more advanced. In the case of GenAI, the goal is not just to interpret the data but to form an understanding that can be used to create something new. Let's dive into these topics in more detail to better understand what sets GenAI apart from other AI approaches.

## What is non-GenAI?

Now that GenAI has taken the world by storm, how can we refer to all of the other AI/ML approaches that existed before it (that is, much of the stuff we already covered in this book)? One of the most popular emerging terms for this is "traditional AI/ML." You may also hear it being referred to as "predictive AI/ML" because the goal is generally to predict something, or "discriminative AI/ML" in the case of classification use cases.

## Diving deeper into GenAI versus non-GenAI

To understand the basic differences between GenAI and non-GenAI, we need to revisit some of the mathematical concepts that underpin AI/ML. Don't worry – we'll just cover the concepts at a level required to define the distinctions between GenAI and non-GenAI.

### *The role of probability*

Remember that mathematical probability is a fundamental concept in many ML use cases. For example, when we ask a model to classify a data point into a specific category, it rarely will do so with 100% certainty. Instead, it will calculate the probability of the data point belonging to each category. Then we, or the model itself, depending on the implementation, can select the category with the highest probability.

One of the key factors that distinguishes GenAI from traditional AI is how probability is used in the learning process. Let's explore this in more detail, starting with traditional AI.

#### Traditional AI and conditional probability

In the case of traditional AI, the models try to estimate the probability that the target variable `Y` contains a specific value based on the values of the predictor variables (or features) in the dataset. This is referred to as the **conditional probability** of the target variable's values based on the values of the input features. Conditional probability is the probability of an event occurring given that another event has already happened, and it is represented by the following formula:

$$P(B|A) = P(A \cap B) / P(A)$$

Here, the following applies:

- P(B|A) is the conditional probability of B given A.
- P(A ∩ B) is the probability of both A and B occurring. This is also referred to as the "joint probability," which we will explore in more detail later.
- P(A) is the probability of A occurring.

To understand conditional probability, imagine a game scenario in which there are three straws hidden from our view. Of the three straws, two are long straws, and one is a short straw. We will take turns picking straws, and whoever picks the short straw loses. Each time it's our turn to pick a straw, there's a specific probability that we will pick the short straw.

Initially, no straws were picked, so in the first turn, the probability of picking the short straw is 1/3 (or approximately 33.3%). Now, let's imagine that we pick a straw, and it turns out to be a long straw. This means that two straws remain, and since we picked a long straw, it means that one long straw remains, as well as the short straw. In the next turn, the probability of picking the short straw is, therefore, 1/2 (or 50%).

The important thing to note here is that since we already picked a long straw, it influences the probability distribution of the overall scenario. This is a very simple example of the concept. In ML, there can be many factors (that is, features) involved, leading to very large numbers of potential combinations of those factors that could influence the outcome. This is why ML models usually need to crunch through large amounts of data to learn patterns in those various combinations of features.

Mapping this back to ML use cases, consider the dataset represented in *Figure 15.1*, which shows a simplified version of the Titanic dataset from OpenML (`https://www.openml.org/search?type=data&id=40945`). The target variable is represented by the `survived` column, which is highlighted by the green box. This represents *B* in our preceding equation. All of the other columns combined constitute the input features, and they represent *A* in our previous equation, as highlighted by the red box. Essentially, this depicts that when an ML model learns to predict the value of the target variable based on the values of the input variables, it is asking the question: "What is the probability of *B* (that is, the target variable), given the values of *A* (that is, the input variables)?" In other words: "What is the probability of *B*, given *A*?":

| pclass | name | sex | age | sibsp | parch | ticket | fare | cabin | survived |
|---|---|---|---|---|---|---|---|---|---|
| 1 | Allen, Miss. Elisabeth Walton | female | 29 | 0 | 0 | 24160 | 211.3375 | B5 | 1 |
| 1 | Allison, Master. Hudson Trevor | male | 0.9167 | 1 | 2 | 113781 | 151.55 | C22 C26 | 1 |
| 1 | Allison, Miss. Helen Loraine | female | 2 | 1 | 2 | 113781 | 151.55 | C22 C26 | 0 |
| 1 | Allison, Mr. Hudson Joshua Cre | male | 30 | 1 | 2 | 113781 | 151.55 | C22 C26 | 0 |
| 1 | Allison, Mrs. Hudson J C (Bessie | female | 25 | 1 | 2 | 113781 | 151.55 | C22 C26 | 0 |
| 1 | Anderson, Mr. Harry | male | 48 | 0 | 0 | 19952 | 26.55 | E12 | 1 |

Figure 15.1: Separation of target variable from input feature

Next, let's look at how probability can be used in a slightly different way in GenAI use cases.

## GenAI and joint probability

In the case of GenAI, the models are designed to learn the **joint probability distribution** of the dataset. As we briefly saw in the preceding equation for conditional probability, joint probability refers to the probability of two or more events happening at the same time. We can use this to understand how different variables or events are connected and how they influence each other. It is represented as the following:

$P(A \cap B) = P(A) \times P(B)$

Here, the following applies:

- P(A) is the probability of A occurring
- P(B) is the probability of B occurring

Let's again use an example to describe this concept in more detail. A common analogy is to imagine rolling two fair six-sided dice and calculating the probability of both dice showing a certain number; for example:

- **Event A**: The first die shows a 3
- **Event B**: The second die shows a 5

Here, the joint probability $P(A \cap B)$ is the probability of both the first die showing a 3 and the second die showing a 5 when they come to rest. Since each die roll is independent and each side has a 1/6 chance of appearing, the joint probability is the following:

$P(A \cap B) = P(A) \times P(B) = 1/6 \text{ x } 1/6 = 1/36$

This means there's a 1/36 chance that both dice will show the exact numbers you predicted. The key thing to understand here is the events are independent, but there's a shared probability distribution that governs the overall outcome.

Mapping this back to ML use cases again, the joint probability distribution in a dataset includes the target variable as well as the input variables, as represented by the red box in *Figure 15.2*:

| pclass | name | sex | age | sibsp | parch | ticket | fare | cabin | survived |
|---|---|---|---|---|---|---|---|---|---|
| 1 | Allen, Miss. Elisabeth Walton | female | 29 | 0 | 0 | 24160 | 211.3375 | B5 | 1 |
| 1 | Allison, Master. Hudson Trevor | male | 0.9167 | 1 | 2 | 113781 | 151.55 | C22 C26 | 1 |
| 1 | Allison, Miss. Helen Loraine | female | 2 | 1 | 2 | 113781 | 151.55 | C22 C26 | 0 |
| 1 | Allison, Mr. Hudson Joshua Cre | male | 30 | 1 | 2 | 113781 | 151.55 | C22 C26 | 0 |
| 1 | Allison, Mrs. Hudson J C (Bessie | female | 25 | 1 | 2 | 113781 | 151.55 | C22 C26 | 0 |
| 1 | Anderson, Mr. Harry | male | 48 | 0 | 0 | 19952 | 26.55 | E12 | 1 |

Figure 15.2: Joining target variable with input features

A major difference here is that, while discriminative models use conditional probability to predict the value of the target variable given the values of the input variables, generative models try to learn the overall joint probability distribution of the dataset, including the target variable.

By learning the overall joint probability distribution of the dataset, the model can develop an understanding of how the dataset is constructed and how all of the features relate to each other, including the target variable. In this manner, by accurately approximating how the training dataset is constructed, it can estimate how to create similar data points that do not already exist in the dataset but are of similar structure and composition – that is, it can generate new, similar content.

In addition to the joint probability distribution, today's generative models learn hidden relationships and structure in the dataset by using the attention mechanism outlined in the famous *Attention Is All You Need* (Vaswani et al., 2017) paper that we've referenced numerous times in this book. Before the attention mechanism was developed, models mostly treated all inputs equally. And, considering that all input features (and related hidden features) carry information, treating all inputs equally results in a low signal-to-noise ratio that limits the effectiveness of the learning and prediction processes. For example, when a model tries to predict the next word in a sentence, not all prior words in the sentence contribute equally to predicting the next word. Instead, the most probable next word may be more dependent on (or more highly correlated with) the existence of specific other words in the sentence. For example, consider the sentence: "I added a pinch of salt and a dash of [BLANK]." When a modern generative model attempts to predict the next word in that sentence, it will likely predict the word "pepper." While the word "pepper" must make sense in the overall context of the sentence (that is, it will have varying degrees of contextual relationships to the other words in the sentence), the word "salt" likely has a higher impact on the prediction than other words, such as "pinch," and the attention mechanism helps the models to learn these kinds of important relationships.

While this section focuses mainly on the differences between GenAI and "traditional AI," I also want to mention that the demarcation is not always so clear, and sometimes the lines may be blurred.

### Blurring the lines

It's important to understand that the demarcations between non-GenAI and GenAI often become blurred because many applications combine both approaches. For example, a model or application that analyzes text and generates a summary may use both discriminative and generative processes.

Consider chatbots as an example. Chatbots commonly generate responses by constructing sentences "on the fly." A popular approach in constructing sentences is to predict the most appropriate next word based on previous words in the sentence. This is an instance of using conditional probability. However, if the application or model only did this, then its abilities would be quite limited. In order to generate complex and coherent responses that adhere to **natural language** (**NL**) constructs, the model needs to have learned an accurate representation of the structure (that is, the joint probability distribution) of the language in which it generates responses.

An additional interesting scenario is the use of GenAI to produce new data that can then be used as features for traditional ML models. An example of this use case, as suggested by my colleague, Jeremy Wortz, would be to ask a generative model to rate songs with a score (let's say between 1 and 5) based on a customer persona, and then make playlists with those generated features from the ratings and outputs from a process called "chain-of-thought reasoning," which essentially asks the LLM to elaborate on its thought process. I will describe the chain-of-thought concept in more detail later.

Now that we've discussed the differences between GenAI and "traditional AI," we will explore the development of techniques that have led to the evolution of GenAI.

# GenAI techniques and evolution

As we've already discussed, while there are some distinctions between GenAI and "traditional AI," they share much of the same history. For example, in *Chapter 1*, we discussed a brief history of AI/ML, and in *Chapters 9* and *14*, we learned about the evolution of various types of NNs, such as **recurrent NNs** (**RNNs**), **long short-term memory** (**LSTM**) networks, and Transformers. All of those concepts and milestones also apply to the history and evolution of GenAI. Considering that GenAI is a relatively newer subset of AI, we can view its evolutionary timeline as an extension of the evolution of AI in general. Therefore, the topics in this section build upon what we've already covered in that regard.

The evolution of GenAI could comprise an entire book by itself, and for the purpose of this current book, it would be an unnecessary level of information to cover all of the major milestones and developments that contributed to the evolution of GenAI in great detail. Instead, I will summarize some of the most prominent milestones and contributory developments at a high level, such as Markov chains and **hidden Markov models** (**HMMs**), **restricted Boltzmann machines** (**RBMs**), **deep belief networks** (**DBNs**), AEs, and GANs.

It's important also to note that some mechanisms that were originally developed primarily for discriminative use cases can be reappropriated for generative use cases. For example, while simple Naïve Bayes classifiers are commonly used to predict a given class based on the features in the dataset (that is, an application of conditional probability for discriminative use cases), the algorithm is also capable of learning an approximation of the joint probability distribution of the dataset during training due to how it applies Bayes' Theorem with the (naïve) assumption that each feature is independent. (To learn more about how Naïve Bayes classifiers work, I recommend reviewing the paper at the following URL: `https://doi.org/10.48550/arXiv.1404.0933`). A similar reappropriation can be done with more complex applications of Bayes' Theorem, such as Bayesian networks.

Before diving into specific GenAI approaches, I want to introduce two important concepts that will form the foundation for many of the topics that we will discuss throughout the remainder of this book, referred to as **embeddings** and **latent space**.

## Embeddings and latent space

In *Chapter 7*, we discussed the topic of dimensionality reduction, and we used mechanisms such as **principal component analysis** (**PCA**) to project features from our datasets into lower-dimensional feature spaces. These lower-dimensional feature spaces can be referred to as "latent spaces," and the representations of our data in the latent space can be referred to as "embeddings." Allow me to explain these important concepts in more detail, starting with embeddings.

### Embeddings

Embeddings are numerical representations of data in a lower-dimensional space (that is, the latent space). They can be seen as numerical "fingerprints" that capture the meaning or characteristics of the original data. For example, word embeddings capture the meaning of words, and if words such as "king" and "queen" have similar embeddings, a language model can infer a relationship between the two. Models can also embed other types of data, such as images, audio, and graphs.

Next, let's dive into the concept of latent space in more detail.

### Latent space

We use the term "latent space" to define the abstract feature space where the intrinsic properties or features of a dataset are represented. This space captures underlying structure and patterns within the data that might not be apparent in its original form (hence the term "latent").

It's important to understand that the latent space and its dimensions map in some way to the original features. This means that relationships among the original features are captured as relationships among the projected features in the latent space (that is, semantic context is represented in some way). These representations and mappings are generally learned by models during the training process and are therefore often not easily interpretable by humans, but there are also methods by which we can explicitly create embeddings for the contents of our datasets, which I will describe in more detail in *Chapter 16*. For now, I'll briefly explain how embeddings and latent spaces are used.

### Using embeddings and latent space

When models have created embeddings for our data, these representations in the latent space can be used in interesting ways. One of the most useful ways to use this data is to measure the distances between embeddings in the latent space, using familiar distance metrics such as Euclidean distance or cosine similarity. Considering that the embeddings in the latent space capture the essence of the concepts they represent, we can identify concepts that may be similar or related to the original space. An example would be products in a retail website's product catalog. By embedding the product details and identifying which ones are near each other in the latent space, our models can get an understanding of which products may be related to each other. A recommender system could then use this information to display insights such as "customers who purchased this item also purchased these other items."

You may ask, "Why not just perform the same kinds of actions in the original space?" The process of encoding embeddings converts the objects to vectors, providing a much more efficient way to represent the concepts. Also, rather than dealing with words and images, ML algorithms and models love to work with vectors, so these vectorized representations are more well suited to ML use cases.

Another example that highlights the efficiency of embeddings is when we use ML models for image-processing use cases. Consider high-definition images that contain millions of pixels, in which each pixel is considered a feature. If we have a dataset consisting of hundreds of millions of images, and each image has millions of features, this could lead to extremely compute-intensive training and inference processing. Instead, mapping the features to a lower-dimensional feature space will significantly optimize processing efficiency. Also, individual pixels often don't convey much meaning; rather, it's often the relationships between pixels and patterns that define what an image represents.

We'll revisit the concepts of embeddings and the latent space many more times throughout the rest of this book because they are such foundational concepts in GenAI. Now that I've introduced these important concepts, we'll begin our journey through various important milestones and approaches that led to the development of GenAI, as outlined at a high level in *Figure 15.3*:

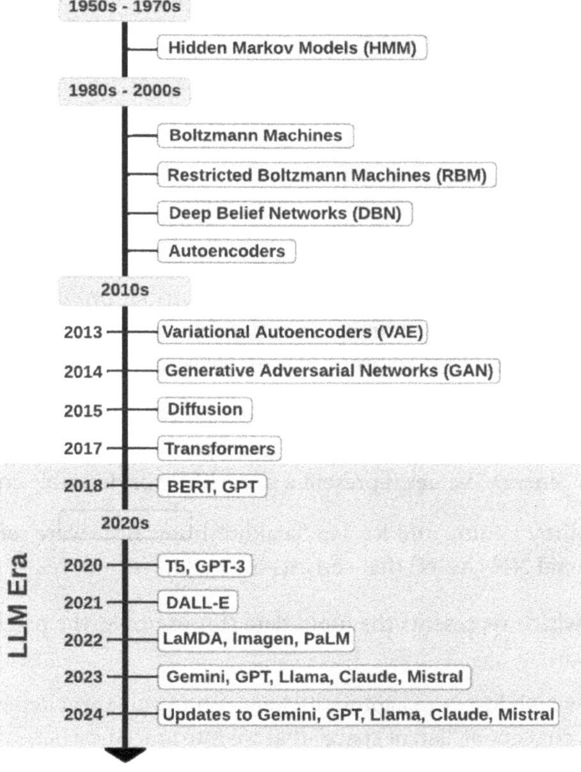


Figure 15.3: Milestones in GenAI evolution

Of course, many additional GenAI models and approaches have been developed and released in the past few decades beyond those shown in *Figure 15.3*, but here I'm focusing on high-level milestones and developments that have most notably influenced the evolution of GenAI. Let's begin diving into each one, starting with Markov chains and HMMs.

### Markov chains and HMMs

The concept of Markov chains was introduced by Andrey Markov in 1906, and they are based on what's referred to as the Markov property, which posits that the next or future state in a sequence or scenario depends only on the current state. They can be used to predict (and therefore generate) the next item in a sequence. HMMs extend this concept to include hidden states that cannot be directly observed in a given scenario. A very simple example would be the potential correlation between current weather conditions and umbrella sales. If we can directly observe the weather, and we see that it is currently raining, then that's an observable state we can use in a Markov chain to predict an increase in umbrella sales. On the other hand, if we cannot observe the weather for some reason (perhaps we're working in a store that's below ground level in a mall) but we notice an increase in umbrella sales, then we could surmise that it's currently raining outside. In this case, the state of the weather is hidden, but we can speculate the probability of rain based on the secondary observable state of increased umbrella sales. This would be a very simple representation of a hidden state in an HMM.

The next concept I'd like to introduce briefly is RBMs.

### RBMs

**Boltzmann machines (BMs)** are a type of **energy-based model (EBM)** that borrows some concepts from physics. "Boltzmann" refers to Ludwig Boltzmann, a physicist associated with statistical mechanics, which uses statistics and probability theory to model the behavior of microscopic particles. The Boltzmann probability distribution (also called the Gibbs distribution) provides the probability of a system being in a particular state based on its energy and temperature. A key concept here is that states with lower energy are more likely to occur than states with higher energy.

Mapping this to data science: Unlike the conditional probability used in traditional AI as described earlier in this chapter, EBMs use an "energy function" to assign a value to every possible configuration of variables, where lower "energy" values represent a more likely or desirable configuration.

Due to the work by Geoffrey Hinton and Ruslan Salakhutdinov, BMs were refined to form RBMs, which are a type of **artificial NN (ANN)** that consists of two layers:

- The visible layer, which represents the input data (for example, the pixels in an image or the words in a sentence).
- The hidden layer, which learns to capture higher-level features and dependencies in the data. This relates to the concept of "latent space" that we introduced earlier.

With RBMs, the learning process involves adjusting the weights between the visible and hidden layers to minimize what's referred to as the **reconstruction error**, which measures how well the RBM can reproduce the original input data. By doing this, the RBM learns to model the probability distribution of the input data.

In the original BM, all of the layers were connected, and the models were quite complex and compute-intensive to train. However, in RBMs, while the visible layers and hidden layers are fully connected to each other, there are no connections within a layer (hence the term "restricted"), which makes them less computationally intensive to train.

RBMs are commonly used for UL use cases, especially for feature extraction and dimensionality reduction, and they can also be seen as a kind of building block for **deep learning** (**DL**), which can be used to build many other kinds of models, including classification and regression. The idea of using RBMs as building blocks brings me to the next topic: DBNs.

### DBNs

DBNs can be viewed as a composition of simple, unsupervised networks such as RBMs, where each sub-network's hidden layer serves as the visible layer for the next sub-network. By stacking multiple RBMs, each layer can learn increasingly abstract and complex representations of the data.

Training DBNs consists of two main phases: pre-training and **fine-tuning** (**FT**). During pre-training, the DBN is trained one layer at a time in an unsupervised manner. Each layer is trained as an RBM, which learns to represent the features passed from the layer below. This layer-wise pre-training helps initialize the weights of the network in a way that makes the subsequent FT phase more effective.

After pre-training, a DBN can sample from the representations it has learned to generate new data similar to the original dataset. We can also use **supervised learning** (**SL**) approaches to fine-tune the network for more specific applications.

Each of the techniques and algorithms I've outlined so far in this chapter are what I would consider to be fundamental milestones and conceptual building blocks that contribute to the development of the more advanced GenAI applications we see today. The next two approaches I will outline are significant steps forward in the evolution of GenAI. The first I will introduce is the concept of AEs.

### AEs

Most – if not all – of the datasets we use for ML use cases represent specific concepts with intrinsic properties (or features); for example, people, cars, medical images, or even more specific concepts, such as people who boarded the Titanic or items purchased on a company's website.

In the case of discriminative ML use cases, the models generally try to predict some kind of output based on the values of the features associated with each instance of the concept being represented. For example, which customers will likely purchase a particular product based on their prior purchasing history?

In the case of GenAI, on the other hand, the models are often expected to generate new data points that validly represent the concept originally represented in the training dataset. For example, having been trained on images of many different concepts, such as cats and motorcycles, the model may be asked to draw a cartoon image of a cat on a motorcycle. To do this, the model needs to "understand" what those concepts are.

Earlier in this chapter, we introduced the topics of embeddings and the latent space, in which latent representations of the data points in our dataset are defined. The process of creating an embedding in the latent space can be referred to as **encoding**, and AEs are a type of ANN used for UL of efficient representations (encodings) for our datasets. At a high level, an AE takes input data, converts it into a smaller, dense representation that captures the essence of the data, and then reconstructs the input data as closely as possible from this representation. Let's dive into how they work.

### How AEs work

AEs are generally made up of three main components: the **encoder**, the latent space (also referred to as the **bottleneck**), and the **decoder**, as depicted in *Figure 15.4*:

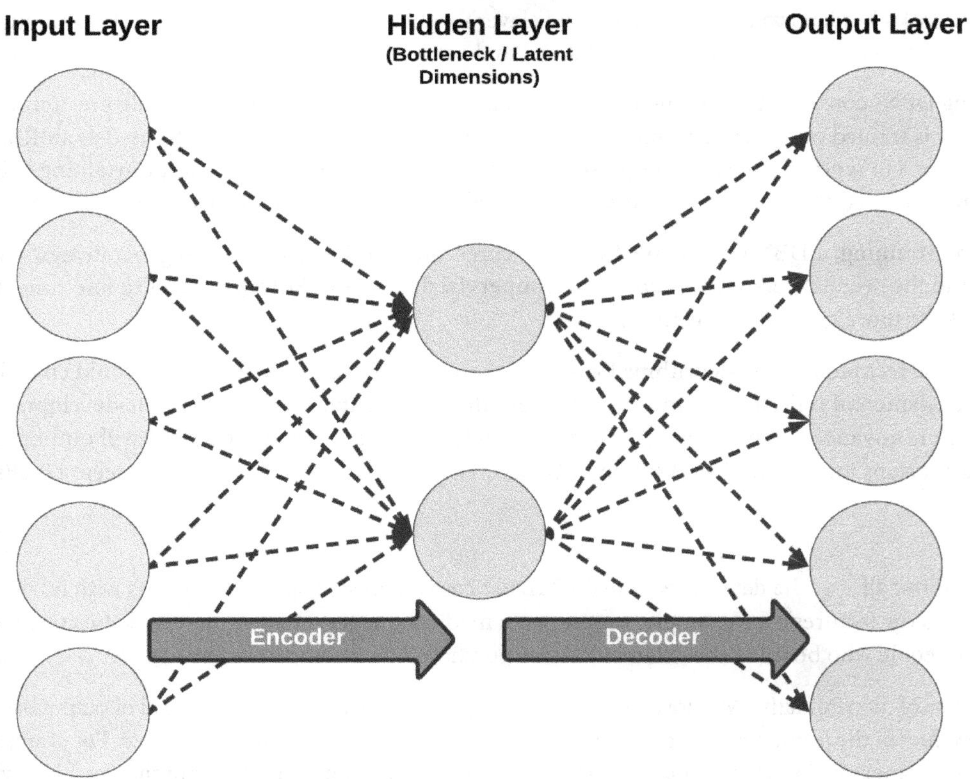


Figure 15.4: AE architecture

During the forward pass in an AE, each input data point is passed through the encoder part of the network. This section of the network compresses the input into a smaller, dense representation in the latent space (hidden layer). The encoded data is then passed through the decoder part of the network, and the decoder tries to reconstruct the original input data from the compressed representation. This is the interesting difference between AEs and traditional, discriminative models – that is, while traditional models generally try to predict an output (*Y*) based on an input (*X*), AEs try to predict (or generate) the original input (*X*).

Just as we did in traditional NNs, we can calculate the difference between the output value and the expected value (*X*), which is referred to as the **reconstruction error**, and then we can use backpropagation to update the network and continue the training cycle as usual. The quality of the reconstruction depends on how well the AE has learned to represent the data in the latent space.

During this training process, the encoder learns to retain only the most relevant features of the data, effectively learning a new way to represent the input data in a smaller-dimensional space. The latent space then contains the compressed knowledge that the network has learned about the data – that is, the essence of the data. To again paraphrase a description by my colleague, Jeremy Wortz, the concept of bottleneck features relates to the quality of the embeddings; since we are typically "squeezing" our most critical information between the encoder and decoder, feature importance is implicitly optimized.

Note, however, that while AEs can learn to reconstruct the original dataset and thereby learn how to perform dimensionality reduction to accurately represent the data in a lower-dimensional space, they are not designed to generate new data. To enable that functionality, we need to bring probability distributions into the image, and this is what has given rise to the development of **variational AEs (VAEs)**, which I'll describe next.

### VAEs

VAEs extend the AE concept by introducing probabilistic approaches to enable the generation of new data points that are similar in structure to the training data. Building on what we've discussed about how AEs work, the easiest way to describe VAEs is to highlight their subtle differences from regular AEs.

As with traditional AEs, VAEs consist of an encoder and a decoder. However, rather than mapping input data to a fixed, deterministic vector as in regular AEs, the encoder in a VAE maps the input data to a probability distribution over the latent space. That's a lot to take in, so let's clarify that further.

As we learned in the previous section, regular AEs encode the input data into a vector that contains latent features representing the original inputs. However, in the case of VAEs, instead of learning a single point in the latent space for each data sample, the encoder learns probability distributions. To be more precise, instead of outputting specific feature values in the latent space, the encoder outputs parameters (such as mean and standard deviation) that describe the probability distribution for features in the latent space. This distribution represents where the encoder believes the input data should be encoded. This approach introduces randomness or variability into the encoding process, which is important for potentially generating new data points that are similar to the input data, and not just reconstructions of the original data.

The following are the high-level steps that are performed when an input is passed through the VAE:

1. The encoder provides the parameters of a probability distribution within the latent space.
2. A point is sampled from this learned distribution.
3. This sampled point then gets passed through the decoder, which tries to reconstruct the input data from this probabilistic encoding.

The goal, again, is to minimize the difference between the original input and its reconstruction, similar to regular AEs.

However, in addition to the reconstruction error, VAEs have an extra term in their loss function referred to as the **Kullback-Leibler** (**KL**) divergence, which measures how much the learned distributions for each input deviate from a standard normal distribution. Without getting into too much detail regarding the mathematics involved, the important thing to understand is that the KL divergence regularization enforces smoothness and continuity in the latent space, which makes the models more robust and can help reduce overfitting.

We covered some complex details in this section, but the key takeaway is that VAEs introduce probabilistic mechanisms that enable them to go beyond simply reconstructing the input data and begin to generate new, similar data points.

We are now firmly getting into the realm of generative models, and the next approach I will outline is that of GANs.

### GANs

We briefly introduced GANs in *Chapter 9*, and in this section, we will explore them in more detail. You may be familiar with the concept of "deepfakes," in which AI is used to create photorealistic images or movies or audiorealistic tracks. By "photorealistic," we mean synthetic data that looks like real photographs or video, and by "audiorealistic," we mean synthetic data that sounds like real audio recordings. GANs are a popular mechanism for generating synthetic data that mimics real-world data. As with any technology, GANs could be used for malicious purposes, such as generating deepfakes of real people without their consent, but they have many useful applications that we will cover shortly. However, let's first dive into what GANs are and how they work.

#### High-level concepts – the generator and the discriminator

The name "generative adversarial networks" (Goodfellow et al., 2014) may sound somewhat abstract, but it perfectly describes the concepts used in this approach to GenAI. In short, GAN implementations consist of two NNs that work in an adversarial manner (that is, they work against each other), in which one of the networks generates synthetic data, and the other network tries to ascertain whether the data is real or not. The network that generates the data is called, not surprisingly, the **generator**, and the network that tries to ascertain the realness of the data is called the **discriminator**. The main premise is that the generator tries to fool the discriminator into believing that the data is real.

One could imagine that having an effective generator is the most important requirement for a GAN, but it is also essential to have an effective discriminator because the discriminator can be seen as the **quality control** (**QC**) mechanism for generated data. If you have an ineffective discriminator, it could easily be fooled by data that does not accurately mimic real data. The more effective your discriminator is at identifying fake data, the harder your generator will need to work to create realistic data. Therefore, both sides of the adversarial partnership need to be trained effectively in order for the GAN to create high-quality data.

The analogy that is often used to describe this process is to imagine a person who wants to create forgeries of expensive art pieces (that is, a forger) and another person who is dedicated to identifying whether a piece of art is real or a forgery (that is, an art expert). In the beginning, the forgeries might be amateur attempts that are easy to identify as fakes. However, as the forger refines their work and learns to create more convincing forgeries, the art expert must become more skilled at identifying minute nuances that distinguish real art from fake art pieces. This cycle then repeats until, ideally, the generator creates data that is indistinguishable from the real data.

## Diving deeper – GAN training process

In a GAN, both networks are usually **convolutional NNs** (**CNNs**). The discriminator is usually a binary classifier that classifies the data as either real or fake. This means that its training process involves many of the same steps that are already familiar to us from earlier chapters in this book, in which we trained classifier models.

As an example, let's imagine that we want to generate images of cats. The discriminator could learn to identify images of cats by being trained on a labeled dataset. In this dataset, some of the inputs are real images of cats, and they are labeled accordingly.

As we briefly mentioned in *Chapter 9*, the generator and the discriminator engage in a kind of adversarial game. In this game, the objective of the discriminator is to *minimize* the error rate for classifying images of cats (that is, it wants to accurately identify images of cats as much as possible). On the other hand, the generator is trying to *maximize* the discriminator's error rate. Due to the opposing objectives of each of the game's participants, where one participant wants to minimize a certain metric and the other participant wants to maximize that same metric, this is referred to as a **minimax** game.

In addition to training the discriminator on the labeled training dataset, the discriminator also receives outputs from the generator and is asked to classify those outputs, as depicted in *Figure 15.5*:

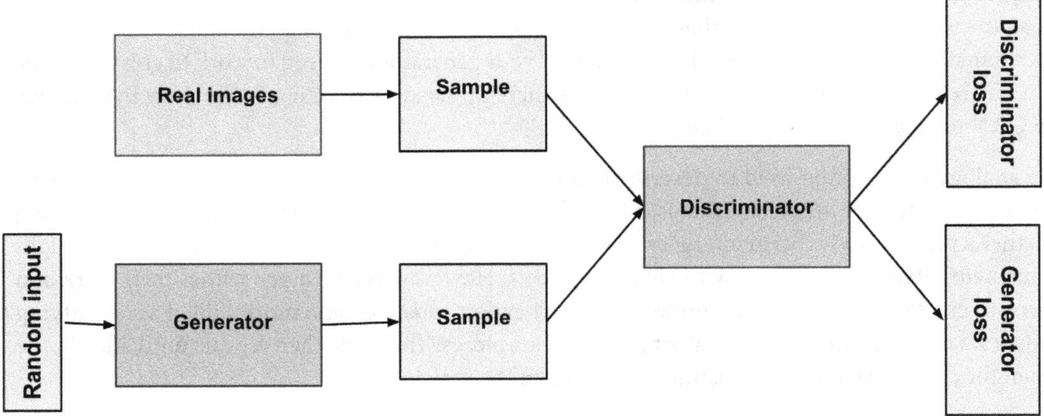


Figure 15.5: GAN (source: https://developers.google.com/machine-learning/gan/gan_structure)

If the discriminator thinks a data point is real when, in fact, it was created by the generator, that counts as an error. This, therefore, increases the error rate for the discriminator, which is a win for the generator. Conversely, if the discriminator accurately classifies the data point as fake, then it reduces the error rate for the discriminator (which is, of course, a win for the discriminator).

While the discriminator uses a typical mechanism such as gradient descent to learn from its mistakes and minimize the loss, the generator can also learn from the discriminator's mistakes, and adjust its weights in a reverse process that aims to increase the discriminator's error rate. This is the novel approach used by GANs, in which the generator doesn't receive direct labels of "good" or "bad" during training, but instead receives feedback on the discriminator's performance. If the discriminator accurately identifies a fake, then the gradients are sent back to the generator so that it can update its weights to create an output that better fools the discriminator in the next round.

During this process, the generator learns the underlying probability distribution of the real data and becomes more adept at generating new data that aligns with that probability distribution (that is, new data with similar properties as the real data). It's also important to understand that each network takes turns in the training process. When the generator is being trained, the discriminator's weights are frozen, and vice versa.

By pitting the models against each other in this fashion, they must continually get better in order to outperform each other. As the discriminator gets better at identifying fakes, the generator learns to generate more realistic data to fool the discriminator, and so on.

GANs gained a lot of popularity since they were first created by Ian Goodfellow and colleagues in 2014, and multiple different types of GAN implementations have emerged, such as **conditional GANs** (**cGANs**), CycleGANs, StyleGANs, **deep convolutional GANs** (**DCGANs**), and progressive GANs. A comprehensive discussion of all of these variants would constitute more detail than is necessary in this book. I encourage you to research these variants further if you have a specific interest in them.

> **Important note – look out for "mode collapse"**
>
> In addition to the typical challenges faced when training many types of NN architectures, such as exploding and vanishing gradients, GANs also often suffer from a specific challenge that is related to their adversarial training process. This problem is called "mode collapse," and it refers to a situation in which the generator may start producing a limited variety of outputs, especially if those outputs successfully trick the discriminator. Since the generator's primary goal is to fool the discriminator, it may learn to generate patterns that are effective in doing so but do not accurately represent the target data distribution. This is an ongoing area of research, in which new mechanisms are being introduced to combat this phenomenon.

Next, let's discuss some common applications of GANs.

### Applications of GANs

In addition to generating photorealistic images and audiorealistic data, GANs can be used for other interesting use cases, such as style transfer and image-to-image translation. For example, some online services enable you to upload a photo and convert it into a cartoon or make it look like an oil painting in the style of a famous artist such as Van Gogh. While these are fun applications, GANs are also utilized in various use cases such as text generation, new drug discovery, and other types of synthetic data generation.

We'll discuss the importance of high-quality synthetic data later, but next, I'd like to introduce another popular GenAI approach, referred to as **diffusion**.

#### *Diffusion*

Diffusion is utilized for many of the same kinds of use cases as GANs, such as image and audio generation, image-to-image translation, style transfer, and even new drug discovery. However, diffusion uses a different approach and often provides more favorable results for some use cases. We'll dive into some of the differences between diffusion models and GANs in this section, but first, as always, let's learn about what diffusion is and how it works.

## High-level concepts – noising and denoising

At a high level, diffusion consists of two main steps:

1. Adding noise to images
2. Reversing the process – that is, removing noise to work back toward the original image

I'll describe these concepts in much more detail, starting with clarifying what it means to add noise to an image. *Figure 15.6* shows an image consisting of noise:

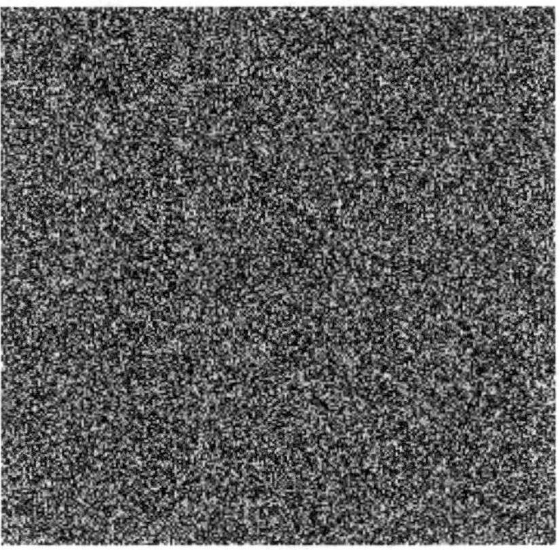

Figure 15.6: Image of noise (source: https://commons.wikimedia.org/wiki/File:256x256_Dissolve_Noise_Texture.png)

Some of you may be familiar with the kind of image shown in *Figure 15.6* if you've ever seen noise displayed on a TV screen when it is not tuned to a channel that provides a image signal. This image consists purely of noise, in which we cannot make out any discernable shapes. However, if we were to take a high-quality image and add some noise to it, then it would become somewhat blurry, but we could still make out what's in the image, as long as not too much noise is added. The more noise we add, the more difficult it becomes to identify the contents of the image.

This concept is used in the **noising** process when training diffusion models. We take images and add small amounts of noise to them to make it progressively more difficult to identify what's in the images. This is also referred to as the **forward diffusion** process.

The **reverse diffusion** or **denoising** process then tries to take a noisy image and work back toward the original image. This might sound a bit pointless – that is, why bother adding noise to images and then trying to train a model to learn how to remove noise to generate the original images? Well, just as in the case of GANs, we are training our model to understand the probability distribution of the original dataset. Diving into a bit more detail, our model actually learns to predict noise in the input. The next step of the process, then, is to simply remove that noise from the input in order to estimate or generate the desired denoised output.

The reason this is called diffusion may be somewhat self-explanatory, but more specifically, the name relates to a concept in the field of non-equilibrium thermodynamics. That's a bit of a mouthful, and, don't worry – we're not going to dive into the physical concept except to briefly introduce why it is named as such.

The analogy often used to explain this concept is to think of a drop of ink being added to a bucket of water. When the ink is added to the water, at first, it occupies a small space in a specific location in the bucket. However, over time, the ink dissipates throughout the bucket of water. Soon, it has spread throughout the bucket, and the bucket then contains a mixture of water and ink. The color of the contents of the bucket may be different from the color that existed before you added the ink, but it's no longer possible to pinpoint the specific location of the ink in the bucket. This is the forward diffusion process, in which the ink diffuses throughout the water in the bucket.

In the physical world, this process of diffusion is usually impossible to reverse – that is, no matter what you try to do, you will not be able to separate the ink from the water and condense the drop of ink back into the original position it occupied when it was first added to the bucket.

In the case of diffusion in ML, however, we try to reverse the process and get back to the original input state.

## Diving deeper

It's important to understand that noise is added to the data in a controlled manner. We don't simply add completely random noise, but instead, we add what is referred to as **Gaussian noise**, meaning the noise is characterized by the "normal" (or "Gaussian") probability distribution. As a refresher, a Gaussian distribution is represented by the familiar "bell curve" as depicted in *Figure 15.7*:

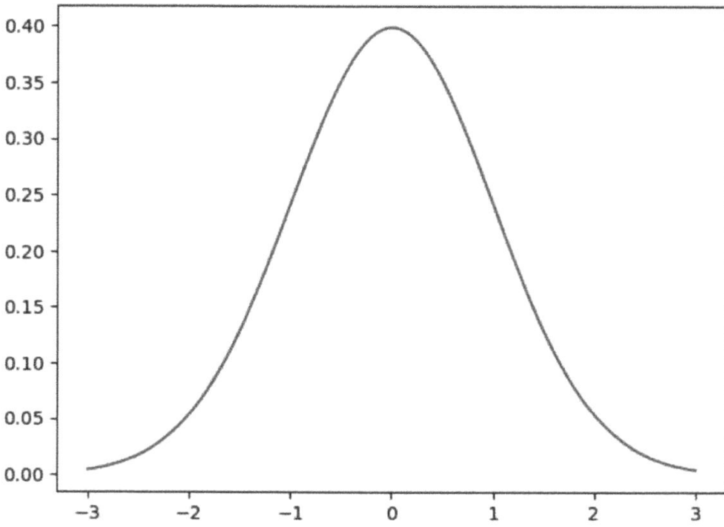


Figure 15.7: Gaussian distribution

As we can see in *Figure 15.7*, the Gaussian distribution is symmetrical, with its mean at the center of the bell curve. The width of the curve represents the variance (or its square root, the standard deviation) of the distribution – that is, how far from the mean we are likely to see data points that fit the distribution. In a Gaussian distribution, data points that are most likely to occur appear closer to the mean, while rarer data points are further from the mean. In *Figure 15.7*, the mean is 0, and the standard deviation is 1, which is a special type of Gaussian distribution referred to as the **standard normal distribution**. By using normally distributed noise, we can easily change the intensity of noise by tweaking the variance of the distribution.

In addition to controlling noise by ensuring it fits the normal distribution, we also introduce noise in a controlled manner by incrementally adding noise to our source images during training, rather than adding too much noise all at once. The addition of noise in this manner is performed according to a **schedule**, in which more noise is added at different time intervals in the process. Remember that, as we discussed in earlier sections of this chapter, randomization is important in generative processes because we don't just want to produce exact images from the original dataset but instead want to produce similar images.

To introduce some controllable randomization into the process, the mean and the variance of noise added in each step are different, and we also add different amounts of noise at various parts of the schedule. Each time we add noise, we form a step in a Markov chain. As you may remember from earlier in this chapter, each step in a Markov chain is dependent only on the step that directly precedes it, and this is important to understand in the context of the reverse diffusion process. As we add noise in each step (during the forward diffusion process), we are training our model to identify (or predict) noise that has been added. Then, during the reverse diffusion process, we start with a noisy image, and we want to try to generate the image in the previous step in the Markov chain and progressively work our way through the chain in that manner until we get back to an approximation of the original image.

To describe this process in more detail, imagine that we start with a image of a cat, and we progressively add noise in each step until we finally end up with a image similar to the image depicted in *Figure 15.6*, which consists almost purely of noise. It would be very difficult to try to jump from such an extremely noisy image back to a image of a cat, so we instead train our model on how to progressively work backward through each step in the Markov chain, predicting the small amount of noise that was added in each step. As our model gets better and better, it can more clearly distinguish between noise and the underlying probability distribution of the original dataset. This understanding of probability distributions allows us to sample new images by starting with pure noise and iteratively denoising them according to what the model has learned.

At the beginning of this section, I promised that I would outline some important differences between diffusion and GANs. Let's take a look at those differences next.

### Differences between diffusion and GANs

I mentioned the concept of mode collapse in the context of training GANs and how that can introduce instability into the training process. Diffusion uses a more stable training process, but diffusion models require many steps to generate data, which can be computationally intensive. Some recent advances aim to reduce the number of required steps and computational costs, but this is still a notable consideration for diffusion models.

Similarly, at inference time, generating samples from diffusion models can be computationally expensive, whereas GANs can generate samples more quickly because they require only a single forward pass through the generator network.

Which approach is best, then? Well, it's a matter of selecting the right tool for the job based on the business requirements for a specific use case.

Next, it's time to move on to discussing perhaps the crowning glory of the GenAI world in recent times: LLMs.

# LLMs

This is another case in which the name of the technology very accurately describes what the technology is; LLMs are large models that are particularly useful for language-based use cases, such as the summarization of large amounts of textual data, or chatbots that can have a conversation with a human.

Behind the scenes, they use statistical methods to process, predict, and generate language based on the context provided to them. They are trained on diverse datasets that include large amounts of textual information, from books and articles to websites and human interactions, enabling them to learn language patterns, grammar, and semantics.

How large are they? Well, some of the latest models at the time of writing this in February 2024 consist of billions or even trillions of parameters. That's pretty huge! Somebody might read this book 20 years from now and laugh at the fact that we considered these sizes to be huge, but these are the biggest models on the planet at the moment, and they constitute an enormous advancement from any models that have existed before now.

Before diving into details on how LLMs are created, let's take a look at some historical milestones that have led to their evolution.

## Evolution of LLMs

The history of language models began with simple rule-based systems that used hand-coded rules to interpret and generate language. It turns out that trying to hand-code all of the complex rules of human language is quite impractical, but the research in this field had to start somewhere.

Next, statistical methods such as N-gram models used probabilities to predict the likelihood of a sequence of words. They were trained on large bodies of text and could capture more language nuances than rule-based systems, but they still struggled with long-term dependencies in text.

The next step forward was to use models such as HMMs and simple NNs for language tasks, and while these models offered better performance by learning more advanced patterns in data, they were still limited in complexity.

Major breakthroughs began when scientists started applying **deep NNs** (**DNNs**) to language use cases. In *Chapter 9* and *Chapter 14*, we discussed the evolution of RNNs, LSTM networks, and Transformers, and how they each enabled progressively more complex language processing. The pivotal event that occurred in the industry was when Google invented the Transformer architecture (Vaswani et al., 2017), which has become the primary technology used in all of the biggest and most advanced LLMs in the industry today.

When we discussed the overall evolution of AI/ML in *Chapter 1*, I mentioned that it was not only the development of more complex ML algorithms that led to the kinds of breakthroughs we've seen in recent decades but also advancements in computing power and the proliferation of available data for training models. These factors all work together to form a kind of ecosystem in which we continue to advance all of these technologies in concert.

Next, we'll dive into how LLMs are created.

## Building LLMs

Let's get one thing out of the way, right from the beginning; it's unlikely that you or I could create our own LLMs from scratch, for two main reasons:

- Training LLMs requires obscenely large amounts of data. Some of the LLMs you or I would interact with are trained on the entire internet, for example. This is one of the things that makes them so powerful and knowledgeable; they are trained on all publicly available data that humans have ever created, in addition to some private and proprietary datasets owned or sourced by the companies that trained them.
- It is generally extremely expensive to train LLMs due to the sheer amount of computing power needed. I'm referring to using thousands of high-performance accelerators, such as the latest and greatest GPUs and TPUs, for months on end to train these models, and those things don't come cheap!

As a result, most people and companies will use LLMs that have already been **pre-trained**, which I will describe next.

### *The LLM training process*

While the companies that build commercially available LLMs almost certainly use a lot of secret magic behind the scenes that they are unlikely to share externally, the following high-level steps are generally involved in the LLM training process:

- Unsupervised or semi-supervised pre-training
- Supervised tuning

Let's take a look at each of these in more detail.

#### Unsupervised or semi-supervised pre-training

Remember from our activities earlier in this book that supervised training requires labeled datasets, which can be cumbersome and expensive to source or create. Bearing in mind that LLMs are often trained on enormous bodies of data such as the entire internet, it would be impossible to label all of that data. Instead, a very clever trick is used to train LLMs in such a way that they can learn from these large bodies of text without the need for explicit labeling, and that trick is called **masked language modeling** (**MLM**).

MLM is actually a very simple yet effective concept; it's essentially a game of "fill in the blanks." You may remember playing this game as a child or encountering it in exam scenarios. Quite simply, we take a sentence and blank out (or mask) one or more of the words, and the task is then to guess what the missing words should be. For example, consider the following sentence:

*"You really hit the nail on the [BLANK] with that last comment."*

What do you think would be the most appropriate word to use for filling in the *[BLANK]* portion of that sentence? If you're familiar with the term, "hit the nail on the head," then you may have guessed that "head" would be the best word to use, and you'd be correct in guessing that.

This approach can be extremely effective in training LLMs because we can feed millions of sentences to the LLM, randomly masking out various words and asking the LLM to predict what the correct words should be. The beauty of this approach is that the masked-out words become the labels, so we don't need to explicitly label the dataset. This is why we refer to this process as UL or **semi-supervised learning** (**SSL**). For example, after the LLM predicts the word to use, we can reveal the word that the sentence actually contained in the masked-out position, and the model can then learn from that. If the model predicted a different word, it would count as an error, and the loss function would reflect this accordingly during training.

The important thing to note is that the model gets to see billions of different sentences during training, and by doing so, it sees words being used in various contexts. Over time, it builds up an understanding of what each word means and which other words it commonly appears alongside. Conceptually, it can build up a graph of how various words relate to each other, as depicted in *Figure 15.8*:

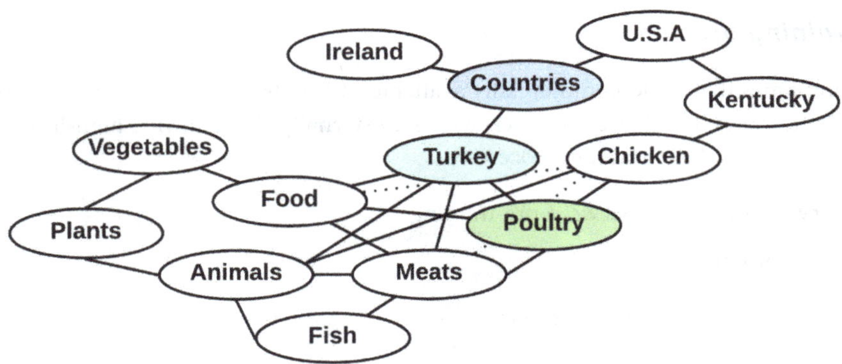


Figure 15.8: Graph of word associations

In *Figure 15.8*, we can see how various words in the English language relate to each other. Notice how some words may be ambiguous, such as the word "Turkey," which could relate to the country, to the animal, or to food. Consider the following sentences:

- "I really like eating turkey."
- "Last year, we went on vacation in Turkey."
- "I really liked the food in Turkey."

Assume that the LLM has seen all of those sentences during training. By seeing these sentences (and billions more) and learning how words are used in various contexts, the LLM forms a probabilistic understanding of the meaning (or **semantic context**) of the words. For example, how does it know that the word "Turkey" might refer to a country based on a sentence about people going on vacation there? Well, from seeing the word "vacation" in millions or billions of sentences, it comes to understand that people go on vacation in places, so in this context, "Turkey" must be a place. It also would have seen other sentences containing the words "Turkey" and "immigration," and it would have developed an understanding that immigration refers not only to places but, more specifically, to countries. It would also have learned what the word "country" means and that its plural form is "countries," based on seeing those words in many other contexts. Similarly, it learns that the word "eat" refers to food, so in the sentence "I really like eating turkey," it understands that "turkey" must be a type of food.

I've presented just a handful of simple sentences here, and we can already see a proliferation of different potential combinations of words and how they relate to each other. Imagine the complexity of the combinations that exist when we bring every sentence on the internet into scope. This is what LLMs learn: extremely complex webs of associations and the underlying meaning of the concepts those words represent. This is an enormous step forward in terms of intelligence that LLMs bring to the AI industry, and this is actually how our own minds learn to understand words, also, which is fascinating. As children, we don't just learn individual words, but we learn words as representations of concepts, usually by forming graphs of associations. For example, imagine you were a young child who was trying to climb up a tree, and your parent said, "Don't climb that tree, or you might fall and injure yourself." You may not have heard all of those words before, but from hearing that sentence, you might learn about the concept of a tree, or of climbing, or falling, or injuring, and you also may intuit that falling and injuring are undesirable things to happen.

In addition to masking words within sentences, another method called **next sentence prediction** (**NSP**) may also be used to train our LLMs to not only predict words but, as the name suggests, also to predict entire sentences based on the provided context. In this case, the LLM is presented with pairs of sentences, and sometimes these sentences have a natural sequential relationship (for example, two consecutive lines from an article), or in other cases, they are unrelated sentences. The LLM's task is to determine if the second sentence logically follows from the first, and by doing this, it learns to discriminate between logical and incoherent sequences of sentences. This pushes the LLM to reason about context across multiple sentences, which helps it generate coherent text in longer responses to requests.

It's important to note that there's an additional layer of abstraction in how the LLMs learn. Earlier in this chapter, we discussed embeddings and latent space. Generally, the kinds of associations I've just outlined occur in the latent space. The latent space is the LLM's representation of the world and all of the various concepts and associations it learns. Concepts with similar meaning or semantic context may be located nearer to each other in the latent space.

After the LLM has learned everything it can learn from the available data via the process I've just described, the pre-training phase is complete. At this point, the LLM has built its latent space representation of the world, and the next step is to teach it how to be useful in responding to requests from humans or other machines. Many different processes can be used for this purpose, but I'll start with the common practice of supervised tuning.

## Supervised tuning of LLMs

As we know by now, supervised training means that we have a dataset that contains labels that can be used to teach a model. For example, if we want our model to perform sentiment analysis, we can label phrases in the dataset with the sentiments expressed by those phrases, such as positive, negative, or neutral. Similarly, for summarization, we could show the model examples of good summarizations, and the model could learn from those examples.

While enormous datasets are usually needed for the pre-training phase, supervised tuning of LLMs requires smaller datasets that are carefully curated and labeled for the particular task at hand. Considering that LLMs usually contain a lot of knowledge from the pre-training phase, surprisingly good tuning results can come from just a few (or perhaps a few hundred) examples in the tuning dataset. Again, this is similar to how humans incrementally learn new skills. For example, imagine there are two people, one of whom has never learned to drive, and another who has been driving cars for 20 years. Now, we want each of those people to learn how to drive a large truck by next week.

Who is most likely to succeed? The person who has been driving cars for 20 years already has a lot of knowledge of the rules of the road and how to operate vehicles, so learning how to drive a large truck will just require a relatively small amount of incremental learning in comparison to the person who has never driven any vehicles before.

It's also important to understand that there are various approaches or levels of tuning that we can apply. For example, **few-shot learning** (**FSL**) refers to a practice in which we provide just a few examples to the LLM. We could even provide these examples along with our prompt (that is, our request to the LLM) rather than needing to provide a dedicated training dataset. This can be effective for some tasks, but, of course, it is quite limited since we are only providing a few examples. When examples are provided along with a prompt, this is an example of prompt engineering, which is an emerging science in itself that focuses on how to craft requests to LLMs in ways that produce the best results. In these cases, the underlying LLM generally does not get retrained on the examples we provide. Instead, the LLM simply uses the examples provided to tailor its responses in alignment with those examples.

On the other side of the spectrum is **full FT** (**FFT**). This level of tuning could require thousands of examples, and the model's parameters get updated based on those examples (that is, the model learns and incorporates the new data into its core structure).

Then, there are levels in between, such as **adapter tuning**, in which **adapter layers** are inserted into the original model, and only those layers get tuned and updated while the original model weights are frozen. Another type of tuning may focus only on updating the output layer of the model while leaving the rest of the model untouched.

On a slightly separate but related note, I want to briefly introduce the topic of **reinforcement learning from human feedback** (**RLHF**). Rather than updating the LLM with more data, this method mainly focuses on teaching the LLM to respond in ways that are more favorable to humans, such as aligning with human values. In this approach, the LLM may provide multiple responses to a prompt, and a human can select (and therefore label) which response is preferred, based on nuances of human communication and culture, such as tone, sentiment, ethical considerations, and many other factors. We introduced the concept of **reinforcement learning** (**RL**) in *Chapter 1*. In the case of RLHF for tuning LLMs, the human feedback is used as the reward signal, teaching the LLM to generate outputs that better align with human preferences, even if there might be multiple technically "correct" ways to respond.

We will explore tuning in more detail in the next chapter, but before we move on to that chapter, let's reflect on what we've learned so far.

## Summary

We started this chapter by introducing the fundamental topics that underpin GenAI, including concepts such as embeddings and latent space. We then described what GenAI is and how it contrasts against "traditional AI," whereby traditional AI typically tries to predict a specific answer, such as a revenue forecast based on historical data or identifying whether a image contains a cat, but GenAI goes beyond those kinds of tasks and creates new content.

We dived into the role of probability in GenAI versus traditional AI, and we discussed how traditional AI often uses conditional probability to predict the values of a target variable based on the values of features in the dataset. On the other hand, GenAI approaches typically try to learn the joint probability distribution of both the features and the target variable.

Next, we explored the evolution of GenAI and various model development milestones that led to the kinds of models we use today. We began this exploration by covering early developments such as Markov chains, HMMs, RBMs, and DBNs. We then covered more recent developments such as AEs, GANs, and diffusion.

Finally, we dived into the world of LLMs. We highlighted major milestones in their evolution by refreshing what we had learned in previous chapters about RNNs, LSTM networks, and Transformers. We then discussed how LLMs are trained, including the kinds of mechanisms used in the pre-training phase, such as MLM and NSP, and how FFT can be used to further train the models for specific use cases.

Although this is an introduction chapter for the GenAI section of our book, we've covered a lot of important concepts in a considerable amount of depth, and this will give us a strong foundation for the remaining chapters.

With that in mind, let's continue our journey into the world of GenAI.

# 16
# Advanced Generative AI Concepts and Use Cases

Now that we've covered the basics of **generative AI (GenAI)**, it's time to start diving deeper. In this chapter, we will cover more advanced topics in the field of GenAI. We'll begin by learning about some techniques to tune and optimize generative models for specific domains or tasks. Then, we'll dive into more detail on the important topics of embeddings and vector databases and how they relate to a relatively new pattern of using **retrieval-augmented generation** (**RAG**) to ground our **large language model** (**LLM**) responses in our own data. Next, we'll discuss multimodal models, how they differ from standard, text-based LLMs, and the kinds of use cases they support. Finally, we will introduce LangChain, which is a popular framework that enables us to design complex applications that can build on the functionalities provided by LLMs. Specifically, this chapter covers the following topics:

- Advanced tuning and optimization techniques
- Embeddings and vector databases
- RAG
- Multimodal models
- GenAI model evaluation
- LangChain

Let's dive right in and start learning about advanced tuning and optimization techniques!

> **Note**
>
> The term *LLM* has become almost synonymous with the field of GenAI in general, but considering that **LLM** stands for **large language model**, the term technically relates them to language processing. It's important to note that there are other types of GenAI models, such as some image generation and multimodal models, that are not strictly language models. Additionally, a more recent concept of **large action models** or **large agentic models** (**LAMs**) has emerged, which combine the language understanding and generation capabilities of LLMs with the ability to perform actions (such as planning and booking a vacation) by using tools and agents to interact with their environment. However, for simplicity, throughout the rest of this book, I will use the terms *LLM* and *GenAI model* interchangeably.

# Advanced tuning and optimization techniques

At the end of the previous chapter, we discussed LLMs and how they are trained and tuned. I mentioned some of the tuning approaches at a high level, and in this section, we will dive deeper into how we can tune LLMs to more adequately address our specific needs. Let's set the stage by outlining how we interact with LLMs in the first place, which we generally do via **prompts**.

> **Definition**
>
> A prompt is a piece of text or instruction that we provide to an LLM to guide its response or output. It tells the LLM what to do and, in some cases, provides guidance on how to do it; for example, "`summarize this financial document, specifically focusing on details relating to company performance in Q4, 2023`."

The first LLM tuning technique we'll explore is **prompt engineering**.

## Prompt engineering

Prompts are the most straightforward method we can use to tune an LLM's outputs to our specific needs. In fact, during the early days of the GenAI popularity explosion that happened in late 2022 and early 2023, you may recall seeing news headlines about a new type of extremely high-demand role that was emerging, called **prompt engineer**. In this section, we'll discuss what prompt engineering is and how it can be used to improve the outputs of GenAI models, which will help provide context on why there was such a sudden increase in demand for talented people in this space.

To begin our discussion, consider the following prompt:

```
Write a poem about nature.
```

This is a very simple prompt, and it does not provide much information to the LLM about what kind of poem we would like it to write for us. If we want a specific type of output, we could include additional instructions in our prompt, such as the following:

```
Write a poem about nature with the following properties:
Format: sonnet
Theme: how the changing seasons affect human emotions
Primary emotion: the poem should focus particularly on the transition
from summer to autumn and the sense of longing that some humans feel
when summer is coming to an end.
```

The latter prompt will produce a much more specific poem that abides by the parameters we've outlined. Feel free to go ahead and try this out for yourself. Visit the following URL, enter each of those prompts, and see how different the responses appear: `https://gemini.google.com`.

> **Note**
>
> You can use the aforementioned URL to test all of the prompts in this book.

This is a rudimentary form of prompt engineering that simply includes additional instructions in the prompt to help the LLM better understand how it should respond to our request. I'll explain additional prompt engineering techniques in this section, beginning by outlining some standard best practices for crafting effective prompts.

### *Core prompt design principles*

The following are some principles we should keep in mind when creating prompts:

- **Be clear**: We should be clear about what we want the LLM to do and avoid using ambiguous language
- **Be specific**: Using specific instructions can help us get better-quality results
- **Provide adequate context**: Without appropriate context, the LLM's responses might not be as relevant to what we are trying to achieve

I included some of those principles in the latter example prompt in the previous section, and I'll outline them more explicitly here. In most cases, it's important for prompts to be as clear and specific as possible due to the sheer and ever-growing versatility of LLMs. For example, if I ask an LLM to write a story about happiness, there's almost no way of knowing what kinds of details it will come up with. While this is one of the wonderful properties of LLMs, it can be challenging when we want an LLM to provide results within specific parameters.

Before I provide additional examples regarding clarity and specificity, consider use cases in which we might not want to be clear and specific, such as when we're not yet sure of our exact requirements, and we want the LLM to help us explore various potential options. In that case, we could start by sending a broad prompt to the LLM, such as the following:

```
Suggest ideas for a new business that will help me make money.
```

In response to a broad prompt such as this one, the LLM will likely come up with all kinds of random ideas, from baking cakes to developing video games that help kids learn mathematics.

After getting some initial outputs from the LLM, we could then iteratively refine the process by diving in on certain ideas more specifically. For example, if we really like the idea of developing video games for kids, we could follow up with a more specific prompt, such as the following:

```
I want to develop a video game that helps kids to learn mathematics.
The target audience will be kids between ten and fifteen years old,
and the game will focus on algebra and precalculus. List and explain
the steps I need to take to start this business.
```

As we can see, this prompt is much more specific than the previous prompt and will likely provide more targeted and actionable responses. This updated prompt abides by all of the principles we've outlined because it's specific in terms of exactly what we want to achieve (starting a business to create a video game), it provides context such as the intended content of the game and the target audience, and it's clear in terms of how we want the LLM to respond (that is, list and explain the steps).

Sometimes, we need to provide large amounts of context, and we will discuss how to do that later in this chapter. Next, however, let's explore how we can use chaining to refine the outputs we get from an LLM.

### *Chaining*

The process I described in the previous section, in which we use the output of the first prompt as an input to a subsequent prompt, is called **prompt chaining**. This can be done via an interactive process such as a chat interface, or in an automated fashion using tools I will describe later, such as LangChain. There's also a framework called ReAct, in which we can chain multiple actions together to achieve a broader goal, which I'll describe next.

#### ReAct and agents

I briefly touched upon the concept of LAMs at the beginning of this chapter, and I'll cover it in more detail here.

**ReAct** stands for **Reasoning and Acting**, and it's a framework for combining the reasoning abilities of LLMs with the ability to take actions and interact with external tools or environments, which helps us build solutions that go beyond simply generating content to achieve more complex goals. Examples of external tools include software, APIs, code interpreters, search engines, or custom extensions.

By using ReAct to go beyond generating content, we can build what are referred to as **agents**. The role of an agent is to interpret the user's desired outcome and decide which tools to utilize to achieve the intended goal. Here, I will again highlight the example of planning and booking a vacation, which might include booking flights, accommodation, restaurant reservations, and excursions, among other tasks.

Chaining is not to be confused with another prompt engineering approach, called **chain-of-thought** (**CoT**) prompting, which we can use to help LLMs work through complex reasoning tasks, which I'll describe next.

## CoT prompting

CoT prompting involves creating prompts in a way that guides the model through a step-by-step reasoning process before arriving at a final answer, much like a human might do when solving a problem. Interestingly, simply adding the words *step by step* to a prompt can cause the model to respond differently and work through a problem in a logical manner, while it might not have done so in the absence of those words. We can also help the model if we describe tasks in a step-by-step manner. For example, consider the following prompt as a baseline, which does not implement CoT prompting:

```
If an investor places $10,000 in a savings account with an annual
interest rate of 5% and the interest is compounded annually, how much
money will be in the account after 3 years?
```

We know that LLMs are large language models and not large mathematics models, but we may be able to teach an LLM how to perform complex mathematical calculations by teaching it the step-by-step process. If the LLM were struggling to provide an accurate answer, we could explain the steps as follows:

```
Annually compounded interest means the interest is added to the
principal amount at the end of each year, and in the next year,
interest is earned on the new principal. Based on that information,
work through the following process step by step: If an investor places
$10,000 in a savings account with an annual interest rate of 5% and
the interest is compounded annually, how much money will be in the
account after 3 years?
```

While the model may not have been able to immediately provide the requested answer, we can help it to get there by teaching it about what should happen in each step, and then it can chain those steps together to get to the end result. To learn more about this fascinating topic, I recommend reading the research paper titled *Chain-of-Thought Prompting Elicits Reasoning in Large Language Models* (Wei et al., 2022), available at the following URL:

`https://doi.org/10.48550/arXiv.2201.11903`

Next, I'll cover another popular prompt engineering technique, which I briefly mentioned in *Chapter 15*, called **few-shot prompting**.

### *Few-shot prompting*

All of the prompts I've included so far in this chapter are examples of **zero-shot prompting** because I simply asked the LLM to do something without providing any referenceable examples of how to do that thing.

Few-shot prompting is a rather simple concept, as it just means that we are providing some examples in our prompt that teach the LLM how to perform the task we are requesting. An example of few-shot prompting would be the following:

```
The following are some examples of product reviews, as well as the
categorized customer sentiment associated with those reviews:
Review: "Sending this back because the strap broke the first time I
used it." Sentiment: Negative. Properties: construction.
Review: "Love the color! It's the perfect size and seems well-made."
Sentiment: Positive. Properties: appearance, construction.
Review: "Disappointed. The zipper gets stuck constantly." Sentiment:
Negative. Properties: construction.
Based on those reviews and their associated sentiment categorizations,
categorize the sentiment of the following review:
"The fabric feels flimsy, and I'm worried it might tear easily."
```

As we can see in the prompt, I've instructed the LLM on how to categorize the sentiment of a review, and I provided detailed examples of the format in which I want the LLM to respond, including specific product properties that the sentiment appears to reference.

In the case of few-shot prompting, we're providing the LLM with a mini "training set" within the prompt itself, and it can use these examples to understand the expected mapping between the inputs and the desired outputs. This helps the LLM to understand what it should focus on and how it should structure its responses, hopefully providing results that more closely match the specific needs of our use case.

Although the examples can be seen as a small training set included in our prompt, it's important to note that the underlying LLM's weights are not changed by the examples we provide. This is true for most prompt engineering techniques, whereby the responses may change significantly based on our inputs, but the underlying LLM's weights remain unchanged.

Few-shot prompting and prompt engineering, in general, can be surprisingly effective, but in some cases, we need to provide many examples to the LLM, and we want to perform incremental training (that is, updating weights), especially for complex tasks. I will outline how we can approach those scenarios shortly, but first, I'll introduce the important emerging field of prompt management.

### *Prompt management practices and tooling*

While new technologies bring new functionality and opportunities, they often also introduce new challenges. As prompt engineering continues to develop, a common challenge companies encounter is the need to store and manage their prompts effectively. For example, if a company's employees are coming up with awesome prompts that provide wonderful results, they will want to track and reuse those prompts. They may also want to update the prompts over time as the LLMs and other components of their solutions evolve, in which case they will want to track different versions of prompts and their results.

In software development, we use versioning systems to track and manage updates to our application code. We can apply the same principles to prompt engineering, using versioning systems and templates to help us efficiently develop, reuse, and share prompts. Companies performing advanced prompt engineering practices will likely end up curating prompt libraries and repositories for these purposes.

**Prompt templates** can be used to standardize the structures of effective prompts so that we don't need to keep using trial and error and reinventing the wheel to create prompts that are best suited for specific use cases. For example, imagine that we work in the marketing department and we run a monthly report to measure the success of our marketing campaigns. We may want to use an LLM to summarize the reports, and there are likely specific pieces of information that certain team members want to review each time. We could create the following prompt template to formulate those requirements:

```
Summarize the [Report Name] report for [Target Audience], considering
that [Target Audience] is particularly interested in reviewing
increases or decreases in [Metric 1] and [Metric 2].
```

Now, the marketing team members simply need to use this prompt template and fill in their desired values for each of the placeholders in the template. As is the case with individual prompts, we could maintain different versions of our prompt templates and evaluate how each version performs.

We will likely see many more prompt engineering and prompt management approaches being developed in the coming years. In addition to this, we can even apply **machine learning** (**ML**) techniques to find or suggest the best prompts for our use cases, either by asking an LLM to suggest the best prompts or by using traditional ML optimization approaches such as classification and regression to find or suggest prompts that will likely lead to the best outcome for a given use case. Expect to continue seeing interesting approaches emerging in this field.

Going beyond prompt engineering and management, the next section describes larger-scale LLM tuning techniques, which can mainly be categorized under the umbrella of **transfer learning** (**TL**).

## TL

In *Chapter 15*, I mentioned at a high level that LLMs usually go through an unsupervised pre-training phase followed by a supervised tuning phase. This section describes in more detail some of the supervised training techniques that are used to fine-tune LLMs. Let's begin with a definition of TL.

**Definition**

TL is an ML approach that uses a model that has already been trained on a certain task (or set of tasks) as a starting point for a new model that will perform a similar task (or set of tasks) with different but somewhat related parameters or data.

An example of TL would be to take a model that was pre-trained on generic image recognition tasks and then fine-tune it to identify objects that are specifically relevant to driving scenarios by incrementally training it on datasets containing road scenes.

TL approaches can be seen as a kind of spectrum, where some TL techniques and use cases only require updating a small portion of the model weights, while others involve much more extensive updates. Different points on this spectrum represent a trade-off between customizability and computational expense. For example, updating a lot of model weights provides more customizability but is more computationally expensive, while updating a small number of model weights offers less customizability but is also less computationally expensive.

Let's begin our discussion at one extreme of the spectrum, in which we will update all or a large portion of the original model's weights, which we refer to as **full fine-tuning**.

### *Full fine-tuning*

In the case of fully fine-tuning LLMs, we could begin with a model that has been pre-trained on an enormous body of data and has learned a broad understanding of the concepts on which it was trained. We then introduce the model to a new dataset specific to the task at hand (for example, understanding the rules of the road). This dataset is usually smaller and more focused than the data used in the initial pre-training phase. During the fine-tuning process, the weights across all layers of the model are updated to minimize the loss in relation to our new task.

A couple of things to bear in mind are that full fine-tuning can require a lot of computational resources, and there's also a risk that the model might forget some of the useful representations it learned during pre-training (sometimes referred to as **catastrophic forgetting**), especially if the new task is considerably different from the pre-trained task.

Remember that LLMs are large—in fact, they're huge! With that in mind, many companies may find full fine-tuning infeasible due to the sheer amount of data, computational resources, and expense it often requires.

Rather than tuning all of the weights in an enormous LLM, research has found that sometimes, we can get effective improvements on specific tasks by only changing some weights. In such cases, we can "freeze" some of the model's weights (or parameters) to ensure they do not get updated while allowing others to be updated. These approaches can be more efficient because they require fewer parameters to be updated and are therefore referred to as **parameter-efficient fine-tuning** (**PEFT**), which I will describe in more detail in the following subsections, starting with **adapter tuning**.

### Adapter tuning

In the case of adapter tuning, the original model layers remain unchanged, but we add **adapter layers** or **adapter modules**, which are small **neural networks** (**NNs**) inserted between the layers of a pre-trained model, consisting of just a few learnable parameters. As input data is fed into the model, it flows through the original pre-trained layers and the newly inserted adapter modules, and the adapters slightly alter the data processing flow, which introduces additional transformation steps in some of the layers.

The loss calculation steps that are performed after the data passes through the network are the same steps we learned about in our chapter on NNs, and the calculated loss is used to compute gradients for the model parameters. However, unlike traditional full-model fine-tuning, the gradients are only applied to update the weights of the adapter modules, while the weights of the pre-trained model itself remain frozen and unchanged from their initial values.

Another popular PEFT technique is **low-rank adaptation** (**LoRA**), which I'll describe next.

### LoRA

LoRA is based on the premise that not all parameters in a NN are equally important for transferring learned knowledge to a new task, and that, by identifying and changing only a small subset of the model's parameters, it's possible to adapt the model for a specific task without completely retraining the model. Instead of modifying the original weight matrices directly, LoRA creates low-rank matrices that can be thought of as a simplified or compressed version of the full matrix that captures the most important properties with fewer parameters.

This is what makes LoRA more parameter-efficient. As with adapter tuning, during backpropagation, only the parameters of the low-rank matrices are updated, and the original model parameters remain unchanged. Since the low-rank matrices have fewer parameters than the full matrices they represent, the training and updating process can be much more efficient.

In addition to improving a model's performance on specific tasks, there are also more subtle performance improvement practices that are just as important, such as aligning with human values and expectations. Let's explore that concept in more detail next.

## Aligning with human values and expectations

Have you ever met somebody who is highly intelligent and talented, but has poor communication skills? Such a person might perform amazingly when it comes to troubleshooting and fixing technology issues, for example, but they may not be a suitable person to lead a customer meeting. They may say things that could be considered rude or inappropriate, or perhaps just a bit odd, due to their poor communication skills. This is the analogy I like to use to explain the concept of tuning GenAI models to align with human values and expectations because such values and expectations can often be more subtle than scientific and, therefore, require tailored approaches. For example, in addition to a model's outputs being accurate, human expectations could require the model's outputs to be friendly, safe, and unbiased, among other subtle, nuanced qualities. I'll refer to this concept from here onward as **alignment**, and in this section I'll describe two approaches that are commonly used for human value alignment today, starting with **reinforcement learning from human feedback** (**RLHF**).

### RLHF

We explored the concept of **reinforcement learning** (**RL**) in *Chapter 1* of this book, and I explained that in RL, the model learns from rewards or penalties based on interactions with its environment as it pursues a specific goal. RLHF is one of those examples in which the name of the technology is highly descriptive and accurately captures what the technology entails. As the name suggests, RLHF is an extension of RL, in which feedback to the model is provided by humans.

In the case of RLHF, we start with an LLM that has already been pre-trained on a large dataset. The model generates multiple possible responses to a prompt, and humans evaluate the responses based on various preferences. The human feedback data is then used to train a separate model called the **reward model**, which learns to predict the kinds of responses humans are likely to prefer, based on the feedback given. This process is intended to capture human preferences in a quantifiable way.

The LLM then uses this feedback (like a reward signal) to update its parameters through RL techniques, making it more likely to generate responses that humans find desirable in the future.

Another technique that can be used to align with human values and expectations is **Direct Preference Optimization** (**DPO**). Let's discuss that next.

### DPO

DPO also uses human feedback to improve a model's performance in aligning with human values and expectations. In the case of DPO, again, the model may provide multiple responses to a prompt, and a human can pick which response they prefer (similar to RLHF). However, DPO does not involve training a separate reward model. Instead, it uses pairwise comparisons as the optimization signal, and it directly optimizes policies based on user preferences rather than predefined reward functions, which is especially useful in cases where it's difficult to define an explicit reward function.

It's important to note that, while RLHF and DPO are valuable and important techniques, there are some challenges that human interaction inherently brings into scope. For example, any process that requires human interaction can be difficult to scale. This means that it can be difficult to gather large amounts of data via feedback from humans. Also, humans can make mistakes or introduce conscious or unconscious bias into the results. These are some factors you would need to monitor and mitigate if you implement an RLHF or DPO solution.

Other LLM tuning techniques continue to emerge in addition to the tuning approaches I've covered here, and a lot of research is being invested in this field. This is another space in which you can expect to continue seeing groundbreaking leaps forward in development.

Next, let's dive deeper into the role of embeddings and vector databases in GenAI.

## Embeddings and vector databases

In *Chapter 15*, we discussed the importance of embeddings and latent spaces, and I explained how they can be created in different ways. One way is when a generative model learns them intrinsically during its training process, and another is when we use specific types of models to create them explicitly.

I also touched on why we would want to explicitly create them since they can be processed more efficiently and are a more suitable format for ML use cases. In this context, when we create an embedding for something, we are simply creating a vector of numeric values to represent it (how we actually do that is a more advanced topic we will cover shortly).

Another concept I briefly touched upon was the importance of relationships between embeddings in the vector space. For example, the proximity of embeddings to each other in the vector space can reflect the similarity between the concepts they represent. Let's examine this relationship in more detail.

### Embeddings and similarity of concepts

A well-constructed embedding contains everything needed to describe and explain the concept it represents—that is, the "meaning" of the concept. That may seem somewhat abstract to us because we don't often think of everything that goes into the meaning of something. For example, when I utter the word "car," a certain image might appear in your mind, and a lot of other information is immediately contextualized by that word. You know that you can drive a car; you know that they are usually relatively expensive; you know that they have wheels and windows, and they're traditionally made from some kind of metal. You know lots of pieces of information about the concept of a car, and this helps you to understand what it is. Imagine that all of the information required to conjure up the idea of a car is stored as a vector in your mind. Now, imagine that you have never heard of a truck before, and I suddenly show you a picture of a truck. It has wheels and windows; it's made from steel; it looks like something you could possibly drive. Although you've never seen anything like it before, you can understand that this new object is similar to a car. That's because the pieces of information (that is, the vector of information) associated with it are very similar to that of a car.

How do we make these kinds of associations? It feels somewhat intuitive, and most lay people (including myself) don't understand how this all really happens in our brains. However, in the case of embeddings, it's a lot easier to understand because our good old friend, mathematics (much loved by computers and ML models), comes to the rescue. As I mentioned already, we can mathematically compare different vectors by calculating the distances between them in their vector space, using well-established distance metrics such as Euclidean distance or cosine similarity.

If concepts are close in the vector space, they are likely close in meaning, in which case, we can say that they are **semantically similar**. For example, in a text-based vector database, the phrases "The cat sat on the mat" and "The feline rested on the rug" would have very similar vector embeddings, even though their exact words differ. A query for one of these vectors is likely to also identify the other one as a highly relevant result. A classic example of this concept is the representation of the words "man," "woman," "king," and "queen," as represented in *Figure 16.1*:

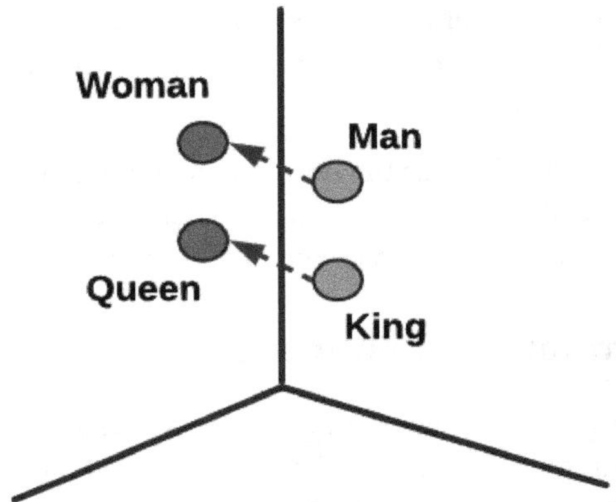


Figure 16.1: Embeddings and semantic similarity

As depicted in *Figure 16.1*, depending on how embeddings of items are captured in a vector space, we may see somewhat consistent projections in the vector representations and distances between vectors. The example shows that the concept of binary gender (or some kind of consistent semantic relationship) is captured in the words "man," "woman," "king," and "queen," and moreover, the distance and direction from "man" to "woman" is similar to that as from "king" to "queen." Additionally, the distance and direction from "man" to "king" is similar to that from "woman" to "queen." In this scenario, hypothetically, you could infer the value "queen" by performing the following mathematical operation in the vector space:

$$king - man + woman = queen$$

That is, if you take the concept of a king, subtract the male gender, and add the female gender, you end up with the concept of a queen.

We can also apply this approach more broadly. For example, in a vector database storing image embeddings, a picture of a tangerine may lie in close proximity to a picture of a clementine. These objects have similar properties, such as size, shape, and color, and a vector representing a banana would likely not be as close to them as they are to each other because a banana does not share as many similarities. However, from a multimodal perspective, which we will describe later in this chapter, considering that they are all types of fruit, the banana vector may still be closer to them than a vector representing a car or an umbrella, for example. This concept is depicted in a simplified manner in *Figure 16.2*:

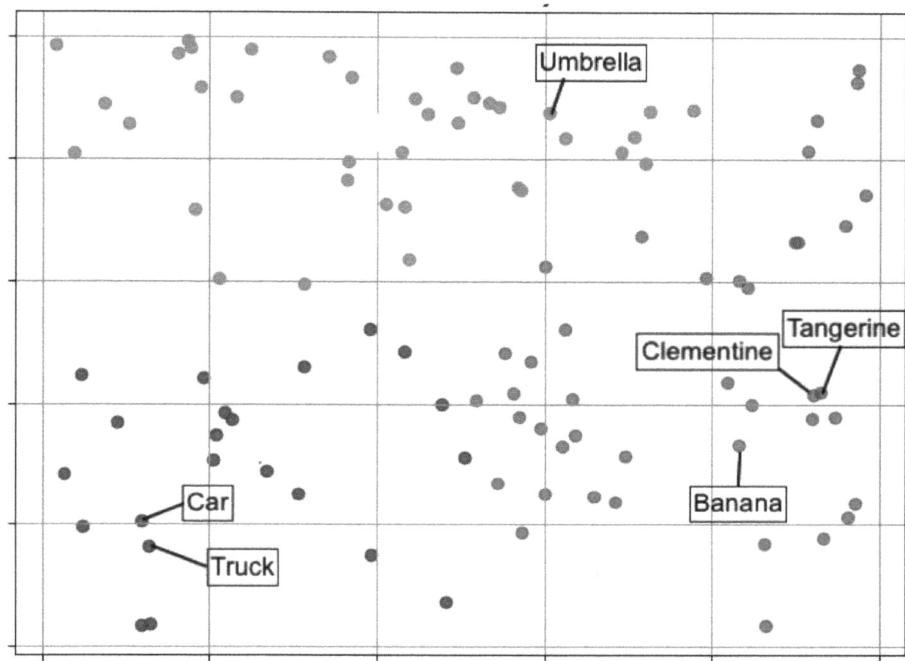


Figure 16.2: Elaborated example of embeddings and semantic similarity

Note that, in reality, the embedding space consists of many abstract dimensions and is not usually visualizable or directly interpretable to humans. Embeddings can be seen as a way to translate the messy complexity of the real world into a numerical language that ML models can understand.

If we create embeddings, we might want somewhere to store them and easily find them when needed. This is where vector databases come into the picture. Let's dive into this in more detail and explain what vector databases are.

## Vector databases

Vector databases are specialized databases designed to store and manage vector embeddings. Unlike traditional databases that focus on exact matches (for example, finding a customer with a specific ID), vector databases are more suitable for use cases such as **similarity search**, in which they use distance metrics to determine how close and, therefore, how similar two vector embeddings are in the latent space they occupy.

This approach significantly changes how we can retrieve and analyze data. While traditional methods rely on keywords or predefined attributes, vector databases allow us to perform semantic search, in which we can query based on meaning, and this enables new applications. If I search for "trousers," for example, the responses can also include jeans, slacks, and chinos because these are all semantically similar concepts. Most modern search engines, such as Google, for example, use sophisticated mechanisms such as semantic search (and a combination of other techniques), rather than simple keyword matching.

Semantic search provides a much better customer experience, as well as a potential increase in revenue for the company implementing it. If you run a retail website, for example, bear in mind that customers will not buy a product if they cannot find it, and their business may go to your competitors instead. By implementing semantic search, we can understand the meaning and the intent of the user and, therefore, present the most relevant products to them.

These new search capabilities are also important in the context of GenAI for reasons I will explain shortly. First, however, let's dive into more detail on how vector databases work.

## How vector databases work

As a precursor to explaining how vector databases work, let's briefly talk about how databases work in general. The main purpose of any database is to store and organize data in a way that makes it easy to find as quickly as possible. Most databases make use of indexes for this purpose, which I'll describe next.

### *Indexing and neighbor search*

To understand the role of indexes, imagine I handed you a book and asked you to find information in the book about "anthropomorphism" as quickly as possible. Without any mechanisms to assist you, you would have to read through every word on every page until you found that word. This, of course, would be a very inefficient process, and it would likely take you a long time to find the information unless it happened to appear near the beginning of the book. This is referred to as a **full table scan** in the database world, which we usually want to avoid, if possible. This is where indexes come into play—if the book I handed you contains an index, you could look in the index to see which pages contain references to "anthropomorphism" and then go directly to those pages, cutting out a lot of inefficient searching.

Relational databases typically use tree-like (for example, B-trees) or hashing-based indexes for rapid lookups, but vector database indexes can be more complex due to the complexities of the embedding space and how each vector is represented in potentially high-dimensional space. Such indexes can resemble graph-like structures that map relationships between embeddings to enable proximity-based searches.

The equivalent of a full table scan in the vector database world is called a **brute-force** search, which borrows its name from cybersecurity and relates to the practice of trying every possible combination of inputs to find the desired result. This is a common way in which bad actors try to guess a person's password—they try every possible combination of characters (within certain parameters).

A longer password makes it difficult for them to guess it via brute force because each additional character exponentially increases the number of potential combinations.

In the case of vector databases, our queries generally try to find items near the queried item in the vector space. If we have a lot of vectors in the database, then brute force would be impractical, especially in high-dimensional spaces where the **curse of dimensionality** (**CoD**) makes exact searches computationally expensive. Fortunately, brute force is usually unnecessary because it seeks to precisely find the vector that is closest to the vector being queried (referred to as the **exact nearest neighbor**), but we don't always need the exact nearest neighbor. For example, use cases such as recommendation engines, which recommend items that are similar to a particular item, generally just need to find vectors that are in the close vicinity of the vector being queried, but it doesn't have to be the exact nearest neighbor of the vector being queried. This is referred to as an **approximate nearest neighbor** (**ANN**) search, and it's extremely useful in the context of vector databases because it allows a trade-off between the accuracy of the search results and the speed of the query – that is, for these kinds of use cases, it's better to quickly get results that are close enough to the queried vector rather than spending a lot of time finding the exact best result.

> **Note – the CoD and vector proximity**
>
> In high-dimensional spaces, points become increasingly spread out, and as the number of dimensions increases, the distance between the nearest and farthest points becomes diluted, which makes it harder to distinguish between truly similar and dissimilar items based on distance metrics.

Some vector databases implement hybrid approaches that combine ANN with other indexing or search strategies, such as hierarchical indexing structures, multi-index hashing, and tree-based methods to improve accuracy, reduce search latency, or optimize resource usage. Graph-based methods such as **Navigable Small World** (**NSW**) graphs, **Hierarchical Navigable Small World** (**HNSW**) graphs, and others create a graph where nodes represent vectors, and queries then traverse the graph to find the nearest neighbors. Some vector databases also use partitioning or clustering algorithms to divide the dataset into smaller, more manageable chunks or clusters, and indexing can then be performed within these partitions or clusters, often using a combination of methods to first identify the relevant partition and then perform an ANN search within it.

Also, we can use approaches such as product quantization or scalar quantization to compress vectors by mapping them to a finite set of reference vectors, which reduces the dimensionality and storage requirements of the vectors before indexing and enables faster searches by approximating distances in the compressed space.

In *Chapter 17*, we'll walk through the various vector database offerings in Google Cloud, but for now, let's move on to discuss how we can create embeddings in the first place.

## Creating embeddings

We've already discussed **autoencoders** (**AEs**) in *Chapter 15*, and we learned how they can be used to create latent space representations of their inputs. There are many more ways in which we can create embeddings, and I'll describe some of the more popular methods in this section, starting with one of the most famous word embedding models, Word2Vec.

### Word2Vec

Word2Vec (short for "word to vector") is a group of algorithms invented by Google (Mikolov et al., 2013) to learn representations of words as vectors. It's often considered to be the grandfather of word embedding approaches, and although it wasn't the first word embedding technique to be invented, it promoted the idea of representing words as dense vectors that capture the semantic meaning and relationships between words in a high-dimensional space. It works by building a vocabulary of unique words and learning a vector representation for each word, where words with similar meanings have similar vectors.

The two main approaches used in Word2Vec are the **Continuous Bag of Words** (**CBOW**) model, which predicts a target word based on its surrounding words, and **skip-gram**, which predicts surrounding words based on a given target word.

While Word2Vec has been a popular and useful approach, newer, transformer-based approaches have emerged that provide more advanced capabilities. Let's take a look at those next.

### Transformers

We've already introduced transformers in previous chapters, and in this section, I will briefly describe transformer-based approaches to create embeddings. While there are many options to choose from, I'll focus on the famous **Bidirectional Encoder Representations from Transformers** (**BERT**) and its derivatives.

#### BERT

Earlier in this book, I mentioned that, every now and then, there's a significant step forward in AI/ML research, and it's well established that the invention of the transformer architecture by Google in 2017 (Vaswani et al., 2017) was one of those significant leaps. BERT, which was invented by Google in 2018 (Devlin et al., 2018), was another significant step.

As indicated in the name, BERT is based on the transformer architecture, and it is pre-trained on a large dataset of text and code. During its training, it learns to model complex patterns and relationships between words and understand nuances such as context, grammar, and how different parts of a sentence relate to each other.

The two main approaches used in BERT are **masked language modeling** (**MLM**), which predicts missing words within a sentence based on the surrounding context, and **next sentence prediction** (**NSP**), which tries to determine if two sentences are logically connected. These are examples of the pre-training phase described in *Chapter 15*. By combining these two approaches, BERT develops an understanding of language structure.

At the heart of BERT are the transformer layers, which take the text we give it and produce word embeddings that capture meaning depending on the surrounding words, which is a major improvement over static embeddings such as Word2Vec.

BERT was such an important breakthrough that many variants have been created since its initial release, such as DistilBERT and ALBERT, which are smaller, distilled versions of BERT with fewer parameters (trading off some accuracy for computational efficiency), as well as domain-specific variants, such as SciBERT (trained on scientific publications) and FinBERT (fine-tuned for financial industry use cases).

There are also more targeted models that build on top of transformer-based models such as BERT. For example, rather than focusing on individual words, **Sentence Transformers** use pooling strategies such as mean pooling or max pooling to create semantically meaningful embeddings of entire sentences.

For image embeddings, while we've discussed **convolutional NNs** (**CNNs**) in previous chapters for use cases such as image classification, it's important to note that CNNs can also be used in creating image embeddings.

Now that we know about some options for creating embeddings, as well as options for storing embeddings in vector databases, let's explore a relatively new pattern that is becoming increasingly popular in the industry and that combines these topics, referred to as RAG.

# RAG

While LLMs are clearly impressive and powerful technologies, they have some limitations. Firstly, LLMs are only as good as the data on which they were trained, and they can be expensive and time-consuming to train, so for most companies, it would not be feasible to retrain them with new information every day. For this reason, in many cases, updated versions of LLMs tend to be released every few months. This means that the information provided in their responses depends on their most recent training dataset. If you want to ask about something that happened yesterday, but the LLM was last updated a month ago, it simply will not have any information on that topic. This can be quite limiting in today's fast-paced business world, where people and applications constantly need up-to-date knowledge at their fingertips.

Secondly, unless you are one of the large corporations that have created the popular LLMs being used today, you most likely did not train the LLM yourself from scratch, so it has not been trained on your specific data. Imagine you are a retail company that wants to use an LLM to enable customers to find out about your products via a chat interface. Since the LLM was not trained specifically on your product catalog, it will not be familiar with the details of your products.

To combat these kinds of challenges, rather than relying solely on the data that was used to train an LLM, we can insert additional data into the context of the prompt being sent to the LLM. We already touched on this in the *Prompt engineering* section of this chapter, in which we provided additional information in the prompts in order to guide the LLM's responses to more accurately match our required outputs. That approach works great for small amounts of data, but if we consider the aforementioned use case of getting the LLM to answer questions about a product catalog, we cannot include the entire product catalog in every prompt. This is where RAG can help.

With RAG, we can *augment* the responses *generated* by LLMs by first *retrieving* relevant information from a data store and then including that information in the prompt being sent to the LLM. That data store can contain whatever information we want, such as the contents of our product catalog. Let's take a look at this pattern in more detail.

## How RAG works

To set the context for explaining how RAG works, let's level-set the discussion by reviewing how LLM interactions typically work without RAG. This is very straightforward and is depicted in *Figure 16.3*:

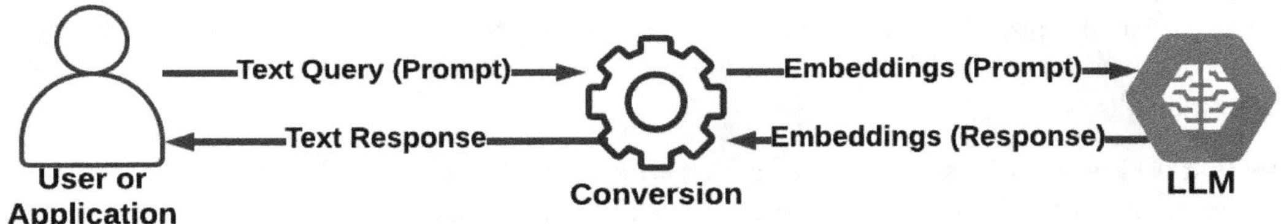


Figure 16.3: LLM prompt and response

As we can see, the typical LLM interaction simply consists of sending a prompt and getting a response. Note that LLMs can only understand numerical values, so behind the scenes, the textual prompt sent by the user is broken down into tokens, and the tokens are encoded into vectors that the LLM can understand. The reverse process is performed when sending the response to the user. Next, let's look at how RAG changes the process, as depicted in *Figure 16.4*:

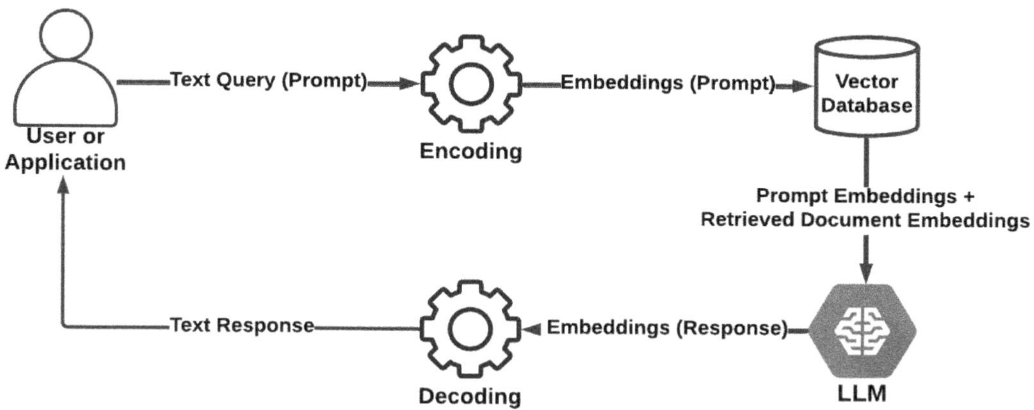


Figure 16.4: RAG

*Figure 16.4* outlines how RAG works at a high level. The steps are as follows:

1. The user or client application sends a textual request or prompt.
2. The contents of the prompt are converted to vectors (embeddings).
3. During the "retrieval" part of the process, the prompt embeddings are used to find similar embeddings in the vector database. These embeddings represent the data that we want to use to augment the prompt (for example, embeddings that represent data from our product catalog).
4. The embeddings from our vector database are combined with the embeddings of the user prompt and sent to the LLM.
5. The LLM uses the combined embeddings of the user prompt and the retrieved embeddings to augment its response, aiming to provide a response that is not only based on the data on which it was originally trained but that is also relevant in the context of the data retrieved from our vector database.
6. The response embeddings are decoded to provide a text response back to the user or client application.

We will implement RAG in a later chapter in this book. Next, let's begin to explore the topic of multimodal models.

# Multimodal models

Given that LLMs, as the name suggests, focus on language, the primary modality for many of the popular LLMs you are likely interacting with at the time of writing this in early 2024 is likely to be text. However, the concept of GenAI also goes beyond text and expands into other modalities, such as pictures, sound, and video.

The term *multimodal* is becoming ever more prominent within the field of GenAI. In this section, we explore what that means, starting with a definition of "modality."

> **Definition**
>
> The Oxford English Dictionary defines modality as:
>
> "*Those aspects of a thing which relate to its mode, or manner or state of being, as distinct from its substance or identity; the non-essential aspect or attributes of a concept or entity. Also: a particular quality or attribute denoting the mode or manner of being of something.*"
>
> *Oxford English Dictionary, s.v. "modality (n.), sense 1.a," December 2023,* `https://doi.org/10.1093/OED/1055677936`.

Considering this formal definition of modalities, unlike traditional models that focus on a single data type, multimodal models are designed to process information from multiple modalities or data types. This is another one of those important leaps forward in AI research because it opens up new use cases and applications for AI. Let's dive into this in more detail.

## Why multimodality matters

In the section on RAG in this chapter, I mentioned an example of a company using an LLM to help customers explore product details via a chat interface and how that would require interacting with data from the company's product catalog. Diving into this example in more detail, consider that items in a product catalog can have many different types of information associated with them. This may consist of structured data with standardized fields such as product dimensions, color, price, and other attributes. It may be associated with semi-structured or unstructured data such as customer reviews as well as product images. Imagine how much of a more detailed understanding our model could gain regarding the items in our catalog if it could ingest and understand all of those data modalities. If the written product description was somehow missing details regarding the color of the product, the model could instead learn that information from pictures of the product. Also, in addition to written customer reviews, some websites allow customers to post videos in their reviews. The model could learn more information from those videos. Overall, this could provide a much richer end-user experience. This is just one example of the power of multimodality, and this concept becomes even more important when creating models that need to interact with the physical world, such as robots and self-driving vehicles.

Multimodal implementations are also important in AI research because one of the main challenges in creating useful and powerful models is getting access to relevant data. Considering that there is a finite amount of text that has ever been created in the world, if text remained the only modality, it would eventually become a limiting factor in AI development. As we can all clearly see on various social media platforms and other websites, video, audio, and pictures have equaled or surpassed text as the primary modalities in user-generated content, and we can expect those trends to continue. We've all heard the expression, "A picture is worth a thousand words," and this becomes ever more true (in addition to other modalities) when training and interacting with large generative models!

Having expressed the importance and advantages of multimodal model implementations, I'll balance the discussion by highlighting some of the related challenges next.

## Multimodality challenges

I'll begin this section with a challenge that consistently emerges in many contexts that we've discussed throughout this book. Generally, creating more powerful models requires more computing resources. Multimodal approaches are no exception here, and training multimodal models often requires significant computational resources.

Another challenge is the complexity—combining data from different modalities can be complex and require careful preprocessing, especially considering that the feature representations in different modalities usually have very different feature spaces, and some modalities have richer or more sparse data than others. For example, text is usually sequential, images are spatial, and audio and video have temporal properties, so aligning all of these for the model to build a unified understanding can be difficult.

In previous chapters, we looked at examples of data quality issues and the kinds of data preparation steps and pipelines required to address those issues. The tricky thing, however, is that data in different modalities has different kinds of potential quality issues, preparation steps, and pipelines that are required to get it ready for training models.

Despite the challenges, multimodality is a rapidly developing space that will continue to get significant attention for the foreseeable future.

While this section focused mainly on training GenAI models (specifically, multimodal approaches), other parts of the model development life cycle also have unique approaches that must be factored in when dealing with GenAI models rather than traditional AI models. The next important topic to highlight is how we evaluate GenAI models and how this can differ in some cases from evaluating traditional AI models.

# GenAI model evaluation

While evaluating traditional AI models usually involves measuring the model's outputs against a ground truth dataset and calculating established, objective metrics such as accuracy, precision, recall, F1 score, **Mean Squared Error** (**MSE**), and others we've covered in this book, evaluating generative models is not always as straightforward.

When evaluating a GenAI model, we might want to focus on different factors, such as the model's ability to produce creative and human-like outputs while staying relevant to the task at hand. An extra challenge in this case is that evaluating these properties can be somewhat vague and subjective. For example, if I ask a generative model to write a poem or create a photorealistic picture of a cat in a meadow, the quality of that poem or the realness of the picture is not always easy to represent with a mathematically calculated number, although some formulaic measurements do exist, which I will describe in this section. However, let's first discuss human evaluation.

## Human evaluation

Perhaps the most straightforward but also slow, cumbersome, and potentially expensive approach is for human evaluators to assess generated outputs for quality, creativity, coherence, and how well they align with the task requirements. There are tools, frameworks, services, and entire companies dedicated to performing these kinds of evaluations. Some common challenges with this approach are that the evaluations are only as good as the instructions provided to the evaluators, so it's necessary to clearly understand and explain how the responses should be evaluated, and the evaluations can be subjective, based on the personal interpretations of the evaluator.

Next, I'll describe some of the formalized methods that can be used in evaluating generative models, starting with metrics for textual language models such as **BiLingual Evaluation Understudy** (**BLEU**) and **Recall-Oriented Understudy for Gisting Evaluation** (ROUGE).

## BLEU

BLEU is used mainly for evaluating machine translation use cases, and it works by measuring how similar the generated text is to human-written text, so when we're using BLEU, we need a set of reference, human-written text examples, which are compared against the generated text. More specifically, BLEU measures something called **n-gram overlap**, which is the overlap in sequences of words between the generated text and the reference text.

## ROUGE

ROUGE also measures the overlap between generated text and human-written reference examples, but it focuses on summarization use cases by measuring the **recall**, which represents how well a generated summary captures important information from the reference summaries.

In addition to overlap-based evaluations, we can also evaluate text generation performance with metrics such as **perplexity** and **negative log-likelihood** (**NLL**), which can be used to evaluate a model's performance in predicting sequences of words, such as predicting the next word in a sentence.

Next, let's consider some approaches for evaluating image generation models, such as the **Inception Score** (**IS**) and **Fréchet Inception Distance** (**FID**).

### IS

IS is used to measure the quality and **diversity** of generated images, in which a good image generation model should be able to generate a wide variety of images, not just variations of the same thing, and the images should look clear and realistic. We can calculate the IS by using a pre-trained image classification model to classify images generated by a generative model.

### FID

While the IS evaluates generated images independently, FID was designed to improve on it by measuring how similar a distribution of generated images is to a distribution of real images (it gets its name from using the Fréchet distance to measure the distance between two distributions).

For the purpose of this chapter, we mainly need to be aware that specific metrics exist for evaluating some generative models, but going into the mathematical detail of how these metrics are calculated would be more information than is required at this point. There are additional approaches and metrics beyond the ones we've covered in this section, but this section describes some popular ones to consider. Next, I'll introduce a somewhat different approach to evaluating generative models, which is to use an **auto-rater**.

## Auto-raters and side-by-side evaluations

At a high level, an auto-rater is an ML model that rates or evaluates the outputs of models. There are different ways in which this can be done, and I will explain the concept in the context of Google Cloud Vertex AI.

Google Cloud Vertex AI recently launched a service called **Automatic side-by-side** (**AutoSxS**), which can be used to evaluate pre-generated predictions or GenAI models in the Vertex AI Model Registry. This means that, in addition to Vertex AI foundation models, we can also use it to evaluate third-party language models.

In order to compare the results of two models against each other, it uses an auto-rater to decide which model gives the better response to a prompt. The same prompt is sent to both models, and the auto-rater evaluates the responses from each model against various criteria, such as relevance, comprehensiveness, the model's ability to follow instructions, and whether the responses are grounded in established facts.

As I mentioned at the beginning of this section, evaluating generative models often requires some subjectivity and more flexibility than the simple objective evaluations of traditional ML models. The various criteria supported by Google Cloud Vertex AI AutoSxS provide such additional flexibility.

Now that we have a high-level understanding of how the evaluation of GenAI models differs in some ways from that of traditional ML models, let's move on to introduce another important topic in GenAI—LangChain.

# LangChain

By sending a prompt to an LLM, we can achieve amazing feats and get useful information. However, we may want to build applications that implement more complex logic than we could achieve in a single prompt, and these applications may require interacting with multiple systems in addition to an LLM.

LangChain is a popular framework for developing applications using LLMs, and it enables us to combine multiple steps into a chain, in which each step implements some logic, such as reading from a data store, sending a prompt to an LLM, taking the outputs from the LLM, and using them in subsequent steps.

One of LangChain's advantages is that it uses a modular approach, so we can build complex workflows by combining smaller, simpler modules of logic. It provides tools for useful tasks such as creating prompt templates and managing context for interactive integrations with LLMs, and we can easily find integrations for accessing information from sources such as Google Search, knowledge bases, or custom data stores.

We will use LangChain later in this book to orchestrate a number of different steps in a workflow. For now, let's summarize what we covered in this chapter before we move on to the next one.

# Summary

In this chapter on advanced GenAI concepts and use cases, we started by diving into techniques for tuning and optimizing LLMs. We learned how prompt engineering practices can affect model outputs, and how tuning approaches such as full fine-tuning, adapter tuning, and LoRA enable pre-trained models to be adapted for specific domains or tasks.

Next, we dived into embeddings and vector databases, including how they represent the meanings of concepts, and enable similarity-based searches. We looked into specific embedding models such as Word2Vec and transformer-based encodings.

We then moved on to describe how RAG can help us to combine information from custom data stores into prompts being sent to an LLM, thereby enabling the LLM to modify its responses in alignment with the contents of our data stores.

After that, we discussed multimodal models and how they can open up additional use cases beyond textual language. We then moved on to discuss how the evaluation of GenAI models differs in some ways from that of traditional ML models, and we introduced some new approaches and metrics for evaluating GenAI models.

Finally, we introduced the important topic of LangChain and how it can help us build applications that implement complex logic by chaining simple modules or building blocks together.

In the next chapter, we will learn about the various GenAI offerings and implementations on Google Cloud.

# 17

# Generative AI on Google Cloud

Now that we've covered many important topics in the world of generative AI, this chapter will explore generative AI specifically in Google Cloud. We will discuss Google's proprietary models, such as the Gemini, PaLM, Codey, Imagen, and MedLM APIs, which are each designed for different kinds of tasks, from language processing to medical analysis.

We will also review open source and third-party models available on Google Cloud via repositories such as Vertex AI Model Garden and Hugging Face.

Continuing from our discussion of vector databases in the previous chapter, we will explore various vector database options in Google Cloud, as well as potential use cases for each option.

Finally, we will use the information we'll cover in this chapter to start building generative AI solutions in Google Cloud.

Specifically, this chapter covers the following topics:

- Overview of generative AI in Google Cloud
- Detailed exploration of Google Cloud generative AI
- Implementing generative AI solutions in Google Cloud

Let's begin with a high-level overview of generative AI in Google Cloud.

## Overview of generative AI in Google Cloud

The pace of generative AI development in Google Cloud is nothing short of astonishing. It seems like almost every day, a new model version, service, or feature is announced. In this section, I'll introduce the various models and products at a high level, which will set the stage for deeper dives later in this chapter.

Overall, the models and products are categorized by modality, such as text, code, image, and video. Let's begin our journey with a discussion of Google's various generative AI models.

## Google's generative AI models

Google has created many generative AI models over the past few years, with a dramatic acceleration of new models and model versions launched in the past year. This section discusses the Google first-party foundation models that are currently available on Google Cloud in early 2024, starting with the quite famous "Gemini."

### Gemini API

As of early 2024, Gemini is Google's largest and most capable series of AI models. It comes in three model families:

- **Gemini Ultra**: This is the biggest model in the Gemini family and is intended for highly complex tasks. At the time of its announced launch in December 2023, it exceeded current state-of-the-art results on 30 of the 32 widely used academic benchmarks in LLM research.
- **Gemini Pro**: While this is not the largest model, it is considered Google's best model for scaling across a wide range of tasks. Its most recent version at the time of writing this in early 2024 is version 1.5, which introduced a one-million-token context window – the largest context window of any model in the industry at the time of its launch. This enables us to send very large amounts of data in a single prompt, opening new use cases and solving significant limitations of earlier LLMs.
- **Gemini Nano**: This is Google's most efficient Gemini model family and is intended to be loaded into the memory of a single device for performing on-device tasks. It was originally launched with two variants of slightly different sizes, Nano-1 and Nano-2.

There are also multimodal variants – for example, **Gemini Ultra Vision** and **Gemini Pro Vision** – that are trained on multiple kinds of input data, including text, code, image, audio, and video. This means that we can mix the modalities in our interactions with Gemini, such as sending an image in our prompt, asking questions (via text) about the contents of the photo, and receiving textual outputs.

### PaLM API models

PaLM is Google's **Pathways Language Model**, based on the "Pathways" system created by Google to build models that could perform more than one task (as opposed to traditional, single-purpose models). The current suite of PaLM models is based on the PaLM 2 release, and there are multiple offerings within this suite, such as PaLM 2 for Text, PaLM 2 for Chat, and Embeddings for Text, each with variants I'll describe next.

### PaLM 2 for Text

As the name suggests, this family of models is designed for text use cases. There are currently three variants of this model family:

- **text-bison**: This can be used for multiple text-based language use cases, such as summarization, classification, and extraction. It can handle a maximum input of 8,192 tokens and a maximum output of 1,024 tokens.
- **text-unicorn**: The most advanced model in the PaLM family for complex natural language tasks. While it's a more advanced model than text-bison, it has the same input and output token limits.
- **text-bison-32k**: This variant is similar to the aforementioned text-bison, but it can handle a maximum output of 8,192 tokens and an overall maximum of 32,768 tokens (combined input and output).

When we send a prompt to **PaLM 2 for Text** models, the model generates text as a response, and each interaction is independent. However, we can build external logic to chain multiple interactions. Next, we'll discuss a set of Google Cloud products that are designed for interactive prompt integrations.

### PaLM 2 for Chat

With **PaLM 2 for Chat**, we can engage in a **multi-turn** conversation with the PaLM models. This product family consists of two variants:

- **chat-bison**: This variant is designed for interactive conversation use cases. It can handle a maximum input of 8,192 tokens and a maximum output of 2,048 tokens.
- **chat-bison-32k**: This variant is similar to the aforementioned chat-bison, but it can handle a maximum output of 8,192 tokens and an overall maximum of 32,768 tokens (combined input and output).

Given the nature of these product variants, they are well suited for use cases in which we want to maintain context across multiple prompts to provide a natural, human-like conversation experience. We can also use PaLM models to generate text embeddings. We'll take a look at those models next.

### PaLM 2 embedding models

We've already generically discussed models that can be used to create embeddings. In this section, we'll look at Google's PaLM 2 embedding models, which come in two variants:

- **textembedding-gecko**: The name "gecko" refers to smaller models. Considering that we are using these models to create embeddings for text, we generally don't need a very large model for that use case, so these smaller and more efficient models make more sense in this context. This particular model focuses on words in the English language.
- **textembedding-gecko-multilingual**: This variant is similar to the aforementioned textembedding-gecko, but this model can support over 100 languages.

So far, all of the PaLM 2 models I've described focus on natural human language. Next, we'll discuss models that are designed for use cases involving computer programming languages.

## Codey API models

Codey is a fun name for models that are designed to implement code use cases. Codey models can generate code based on natural language requests and can even convert code from one programming language into another. Like the text-based PaLM 2 models, Codey models come in different variants based on whether we simply want the models to generate distinct responses to independent prompts, or we want to implement an interactive, chat-based use case. Let's dive into each in more detail.

### Codey for Code Generation

This set of offerings is designed for independent prompts (although, again, we can build logic to chain them together if we wish), and they come in two variants:

- **code-bison**: This model can generate code based on a natural language prompt – for example, `write a Java function to fetch the customer_id and account_balance fields from the 'accounts' table in the 'customer' MySQL database`. This variant can handle a maximum input of 6,144 tokens and a maximum output of 1,024 tokens.
- **code-bison-32k**: This variant is similar to the aforementioned code-bison, but it can handle a maximum output of 8,192 tokens and an overall maximum of 32,768 tokens (combined input and output).

Next, let's discuss model variants for interactive coding use cases.

### Codey for Code Chat

This product family enables us to engage in multi-turn conversations regarding code. These models also come in two variants:

- **codechat-bison**: This model can be used to implement chatbot conversations involving code-related questions. Like the code-bison model, it can handle a maximum input of 6,144 tokens and a maximum output of 1,024 tokens.
- **codechat-bison-32k**: This is similar to the aforementioned codechat-bison, but it can handle a maximum output of 8,192 tokens and an overall maximum of 32,768 tokens (combined input and output).

In addition to the Code Generation and Code Chat models, there is also a version of Codey that's used for code completion, which I'll describe next.

### Codey for Code Completion

The code completion variant of Codey is intended to be used in **integrated development environments** (**IDEs**). There is one variant of this model, called **code-gecko**, which is designed to act as a helpful coding assistant to help developers write effective code in real time. This means that it can suggest snippets of code within the IDE as developers are writing their code. We're all familiar with predictive text and auto-correct features on our mobile phones. This is a similar concept, but for code, and it helps developers get their work done more quickly and effectively.

Now, let's switch our discussion to another modality: images.

## *Imagen API models*

Imagen is the name of the suite of Google models that are designed for working with images, where each model in the suite can be used for different types of image-related use cases, such as the following:

- **Image generation**: The **imagegeneration** model, as its name suggests, can be used to generate images based on natural language prompts. For example, the image in *Figure 17.1* was generated by the prompt, "an impressionistic painting of a woman fishing late on a summer evening in a small boat on a placid river with reeds and trees in the background." We can also interactively edit our pictures to refine them based on our desired outcomes:

Figure 17.1: Image generated by Imagen

- **Image captioning**: We can send an image in our prompt to the **imagetext** model, and it will generate descriptive text based on the contents of that image. This can be useful for many kinds of business use cases, such as generating product descriptions based on images in a product catalog on a retail website, generating captions for images in news articles, or generating "alt text" for images on websites.
- **Visual Question Answering (VQA)**: In addition to generating captions for our pictures, the **imagetext** model also allows us to interactively ask questions about the contents of images.
- **Multimodal embeddings**: While we can use the **textembedding-gecko** models to generate text embeddings, the **multimodalembedding** model can generate embeddings for both images and text.

In addition to the general-purpose models I've just described, Google also provides models that have been designed to focus specifically on medical use cases. The next section briefly describes those models.

### MedLM API models

There are two model variants in this suite of models, **medlm-medium** and **medlm-large**, both of which are HIPAA-compliant and can be used for summarizing medical documents and helping healthcare practitioners with medical questions.

We can expect many more models and variants to be added by Google continuously to support an ever-growing plethora of use cases. In addition to Google's first-party models described in the previous subsections, Google Cloud supports and embraces open source development, something we will explore next.

## Open source and third-party generative AI models on Google Cloud

Google is a well-established contributor to open source communities, having created and contributed critically important inventions such as Android, Angular, Apache Beam, Go (programming language), Kubernetes, TensorFlow, and many others. In this section, we will explore Google's open source generative AI models, as well as third-party (both open source and proprietary) generative AI models that we can easily use on Google Cloud. We'll begin by discussing the Google Cloud Vertex AI Model Garden, which enables us to access such models.

### Google Cloud Vertex AI Model Garden

Vertex AI Model Garden is a centralized library that provides a "one-stop shop" that makes it easy for us to find, customize, and deploy pre-trained AI models.

Within Model Garden, we can access foundation models, such as Gemini and PaLM 2, as well as open source models, such as Gemma (described shortly) and Llama 2, and third-party models, such as Anthropic's Claude 3. We can also access task-specific models that cater to use cases such as content classification and sentiment analysis, among many more.

Vertex AI Model Garden is closely integrated with the rest of the Google Cloud and Vertex AI ecosystem, which enables us to easily build enterprise-grade solutions by providing access to all of the data processing and solution-building tools we've discussed in previous chapters, as well as many more that are beyond the scope of this book.

In addition to making models accessible via Vertex AI Model Garden, Google Cloud has established a strategic partnership with Hugging Face, which I'll describe next.

### *Hugging Face*

Hugging Face is both a company and a community that makes it easy to share ML models, tools, and datasets. While it started as a chatbot company, it rapidly grew in popularity due to its Model Hub and Transformer libraries, which amassed a broad community of contributors and made it easy to access and use large, pre-trained models. With the relatively new direct partnership between Google Cloud and Hugging Face, Google Cloud customers can now easily avail of the immense variety of models, tools, and datasets from within their Google Cloud environments running on services such as Vertex AI and **Google Kubernetes Engine** (**GKE**).

Considering that this is a book about Google ML and generative AI, the descriptive discussions here will focus on Google's models; I'll refer you to the external documentation for the non-Google open source and third-party models. With that in mind, in the next section, I will introduce Google's suite of open source models, named "Gemma."

### *Gemma*

Gemma is a family of state-of-the-art and open source, lightweight models from Google (DOI citation: 10.34740/KAGGLE/M/3301) that were built from the same technology and research that was used to create the Gemini models. These are decoder-only, text-to-text LLMs with pre-trained, instruction-tuned variants and open weights, which were trained on data from a wide variety of sources, including web pages, documents, code, and mathematical texts. They are suitable for many different kinds of text-generation use cases, such as summarization or question answering, and their open weights mean that they can be customized for specific use cases. Also, since they are relatively lightweight, they don't require specialized or large-scale computing resources to run, so we can use them in many environments, including a local laptop, for example, which makes it easy for developers to start experimenting with them.

**Fun fact**

Gemma happens to be my sister's name, so I was pleasantly surprised when that name was chosen for this suite of models, although I had no involvement whatsoever in the naming of these models.

We will look at the Gemma models in more detail later. In *Chapter 15*, I described the importance of embeddings and vector databases in generative AI. Now that we've explored the various generative AI models in Google Cloud, let's discuss vector databases in Google Cloud.

## Vector databases in Google Cloud

This section provides a brief introduction to vector databases in Google Cloud, and we'll dive into them in more detail later in this chapter. I'll begin this section with a simple question: "*Which Google Cloud database service provides vector database functionality?*" Quite simply, the answer is, "*Pretty much all of them, with just a few exceptions!*"

The next question, then, might be, "*Which one should I use?*" The answer to that question is a bit more nuanced. We'll explore the options in more detail in this chapter, starting with a couple of easy decisions. Firstly, the reason almost all Google Cloud database services provide vector database functionality is that Google wants to make it as easy as possible for you to access this functionality. If you already use AlloyDB to manage your application's operational data, you can easily go ahead and use AlloyDB AI for your vector database needs. If you use BigQuery for your analytical needs, go ahead and use BigQuery Vector Search, although bear in mind that BigQuery is designed primarily for large-scale data processing rather than for optimizing latency. If you have strict low-latency requirements, your workload may be better suited for one of the other Google Cloud options we'll cover later.

If you're starting fresh in Google Cloud and want to set up a vector database, begin your journey by exploring the offerings under Vertex AI. For a fully managed platform that enables developers to build Google-quality search experiences for websites, structured and unstructured data, or to integrate your applications with generative AI and search functionality that's grounded in your enterprise data, start with Vertex AI Search and Conversation. If you want to create and store your vectors and rapidly search billions of semantically related items, look into Vertex AI Vector Search.

In the next section, we will dive deeper into Google Cloud's generative AI offerings, starting with a hands-on exploration of the models, and then covering the various vector database offerings in more detail.

# A detailed exploration of Google Cloud generative AI

In this section, we'll start interacting with Google Cloud's generative AI offerings. To set the stage, I'll begin by briefly introducing Google Cloud Vertex AI Studio.

## Google Cloud Vertex AI Studio

Vertex AI Studio can be accessed within the Google Cloud console UI, and it provides an interface to easily start using all of the generative AI models I described in the previous section.

To access the Google Cloud Vertex AI Studio UI, perform the following steps:

1. In the Google Cloud console, navigate to the Google Cloud services menu and choose **Vertex AI**.
2. Under the **Get started with Vertex AI** section, click **ENABLE ALL RECOMMENDED API**, as depicted in *Figure 17.2*:

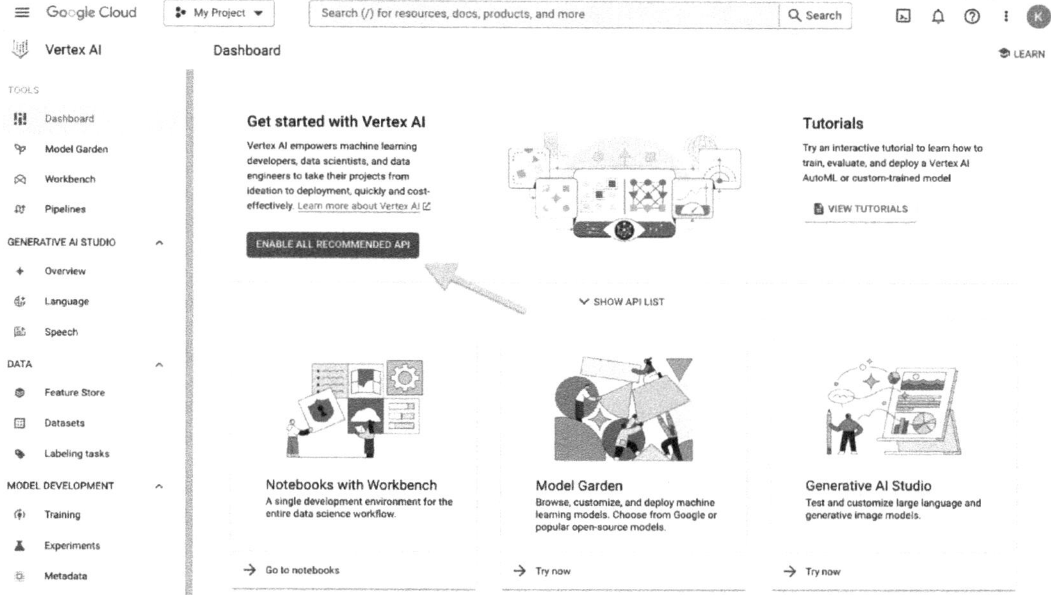


Figure 17.2: Enabling the recommended APIs

3. Give it a few minutes for the APIs to enable. Once the APIs have been enabled, you can start using them. You can interact with each of the models I described in the previous section by clicking on each of the modalities – such as **Language**, **Vision**, **Speech**, and **Multimodal** – in the menu on the left-hand side of the screen.

4. For example, if we click on **Language**, a screen similar to what's shown in *Figure 17.3* will be displayed:

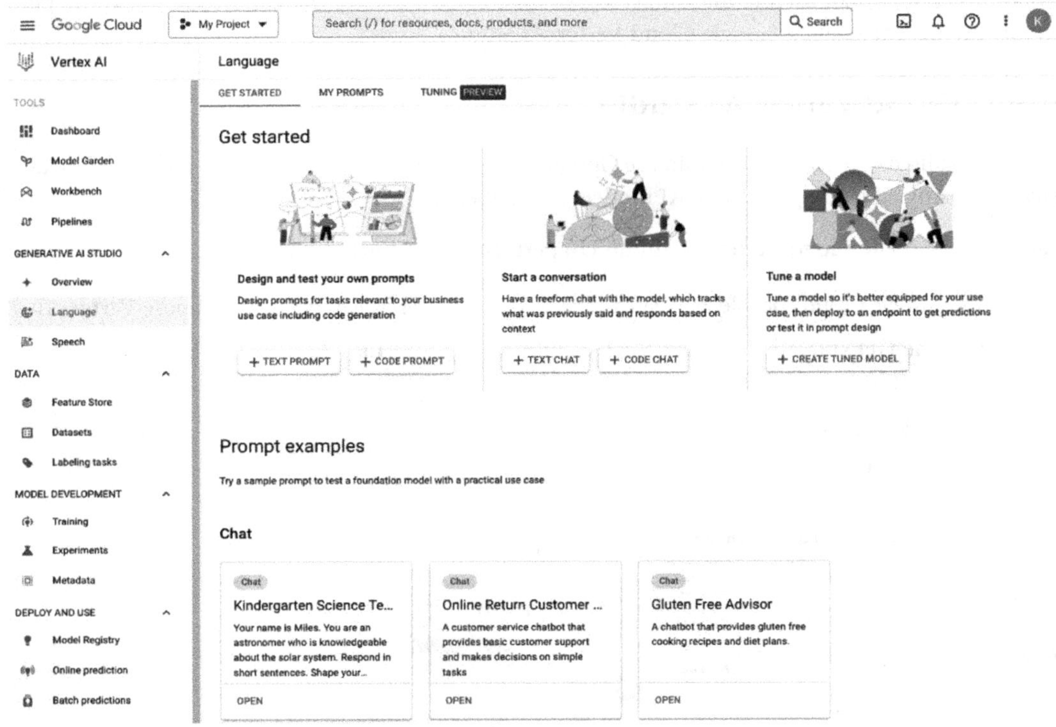


Figure 17.3: The Language section

5. From here, we can click on the various links and buttons to try out the different models. For example, clicking the **TEXT CHAT** button will present us with a chat interface into which we can type our prompts, as shown in *Figure 17.4*:

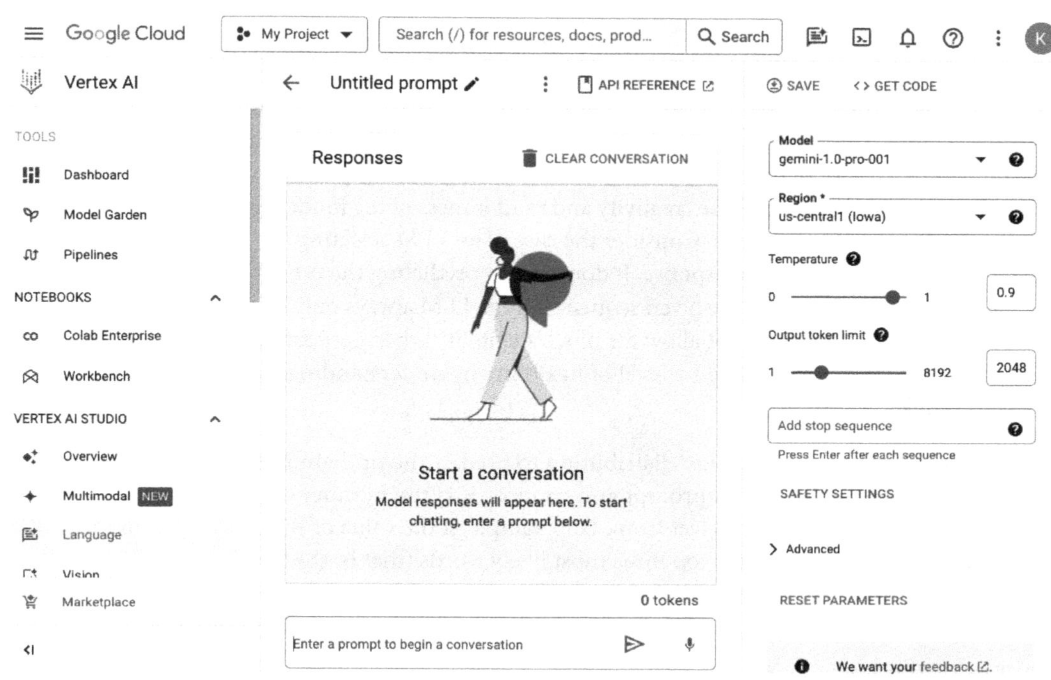


Figure 17.4: Text chat

The parameters we can configure can be found on the right-hand side of the screen. Let's take a closer look:

- **Model**: The model to which we want to send a prompt.
- **Region**: The region in which we want to run our prompt.
- **Temperature**: This parameter configures the level of creativity or randomness in the model's responses. This can be seen as the level of imagination we want the model to use when generating responses. If we consider the use case of an LLM predicting the next word in a sequence, the temperature parameter influences the overall probability distribution of the next word, such that a higher temperature can result in more creative outputs, while a lower temperature guides the model to provide more conservative and predictable outcomes. If we want the model to create imaginative art or text, for example, we could configure a high temperature, whereas if we want more formal, factual responses, we would configure a low temperature.
- **Output token limit**: The maximum number of tokens we want the model to generate in its response.

We can click on the **Advanced** section at the bottom of the screen to expand it. Here, we will see additional parameters, such as **Max responses**, which configures the maximum number of example responses we want the model to return (note that this parameter is only valid for some types of interactions – for example, it is not applicable to chat interactions because chat will always provide one response in each turn of the conversation), as well as **Top-K** and **Top-P**. Like **Temperature**, **Top-K** and **Top-P** can be used to control the creativity and randomness of the model's generated outputs, but they work in slightly different ways. Consider the case of an LLM selecting the next word (or token) in a sequence when generating a response. It does this by predicting the probability of the next word being the most likely to occur in the given sequence. If the LLM always only selects the word with the highest probability, then it does not allow for much flexibility when generating responses. However, we can use **Top-K** and **Top-P** to add a level of flexibility by understanding how they work. So, let's take a closer look:

- **Top-K**: This uses a probability distribution to predict the probabilities for all possible next words based on the current prompt and context. K is the number of top-probability words that we want the model to select from. For example, if the value of K is 3, then the model will pick the next word from the top three most likely words (that is, the top three words with the highest probabilities). If the value of K is 1, then the model will pick only the most likely (that is, highest probability) word. This is a specific case we refer to as **greedy selection**.
- **Top-P**: This is a bit more complex to explain, but only slightly. Rather than picking a specific number of the highest-probability words to choose from, the model will calculate the **cumulative** probability for all possible next words – that is, it will add up the probabilities of the most likely words until the sum of the probabilities reaches the specified threshold (P), at which point it will randomly select a word from the pool of words that fall under the probability threshold.

In the case of both **Top-K** and **Top-P**, lower values guide the outputs toward conservative responses, while higher values allow some wiggle room for more creative responses, similar to (and in conjunction with) the **Temperature** parameter.

In the **Advanced** section, for some models, we can also enable **Grounding**, in which case we can ground our responses based on data we have stored in Vertex AI Search and Conversation. We also have the option to stream responses, which refers to printing responses as they are generated.

I encourage you to explore the different modalities and types of models in Vertex AI Studio. If you need help thinking of creative prompts to write, you can ask one of the text-based models to suggest some examples for you to use!

In addition to accessing the various models through the Vertex AI Studio UI, you can also interact with the models programmatically via the REST APIs and SDK.

Next, let's explore the vector database options available in Google Cloud.

## A detailed exploration of Google Cloud vector database options

Earlier in this chapter, I mentioned that almost all of Google Cloud's database services provide vector database support. In this section, we'll take a look at each one in more detail, starting with Vertex AI Search and Conversation.

### *Vertex AI Search and Conversation*

While the main purpose of the Vertex AI Search and Conversation product suite, as its name suggests, is to build search and conversation applications, we can also use it as a vector database for implementing **retrieval augmented generation** (**RAG**) solutions. As we discussed earlier in this chapter, if you're getting started with RAG on Google Cloud, unless you have a specific need to control the chunking, embedding, and indexing processes, or you have a specific operational need to link all of your data with another Google Cloud database service, Vertex AI Search and Conversation should be your first choice to consider. This is because it performs all of the chunking, embedding, and indexing processes for you, and it abstracts away all of those steps behind a simple and convenient orchestration interface, saving you a lot of time and effort. You can also use data connectors to ingest data from third-party applications, such as JIRA, Salesforce, and Confluence.

If you need to specifically control the chunking, embedding, and indexing processes, you can choose from other Google Cloud database offerings, as described next.

### *Vertex AI Vector Search*

As we've seen throughout this book, Vertex AI provides an entire ecosystem of tools and services for pretty much everything we could wish to do in the realm of ML and artificial intelligence. So, it makes sense that it would include a vector database, and that vector database is called Vertex AI Vector Search. It enables us to store and search through massive collections of embeddings to find the most similar or relevant items with high speed and low latency.

Although we need to create the chunks, embeddings, and indexes (unlike when using Vertex AI Search and Conversation), it still provides a fully managed service in that we don't need to worry about infrastructure or scaling because Vertex AI handles all of that for us. The fact that we need to create the chunks, embeddings, and indexes means that we can implement more granular control over those processes if needed, such as the models used to create our embeddings (note that this is also true for all of the other vector database options that we will discuss in the remainder of this chapter).

Vertex AI Vector Search also integrates closely with the rest of the Vertex AI and Google Cloud ecosystem to serve as part of larger orchestration solutions requiring the use of embeddings. On the topic of the broader ecosystem, let's consider additional Google Cloud vector databases beyond Vertex AI, starting with BigQuery.

### BigQuery Vector Search

BigQuery Vector Search is a feature within Google Cloud BigQuery that allows us to store and find embeddings. We can send a query embedding to BigQuery by using the BigQuery **VECTOR_SEARCH** function, which will quickly find similar embeddings within the vector database. This is a major benefit for users who are familiar with SQL syntax, and it enables semantic search and similarity-based analysis directly within your BigQuery data warehouse.

We can also choose to create a vector index to speed up the embedding retrieval process. When we use a vector index, an **approximate nearest neighbor** (**ANN**) search is used to return approximate results, which is much quicker than doing a brute-force search to find an exact match. If we do not create a vector index, then a brute-force search will be performed. We also have the option to explicitly implement a brute-force search even when a vector index exists if we want to get an exact result.

BigQuery Vector Search is a great choice for data science teams who already store a lot of data in BigQuery, and it also provides the benefits of BigQuery's automatic and enormous scalability. Continuing the theme of scalability, another database service in Google Cloud that is renowned for its impressive scalability is Google Cloud Spanner. Let's take a closer look.

### Spanner Vector Search

Google Cloud Spanner is a fully managed and highly performant database service that is capable of global scale. It's often referred to as a "NewSQL" database because it combines the scalability and flexibility of non-relational (NoSQL) databases with the familiar SQL interface of relational databases. It can easily handle petabytes of data, and unlike many NoSQL databases, it can guarantee strong consistency across transactions, even when distributed globally.

With all of this in mind, Spanner is suitable for applications that require highly distributed and strongly consistent data, such as online banking use cases. In early 2024, Spanner also launched support for cosine distance and Euclidean distance comparisons, and we can use these vector distance functions to perform **K-nearest neighbors** (**KNN**) vector searches for use cases such as similarity search or RAG. This means that this highly distributed database service that is commonly used for enterprise-scale, business-critical workloads now also supports vector database functionality.

While Spanner combines the best of NoSQL and relational database functionality, I'll describe relational database options next before covering NoSQL options explicitly.

### Cloud SQL and pgvector

Cloud SQL is a fully managed relational database service within Google Cloud that provides support for PostgreSQL, MySQL, and SQL Server. According to the Google Cloud documentation, "*More than 95% of Google Cloud's top 100 customers use Cloud SQL to run their businesses.*" It is often seen as the default relational database service to use in Google Cloud, and considering that it's fully managed, we don't need to worry about server provisioning, operating system updates, or database patches because Google handles all of those activities for us.

While Cloud SQL doesn't natively support vector database capabilities, **pgvector** is an open source PostgreSQL extension that's designed to provide vector similarity search functionality. It includes new data types, operators, and functions that enable us to store, search, and analyze high-dimensional vectors directly within our PostgreSQL database. It also includes specialized index types such as **Hierarchical Navigable Small Worlds** (**HNSW**) for fast ANN searches, and it Includes functions for vector addition, subtraction, normalization, and other operations, allowing us to manipulate embeddings within our database, as well as a similarity operator (<->) to calculate distance metrics such as cosine similarity between vectors to help us find the most similar embeddings. What's great is that it allows us to use SQL for all of those vector operations, which makes it easy to use for people who are familiar with PostgreSQL.

In addition to using `pgvector` with Cloud SQL for PostgreSQL, we can also use it with AlloyDB, which I'll describe next.

### *AlloyDB AI Vector Search*

AlloyDB is a fully managed PostgreSQL-compatible database service on Google Cloud that's designed for high-performance enterprise workloads. It significantly improves performance and scalability compared to standard PostgreSQL and provides advanced features such as auto-scaling and automatic failover, as well as built-in database backups and updates. It's also suitable for hybrid transactional and analytical workloads, and recently, it has added multiple ML-related features.

AlloyDB AI now also includes vector similarity search as it uses `pgvector` natively within the database.

Next, we'll discuss NoSQL database options for vector similarity search in Google Cloud.

### *NoSQL database options for vector similarity search in Google Cloud*

Firestore is a NoSQL database within the Google Firebase ecosystem. Firebase itself is a Google Cloud framework that provides a collection of tools and services for developing mobile and web applications. In addition to the first-party functionality provided by Firebase, there's also **Firebase Extensions Hub**, which provides a way for users to add more functionality to their solutions. While Firestore does not provide vector database support out-of-the-box, that functionality can be added via an extension named **Semantic Search with Vertex AI**, which adds text similarity search to a Firestore application by integrating with Vertex AI Vector Search.

Another highly popular NoSQL database in Google Cloud is Bigtable. Similar to Firestore, Bigtable does not currently provide native support for vector similarity search use cases, but it can integrate with Vertex AI Vector Search to help build a solution for that purpose.

Now that we've covered both the relational and NoSQL database options that provide vector database support in Google Cloud, it's important to describe one additional type of data store, called **Memorystore**, which is used to implement caching solutions for extremely low-latency workloads.

### *Memorystore for Redis*

Redis is an open source in-memory data store, meaning that, unlike traditional databases that store data on disk, Redis primarily stores data in RAM, giving it ultra-fast performance. Memorystore for Redis is a fully managed Redis service in Google Cloud that provides scalable, highly available, and secure Redis instances without requiring us to manage the instance or the related infrastructure.

Memorystore for Redis now also supports ANN and KNN vector search, enabling us to implement an in-memory vector store cache for applications that require extremely low-latency responses.

At this point, we've covered most of the main generative AI offerings on Google Cloud that are available. Now, it's time to put some of our knowledge into action as we begin to implement generative AI solutions on Google Cloud.

## Implementing generative AI solutions on Google Cloud

In this section, we will create a Vertex AI Search and Conversation application that will enable us to ask questions about the contents of a set of documents. We will use some publicly available medical study reports as examples, but this pattern can be applied to many different kinds of use cases. The following are some other popular applications of this pattern:

- Enabling users of a retail website to get information about products on the site via a question-and-answer interface, in which they can ask human language questions and get factual responses about products in the catalog.
- Asking questions about financial documents. For example, employees in a company's finance department could upload the company's financial documents, such as quarterly earnings reports, and ask natural language questions about the contents of the reports to identify trends or other important information.
- Providing natural language search over a company's vast corpus of internal documentation. For example, given a simple search interface, an employee could ask, "How can I modify my retirement contributions?" and they can get a response that tells them how to do so, along with a link to the internal documentation that describes the process in detail.

We will use Vertex AI Search and Conversation to build this application, as described in the following subsections.

## Building a Vertex AI Search and Conversation application

At a high level, we will perform three main activities to build our Vertex AI Search and Conversation application:

1. Enable the Vertex AI Search and Conversation API.
2. Create a data store to store the data we will use in our application.
3. Create the Vertex AI Search and Conversation application so that it can interact with our data store.

Let's begin by enabling the Vertex AI Search and Conversation API.

### *Enabling the Vertex AI Search and Conversation API*

To enable the Vertex AI Search and Conversation API, perform the following steps:

1. In the Google Cloud console, type `search` into the search box and select **Search and Conversation**.
2. If this is your first time using this service, a page similar to what's shown in *Figure 17.5* will be displayed:

Welcome to Vertex AI Search and Conversation
Vertex AI Search and Conversation allows developers to quickly build new experiences such as custom search engines and conversational apps via out-of-the-box templates and APIs.
Improve the quality and the performance of your Vertex AI Search and Conversation models, and diagnose issues faster by allowing Google to selectively sample model inputs and results. See Terms
We do not share model weights or Customer Data cross customers.
CONTINUE AND ACTIVATE THE API

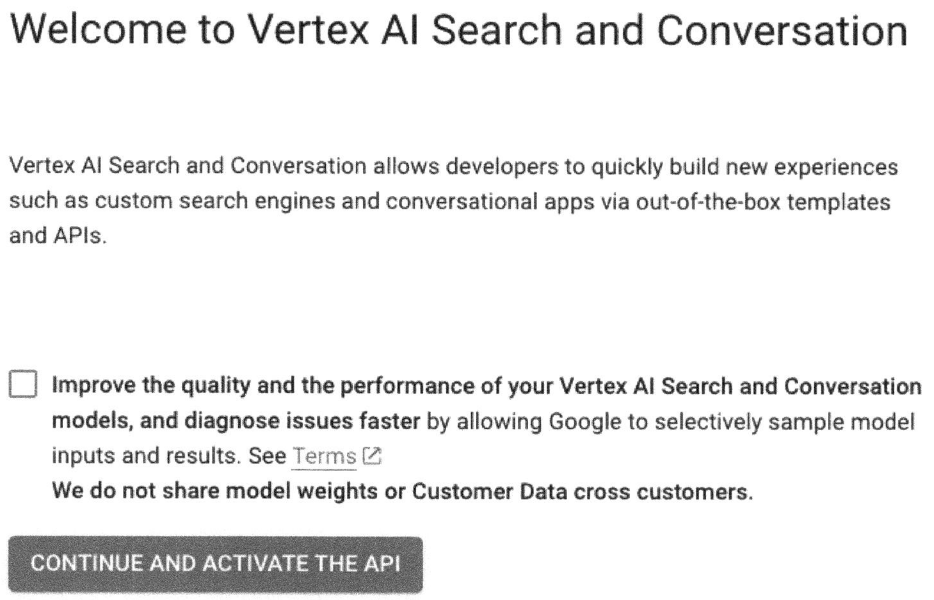

Figure 17.5: Activating Vertex AI Search and Conversation

3. Click **CONTINUE AND ACTIVATE THE API** (this checkbox is optional – you can either select it or leave it blank).
4. After a few seconds, your environment will be created. A page similar to what's shown in *Figure 17.6* will be displayed:

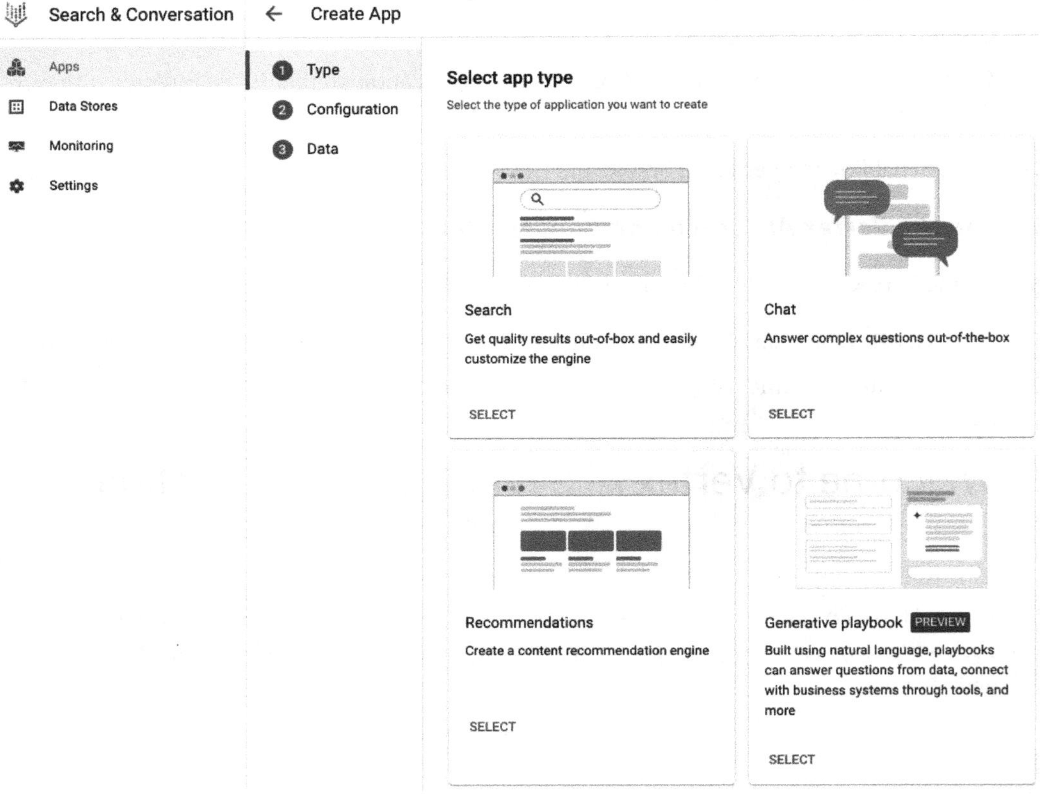


Figure 17.6: Vertex AI Search and Conversation environment page

Now that we've enabled the API, we can start the next prerequisite step to set up our application, which is to create a data store.

### *Creating a data store for Vertex AI Search and Conversation*

At the time of writing this, in March 2024, Vertex AI Search and Conversation can support the following data store sources:

- **Website URLs**: Automatically crawl website content from a list of domains you define
- **BigQuery**: Import data from your BigQuery table

- **Cloud Storage**: Import data from your storage bucket
- **API**: Import data manually by calling the API

We're going to use Google Cloud Storage as our data store, so we'll begin by setting up a location and uploading our data there. The next subsection describes the process.

> **Note**
>
> In *Chapter 4* of this book, we created a local directory in our Cloud Shell environment and cloned our GitHub repository into that directory. If you did not perform those steps, please reference those instructions now. They can be found in the *Creating a directory and cloning our GitHub repository* section.
>
> Once you have ensured that the GitHub repository for this book has been cloned into the local directory in your Cloud Shell environment, continue with the following steps.

### Storing our data in Google Cloud Storage

The easiest way to stage our data is to use Google Cloud Shell. To set up our data store, perform the following steps:

1. Click the **Cloud Shell** symbol in the top-right corner of the screen, as shown in *Figure 17.7*:

Figure 17.7: Activating Cloud Shell

2. This will activate the Cloud Shell environment, which will appear at the bottom of your screen. It will take a few seconds for your environment to activate. Once your environment has been activated, you can paste and run the commands mentioned in the following steps into Cloud Shell.
3. Run the following commands to set up environment variables so that you can store your preferred region, your Cloud Storage bucket name, and the data path (**replace YOUR-REGION with your preferred region, such as us-central1, and YOUR-BUCKET-NAME with your bucket name**):

```
export REGION=YOUR-REGION
export BUCKET=YOUR-BUCKET-NAME
export BUCKET_PATH=${BUCKET}/data/Chapter-17/
```

4. Verify the bucket path (you should also copy the contents of the response from this command and keep it for reference to be used in a later step):

```
echo $BUCKET_PATH
```

5. Create the bucket if it doesn't already exist:

```
gsutil mb -l $REGION gs://${BUCKET}
```

6. Change directories to the location at which the data for this chapter is stored:

```
cd ~/packt-ml-sa/Google-Machine-Learning-for-Solutions-
Architects/Chapter-17/data
```

7. Upload the files to the bucket (this also creates the path within the bucket):

```
for i in *; do gsutil cp $i gs://${BUCKET_PATH}; done
```

8. Verify that the files have been uploaded (the following command should list the files):

```
gsutil ls gs://${BUCKET_PATH}
```

Now that we've uploaded the files to our Cloud Storage bucket, we can create the Vertex AI Search and Conversation data store for our application, as described in the next subsection.

> **Note**
>
> The citations for the documents used in this exercise can be found in the `document_citations.txt` file in the `Chapter-17` directory of this book's GitHub repository: `https://github.com/PacktPublishing/Google-Machine-Learning-for-Solutions-Architects/blob/main/Chapter-17/document_citations.txt`.

### Creating the Vertex AI Search and Conversation data store

Within Vertex AI Search and Conversation, we will define a data store associated with the files we uploaded to Cloud Storage. To do that, perform the following steps:

1. In the Vertex AI Search and Conversation console UI, select **Data Stores** | **Create data store** | **Cloud Storage**.
2. A screen similar to the one shown in *Figure 17.8* will be displayed:

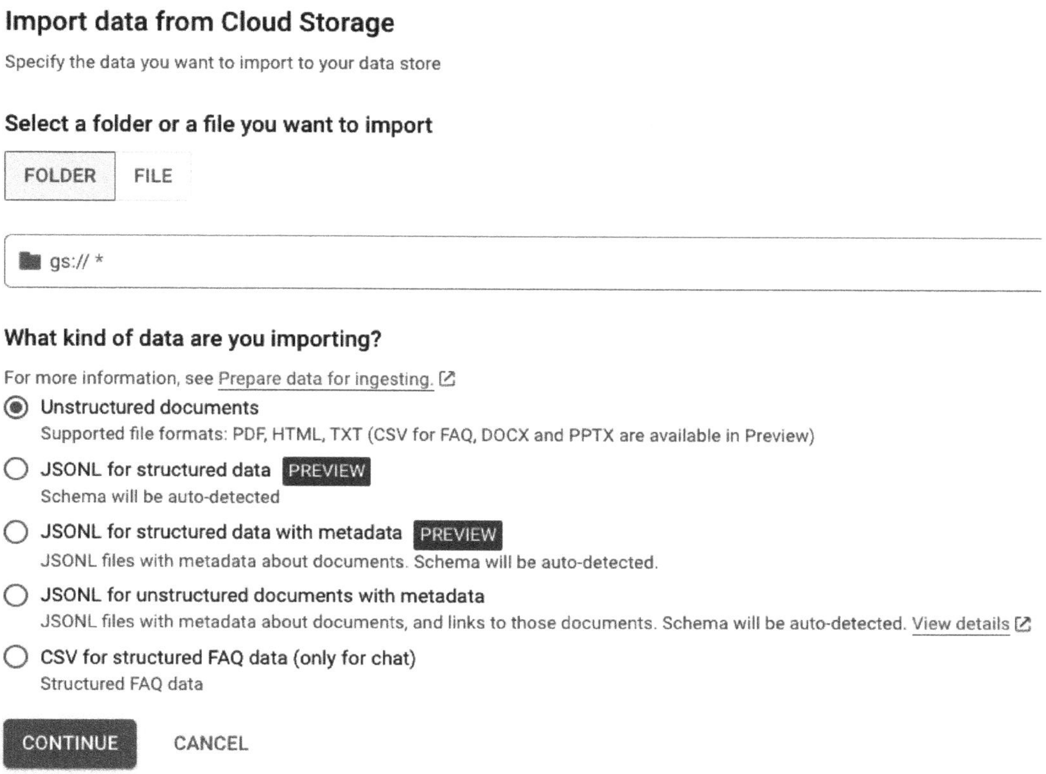

Import data from Cloud Storage

Specify the data you want to import to your data store

Select a folder or a file you want to import

FOLDER FILE

gs:// *

What kind of data are you importing?

For more information, see Prepare data for ingesting.

Unstructured documents
Supported file formats: PDF, HTML, TXT (CSV for FAQ, DOCX and PPTX are available in Preview)

JSONL for structured data PREVIEW
Schema will be auto-detected

JSONL for structured data with metadata PREVIEW
JSONL files with metadata about documents. Schema will be auto-detected.

JSONL for unstructured documents with metadata
JSONL files with metadata about documents, and links to those documents. Schema will be auto-detected. View details

CSV for structured FAQ data (only for chat)
Structured FAQ data

CONTINUE CANCEL

Figure 17.8: Import data from Cloud Storage

3. Enter the path to the files you uploaded (this is the value you specified for the `BUCKET_PATH` environment variable in your Cloud Shell environment in the previous section).
4. We will use unstructured documents in this example, so the **Unstructured documents** option should remain selected.
5. Click **Continue**. A screen similar to what's shown in *Figure 17.9* will be displayed:

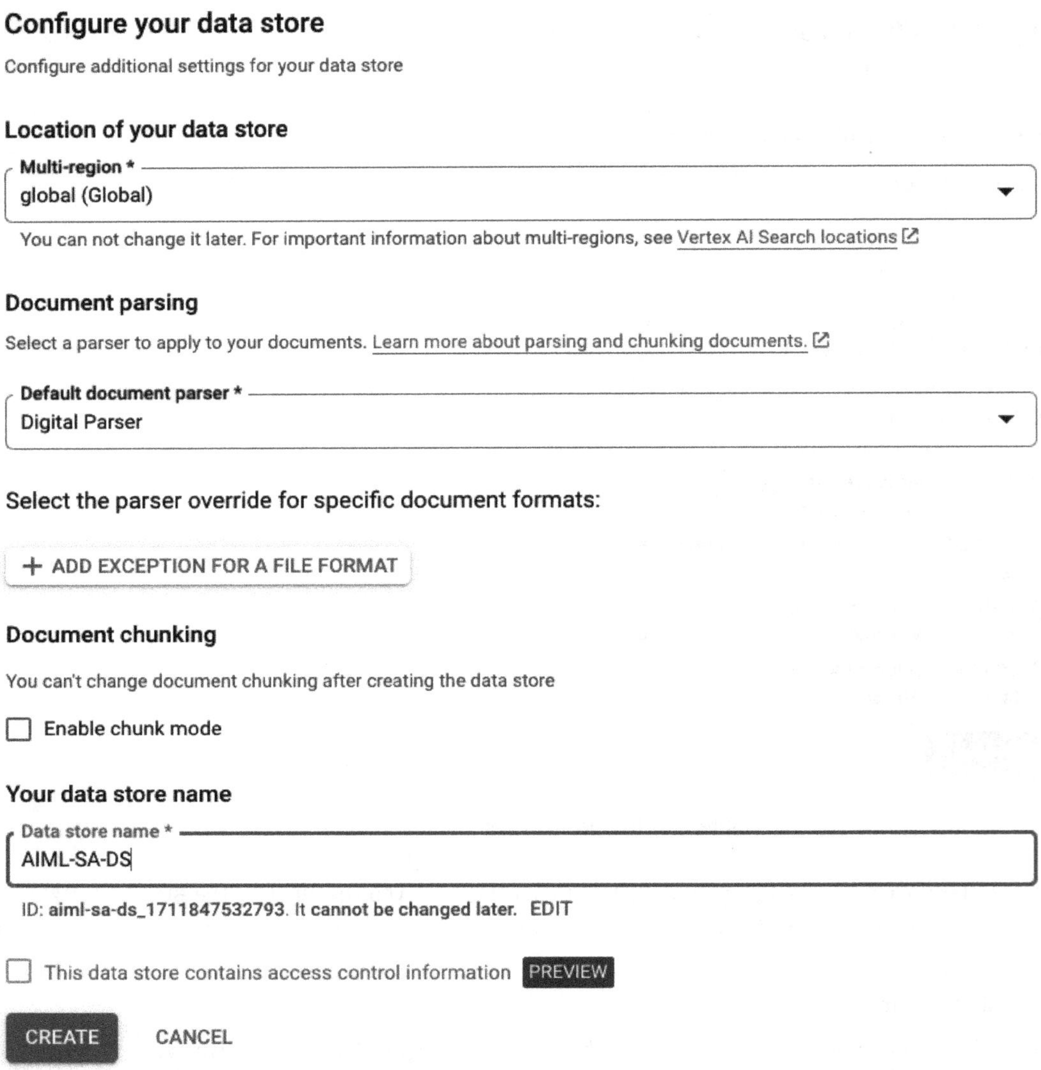

Figure 17.9: Configure your data store

6. Select **global (Global)** under **Multi-region**.
7. Select **Digital Parser** under **Default document parser**.
8. Enter a name for the data store, such as `AIML-SA-DS`, and click **CREATE**.

Now that the data store has been created, it's time to create the Vertex AI Search and Conversation application.

### *Creating the Vertex AI Search and Conversation application*

To create the Vertex AI Search and Conversation application, perform the following steps:

1. In the Vertex AI Search and Conversation console UI, select **Apps** | **Create a new app** | **Search**.
2. On the next screen that appears, ensure that the **Enterprise edition features** and **Advanced LLM features** checkboxes are enabled, as shown in *Figure 17.10* (read the descriptions for each option to understand their features):

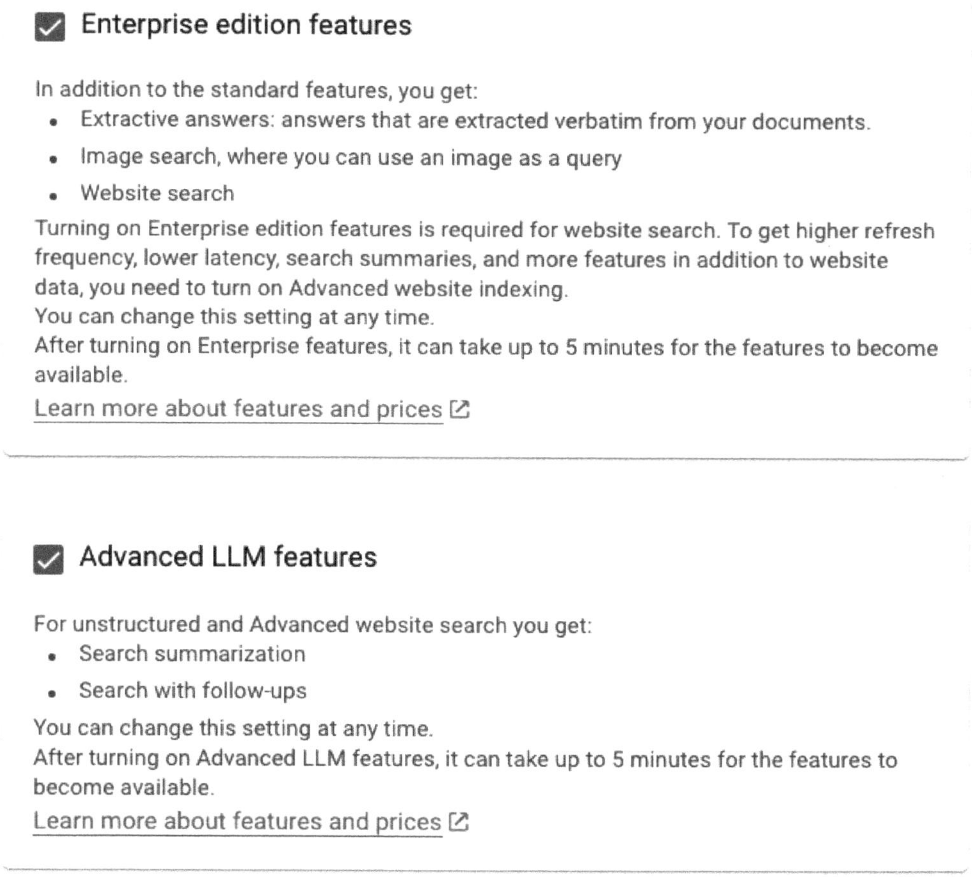

Enterprise edition features

In addition to the standard features, you get:

- Extractive answers: answers that are extracted verbatim from your documents.
- Image search, where you can use an image as a query
- Website search

Turning on Enterprise edition features is required for website search. To get higher refresh frequency, lower latency, search summaries, and more features in addition to website data, you need to turn on Advanced website indexing.
You can change this setting at any time.
After turning on Enterprise features, it can take up to 5 minutes for the features to become available.
Learn more about features and prices

Advanced LLM features

For unstructured and Advanced website search you get:

- Search summarization
- Search with follow-ups

You can change this setting at any time.
After turning on Advanced LLM features, it can take up to 5 minutes for the features to become available.
Learn more about features and prices

Figure 17.10: Enabling Search and Conversation features

3. Next, enter an application name and a company name, select the location for the application, and select **CONTINUE**, as shown in *Figure 17.11* (note the recommendation to choose a global location if you do not have compliance or regulatory reasons to locate your data in a particular multi-region):

**Your app name**

App name *
test-search-app

ID: test-search-app_1711828284169. It cannot be changed later. EDIT

**External name of your company or organization**

Company name *
Google

Providing your company name helps the model provide higher-quality responses

**Location of your app**

We recommend that you choose the **global location**, if you do not have compliance or regulatory reasons to locate your data in a particular multi-region. (EU and US regions are currently in preview)

Multi-region *
global (Global)

You can not change it later. For important information about multi-regions, see Vertex AI Search locations

CONTINUE CANCEL

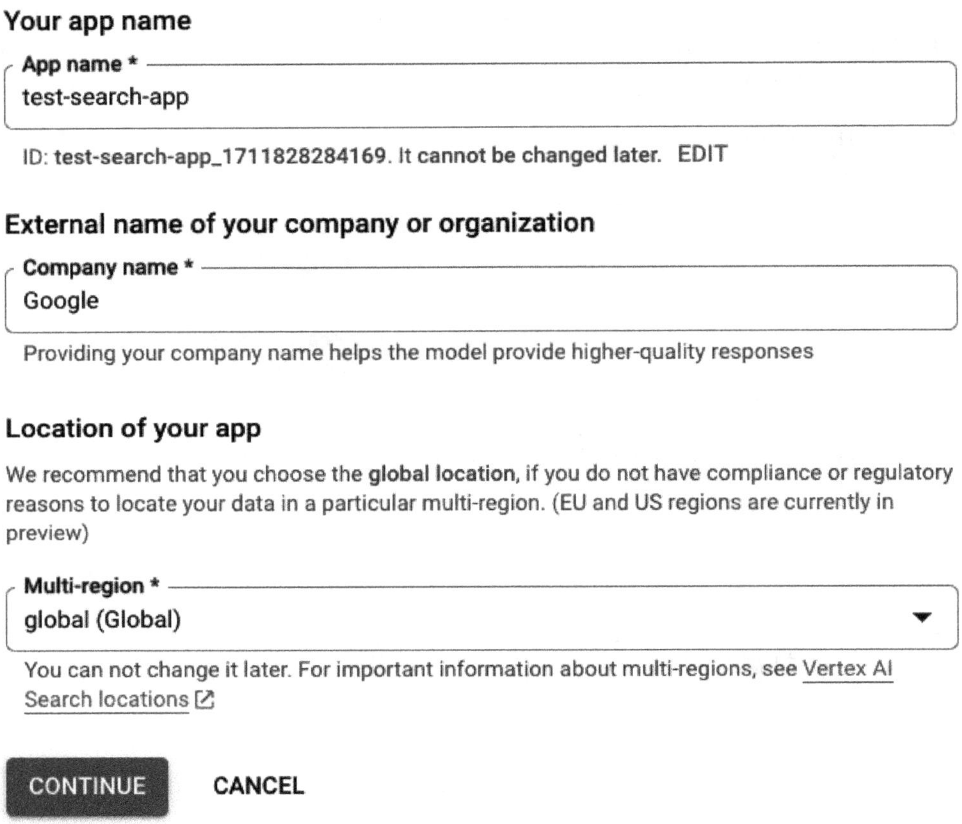

Figure 17.11: Application details

4. Next, select the data store we created that you wish to integrate with our application, as shown in *Figure 17.12*:

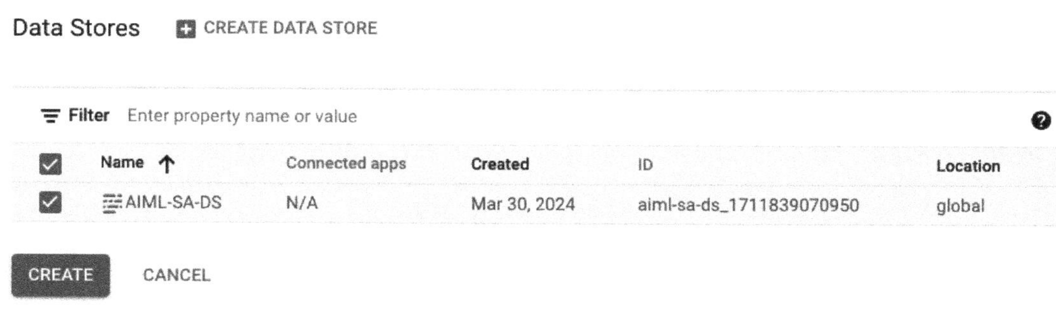


Figure 17.12: Selecting a data store

It will take some time for the data for our application to be processed. Periodically refresh the page to check its status. Once it's complete, a list of files will be displayed. At that point, we are ready to configure how we want our application to behave, as described in the following subsection.

### *Configuring our newly created application*

There are numerous ways in which our new Search and Conversation application can work, and we can specify how we want it to work by modifying its configuration. To do that, perform the following steps:

1. In the Vertex AI Search and Conversation console UI, select **Apps** from the menu on the left-hand side of the page.
2. Click on our newly created app.
3. Select **Configurations** from the menu on the left-hand side of the page.

   For the **Search type** configuration, there are three options:

   A. **Search**: Simply respond to the search query with a list of relevant results.

   B. **Search with an answer**: Provide a generative summary above the list of search results.

   C. **Search with follow-ups**: Provide conversational search with generative summaries and support for follow-up questions.

4. Select the **Search with follow-ups** option.
5. Scroll down to the **Large Language Models for summarization** section and select the latest version of Gemini.
6. You can leave all other options at their default values unless you have any specific requirements to change them.
7. Click **SAVE AND PUBLISH**.

At this point, we are ready to start using our application, as described in the following subsection.

### *Using our newly created application*

Now that we've created our application, we can start using it. To do that, perform the following steps:

1. In the Vertex AI Search and Conversation console UI, select the **Preview** section for our newly created application.
2. We will be presented with a search box in which we can ask questions about the contents of the documents in our data store.
3. Begin with the following question: `What are the effects of cinnamon on health?`
4. Click **Ask a follow-up** and enter the following question: `What specific properties contribute to its ability to regulate glucose levels?`

   Note that the follow-up question simply mentions "it" and does not mention cinnamon directly. However, the model uses the context from the previous question to understand our intent.

   Also, note that the generated answers may contain references, and the referenced documents that were used to generate the answer are listed in the response.
5. Try out the following additional questions:
   A. `How does the daily dosage of cinnamon used in the study compare to typical dietary cinnamon intake?`
   B. `Were there any observed long-term effects of cinnamon supplementation beyond the 4-week trial period?`
   C. `How might cinnamon supplementation interact with other dietary or lifestyle interventions for prediabetes management?`
   D. `Could the study's findings on cinnamon's glucose-regulating effects extend to individuals with types 1 or 2 diabetes?`
6. Next, let's ask some questions that will cause the model to respond based on other documents in the collection. Try out the following questions:
   A. `What are the main challenges FIM programs face when attempting to use EHR data for evaluation?`
   B. `How can FIM programs overcome the barriers to accessing and utilizing EHR data effectively?`
   C. `What alternative data sources can FIM programs use to evaluate health outcomes and healthcare utilization, apart from EHR data?`

D. `How do the privacy and liability concerns of healthcare partners impact the sharing of EHR data with FIM programs?`

E. `What role does albumin play in the nutritional status of children, and how are red bean cookies effective in improving it?`

F. `What specific dietary interventions have shown promise in modifying the gut microbiome to improve outcomes for patients with diseases?`

G. `How does the gut microbiome's interaction with the body impact mental health disorders, and what mechanisms are involved?`

H. `Can changes in the gut microbiome serve as early indicators for the development of chronic kidney disease?`

These are just some examples – feel free to play around with additional documents and questions, and think about how this pattern can be extended to pretty much any use case due to the vast knowledge of these wondrous models.

That's it! You have successfully built your first generative AI application in Google Cloud! As you can see, Vertex AI Search and Conversation makes it very easy for us to do this. Behind the scenes, this can be seen as a RAG solution because we are interacting with the Gemini model and getting it to generate responses that are grounded in the contents of the documents we uploaded to our data store, although Vertex AI Search and Conversation abstracts away and manages all of the complexities and steps required to implement the solution.

In the next chapter, we will build some additional and more complex use cases. First, however, let's summarize what we've learned in this chapter.

## Summary

In this chapter, we dived into generative AI in Google Cloud, exploring Google's native generative AI models, such as the Gemini, PaLM, Codey, Imagen, and MedLM APIs. We discussed the multiple versions of each model and some example use cases for each. Then, we introduced Vertex AI Studio and discussed open source and third-party models available on Google Cloud via repositories such as Vertex AI Model Garden and Hugging Face.

Next, we discussed vector databases in Google Cloud, covering various options available, such as Vertex AI Search and Conversation, Vertex AI Vector Search, BigQuery Vector Search, Spanner Vector Search, `pgvector`, and AlloyDB AI Vector Search, including some decision factors for choosing one solution over another. It is these kinds of decision points that are often most important in the role of a solutions architect, and the decisions will vary based on the specific needs of the customer or project, including the cost of each solution. I recommend always consulting the latest pricing information for each product and factoring that into your decision process.

Finally, we put some of this chapter's topics into action and built a generative AI application in Google Cloud – specifically, we built a Vertex AI Search and Conversation application that enabled us to ask natural language questions about the contents of a collection of documents.

In the next chapter, we will continue with this theme of using the topics we've covered throughout this book to build solutions in Google Cloud. Join me there to start diving in further!

# 18

# Bringing It All Together: Building ML Solutions with Google Cloud and Vertex AI

You've finally made it! This is the last chapter in our book. In this chapter, we will bring together the topics we've learned in this book by building a couple of example solutions that combine multiple concepts and Google Cloud services we discussed throughout the book. This chapter focuses mainly on building solutions, so it will be short on textual content, and the primary activities will be outlined in the Jupyter Notebook files accompanying the chapter.

We will build two solutions in the chapter: the first will be a deeper dive into **retrieval-augmented generation** (**RAG**), and the second will be an end-to-end solution that combines traditional **machine learning** (**ML**), MLOps, and generative AI.

The topics in this chapter are as follows:

- Building a RAG implementation piece by piece
- An example business use case and reference architecture
- Building and implementing the use case
- Recap and next steps

Before we dive into the architectural details, we need to perform a quick prerequisite step to set up the required environment settings in our Google Cloud project.

# Prerequisites

This section describes the steps we need to perform to set up our environment for the implementation steps later in this chapter.

> **Note**
>
> In *Chapter 4* of this book, we created a local directory in our Cloud Shell environment and cloned our GitHub repository into that directory. If you did not perform those steps, please reference those instructions now, in the section named *Create a directory and clone our GitHub repository* in *Chapter 4*.
>
> When you have ensured that the GitHub repository for this book has been cloned into the local directory in your Cloud Shell environment, continue with the prerequisite steps in this section.

## Using the Google Cloud Shell

We will use the Google Cloud Shell to perform the activities in this section. You can access it by clicking the Cloud Shell symbol in the top-right corner of the Google Cloud console screen, as shown in *Figure 18.1*.

Figure 18.1: Activating the Cloud Shell

Open the Google Cloud Shell as shown in *Figure 18.1*, and perform the following steps:

1. Change the directory to the location at which the code for this chapter is stored:

```
cd ~/packt-ml-sa/Google-Machine-Learning-for-Solutions-
Architects/Chapter-18/prereqs/
```

2. Run the `chapter-18-prereqs.sh` script:

```
bash chapter-18-prereqs.sh
```

3. If you are prompted to authorize the Cloud Shell, click **Authorize**.

The script performs the following steps on our behalf:

I. Grant the **Eventarc Event Receiver** IAM role to our project's default Compute Engine service account (our Jupyter Notebook, Cloud Functions, and other components use this service account). If you want to use a service account other than the default Compute Engine service account, you would need to grant the **Eventarc Event Receiver** IAM role to that service account instead.

II. Grant the **Service Account Token Creator** IAM role to our project's Pub/Sub service agent. This is required for when we use Pub/Sub later in this chapter.

Now that the prerequisites have been completed, we can continue with the main content of the chapter. The first solution we will focus on is building a RAG implementation piece by piece.

# Building a RAG implementation piece by piece

In *Chapter 17*, we implemented a RAG solution using Vertex AI Search. I explained that Vertex AI Search makes the process very easy for us because it abstracts away the steps in the process, such as chunking and embedding our content, and it performs all of those steps for us behind the scenes. There are also popular frameworks, such as LlamaIndex, that help to simplify RAG implementations, and Vertex AI has also launched a grounding service (Vertex AI Grounding) that you can use to ground responses from a generative model with results from Google Search or with your own data (using the aforementioned Vertex AI Search solution). In this section, we will dive deeper into the process and build a RAG solution, piece by piece. Before we dive into the solution architecture, we'll cover some of the concepts that had been abstracted away in *Chapter 17*, most notably regarding **tokens** and **chunks**.

## Tokens and chunks

In this section, we'll discuss the concepts of tokens and chunks, beginning with tokens.

### *Tokens*

LLMs generally work with tokens rather than words. For example, when dealing with text, a token often represents subsections of words, and tokenization can be done in different ways, such as breaking text up by characters, or using subword-based tokenization (the word "unbelievable" could be split into subwords such as "un," "believe," and "able").

The exact size and definition of a token can vary based on different tokenization methods, models, languages, and other factors, but a general rule of thumb for English text using subword tokenizers is around four characters per token on average.

Next, let's discuss the concept of chunks.

### Chunks

When creating embeddings, we usually break a document into chunks and then create embeddings of those chunks. Again, this can be done in different ways, using different tools. In the Jupyter Notebook file accompanying this chapter, we use Google Cloud Document AI to break our documents into chunks.

For this purpose, one of the parameters we need to specify for our Document AI processor is the chunk size to use, which is measured (in this case) by the number of tokens per chunk. You may need to experiment with the value of this parameter to find the chunk size that works best for your use case (e.g., based on the length and structure of the document sections). We generally want our chunks to capture some level of semantic granularity, but there are trade-offs in terms of this granularity. For example, smaller chunks can capture more granular semantic context and provide more precise search results but can be less efficient (computationally) to process. We also need to ensure chunk sizes are within the input length limits of the embedding model we're using in order to avoid possible truncation. A good practice is to start with a moderate chunk size and adjust it based on how well it fits our needs.

Fortunately, Document AI can automatically handle chunking based on layout, even if you don't specify a preconfigured chunk size, which can be helpful if you don't know what chunk size to use.

Now that we've covered those concepts, let's dive into the architecture of the RAG implementation we will build in this chapter.

## Example RAG solution architecture

In this section, we will review the architecture of our example RAG implementation. The proposed solution architecture is shown in *Figure 18.2*.

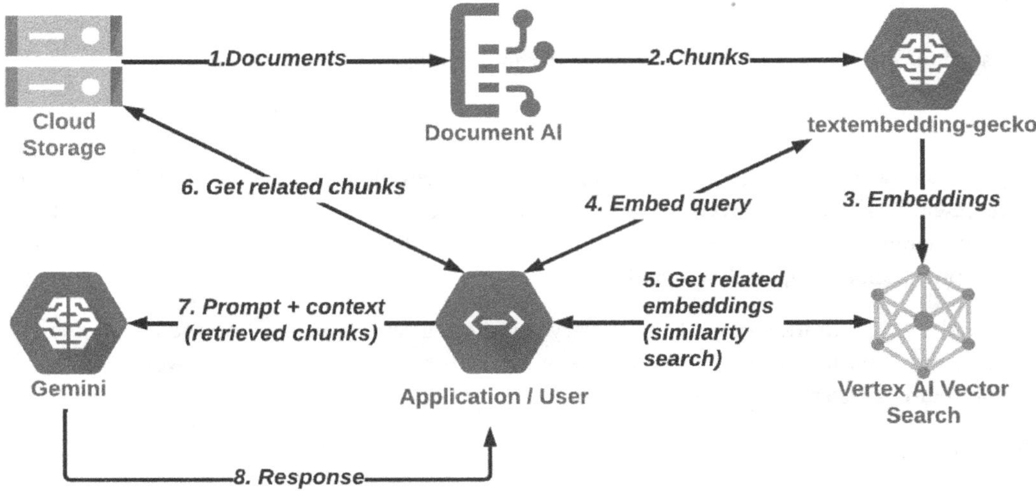


Figure 18.2: RAG solution piece by piece on Google Cloud

You can see that between some of the steps depicted in *Figure 18.2*, **Google Cloud Storage** (**GCS**) is used to store the inputs and outputs of each step, but those intermediary processes are omitted to make the diagram more readable. Also, when we implement this solution in the Jupyter Notebook that accompanies this chapter, the notebook is the application/user that coordinates each of the steps in the overall process.

The steps in the solution depicted in *Figure 18.2* are described as follows:

1. Our documents, which are stored in GCS, are sent to Google Cloud Document AI for chunking. As the name suggests, the chunking process breaks the documents into chunks, which are smaller sections of the document. This is required in order to create standard-sized chunks that serve as inputs to the embedding process that comes in the next step. The size of the chunks is configurable in Document AI, and further details on this process are described later.
2. The chunks are then sent to the Google Cloud text embedding LLM to create embeddings for the chunks. The resulting embeddings are stored in GCS, alongside their respective chunks (this step is omitted from the diagram).
3. We create a Vertex AI Vector Search index, and the embeddings are ingested from GCS to the Vertex AI Vector Search index (the GCS intermediary step is omitted from the diagram).
4. Next, the application/user asks a question that relates to the contents of our documents. The question is sent as a query to the Google Cloud text embedding LLM to be embedded/vectorized.
5. The vectorized query is then used as input in a request to Vertex AI Vector Search, which searches our index to find similar embeddings. Remember that the embeddings represent an element of semantic meaning, so similar embeddings have similar meanings. This is how we can perform a semantic search to find embeddings that are similar to our query.
6. Next, we take the embeddings returned from our Vertex AI Vector Search query and find the chunks in GCS that relate to those embeddings (remember that *Step 2* in our solution created a stored association of chunks and embeddings).
7. Now, it's finally time to send a prompt to Gemini. The retrieved document chunks from *Step 6* serve as the context for the prompt. This helps Gemini respond to our prompt based on the relevant content from our documents and not just from its pre-trained knowledge.

   Gemini responds to the prompt.

Now that we've walked through the steps in the process, let's go ahead and implement this solution!

## Building the RAG implementation on Google Cloud

To build our solution, open the Jupyter Notebook file that accompanies this chapter, and perform the activities described in that file. We can use the same Vertex AI Workbench instance that we created in *Chapter 14* for this purpose. Please open JupyterLab on that notebook instance. In the directory explorer on the left side of the screen, navigate to the `Chapter-18` directory and open the `rag.ipynb` notebook file. You can choose Python (local) as the kernel. Again, you can run each cell in the notebook by selecting the cell and pressing *Shift* + *Enter* on your keyboard. In addition to the relevant code, the notebook file contains markdown text that describes what the code is doing.

When you have completed the steps in the notebook, you will have officially built your own RAG solution—nice work!

Note that in this case, we are using documents stored in GCS as our source of truth, but we could also use other data, such as data stored in BigQuery.

> **Document citation**
>
> For congruity and comparability, we will use one of the same reference documents that we used in our RAG implementation in *Chapter 17*, cited as follows:
>
> *Hila Zelicha, Jieping Yang, Susanne M Henning, Jianjun Huang, Ru-Po Lee, Gail Thames, Edward H Livingston, David Heber, and Zhaoping Li, 2024. Effect of cinnamon spice on continuously monitored glycemic response in adults with prediabetes: a 4-week randomized controlled crossover trial. DOI:https://doi.org/10.1016/j.ajcnut.2024.01.008*

Next, we will build a broader solution that brings together many topics from throughout this entire book. I'll begin by explaining the use case.

# An example business use case and reference architecture

We will focus on a **computer vision** (**CV**) use case to identify objects in images. Extending the CV use case we implemented in *Chapter 15*, we will build our CV model via an MLOps pipeline that incorporates the majority of the main topics from the earlier chapters in this book, such as the following:

- Data preparation, cleaning, and transformation
- Model training, deployment, inference, and evaluation

The novel approach we are using here is to utilize generative AI to generate the data that will be used to test and evaluate our model. Before we start building, I'll present additional context on why I chose to include this use case as an example.

## Additional background on our use case

Throughout this book, one topic has come up more frequently than any other: data is often the most important factor in training high-quality ML models. Without adequate data, your ML project will not succeed. We've discussed that getting access to sufficient data often proves difficult, and this challenge continues to become ever more prevalent as the average size of models increases (today's models with trillions of parameters require enormous amounts of data). We also discussed that there's a limited amount of data in the world, but one mechanism that could help address this challenge is the creation of synthetic data and the use of generative AI models to generate new data. The use case I outline in this section combines all of the steps in the traditional ML model development life cycle and extends the process to include generating data that can be used to test our model. In the next subsection, I'll describe the architecture for this solution.

## The reference architecture

This section outlines the components of the solution architecture we will build. In addition to Vertex AI components, the reference architecture brings other Google Cloud services into scope to build the overall solution, such as Google Cloud Functions, Pub/Sub, Eventarc, and Imagen. We originally introduced and described all of those Google Cloud services in *Chapter 3* of this book. If you are not familiar with them, I recommend refreshing your knowledge from that chapter.

Our solution begins with an MLOps pipeline that will train and deploy our CV model, which I will describe in detail shortly. The resulting model is then used in a broader architecture to implement the overall solution. I'll begin by outlining the MLOps pipeline.

> **A note on the dataset (CIFAR-10)**
>
> In our model training and evaluation code, we're using something called the **CIFAR-10** (**Canadian Institute For Advanced Research, 10 classes**) dataset, which is a commonly used benchmark dataset for CV and image classification. It contains 60,000 color images belonging to ten different classes: Airplane, Automobile, Bird, Cat, Deer, Dog, Frog, Horse, Ship, and Truck. Each image has a fixed size of 32x32 pixels, and the images are stored in the RGB color space, meaning each pixel is represented by three values corresponding to the red, green, and blue color channels. It's important to keep these details in mind when we want to send generated data to our model at inference time.
>
> The CIFAR-10 dataset is so commonly used for benchmarking CV models that it is included in the built-in datasets module in Tensorflow/Keras. This means that we can use it in our code by simply importing that module rather than needing to download the data from an external source.

### Our MLOps pipeline

This section will walk through the steps implemented in our MLOps pipeline, as depicted in *Figure 18.3*.

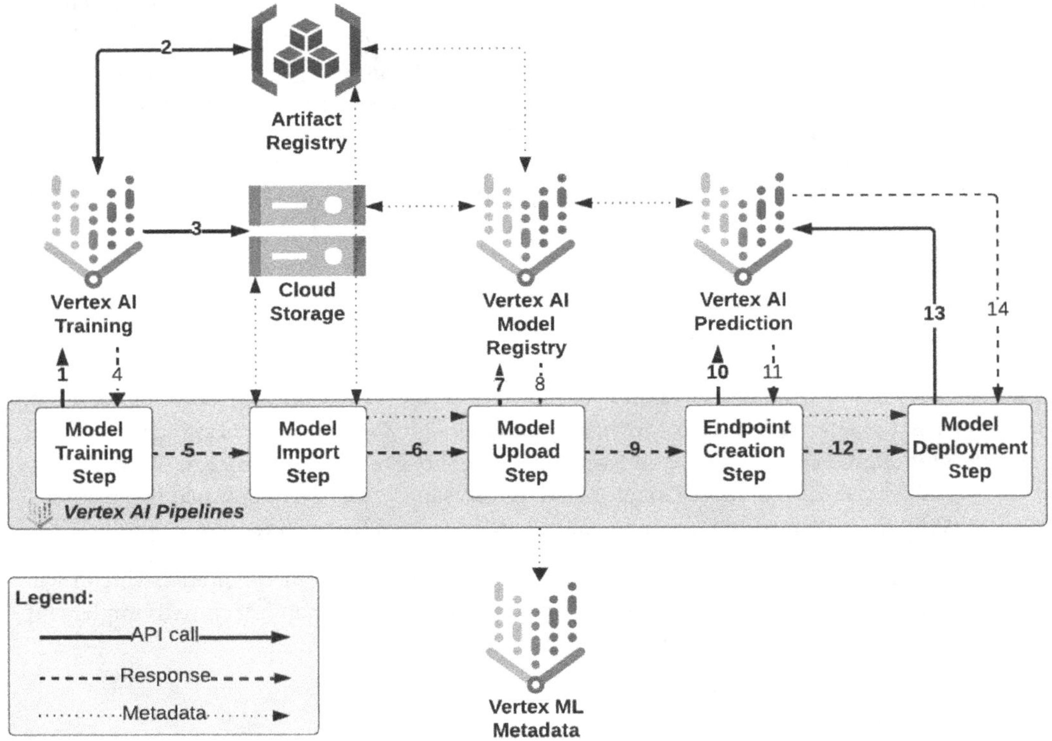


Figure 18.3: MLOps pipeline for CV model

In *Figure 18.3*, the process works as follows:

1. The first step in our pipeline—the model training step—is invoked. While the MLOps pipeline we built for our tabular Titanic dataset in *Chapter 11* started with distinct data preprocessing steps using Serverless Spark in Dataproc, in our pipeline in this chapter, the data ingestion and preparation steps are handled directly in the code of our model training job. Also, as noted, in this case, we are using the built-in CIFAR-10 image dataset in Tensorflow/Keras rather than fetching a dataset from an external source. Vertex AI Pipelines starts the model training process by submitting a model training job to the Vertex AI training service.
2. In order to execute our custom training job, the Vertex AI training service fetches our custom Docker container from Google Artifact Registry.

When our model has been trained, the trained model artifacts are saved in GCS.

The model training job status is complete.

3. The next step in our pipeline—the model import step—is invoked. This is an intermediate step that prepares the model metadata to be referenced in later components of our pipeline. The relevant metadata in this case consists of the location of the model artifacts in GCS and the specification of the Docker container image in Google Artifact Registry that will be used to serve our model.
4. The next step in our pipeline—the model upload step—is invoked. This step references the metadata from the model import step.
5. The model metadata is used to register the model in the Vertex AI Model Registry. This makes it easy to deploy our model for serving traffic in Vertex AI.

   The model upload job status is complete.
6. The next step in our pipeline—the endpoint creation step—is invoked.
7. An endpoint is created in the Vertex AI prediction service. This endpoint will be used to host our model.

   The endpoint creation job status is complete.
8. The next step in our pipeline—the model deployment step—is invoked.
9. Our model is deployed to our endpoint in the Vertex AI prediction service. This step references the metadata of the endpoint that has just been created by our pipeline, as well as the metadata of our model in the Vertex AI Model Registry.

   The model deployment job status is complete.

Now that we've walked through the steps in our model training and deployment pipeline, I will begin to outline the broader solution architecture we will build in this chapter, of which our MLOps pipeline is just a subset.

### The end-to-end solution

We will build an event-driven architecture, which is a popular pattern for developing serverless solutions intended to be implemented automatically in response to an event or a set of events. This is the end-to-end solution that incorporates the majority of the topics we've covered throughout this book, as depicted in *Figure 18.4*.

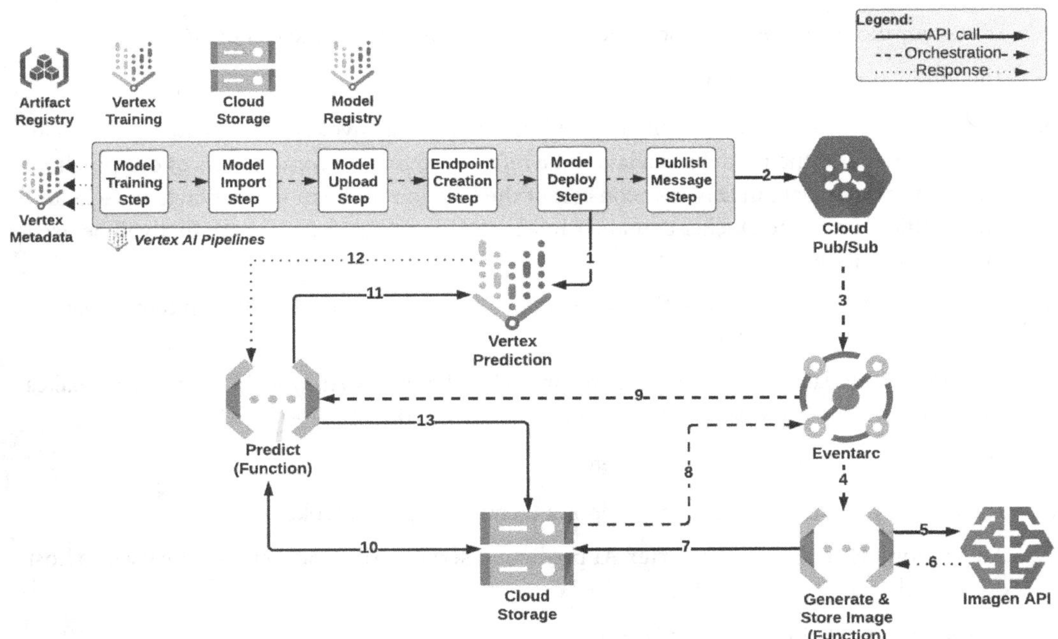


Figure 18.4: The end-to-end solution

In *Figure 18.4*, our MLOps pipeline is simplified in the top-left corner of the diagram. It still implements all of the same steps we discussed in the previous section, but the diagram is simplified so we can focus our discussion on the broader, end-to-end solution. In this context, the MLOps pipeline is represented as a single step in the overall process.

Notice that Eventarc features prominently in our solution, as it is the primary mechanism for orchestrating the steps in our event-based architecture. In the Jupyter Notebook file that accompanies this chapter, I will explain in more detail exactly how Eventarc is being configured behind the scenes.

The following set of steps describes the architecture implemented in *Figure 18.4*:

1. Our MLOps pipeline trains and deploys our CV model.
2. When the MLOps pipeline completes, it publishes a message to a Pub/Sub topic we created for that purpose.
3. Eventarc detects that a message has been published to the Pub/Sub topic.
4. Eventarc triggers the Cloud Function we've created to generate an image.
5. The code in our image generation function makes a call to the Imagen API with a prompt to generate an image containing one of the types of objects our model was trained to recognize (a type of object supported by the CIFAR-10 dataset).
6. Imagen generates an image and returns it to our function.

7. Our function stores the new image in GCS.
8. GCS emits an event indicating that a new object has been uploaded to our bucket. Eventarc detects this event.
9. Eventarc invokes our next Cloud Function and passes the GCS event metadata to our function. This metadata includes details such as the identifiers of the bucket and the object in question.
10. Our prediction function takes the details regarding the bucket and the object in question from the event metadata and uses those details to fetch the newly created object (i.e., the newly generated image from Imagen).
11. Our prediction function then performs some preprocessing on the image to transform it into a format that is expected by our model (i.e., similar to the format of the CIFAR-10 data the model was trained on). Our function then sends the transformed data as a prediction request to the Vertex AI endpoint that hosts our model.
12. Our model predicts what type of object is in the image, and sends a prediction response to our Cloud Function.
13. Our Cloud Function saves the prediction response in GCS.

When the process has been completed, you can view the generated image and the resulting prediction from our model in GCS.

Notice that all of the steps in the solution are implemented automatically and without the need to provision any servers. This is a fully serverless, event-driven solution architecture.

An interesting side effect of this solution is that, although the primary intention is to test our newly trained model on generated data, this solution could also be applied to the inverse use case. That is, if we are confident that our model has been trained effectively and provides consistently accurate results, we could use it to evaluate the quality of the generated data. For example, if our model predicts that the generated data contains a particular type of object with a probability of 99.8%, we can interpret this as a reflection of the quality of the generated data.

Now that we've discussed the various steps in the process, let's start building it!

# Building and implementing the use case on Google Cloud

To build our solution, open the Jupyter Notebook file that accompanies this chapter, and perform the activities described in that file. We can use the same Vertex AI Workbench instance that we created in *Chapter 14* for this purpose. Please open JupyterLab on that notebook instance. In the directory explorer on the left side of the screen, navigate to the `Chapter-18` directory and open the `end-to-end-mlops-genai.ipynb` notebook file. You can choose Python (local) as the kernel. Again, you can run each cell in the notebook by selecting the cell and pressing *Shift* + *Enter* on your keyboard. In addition to the relevant code, the notebook file contains markdown text that describes what the code is doing.

If you have followed all of the practical steps and successfully built the solution, then it's time to give yourself a good pat on the back. Let's take a moment to reflect on what you have done here. You have successfully built and executed an MLOps pipeline that trains and deploys a Keras convolutional neural network implementing a CV use case. You have then built a completely serverless, event-driven solution architecture using a combination of many Google Cloud services. That is absolutely something to be very proud of!

As we approach the end of this chapter and book, let's summarize what we've learned and discuss how to keep learning in this space.

# Recap and next steps

In this chapter, we combined the majority of the important concepts that we've covered throughout this book and explained how they all tie together. We did this by building solutions that incorporated many of the topics and Google Cloud products we've discussed.

First, you built a RAG implementation, piece by piece, focusing on the combination of various generative AI concepts in that process, such as using Google Cloud Document AI to break a document into chunks in a manner that preserves the hierarchical structure of the original document. This solution also included using the Google Cloud text embedding LLM to create embeddings for the document chunks and using Vertex AI Vector Search to store and index those embeddings. You used the resulting solution to implement a question-answering use case with Gemini, grounding the answers in the contents of the document. To do this, you used the Google Cloud text embedding LLM to create embeddings of the prompted question and used those embeddings to perform a semantic similarity search in Vertex AI Vector Search to find similar embeddings relating to specific chunks from the document. Those chunks were then provided as context when sending the request to Gemini.

Next, you built a solution that combined many topics from this entire book, including all of the steps in a traditional model development life cycle, as well as generative AI and broader solution architecture concepts. This resulted in a fully automated, serverless, event-driven solution that incorporated many Google Cloud services, such as Google Cloud Functions, Pub/Sub, Eventarc, and Imagen.

We've certainly covered a lot of topics in this book, and the learning journey never ends. Let's wrap up by discussing how you can continue your learning journey in the fields of Google Cloud solutions architecture, ML, and generative AI.

## Next steps

The technology fields we've discussed in this book continue to evolve at an accelerating pace, with new frameworks and patterns emerging daily, and generative AI is arguably the fastest-evolving field of them all. For example, in addition to the popular LangChain framework we've discussed in this book, which can also be implemented as a managed experience in the Vertex AI Reasoning Engine, extensions of those frameworks, such as LangGraph, LangSmith, and many others continue to emerge, providing additional functionality and flexibility for building generative AI solutions and applications. Due to the accelerating pace of development and overall excitement in this space, new learning resources will continue to emerge. In addition to the GitHub repository associated with this book, many other example repositories provide valuable tutorials on how to implement various patterns. Perhaps the most valuable of these is, of course, the official Vertex AI example repository, which can be accessed at the following link:

`https://github.com/GoogleCloudPlatform/vertex-ai-samples`

I also find the information available at `langchain.com` to be a useful learning resource, and I highly recommend referencing the example architecture use cases in the Google Cloud Solutions Center, accessible at the following link:

`https://solutions.cloud.google.com/`

As we wrap up, remember that the practical exercises and notebooks we used throughout this book create resources on Google Cloud, which can incur costs. For that reason, I recommend reviewing each of the practical steps we performed in this book and ensuring that all resources are deleted if you no longer plan to use them. The GitHub repository associated with this book contains steps for deleting resources such as models, endpoints, and data stores. For other types of resources, please refer to the Google Cloud documentation to ensure that the relevant resources are deleted.

I write this final section with a sense of sadness that our journey together is coming to an end, and I want to thank you for embarking on this journey with me. If you've made it this far in the book, then you have gained expert knowledge in AI/ML, generative AI, Google Cloud, and solutions architecture, and you now know a lot more about all of those topics than most people in the world! Congratulations, and well done! I wish you the best of luck in your future AI adventures, and I hope you will use your knowledge and skills to achieve wonderful things!

# Index

## Symbols



## A

# B

# D

# F

# G

# H

## I



## J



## K



## L

# M

## N

# Q



# R

## S

## W



## X



## Z

packtpub.com

Subscribe to our online digital library for full access to over 7,000 books and videos, as well as industry leading tools to help you plan your personal development and advance your career. For more information, please visit our website.

## Why subscribe?

- Spend less time learning and more time coding with practical eBooks and Videos from over 4,000 industry professionals
- Improve your learning with Skill Plans built especially for you
- Get a free eBook or video every month
- Fully searchable for easy access to vital information
- Copy and paste, print, and bookmark content

Did you know that Packt offers eBook versions of every book published, with PDF and ePub files available? You can upgrade to the eBook version at packtpub.com and as a print book customer, you are entitled to a discount on the eBook copy. Get in touch with us at customercare@packtpub.com for more details.

At www.packtpub.com, you can also read a collection of free technical articles, sign up for a range of free newsletters, and receive exclusive discounts and offers on Packt books and eBooks.

# Other Books You May Enjoy

If you enjoyed this book, you may be interested in these other books by Packt:

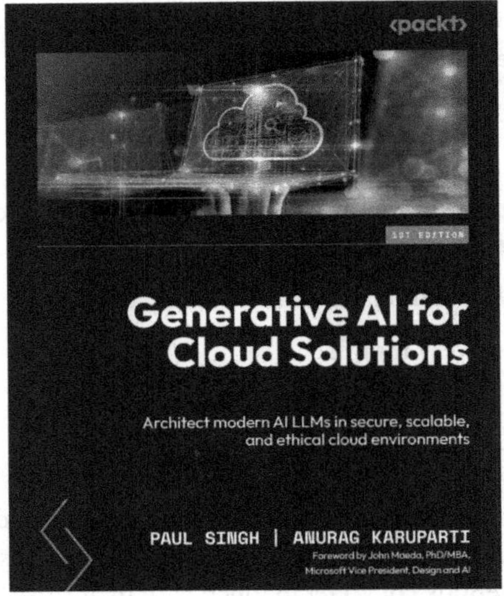


**Generative AI for Cloud Solutions**

Paul Singh, Anurag Karuparti

ISBN: 978-1-83508-478-6

- Get started with the essentials of generative AI, LLMs, and ChatGPT, and understand how they function together
- Understand how we started applying NLP to concepts like transformers
- Grasp the process of fine-tuning and developing apps based on RAG
- Explore effective prompt engineering strategies
- Acquire insights into the app development frameworks and lifecycles of LLMs, including important aspects of LLMOps, autonomous agents, and Assistants APIs
- Discover how to scale and secure GenAI systems, while understanding the principles of responsible AI

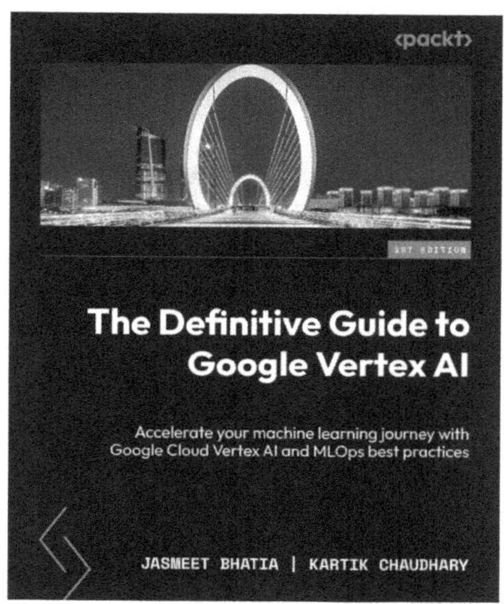


**The Definitive Guide to Google Vertex AI**

Jasmeet Bhatia, Kartik Chaudhary

ISBN: 978-1-80181-526-0

- Understand the ML lifecycle, challenges, and importance of MLOps
- Get started with ML model development quickly using Google Vertex AI
- Manage datasets, artifacts, and experiments
- Develop no-code, low-code, and custom AI solution on Google Cloud
- Implement advanced model optimization techniques and tooling
- Understand pre-built and turnkey AI solution offerings from Google
- Build and deploy custom ML models for real-world applications
- Explore the latest generative AI tools within Vertex AI

## Packt is searching for authors like you

If you're interested in becoming an author for Packt, please visit `authors.packtpub.com` and apply today. We have worked with thousands of developers and tech professionals, just like you, to help them share their insight with the global tech community. You can make a general application, apply for a specific hot topic that we are recruiting an author for, or submit your own idea.

## Share Your Thoughts

Now you've finished *Google Machine Learning and Generative AI for Solutions Architects*, we'd love to hear your thoughts! Scan the QR code below to go straight to the Amazon review page for this book and share your feedback or leave a review on the site that you purchased it from.

`https://packt.link/r/1-803-24527-1`

Your review is important to us and the tech community and will help us make sure we're delivering excellent quality content.

## Download a free PDF copy of this book

Thanks for purchasing this book!

Do you like to read on the go but are unable to carry your print books everywhere?

Is your eBook purchase not compatible with the device of your choice?

Don't worry, now with every Packt book you get a DRM-free PDF version of that book at no cost.

Read anywhere, any place, on any device. Search, copy, and paste code from your favorite technical books directly into your application.

The perks don't stop there, you can get exclusive access to discounts, newsletters, and great free content in your inbox daily

Follow these simple steps to get the benefits:

1. Scan the QR code or visit the link below

`https://packt.link/free-ebook/978-1-80324-527-0`

2. Submit your proof of purchase
3. That's it! We'll send your free PDF and other benefits to your email directly

www.ingramcontent.com/pod-product-compliance
Lightning Source LLC
Chambersburg PA
CBHW060637060326
40690CB00020B/4434

* 9 7 8 1 8 0 3 2 4 5 2 7 0 *